# TURBULENCE IN POROUS MEDIA

## MODELING AND APPLICATIONS

# TURBULENCE IN POROUS MEDIA

## MODELING AND APPLICATIONS

Marcelo J.S. de Lemos

*Instituto Tecnológico de Aeronáutica – ITA*
*Brazil*

ELSEVIER

AMSTERDAM • BOSTON • HEIDELBERG • LONDON • NEW YORK • OXFORD
PARIS • SAN DIEGO • SAN FRANCISCO • SINGAPORE • SYDNEY • TOKYO

ELSEVIER B.V.
Radarweg 29
P.O. Box 211, 1000 AE Amsterdam
The Netherlands

ELSEVIER Inc.
525 B Street, Suite 1900
San Diego, CA 92101-4495
USA

**ELSEVIER Ltd.**
**The Boulevard, Langford Lane**
**Kidlington, Oxford OX5 1GB**
**UK**

ELSEVIER Ltd.
84 Theobald's Road
London WC1X 8RR
UK



First edition 2006

ISBN-13: 978-0-080-44491-8
ISBN-10: 0-080-44491-1

Printed in Great Britain

06 07 08 09 10 10 9 8 7 6 5 4 3 2 1

*to the memory of Floriano Eduardo de Lemos*

*to Magaly, Pedro and Isabela*

# Preface

The main idea of this monograph is to introduce the reader to the possible characterizations of turbulent flow, and heat and mass transport in permeable media, including analytical, numerical and a review of available experimental data. Such transport processes, occurring at relatively high velocity in permeable media, are present in a number of engineering and natural flows. Therefore, a book like this one that compiles, details, compares and evaluates the available methodologies in the literature for modeling and simulating such flows can be useful to different fields of knowledge, covering engineering (mechanical, chemical, aerospace and petroleum) and basic sciences (thermal sciences, chemistry, physics, applied mathematics, and geological and environmental sciences).

This work has as its primary audience graduated students, engineers and scientists involved in design and analyses of: (a) modern engineering system such as fuel cells, porous combustors and advanced heat exchangers, (b) studies of biomechanical problems such as air flow in lungs and (c) modeling of natural and environmental flows such as atmospheric boundary layer over vegetation, fire in forests and spread of contaminants near rivers and bays.

In regard to already published works, it looks to the author that a book on the specific topic of "turbulence in porous media" is still not yet available as this topic is still quite new and unexplored. Only a few years ago papers on this topic have started to appear in scientific journals.

This book was written for use in graduate courses on turbulent transport phenomena in engineering, chemical and environmental programs which deal with research areas involving this theme.

My former and present graduate students have all contributed to the completion of this monograph. In the book, I also tried to collect, review and compile some of our joint publications and reports, which were completed during the last fifteen years or so. My sincere thanks to all of them, which includes Drs Pedras, M.H.J., Rocamora Jr., F.D., Assato, M., Braga, E.J., Mesquita, M.S., doctoral candidates Silva, R.A., Saito, M.B., Tofaneli, L.A., master graduates Magro, V.T., Graminho, D.R. and Santos, N.B.

Our research sponsors, CNPq and FAPESP (Brazilian Federal and São Paulo State Research Funding Agencies, respectively) have also greatly contributed to our "existence" as a research group. Most of my graduate students so far have been funded by them. To CNPq and FAPESP, my sincere thanks and my hopes that they keep on believing in our work.

In my nearly twenty years of teaching at Instituto Tecnológico de Aeronáutica (ITA), Brazil, I have always had support from colleagues and the Institute administration. None of our research goals would have been accomplished if we had been exposed to an "adverse pressure gradient" during these years. ITA, consistently and continually, has provided a fruitful and open environment were new ideas find fertile ground to grow and develop. I wish to express my sincere thanks to ITA and all the colleagues of the Mechanical Engineering Division.

A prominent professor in Brazil recently said that in the post-war era the international language spoken in most meetings and conferences around the world is "Bad English". Transposing this notion to written material, I wish to convey my deep apologies to all native English speakers for doing so much damage to Shakespeare's language.

The idea of writing this book came from a conversation I had with the senior publishing editor of Elsevier Arno Schouwenburg, who trusted that I had adequate scholarship for putting forward this project. I am grateful to him and to Vicki Wetherell, Editorial Assistant, for their invaluable help during the preparation of this work.

Finally, I hope that this work does not disappoint the reader who wishes to keep abreast with the latest developments in the interesting and innovative field of modeling turbulence in permeable structures.

Marcelo J.S. de Lemos
São José dos Campos, September 2005

This photo, with my father[1] and son, is to me an unmistakable evidence that, in our short life, uncertainty and fear lie in between youth hopefulness and old age wisdom.

[1] During the production stage of this book, Prof. Dr. Floriano Eduardo de Lemos passed away.

# Overview

The main scope of this book is to present in an organized, self-contained and systematic format new engineering techniques and novel applications of turbulent transport modeling for flow, heat and mass transfer in porous media. The motivation for writing this text is the fact that modern engineering equipment design and environmental impact analyses can benefit from appropriate modeling of turbulent flow in permeable structures. Examples span from flow in advanced porous combustors to atmospheric boundary layers over thick and dense rain forests. Besides, this is a very new topic because it involves disciplines that, traditionally, have been developed apart from each other such as *turbulence*, mostly associated with clear (unobstructed) flows and *porous media*, which is usually related to low speed currents, in laminar regime, through beds or materials containing interconnected pores.

Accordingly, a number of natural and engineering systems can be characterized by some sort of porous structure through which a working fluid permeates in turbulent regime. That is the case of many transport systems in environmental, petroleum, chemical, mechanical and aerospace technologies, including flows and fires in rain forests, for example. Here, forests are modeled as porous layers through which atmospheric air flows. In fact, forest fire devastates huge amounts of native land and occurs frequently in many countries such as Canada and Brazil, to mention a few. The ability to more precisely predict the spread of fire fronts, for example, might help authorities to better manage resources in order to minimize risk to human life. For analyzing such systems in a realistic and useful way, one has to consider the heterogeneity of the medium (different phases) and the fact that the flow fluctuates with time.

Another important application is the analysis of flow and heat transfer in advanced energy systems such as fuel cells and porous combustors. In addition, many transport processes in chemical engineering equipment also occur at a high velocity within a porous bed.

The two independent characteristics mentioned above, namely the medium heterogeneity and flow turbulence, have never been treated in a book and, as mentioned, this new knowledge could be useful to scientists, engineers and environmentalist dealing with natural disasters and working on the development of advanced energy systems.

As such, for practical, simple and tractable analyses, engineers, scientists and environmentalists tend to look at these systems as if the medium was made by a unique material (*after application of some volumetric average*) and did not present high frequency variation in its fluid phase velocity (*by using a model for handling turbulence effects*). The advantages of having an easy-to-use tool based on this macroscopic treatment are many, such as unveiling important overall flow characteristics without having to resort to sole experimental analysis which, in turn, can be time-consuming and expensive.

Turbulence models proposed for such flows depend on the order of application of time and volume average operators. Two developed methodologies, following the two orders of integration, lead to different governing equations for the statistical quantities. This book will review recently published methodologies to mathematically characterize turbulent transport in porous media.

The concept of *double-decomposition* is discussed in detail and models are classified in terms of the order of application of time and volume average operators, among other peculiarities. A total of four major classes of models have been identified and a general discussion on their main characteristics is carried out.

For hybrid media, involving both a porous structure and a clear flow region, difficulties arise due to the proper mathematical treatment given at the interface. This book also presents and discusses numerical solutions for such hybrid medium, considering here a channel partially filled with a porous layer through which fluid flows in turbulent regime. In addition, macroscopic forms of buoyancy terms were also considered in both the mean and the turbulent fields. Cases reviewed include heat transfer in porous square enclosures as well as cavities partially and totally filled with porous material.

In summary, in this book an overview on both porous media modeling and turbulence modeling is first presented with the aim of positioning this work in relation to these two classical analyses. In the next chapter, governing equations are reviewed for clear flow before the averaging operations are applied to them. The *double-decomposition* concept is presented and thoroughly discussed prior the derivation of macroscopic governing equations. Equations for turbulent momentum transport follow, showing detailed derivation for the mean and turbulent field quantities. The statistical $k$–$\varepsilon$ model for clear domains, used to model turbulence effects, serves also as the basis for modeling. Turbulent heat transport in porous matrices is further reviewed in light of the double-decomposition concept. The chapter on numerical modeling and algorithms details major methodologies and techniques used for numerically solving the flow governing equations. A final chapter on applications in hybrid media covers forced flows in composite channels, channels with porous and solid baffles, turbulent impinging jet onto a porous layer, buoyant flows and flow and heat transfer in a back-step.

# Contents

# List of Figures

# List of Tables

# Nomenclature

## LATIN CHARACTERS

| | |
|---|---|
| $A_i$ | Macroscopic interface area between the porous region and the clear flow |
| $A_i'''$ | Microscopic interfacial area between the solid and the liquid phases |
| $A_{nb}$ | Area of volume face associated with neighbor point in the grid |
| $A$ | Coefficient for morphology of porous medium in Ergun's equation |
| $a$ | Porous Insert thickness |
| $B_\gamma$ | Formation factor of phase $\gamma$ |
| $c_{1NL}$, $c_{2NL}$, $c_{3NL}$ | Coefficients of non-linear model |
| $c_\mu^*$ | Constant in turbulence model |
| $c_1$, $c_2$ | Constants in Equation (9.25) |
| $c_{1\varepsilon}$, $c_{2\varepsilon}$ | Turbulence model constants |
| $c_p$ | Fluid specific heat |
| $c_F$ | Forchheimer coefficient in Equation (9.7), $c_F = 0.55$ |
| $c_\mu$ | Turbulence model constant in Equation (4.20) |
| $c_k$ | Turbulence model constant in Equation (9.24) |
| $c_D$ | Constant in one-equation turbulence model |
| $C_\ell$ | Volumetric molar concentration of species $\ell$ |
| $D$ | Diffusion coefficient; diameter |
| $D_\ell$ | Diffusion coefficient of species $\ell$ |
| $D_j$ | Diameter of jet nozzle |
| $D_p$ | (1) Square rod size; (2) Particle diameter, $D_p = \sqrt{\frac{144K(1-\phi)^2}{\phi^3}}$ |
| $d$ | Particle or pore diameter |
| $Da$ | Darcy number, $Da = \frac{K}{H^2}$ |
| $Da_{eq}$ | Equivalent Darcy number using $K_{eq}$ given by Equation (1.9); $Da_{eq} = \frac{K_{eq}}{H^2}$ |
| $\mathbf{D}$ | Deformation rate **tensor**, $\mathbf{D} = [\nabla\mathbf{u} + (\nabla\mathbf{u})^T]/2$ |
| $\mathbf{D}_{disp}$ | Mass dispersion **tensor** |
| $\mathbf{D}_{disp,t}$ | Turbulent mass dispersion **tensor** |
| $\mathbf{D}_t$ | Turbulent mass flux **tensor** |
| $F$ | (1) Inertia coefficient; (2) Coefficients in discretized continuity equation |
| $F_{in}$ | Total mass flux entering the control volume |

| | |
|---|---|
| $F_{\text{out}}$ | Total mass flux leaving the control volume |
| $f$ | Friction factor |
| $G^{\text{i}}$ | Production rate of $k$ due to the porous matrix, $G^{\text{i}} = c_k \rho \phi \langle k \rangle^{\text{i}} \lvert \bar{\mathbf{u}}_{\text{D}} \rvert / \sqrt{K}$ |
| $\mathbf{g}$ | Gravity acceleration **vector** |
| $g$ | Gravity acceleration value |
| $g_z$ | $z$-component of gravity vector |
| $H$ | (1) Clear medium height; (2) Distance between the channel walls; (3) Channel height, $H = 40\,\text{mm}$; (4) Square cavity height; (5) Backward step height |
| $h_{\text{p}}$ | Porous medium height |
| $h$ | (1) Heat transfer coefficient; (2) Baffle height, $h = 20\,\text{mm}$ |
| $\mathbf{I}$ | Unity **tensor** |
| $\vec{J} = \mathbf{J}$ | Diffusion flux vector |
| $\mathbf{J}_\ell$ | Mass diffusion flux of species $\ell$ |
| $J$ | Jacobian |
| $\mathbf{K}_{\text{disp}}$ | Conductivity **tensor** due to dispersion |
| $\mathbf{K}_{\text{tor}}$ | Conductivity **tensor** due to tortuosity |
| $\mathbf{K}_{\text{disp,t}}$ | Conductivity **tensor** due to turbulent dispersion |
| $\mathbf{K}_{\text{t}}$ | Conductivity **tensor** due to turbulent heat flux |
| $K_{\text{eq}}$ | Equivalent permeability for the continuum model, $K_{\text{eq}} = \frac{\phi^3 D_{\text{p}}^2}{A(1-\phi)^2}$ |
| $k_{\text{f}}$ | Fluid thermal conductivity |
| $\langle k \rangle^{\text{i}}$ | Intrinsic (fluid) average of $k$ |
| $K$ | Permeability of porous medium, $K = \frac{\phi^3 D_{\text{p}}^2}{A(1-\phi)^2}$ |
| $K_{\text{r}\gamma}$ | Relative permeability of phase $\gamma$ |
| $k_{\text{s}}$ | Solid thermal conductivity |
| $k$ | Turbulent kinetic energy per unit mass, $k = \overline{\mathbf{u}' \cdot \mathbf{u}'}/2$ |
| $\langle k \rangle^{\text{v}}$ | Volume (fluid + solid) average of $k$ |
| $L$ | (1) Axial length of periodic section of channel; (2) Cavity width; (3) Cell length; (4) Recirculation length |
| $L_{\text{c}}$ | Test section length, $L_{\text{c}} = 600\,\text{mm}$ |
| $L_{\text{i}}$ | Entry length, $L_{\text{i}} = 400\,\text{mm}$ |
| $L_{\text{t}}$ | Channel length, $L_{\text{t}} = 1400\,\text{mm}$ |
| $l$ | Distance from the lower wall to the center of the porous medium |
| $\ell$ | Chemical species identifier |

| | |
|---|---|
| $\ell_e$ | Energy containing length scale |
| $\ell_m$ | Mixing length |
| $m_\ell$ | Mass fraction of component $\ell$ |
| $M_\ell$ | Molar weight of component $\ell$ |
| $M$ | Number of cells in $z$-direction |
| | |
| $N$ | (1) Number of neighbor grid points; (2) Number of obstacles; (3) Number of cells in $y$-direction |
| $Nu$ | (1) Nusselt number based on $H$ and $k$; $Nu = hH/k$; (2) Nusselt number based on $H$ and $k_{\text{eff}}$, $Nu = hH/k_{\text{eff}}$; (3) Nusselt number based on $L$ and $k_{\text{eff}}$, $Nu = hL/k_{\text{eff}}$ |
| $n$ | Coordinate normal to the interface |
| $\mathbf{n}$ | Unit vector normal to the interface |
| | |
| $\langle \overline{p} \rangle^{\text{i}}$ | Intrinsic (fluid) average of pressure $p$ |
| $p$ | (1) Thermodynamic pressure; (2) Pressure based on total area in a porous medium |
| $P$ | Pressure |
| $P^{\text{i}}$ | Production rate of $\langle k \rangle^{\text{i}}$ due to mean gradients of $\mathbf{u}_{\text{D}}$, $P^{\text{i}} = -\rho \langle \overline{\mathbf{u}'\mathbf{u}'} \rangle^{\text{i}} : \nabla \overline{\mathbf{u}}_{\text{D}}$ |
| $\mathbf{p}$ | Unit vector parallel to the interface |
| $Pr$ | Prandtl number, $Pr = \nu/\alpha_{\text{eff}}$ |
| $Pe$ | Péclet number |
| $Pe_{\text{D}}$ | Modified Péclet number, $Pe_{\text{D}} = Pe\,(1-\phi)^{1/2}$ |
| $P_{\text{k}}$ | Shear production rate of turbulent kinetic energy $k$ |
| | |
| $q$ | Source term referent to mass injection |
| $q_\lambda$ | Source term of phase $\gamma$ |
| $Q_{\text{cell}}$ | Heat transferred to a single cell |
| | |
| $R_{\text{abs}}$ | Absolute residue for mass continuity equation |
| $R_{\text{rel}}$ | Relative residue for mass continuity equation |
| $R_{\text{T}}$ | Residue for energy equation |
| $R$ | Model combustor radius |
| $\overline{\mathbf{R}}$ | Time average of total drag per unit volume |
| $\mathbf{R}$ | Total drag per unit volume |
| $r/R$ | Non-dimensional radial position |
| $Ra_{\text{cr}}$ | Critical Rayleigh number |
| $Ra_{\text{m}}$ | (1) Darcy-Rayleigh number for continuum model (solid and fluid phases), $Ra_m = Ra \cdot Da_{\text{eq}} = Ra_\phi \cdot Da$; (2) Darcy-Rayleigh number for porous-continuum model (one phase), $Ra_{\text{m}} = Ra_{\text{f}} \cdot Da = g\beta_\phi H \Delta T K / \nu_{\text{f}} \alpha_{\text{eff}}$ |

| | |
|---|---|
| $Ra$ | (1) Fluid Rayleigh number based on $\nu_f$, $Ra = g\beta H^3 \Delta T / \nu_f \alpha$; (2) Fluid Rayleigh number based on $\nu$, $Ra = g\beta H^3 \Delta T / \nu\alpha$ |
| $Ra_f$ | Fluid Rayleigh number based on $\nu_f$ and $\alpha_f$, $Ra = g\beta H^3 \Delta T / \nu_f \alpha_f$ |
| $Ra_\phi$ | Volume averaged Rayleigh number, $Ra_\phi = g\beta_\phi H^3 \Delta T / \nu_f \alpha_{eff}$ |
| $Re$ | (1) Reynolds number; (2) Reynolds number based on average surface velocity $\mathbf{u}_D^*$, $Re = 2H\rho \mathbf{u}_D^* / \mu$ |
| $Re_H$ | Reynolds number based on the channel height, $Re_H = \rho \lvert \mathbf{u}_D \rvert H / \mu$ |
| $Re_p$ | Reynolds number based on the pore diameter, $Re_P = \lvert \mathbf{u}_D \rvert \, d / \nu_f$ |
| | |
| $s$ | (1) Clearance for unobstructed flow; (2) Porous medium thickness |
| $S_\gamma$ | Saturation of phase $\gamma$ |
| $S_\varphi$ | Source term for general variable $\varphi$; $\varphi = U, W, T$ |
| $Sc_t$ | Turbulent Schmidt number |
| $S$ | Swirl parameter |
| | |
| $T$ | Temperature |
| $T_1$ | Hot wall temperature |
| $T_0$ | Cold wall temperature |
| $T_b$ | Bulk temperature |
| $T_f$ | Fluid temperature |
| $T_s$ | Solid temperature |
| $t$ | Baffle thickness, $t = 1.5\,\text{mm}$ |
| | |
| $\mathbf{u}$ | (1) Microscopic (local) velocity vector for single component; (2) Mass-averaged velocity of a mixture, $\mathbf{u} = \sum_\ell m_\ell \mathbf{u}_\ell$ |
| $\overline{\mathbf{u}}$ | Microscopic time-averaged velocity vector |
| $\mathbf{u}_D = \vec{u}_D$ | Instantaneous Darcy, superficial or seepage velocity (volume average of $\mathbf{u}$), $\mathbf{u}_D = \phi \langle \mathbf{u} \rangle^i$ |
| $\overline{\mathbf{u}}_D$ | Time mean Darcy velocity vector, $\overline{\mathbf{u}}_D = \phi \langle \overline{\mathbf{u}} \rangle^i$ |
| $\mathbf{u}_\ell$ | Velocity of species $\ell$ |
| $\mathbf{u}_{D_i}$ | Darcy velocity **vector** at the interface |
| $\overline{\mathbf{u}}_{D_i}$ | Time mean Darcy velocity **vector** at the interface |
| $\mathbf{u}_{D_p}$ | Darcy velocity **vector** parallel to the interface |
| $\overline{\mathbf{u}}_{D_p}$ | Time mean Darcy velocity **vector** parallel to the interface |
| $\mathbf{u}_D^*$ | Average surface velocity, $\mathbf{u}_D^* = \frac{1}{H} \int_0^H \mathbf{u}_D \, dy$ |
| $\langle \mathbf{u} \rangle^i$ | Intrinsic (fluid) average of $\mathbf{u}$ |
| $u_{D_n}, u_{D_p}$ | Components of Darcy velocity at interface along $\eta$ (normal) and $\xi$ (parallel) directions, respectively |
| $u_{D_i}, v_{D_i}$ | Components of Darcy velocity at interface along $x$ and $y$, respectively |

| | |
|---|---|
| $U_i$ | Fluid inlet velocity |
| $U$ | Velocity component in $y$-direction |
| $u_i$ | Velocity component in $i$-direction |
| $U_{D\|\|}$ | Average surface velocity component parallel to the interface |
| $V$ | Tangential velocity component |
| $w$ | Exit flow ring length |
| $W$ | Velocity component in $z$-direction |
| $W_{in}$ | Inlet axial velocity |
| $x$, $y$ | Cartesian coordinates |
| $x_R$ | Reattachment length |
| $y$ | Coordinate along cavity width $L$ |
| $z$ | (1) Axial distance; (2) Coordinate along cavity height $H$ |
| $z/H$ | Non-dimensional axial distance from the collision plate |

## GREEK SYMBOLS

| | |
|---|---|
| $\delta$ | (1) Boundary Layer thickness; (2) Coefficient for grid influence on friction factor, $\delta = [(f\|_{ref} - f)/ f\|_{ref}] \times 100$ |
| $\delta r_{nb}$ | Distance between nodal and neighbor points in grid |
| $\rho$ | (1) Fluid density; (2) Bulk density of mixture, $\rho = \sum_{\ell} \rho_{\ell}$ |
| $\rho_{\ell}$ | Mass density of species $\ell$ |
| $\Delta T$ | Temperature drop across cavity width $L$, $\Delta T = (T_1 - T_0)$ |
| $\eta$ | (1) Kolmogorov length scale; (2) Coefficient for heat transfer comparison, $\eta = [(\overline{Nu}^*\|_{porous} - \overline{Nu}^*\|_{solid})/\overline{Nu}^*\|_{solid}] \times 100$ |
| $\Gamma_{\varphi}$ | Transport coefficient of exchange for general variable $\varphi$; $\varphi = U, W, T$ |
| $\ell$ | Component identifier |
| $\alpha$ | (1) Dispersion coefficient; (2) Cavity tilt angle |
| $\varepsilon$ | (1) Dissipation rate of turbulence kinetic energy $k$, $\varepsilon = \mu \overline{\nabla \mathbf{u}' : (\nabla \mathbf{u}')^T} / \rho$; 2) Coefficient for grid influence on Nusselt number, $\varepsilon = [(\overline{Nu}\|_{ref} - \overline{Nu})/ \overline{Nu}\|_{ref}] \times 100$ |
| $\langle \varepsilon \rangle^i$ | Intrinsic (fluid) average of $\varepsilon$ |
| $\mu$ | (1) Fluid dynamic viscosity; (2) Mixture dynamic viscosity |
| $\mu_t$ | Microscopic turbulent viscosity |
| $\mu_{t_\phi}$ | Macroscopic turbulent viscosity |

| | |
|---|---|
| $\mu_{\text{eff}}$ | Effective viscosity for a porous medium |
| $\lambda$ | (1) Fluid thermal conductivity; (2) Fin-conductance parameter, $\lambda = k_s t/(k_f L)$ |
| $\lambda_\gamma$ | Mobility of phase $\gamma$ |
| $\Theta$ | (1) Mean temperature; (2) Non-dimensional temperature: $\Theta = (T - T_0)/(T_1 - T_0)$ |
| $\theta$ | Fluctuating temperature |
| $\nu$ | Fluid kinematic viscosity |
| $\Delta V$ | Representative elementary volume |
| $\Delta V_f$ | Fluid volume inside representative elementary volume |
| $\varphi$ | General dependent variable (*scalar*) |
| $\boldsymbol{\varphi}$ | General dependent variable (**vector**) |
| $\eta, \xi$ | Generalized coordinates |
| $\beta$ | (1) Interface stress jump coefficient; (2) Thermal expansion coefficient; (3) Compressibility factor |
| $\beta_\phi$ | Macroscopic thermal expansion coefficient |
| $\beta_C$ | Salute expansion coefficient |
| $\beta_{C_\phi}$ | Macroscopic salute expansion coefficient |
| $\Psi_0$ | Maximum value of normalized stream function in the main recirculating eddy |
| $\sigma_t$ | Turbulent Prandtl number |
| $\sigma_k$ | Turbulent Prandtl number for $k$ |
| $\sigma_\varepsilon$ | Turbulent Prandtl number for $\varepsilon$ |
| $\gamma$ | Phase identifier |
| $\gamma_\varphi$ | Relaxation parameter for $\varphi = U, W, P, T$ |
| $\phi$ | Porosity, $\phi = \Delta V_f / \Delta V$ |
| $\tau_w$ | Wall shear stress, $\tau_w = \mu \frac{du}{dy}$ |

## SPECIAL CHARACTERS

| | |
|---|---|
| $\lvert\varphi\rvert$ | Absolute value (Abs) |
| $\varphi_{\text{eff}}$ | Effective value, $\varphi_{\text{eff}} = \phi\varphi_f + (1-\phi)\varphi_s$ |
| $(\varphi)_{H,C}$ | Hot/cold |
| $\langle\varphi\rangle^i$ | Intrinsic average |
| ${}^i\varphi$ | Spatial deviation |
| $(\varphi)_\phi$ | Macroscopic or porous continuum value |
| $\varphi_{s,f}$ | Solid/fluid |
| $\overline{\varphi}$ | Time average |
| $\varphi'$ | Time fluctuation |
| $(\ )^T$ | Transpose |
| $\langle\varphi\rangle^v$ | Volume average |

## SUBSCRIPTS

| | |
|---|---|
| 0 | Parallel-plate channel |
| f | Fluid |
| D | Darcy |
| b | Bulk |
| s | Solid |
| ref | Reference |
| $\beta$ | Buoyancy |
| $\ell$ | Chemical species |
| t | Turbulent |
| $\phi$ | Macroscopic |
| C | Concentration |

## SUPERSCRIPTS

| | |
|---|---|
| i | Intrinsic (fluid) average |
| v | Volume (fluid + solid) average |
| k | Turbulent kinetic energy |

# PART ONE

# Modeling

Chapter 1

# Introduction

## 1.1. OVERVIEW OF POROUS MEDIA MODELING

Due to its ever-broader range of applications in science and industry, the study of flow through porous media has gained extensive attention lately. Engineering systems based on fluidized bed combustion, enhanced oil reservoir recovery, underground spreading of chemical waste, enhanced natural gas combustion in an inert porous matrix and chemical catalytic reactors are just a few examples of applications of this interdisciplinary field. In a broader sense, the study of porous media embraces fluid and thermal sciences, and materials, chemical, geothermal, petroleum and combustion engineering.

Accordingly, applications that are more complex usually require appropriate and, in most cases, more sophisticated mathematical and numerical modeling. Obtaining the final numerical results, however, may require the solution of a set of coupled partial differential equations involving many coupled variables in a complex geometry. This book shall review important aspects of numerical methods, including treatment of multidimensional flow equations, discretization schemes for accurate solution, algorithms for pointwise and block-implicit solutions, algorithms for high performance computing and turbulence modeling. These subjects shall be grouped into major sections covering numerical formulation and algorithms, geometry and turbulence.

### 1.1.1. General Remarks

During the past few decades, a number of textbooks have been written on the subject of porous media. Among them one can mention the works referred to in Muskat (1946), Carman (1956), Houpeurt (1957), Collins (1961), DeWiest (1969), Scheidegger (1974), Dullien (1979), Bear and Bachmat (1990) and Kaviany (1991) . Advanced models documented in recent literature try to simulate additional effects such as variable porosity, anisotropy of medium permeability, non-conventional boundary conditions, flow dimension, geometry complexity, non-linear effects and turbulence. Not all these flow complexities can be analyzed with the early unidimensional Darcy flow model. Recognizing the importance of these applications, the literature has been ingenious in proposing a number of extended theoretical approaches. Below is a short review of basic equations governing fluid flow followed by a summary of some of the classical models for analyzing transport phenomena in porous media.

### 1.1.2. Fundamental Conservation Equations

The basic conservation equations describing the flow of a fluid through an infinitesimal volume can be written in a compact form as

$$\frac{\partial}{\partial t}(\rho\varphi)+\frac{\partial}{\partial x_j}\left[\rho u_j\varphi-\Gamma_\varphi\frac{\partial\varphi}{\partial x_j}\right]=S_\varphi \tag{1.1}$$

where $\varphi$ is the general variable (not to confuse with the porosity $\phi$ to be introduced later), $u_j$ is the $j$-th velocity component, $\rho$ is the density, and $\Gamma_\varphi$ and $S_\varphi$ are the diffusion coefficient and source terms, respectively. The value of $\varphi$ and its corresponding parameters ($\Gamma_\varphi$ and $S_\varphi$) take different forms according to the conserved quantity (mass, momentum, energy, chemical species, turbulent kinetic energy, etc.). The conservation laws recast into the form of Equation (1.1) appear commonly in many texts devoted to the use of the control-volume approach. It is a convenient way to represent all transport phenomena occurring in a certain flow.

When (1.1) is written in Cartesian coordinates all in three dimensions for the case of mass conservation ($\varphi=1, \Gamma_\varphi=S_\varphi=0$), one has:

$$\frac{\partial\rho}{\partial t}+\frac{\partial(\rho u)}{\partial x}+\frac{\partial(\rho v)}{\partial y}+\frac{\partial(\rho w)}{\partial z}=0 \tag{1.2}$$

For a general variable, Equation (1.1) can be written in Cartesian coordinates as,

$$\frac{\partial(\rho\varphi)}{\partial t}+\frac{\partial}{\partial x}\left(\rho u\varphi-\mu\frac{\partial\varphi}{\partial x}\right)+\frac{\partial}{\partial y}\left(\rho v\varphi-\mu\frac{\partial\varphi}{\partial y}\right)+\frac{\partial}{\partial z}\left(\rho w\varphi-\mu\frac{\partial\varphi}{\partial z}\right)=S_\varphi(x,y,z) \tag{1.3}$$

where all physical mechanisms and extra terms not included in the total flux components on the rhs (right hand side) are treated by the source $S_\varphi$.

### 1.1.3. Basic Models for Flow in Porous Media

When the balance equations compacted in (1.1) are written for flow through porous media, the medium porosity has to be accounted for. Therefore, the continuity Equation (1.2) can be modified and expressed in vector form as:

$$\frac{\partial(\phi\rho)}{\partial t}=-\nabla\cdot\rho\vec{u}_{\mathrm{D}} \tag{1.4}$$

where $\phi$ is the medium porosity defined as the ratio of pore volume to total (fluid plus solid) volume, $\rho$ is the density based on total volume and $\vec{u}$ is the superficial velocity defined as the volumetric flow rate divided by unit of total cross-sectional area. The well-known Darcy (1856) law of motion can be further given as,

$$\vec{u}_{\mathrm{D}}=-\frac{K}{\mu}(\nabla p-\rho\vec{g}) \tag{1.5}$$

where $p$ is the pressure based on total area. The quantity $K$ is referred to as the medium **permeability**, the unit of which is Darcy, defined as the permeability of a porous medium to viscous flow for the flow of 1 ml of a liquid of 1 centipoise viscosity under a pressure gradient of 1 atm/cm across 1 cm$^2$ in 1 sec. If in Equation (1.4) the porosity $\phi$ is assumed invariant with time, one has,

$$\phi \frac{\partial \rho}{\partial t} = -\nabla \cdot \rho \vec{u}_{\mathrm{D}} \tag{1.6}$$

The empirical modification owned to Brinkman (1947) replaces Equation (1.5) with,

$$\vec{u}_{\mathrm{D}} = -\frac{K}{\mu} \left( \nabla p - \rho \vec{g} \right) + K \nabla^2 \vec{u}_{\mathrm{D}} \tag{1.7}$$

where the extra term is intended to account for distortion of the velocity profiles near containing walls.

The flow of fluids through consolidated porous media and through beds of granular solids are similar, both having the general function of pressure drop vs. flow rate alike in form, i.e., transition from laminar flow to turbulent flow is gradual (Perry and Chilton, 1973). For this reason, this function must include a *viscous* term and an *inertia* term. Therefore, an extension to (1.7) in order to account for the inertia effects was proposed by Forchheimer (1901) in the form,

$$\vec{u}_{\mathrm{D}} = -\frac{K}{\mu} \left( \nabla p - \rho \vec{g} \right) + K \nabla^2 \vec{u}_{\mathrm{D}} - \frac{F \left| \vec{u}_{\mathrm{D}} \right| \vec{u}_{\mathrm{D}}}{\nu} \tag{1.8}$$

where $F$ is known as the *inertia coefficient*. For the sake of using a coherent nomenclature throughout this book, another coefficient $c_{\mathrm{F}}$, which is related to $F$, is the *Forchheimer coefficient*. The proposition of writing the *inertia coefficient* as a function of $c_{\mathrm{F}}$ is a tentative to separate the dependence of the medium morphology on experimental values of $F$ for a variety of porous media. The relationship between $F$ and $c_{\mathrm{F}}$ will be shown below and, in all simulations to be shown is this book, $c_{\mathrm{F}}$ is taken to be a constant with value 0.55. The treatment followed in this book, i.e., the consideration of an additional mechanism for transport in porous media, namely turbulence, is distinct from the approach taken by many authors in the literature who proposed expressions for $c_{\mathrm{F}}$ in order to fit Equation (1.8) to experimental data for high values of Reynolds number (*Re*) (Bhattacharya *et al.*, 2002). Here, on the contrary, $c_{\mathrm{F}}$ is assumed a constant. The mechanism of *macroscopic turbulent transport*, explicitly appearing after simultaneously time and volume averaging the full convective term in (1.8) (not shown there), is modeled in separate giving an alternative way to account for discrepancies when (1.8) is applied to high Reynolds flow.

Additionally, the non-linear character of Equation (1.8) has direct implications on its numerical solution. For purely viscous flow, away from any containing walls, the last two terms on the right-hand side of Equation (1.8) become negligible and the resulting equation is again Darcy's Equation (1.5). Values of $K$ and $F$ are usually determined experimentally for each type of porous medium. For sphere-pack beds, Ergun (1952) has proposed the following empirical expression:

$$K = \frac{d^2\phi^3}{A(1-\phi)^2}; \quad F = \frac{1.75\,d}{150(1-\phi)} \tag{1.9}$$

where $d$ is the diameter of particles or pores in the bed and $A$ is a parameter that depends on the medium morphology. For a medium formed by an array of circular rods displaced in square arrangements, its value is given in Kuwahara and Nakayama (1998) as $A = 140$. Data on pressure drop as a function of flow rate for various fluids are generally available from the manufacturer of such porous media.

### 1.1.4. Extended Models for Flow in Porous Media

Since 1980s, several studies were published concerned with the application and extension, for several geometries and different processes, of the models embodied in Equation (1.8). A complete review of all these works would be outside the scope of the present text and for that only a few of then will be mentioned.

Boundary and initial effects have been investigated by Vafai (1981) and variable medium porosity studies are presented in Vafai (1984). The flow through packed beds has been accounted for by Hunt and Tien (1988) and by Adnani *et al.* (1995). The limitations imposed by the use of Equation (1.8) when applied to several flows of engineering interest have been the subject of the reports by Nield (1991) and by Vafai (1995). Knupp and Lage (1995) have extended the early Forchheimer's ideas to the tensor permeability case. Studies on natural convection systems in a porous saturated medium with both vertical and horizontal temperature gradients are reported in Manole and Lage (1995).

The advantages of having a combustion process inside an inert porous matrix are today well recognized. Hsu *et al.* (1993) points out some of its benefits including higher burning speed and volumetric energy release rates, higher combustion stability and the ability to burn gases of a low energy content. Driven by this motivation, the effects on porous ceramic inserts have been investigated in Peard *et al.* (1993). Turbulence modeling of combustion within inert porous media has been conducted by Lim and Matthews (1993) on the basis of an extension of the standard $k$–$\varepsilon$ model of Jones and Launder (1972). Work on direct simulation of turbulence in premixed flames, for the case when the porous dimension is of the order of the flame thickness, has also been reported in Sahraoui and Kaviany (1995).

Being a multidisciplinary area, studies on flow, heat and mass transfer in porous media have attracted the attention of many research groups around the world, most of which concerned with different applications, processes and systems configurations. With the recent explosion of the World Wide Web (WWW) on the Internet, the reader is encouraged to *navigate* through interesting sites which summarize the research being conducted at different locations. To mention just a few of them would not do justice to all those important work being presently carried out. The electronic addresses to these sites are readily available through the employment of pertinent key words used in conjunction with the many *search engines* (database-search facilities) available online.

### 1.1.5. Models for Petroleum Reservoir Simulation

The recovery and better use of existing oil fields is being considered lately by many countries due to its impact on internal economies. With exploration of new fields becoming prohibitively expensive and considering further that full experimentation in laboratories is an extremely difficult task, the subject of enhanced oil recovery (EOR) has stimulated many research efforts toward the development of mathematical and numerical tools able to analyze existing oil reserves. Simulation of the movement of a different phase (such as water) or of a different component (such as a miscible tracer) introduced through an injection well (see Figure 1.1) can provide important technical information to oil companies, aiding their decision-making process to pursue any further extraction on existing wells. Below is a summary of two important mathematical frameworks for analyzing the flow of miscible components and different phases through a porous rock embedded in oil.

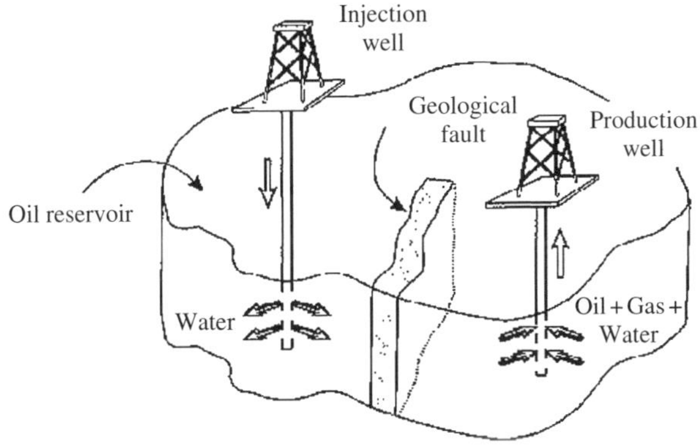


**Figure 1.1.** Examples of enhanced oil recovery system.

### 1.1.5.1. *Miscible Fluids*

If one considers the injection into the soil of a single-phase flow of two miscible fluids, let us say water and a tracer, Equation (1.4) for the overall mass conservation equation can be rewritten as

$$\frac{\partial(\phi\rho)}{\partial t} + \nabla\cdot\rho\vec{u}_{\mathrm{D}} = q \tag{1.10}$$

where $q$ is the source term referent to the mass injection at the point in question, i.e., the total mass flow rate of the mixture of component $\ell$ and water. Using the compressibility coefficient $\beta$, which is defined as,

$$\beta = \frac{1}{\rho}\left(\frac{\partial\rho}{\partial p}\right) \tag{1.11}$$

and considering negligible fluid weight, the substitution of (1.5) into (1.10) gives an equation for the pressure field of the form,

$$\nabla(\nabla p) = \frac{\mu\phi\beta}{\mathrm{K}}\frac{\partial p}{\partial t} - \frac{\mu}{\mathrm{K}}q \tag{1.12}$$

Equation (1.12), once solved with the appropriate boundary conditions, gives the distribution of pressure which, when applied in the Darcy's Equation (1.5), will give the velocity field.

To obtain the mass density distribution for the component $\ell$, one needs to perform the mass conservation for such tracer, and the governing equation is

$$\frac{\partial(\phi P_\ell)}{\partial t} + \nabla\cdot(\vec{J} + \vec{u}_{\mathrm{D}}P_\ell) = \mathrm{q}P_\ell^* \tag{1.13}$$

where $P_\ell$ is the component density (mass of component $\ell$ over void volume), $P_\ell^*$ its value at the injection well and $\vec{J}$ the diffusion flux given by

$$\vec{J} = -\phi D\Delta P_\ell \tag{1.14}$$

where $D$ is the diffusion coefficient. A common application of this model is the study of oil reservoirs where a radioactive tracer, mixed with water, is injected into the ground. Its concentration is monitored at production wells, giving important information on soil characteristics and oil availability. Once the velocity field is determinate by means of (1.5) and (1.12), Equations (1.13) and (1.14) give the distribution of the tracer throughout the field.

### 1.1.5.2. *Two-Phase Flow*

When distinct phases are considered, for oil and water, the so-called "black-oil model" can be applied (Sharpe, 1993). Not considering capillarity effects and body forces, the unknowns in this model are the water and oil saturations in addition to pressure. Calling $S_\gamma$ the phase saturation, $B_\gamma$ the phase formation factor and $q_\gamma$ the source term for the phase $\gamma$, the governing equations are:

$$\frac{\partial}{\partial t}\left(\frac{S_\gamma}{B_\gamma}\phi\right) = \nabla\cdot\left[\nabla\lambda_\gamma p\right] - q_\gamma \tag{1.15}$$

and

$$\sum_\gamma S_\gamma = 1 \tag{1.16}$$

where $\lambda_\gamma$ is the mobility of phase $\gamma$ given by,

$$\lambda_\gamma = \frac{KK_{r\gamma}}{B_\gamma \mu_\gamma} \tag{1.17}$$

In Equation (1.17), $K_{r\gamma}$ is the relative permeability of phase $\gamma$. Since one is using the black-oil model with no gas phase, it is irrelevant to specify both the component and the phase, since there is only oil in the oil phase and only water in the water phase. Equation (1.15) written for a general phase $\gamma$ in the Cartesian coordinate system is:

$$\frac{\partial}{\partial t}\left(\frac{S_\gamma}{B_\gamma}\phi\right) = \frac{\partial}{\partial x}\left(\lambda_x^\gamma \frac{\partial p}{\partial x}\right) + \frac{\partial}{\partial y}\left(\lambda_y^\gamma \frac{\partial p}{\partial y}\right) + \frac{\partial}{\partial z}\left(\lambda_z^\gamma \frac{\partial p}{\partial z}\right) - q_\gamma \tag{1.18}$$

The solution of the above equation set is usually found by resolving for pressure in an implicit fashion and then updating the saturation until final convergence is achieved (IMPES-IMplicit Pressure Explicit Saturation). More details on the numerics of this solution method can be found in Aziz and Setari (1979).

## 1.2. OVERVIEW OF TURBULENCE MODELING

### 1.2.1. General Remarks

Many research groups around the world are now investigating situations where turbulence might play an important role in the flow of a fluid inside a porous media. As an example, turbulence modeling of combustion within an inert porous matrix has been conducted by Lim and Matthews (1993). In that work the overall effect of the porous media on the turbulence level has been considered by an extension of well-established standard $k$–$\varepsilon$ model of Jones and Launder (1972). Other studies also appraise the direct simulation of turbulence in combustion systems, as reported in Sahraoui and Kaviany (1995).

Adapting turbulence models based on time averaged equations, which govern the continuum fluid phase, seems to be an emerging field of research applied to more realistic simulation of flow in porous media. With this motivation, the following is an overview of single-phase turbulence modeling. The interested reader is referred to more complete reviews considering the suitability of several classes of models (AGARD, 1991), recirculating flows (Leschziner, 1989) and second-moment closures (Launder, 1975).

### 1.2.2. Turbulence Phenomena

Before presenting turbulence models and their use, it is interesting to review a few concepts on turbulence phenomena. Only an overview of basic ideas is here presented, and, to the enthusiastic reader, the classical textbooks of Tennekes and Lumley (1972), Hinze (1975), Monin and Yaglom (1975) and Bradshaw (1978) are suggested for further study. In the last few years, new titles have come such as Warsi (1993), Libby (1996), Lesieur (1997), Chen and Jaw (1998) and Davidson (2004), for example.

Turbulent motion can be visualized as the superimposed movement of eddies of fluid of a wide range of dimensions, having each of those sizes a corresponding fluctuating frequency. The eddies have time-varying vorticities in random directions. The largest eddies in the flow, and consequently their associated frequencies, have sizes limited by the characteristic flow dimensions. For instance, the largest eddies in a pipe will be of the order of the pipe diameter. On the other hand, the smallest eddies, associated with the highest frequencies, will be determined by the fluid molecular viscosity. The larger the Reynolds number, the wider this spectrum will be.

Turbulence can also be characterized by means of a fluctuating field, giving rise to the correlation terms after the time averaging process. In addition, the turbulence *energy spectrum* represents the distribution of kinetic energy carried by eddies of different sizes. The high-energy eddies will carry the most momentum, energy and mass, and therefore will be most responsible for the total value of the Reynolds stresses/fluxes. It is exactly an eddy with this dimension that turbulence models try to simulate. It is interesting to emphasize that the great majority of turbulence models for practical applications make use of only *one* value for the turbulent kinetic energy and for the associated characteristic length, even though the energy spectrum is a *continuous* function of the eddy size. For engineering flows of practical interest, however, only one typical value will be a good approximation for calculation purposes.

The transfer process passing down energy to eddies of smaller sizes is known in the literature as *cascade of energy*. The low-frequency fluctuations extract their energy from the mean flow. The random and rotational motion of those fluid elements tends to pull each other, decreasing their size and therefore feeding high-frequency fluctuations with kinetic energy. This process keeps transferring energy to even smaller eddies, until the energy is finally converted into heat by viscous action. As energy is being fed to eddies of smaller size by the *vortex-stretching* seen above, directional effects became of a lesser intensity. Consequently, for a sufficiently high *Re*, the fine structure of turbulence is said

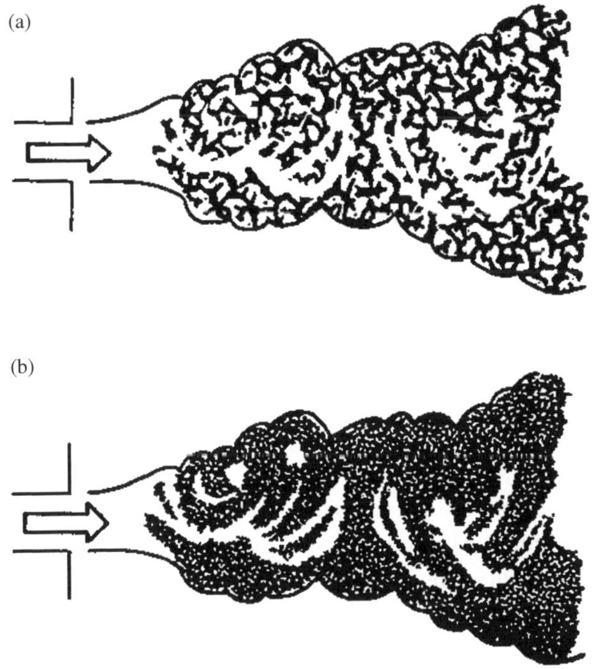


**Figure 1.2.** Influence of Reynolds number of fine turbulence structure.

to have achieved the state of local isotropy. In this situation, energy is transformed into heat for eddies with dimension $\eta = (\nu/\varepsilon)^{1/4}$ (Kolmogorov, 1942).

It is very interesting to point out that the viscosity $\nu$ does not determine the rate at which energy is dissipated, but rather the size of the dissipation eddies. For instance, if in a certain flow $\varepsilon$ is increased through increasing *Re*, eddies of size $\eta$ will have their dimensions reduced in order to balance the energy production and dissipation rates. Figure 1.2 taken from Tennekes and Lumley (1972) illustrates this idea, where a darker pattern for a higher *Re* indicates a decrease in the turbulence fine structure. It is also worth noting that the denser shade shows a more isotropic pattern than before.

Next section presents a classification of modeling techniques according to the most common nomenclature found in the literature.

### 1.2.3. Traditional Classification of Turbulence Models

#### *1.2.3.1. Basic Concepts*

Two important concepts are strictly linked to turbulence modeling. The first, known as *turbulent viscosity*, $\mu_t$, has been widely used over the years, and the second, the *mixing length*, $\ell_m$, is included here for historical reasons.

In 1877, the French scientist Boussinesq (1877) supposed that, in a turbulent flow, the Reynolds stresses acting on an element of fluid could be approximated as proportional to the element deformation rate, in a perfect analogy to molecular momentum transfer. This idea was grounded on the hypothesis that eddies, like molecules, would exchange momentum and heat after interacting with each other. For an incompressible fluid, this concept can be written as:

$$-\rho\overline{u_i u_j} = \mu_t \left[ \frac{\partial U_i}{\partial x_j} + \frac{\partial U_j}{\partial x_i} \right] - \frac{2}{3}\rho k \delta_{ij} \tag{1.19}$$

and for the turbulent heat or mass transfer this approximation reads:

$$-\rho\overline{u_j\theta} = \frac{\mu_t}{\sigma_t}\left( \frac{\partial \Theta}{\partial x_j} \right) \tag{1.20}$$

where $\theta$ is the fluctuating part of $\Theta$ (temperature or concentration) and $\sigma_t$ is known as the turbulent Prandtl number for turbulent energy/mass transfer.

In spite of the simplicity embodied in (1.19) and (1.20), a perfect analogy between laminar and turbulent flow is not possible, because: (a) in contrast with molecules, eddies carrying the most energy have dimensions of the order of the flow domain; (b) eddies cannot be seen as rigid bodies, loosing their identity after colliding with each other; and (c) $\mu_t$ is a local and flow-dependent parameter, presenting sometimes a strong variation with the transport direction.

It should be pointed out that the advantages in using (1.19) and (1.20) greatly overcomes the absence of a more-elaborate physical interpretation of $\mu_t$. It is also worth noting that (1.19) and (1.20) do not constitute a *model* for the stresses/fluxes, since information on $\mu_t$ has still to be supplied. The second important concept here presented constitutes, in fact, the first turbulence model reported in the literature.

In 1925, the German scientist Prandtl (1925), using an argument based on the kinetic theory of gases, assumed that $\mu_t$ could be calculated as proportional to a fluctuating velocity and to a characteristic length, which he named *mixing length*, $\ell_m$. Prandtl further suggested that the characteristic velocity could be taken as the product of $\ell_m$ and the gradient of the mean velocity, given for $\mu_t$:

$$\mu_t = \ell_m \left| \frac{\partial U}{\partial y} \right| \tag{1.21}$$

The physical interpretation of $\ell_m$ is, therefore, "the distance traveled by an eddy corresponding to an average change in the local mean velocity equals $\partial U/\partial y$ times $\ell_m$". With the concepts of $\mu_t$ and $\ell_m$, turbulence models can be more clearly presented and discussed upon.

*Zero-Equation Models* This class of models is also known in the literature as First Order of Phenomenological. In this method, the eddy behavior is explained by linking the Reynolds stresses to the mean flow through the use of the above-presented mixing length concept. Transport equations are then solved only for the mean flow, and for general three-dimensional situations, the Reynolds stresses are given by:

$$\rho\overline{u_i u_j} = \rho l_{\mathrm{m}}^2 \left\{ \left[ \frac{\partial U_i}{\partial x_j} + \frac{\partial U_j}{\partial x_i} \right] \frac{\partial U_i}{\partial x_j} \right\}^{1/2} \left[ \frac{\partial U_i}{\partial x_j} + \frac{\partial U_j}{\partial x y_i} \right] + \frac{2}{3}\rho k \delta_{ij} \tag{1.22}$$

Equation (1.22) is of little use, since specification of $\ell_{\mathrm{m}}$ for general three-dimensional flow is of a certain difficulty. Nevertheless, for two-dimensional shear layers, Equation (1.22) gives a simple formula for the shear stress $\overline{uv}$, as:

$$-\overline{uv} = \ell_{\mathrm{m}}^2 \frac{\partial U}{\partial y} \left| \frac{\partial U}{\partial y} \right| \tag{1.23}$$

For free shear layers, $\ell_{\mathrm{m}}$ is usually taken as proportional to the layer thickness $\delta$. For planar flow, $\delta$ is defined as the distance between the points where the local velocities differ by 1% of $U$, the total velocity variation across the layer. For axi-symmetric situations, $\delta$ is the distance from the symmetry axis to the point where the velocity is $0.99U$.

*One-Equation Models* One of the major drawbacks of the zero-equation models is the assumption that the characteristic velocity and length scales are not transported throughout the flow. Transport processes by convection and diffusion mechanisms are particularly important where the velocity gradient across the layer is small, in some recirculating flows and when there is a rapid development of the flow conditions.

One-equation models solve, then, an additional transport equation for turbulent kinetic energy $k$, giving the turbulent viscosity the formula:

$$\mu_{\mathrm{t}} = c_\mu^* \sqrt{k} \ell_{\mathrm{e}} \tag{1.24}$$

where $c_\mu^*$ is a constant and $\ell_{\mathrm{e}}$ is a length scale associated with the energy containing eddies. Equation (1.24) is known in the literature as the Kolmogorov–Prandtl expression, being introduced independently by Kolmogorov in 1942 and by Prandtl in 1945.

The accompanying equations for $k$ is suggested in the literature as:

$$\frac{D(\rho k)}{Dt} = \frac{\partial}{\partial x_j} \left[ \frac{\mu_{\mathrm{t}}}{\sigma_k} \frac{\partial k}{\partial x_j} \right] + \mu_{\mathrm{t}} \left[ \frac{\partial U_i}{\partial x_j} + \frac{\partial U_j}{\partial x_i} \right] \frac{\partial U_i}{\partial x_j} + \rho c_{\mathrm{D}} k^{3/2} \ell_{\mathrm{e}} \tag{1.25}$$

where $c_{\mathrm{D}}$ is a constant.

*Two-Equation Models* In calculating several flows of engineering interest, it was noticed that although the use of (1.25) could well represent, in some cases, the transport of the characteristic velocity $k^{1/2}$, the application of empirical formulae for the length scale could not do the same for predicting the development of along the flow. In other words, the length scale $\ell_e$, similar to $k^{1/2}$, is also subjected to the processes of convection and diffusion, and a second transport equation for $\ell_e$ was then made mandatory in those cases.

The length scale $\ell_e$ need not necessarily be the dependent variable for this second equation. In fact, after numerous tests in the literature, it has been found that the dissipation rate of $k$, namely $\varepsilon$, gave remarkably good results in a wide range of flows. The dissipation rate $\varepsilon$ can be related to the length $\ell_e$ as:

$$\varepsilon = \frac{k^{3/2}}{\ell_e} \tag{1.26}$$

giving for the turbulent viscosity $\mu_t$:

$$\mu_t = \rho c_\mu \frac{k^2}{\varepsilon} \tag{1.27}$$

where $c_\mu$ is a constant.

An equation for $\varepsilon$ is suggested by Jones and Launder (1972) as:

$$\frac{d\rho\varepsilon}{dt} = \frac{\partial}{\partial x_j}\left[\frac{\mu_t}{\sigma_\varepsilon}\frac{\partial \varepsilon}{\partial x_j}\right] + c_{1\varepsilon}\frac{\varepsilon}{k}\mu_t\left[\frac{\partial U_i}{\partial x_j} + \frac{\partial U_j}{\partial x_i}\right]\frac{\partial U_i}{\partial x_j} + \rho c_{2\varepsilon}\frac{\varepsilon^2}{k} \tag{1.28}$$

In Equation (1.28), the $c$'s and $s_e$ are constants. The basic advantage of the $k$–$\varepsilon$ model over simpler methods is its ability to predict the length scale variation along the flow.

*Reynolds Stress Models* Although the $k$–$\varepsilon$ is, without any question, the most widely tested turbulence model in the literature, there are a few situations where its results are still considered to be poor.

In all the models seen so far, it was assumed the turbulence could be well simulated by knowing only *one* velocity and *one* length scale, and that all stresses could be represented by Equations (1.19) and (1.20). These equations imply that all stresses/fluxes are equally considered, regardless of their geometric plane of action. For predicting turbulence-driven secondary motion and effects caused by buoyancy forces, rotation and wall proximity, the use of (1.1) and (1.2) will not satisfactorily simulate the correct directional influence caused by those factors.

With the aim to describe, for each stress, the effects that certain factors have upon them, methods called *Reynolds stress models* (RSM) were developed. In this case, individual equations are solved containing all information pertaining to that particular component.

A modeled equation for the stress $\rho\overline{u_i u_j}$ can be found in the literature as (Launder *et al.*, 1975);

$$\frac{d\overline{u_i u_j}}{dt} = P_{ij} + G_{ij} + \pi_{ij} - \frac{2}{3}\delta_{ij}\,\varepsilon + \mathrm{D}_{ij} \tag{1.29}$$

where the terms on the rhs represent the processes of production by the mean flow, production by buoyancy forces, distribution by pressure fluctuations, dissipation by molecular effects and diffusion by turbulent and viscous interactions. A complete discussion on the modeling steps necessary to obtain all the terms in (1.29) is beyond the scope of the present work, and to the interested reader Launder *et al.* (1975) is suggested.

With exception of simple cases, the solution of (1.29) is extremely tedious, and its use is only acceptable when simpler theories cannot correctly predict the total value of $\rho\overline{u_i u_j}$.

*Algebraic Stress Models (ASM)* The use of (1.29) implies, for a three-dimensional case with constant temperature, finding the solution of six equations in addition to the solution of the mean flow. This is obviously a formidable task, even considering the present stage of development of digital computers. By changing the character of (1.29) into a more easy-to-handle algebraic relation, Rodi (1972) first suggested that all terms governing the total budget of $\rho\overline{u_i u_j}$ could still be accounted for. His well-known expression reads:

$$\frac{D\overline{u_i u_j}}{Dt} - D_{ij} = \frac{\overline{u_i u_j}}{k}\left(P_{\mathrm{k}} - \varepsilon\right) \tag{1.30}$$

Rodi and a number of coworkers have used (1.30) in a wide range of flow configurations. An analogous expression for turbulent fluxes has also been proposed by Gibson and Launder (1976):

$$\frac{D\overline{u_j\theta}}{Dt} - D_{j\theta} = 0 \tag{1.31}$$

where local equilibrium for the thermal field is assumed.

The major advantage of such method is the elimination of the hypothesis of a *scalar* turbulent viscosity embodied in Equations (1.19) and (1.20) (Rodi, 1972). Among many geometries, ASMs have been used successfully in vertical buoyant jets (Lujboja and Rodi, 1981) and in liquid-metal pipe flow (de Lemos and Sesonske, 1985).

*Large Eddy Simulation and Direct Numerical Simulation* The fundamental idea of all closures seen above is the determination of the extra unknowns appearing in the time-averaged Navier-Stokes and energy equations. A recent method that overcomes the need of the time averaging process is called large eddy simulation (LES). This approach consists of the computation of the instantaneous form of governing equations. A model for the eddies with sizes of the order of the computational grid provides the additional information needed for closure. Solutions of this kind require substantial computer time and storage and only recently started to be used in practical engineering calculations. A review on the several LES techniques available today is beyond the scope of this text. As a suggestion to the interested reader, some early and pioneering information on this method can be found in Herring (1977) and Schumann *et al.* (1980).

## 1.3. TURBULENT FLOW IN PERMEABLE STRUCTURES

The two classical analyses seen above, namely porous media modeling and turbulence modeling, can be combined in order to broaden our ability to simulate real engineering systems, such as permeable structures.

A permeable structure can be seen as a multiphase system with one of the phases being the solid matrix. The flowing fluid can be either a single-phase substance composed of a mixture of distinct chemical species (see section "Miscible Fluids", p. 8) or a two-phase mixture (see "Two-Phase Flow" on p. 8). A large number of physical systems can be seen as a porous or permeable medium. In the past decades, macroscopic equations have been used to analyze innumerous engineering and naturally formed porous media, spanning from underground flow in soils to atmospheric boundary layer over crops and forests. As such, we should here be careful in characterizing what category of flows this book aims at modeling.

Figure 1.3 illustrates the class of flow considered in this book. The Reynolds number based on the statistical value of the void size is sufficiently high for turbulence to be established within the void space. In addition, the system can be seen at a macroscopic

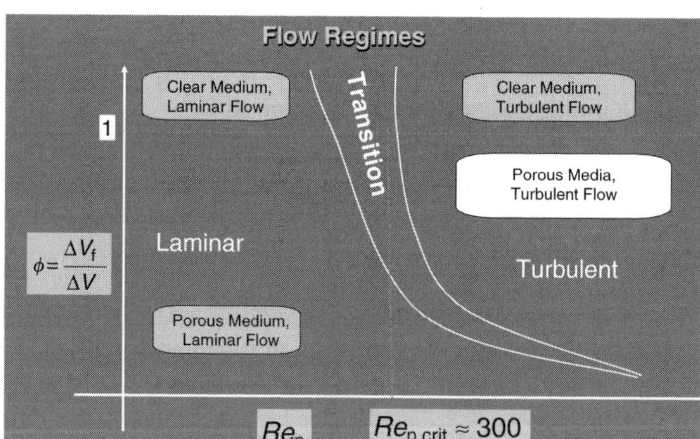


**Figure 1.3.** Flow regimes over permeable structures.

level, and the two phases involved, namely the solid (porous matrix) and the fluid phases, are modeled as a unique phase after the homogenization process. In Figure 1.4 a heat exchanger is seen as a porous structure through which the working fluid permeates.

Environmental flows also benefit from such macroscopic views. Figure 1.5 shows the atmospheric boundary layer over a thick rain forest seen as a layer of porous media covering the earth crust. As such, macroscopic exchange rates of energy and mass can be investigated using an upscaling technique.

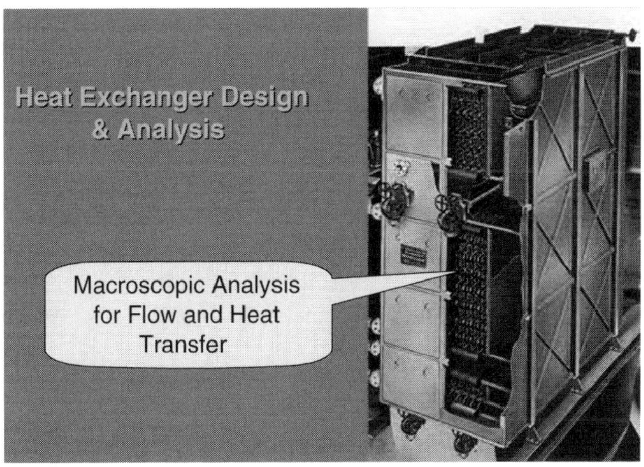


**Figure 1.4.** Macroscopic analysis of heat exchangers.

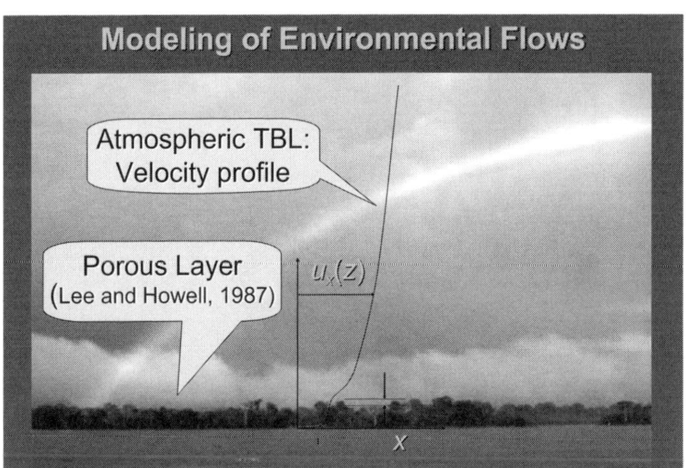


**Figure 1.5.** Macroscopic view of flow over rain forests.

Already, a few reference books (Ingham and Pop, 1998, 2002, 2005; Vafai, 2005) have chapters devoted to the analysis of transitional flow, from Darcy regime to fully turbulent flow, in porous media. This recent interest in the literature reflects the many advantages in having an appropriate mathematical framework for analysis of turbulent flow in permeable media. The chapters to follow are devoted to expose one of such views in which both spatial deviation and time fluctuation of flow variables are simultaneously considered.

Chapter 2

# Governing Equations

## 2.1. LOCAL INSTANTANEOUS GOVERNING EQUATIONS

The steady-state local or microscopic instantaneous transport equations for an incompressible fluid with constant properties are given by:

$$\nabla \cdot \mathbf{u} = 0 \tag{2.1}$$

$$\rho \nabla \cdot (\mathbf{u}\mathbf{u}) = -\nabla p + \mu \nabla^2 \mathbf{u} + \rho \mathbf{g} \tag{2.2}$$

$$(\rho c_{\mathrm{p}}) \nabla \cdot (\mathbf{u} T) = \nabla \cdot (\lambda \nabla T) \tag{2.3}$$

where $\mathbf{u}$ is the velocity vector, $\rho$ is the density, $p$ is the pressure, $\mu$ is the fluid viscosity, $c_{\mathrm{p}}$ is the specific heat, $T$ is the temperature, $\lambda$ is the fluid thermal conductivity and $\mathbf{g}$ is the gravity acceleration vector.

The transient form of the microscopic momentum Equation (2.2) for a fluid with constant properties is given by the Navier-Stokes equation as follows:

$$\rho \left[ \frac{\partial \mathbf{u}}{\partial t} + \nabla \cdot (\mathbf{u}\mathbf{u}) \right] = -\nabla p + \mu \nabla^2 \mathbf{u} + \rho \mathbf{g} \tag{2.4}$$

In addition, a steady-state form of (1.13) for clear medium ($\phi = 1$) using vector notation, giving the mass fraction distribution for the chemical species $\ell$, is governed by the following transport equation:

$$\nabla \cdot (\rho \mathbf{u} m_\ell + \mathbf{J}_\ell) = \rho R_\ell \tag{2.5}$$

where $m_\ell$ is the mass fraction of the species $\ell$, defined as $m_\ell = \rho_\ell / \rho$, $\rho_\ell$ is the mass density of species $\ell$ (mass of $\ell$ over total mixture volume), $\mathbf{u}$ is the mass-averaged velocity of the mixture, $\mathbf{u} = \sum_\ell m_\ell \mathbf{u}_\ell$, $\mathbf{u}_\ell$ is the velocity of species $\ell$, $\rho$ is the bulk density of the mixture ($\rho = \sum_\ell \rho_\ell$) and $\mathbf{J}_\ell$ is the mass diffusion flux of $\ell$. The generation rate of species $\ell$ per unit mass of mixture is given in (2.4) by $R_\ell$.

Further, the mass diffusion flux $\mathbf{J}_\ell$ in Equation (2.5) is due to the velocity slip of species $\ell$, and is given by:

$$\mathbf{J}_\ell = \rho_\ell (\mathbf{u}_\ell - \mathbf{u}) = -\rho D_\ell \nabla m_\ell \tag{2.6}$$

where $D_\ell$ is the coefficient of species $\ell$ for diffusion into the mixture. The second equality in Equation (2.6) is known as Fick's Law, which is a constitutive equation strictly valid for binary mixtures in the absence of any additional driving mechanisms for mass transfer (Ingham and Pop, 2002). Therefore, no *Soret* or *Dufour* effects are here considered.

An alternative way of writing the mass transport equation is using the volumetric molar concentration $C_\ell$ (mol of species $\ell$ over total mixture volume), the molar weight $M_\ell$ (g/mol of species $\ell$) and the molar generation/destruction rate $R_\ell^*$ (mol of species $\ell$ produced or consumed/total mixture volume), giving:

$$M_\ell \nabla \cdot (\mathbf{u} C_\ell + \mathbf{J}_\ell) = M_\ell R_\ell^* \tag{2.7}$$

The mass diffusion flux $\mathbf{J}_\ell$ (mass of $\ell$ per unit area per unit time) in (2.5) or (2.7) can then be written as:

$$\mathbf{J}_\ell = \rho_\ell (\mathbf{u}_\ell - \mathbf{u}) = -\rho_\ell D_\ell \nabla m_\ell = -M_\ell D_\ell \nabla C_\ell \tag{2.8}$$

Rearranging (2.7) for an inert species, dividing it by $M_\ell$ and dropping the index $\ell$ for a simple binary mixture, one has

$$\nabla \cdot (\mathbf{u} C) = \nabla \cdot (D \nabla C) \tag{2.9}$$

If one considers that the density in the last term of (2.2) varies with temperature and concentration, for natural convection flow, the Boussinesq hypothesis reads, after renaming this density $\rho_\mathrm{T}$:

$$\rho_\mathrm{T} \cong \rho[1 - \beta(T - T_\mathrm{ref}) - \beta_\mathrm{C}(C - C_\mathrm{ref})] \tag{2.10}$$

where the subscript "ref" indicates a reference value, and $\beta$ and $\beta_\mathrm{C}$ are the thermal and salute expansion coefficients, respectively, defined by:

$$\beta = -\frac{1}{\rho} \frac{\partial \rho}{\partial T}\bigg|_{\mathrm{p,C}}, \quad \beta_\mathrm{C} = -\frac{1}{\rho} \frac{\partial \rho}{\partial C}\bigg|_{\mathrm{p,T}} \tag{2.11}$$

Equation (2.10) is an approximation of (2.11) and shows how density varies with temperature and concentration in the body force term of the momentum equation.

Further, substituting (2.10) into (2.2), one has:

$$\rho \nabla \cdot (\mathbf{u}\mathbf{u}) = -\nabla p + \mu \nabla^2 \mathbf{u} + \rho \mathbf{g}[1 - \beta(T - T_\mathrm{ref}) - \beta(C - C_\mathrm{ref})] \tag{2.12}$$

Thus, the momentum equation becomes,

$$\rho\nabla\cdot(\mathbf{uu}) = -(\nabla p)^* + \mu\nabla^2\mathbf{u} - \rho\mathbf{g}[(\beta(T - T_{\text{ref}}) + \beta_C(C - C_{\text{ref}})] \tag{2.13}$$

where $(\nabla p)^* = \nabla p - \rho\mathbf{g}$ is a modified pressure gradient.

For fluid and solid phases with heat sources, Equation (2.3) becomes:

$$(\rho c_p)_f\nabla\cdot(\mathbf{u}T_f) = \nabla\cdot(\lambda_f\nabla T_f) + S_f \tag{2.14}$$

$$0 = \nabla\cdot(\lambda_s\nabla T_s) + S_s \tag{2.15}$$

where the subscripts f and s refer to each phase, respectively. Equation (2.15) corresponds to the porous matrix of the solid phase. If there is no heat generation either in the solid or in the fluid phase, we obtain,

$$S_f = S_s = 0 \tag{2.16}$$

As mentioned, there are, in principle, two ways that one can follow in order to treat turbulent flow in porous media. The first method applies a time average operator to the governing Equations (2.1)–(2.5) before the volume average procedure is applied. In the second approach, the order of application of the two average operators is reversed. Both techniques aim at derivation of suitable macroscopic transport equations.

Volume averaging in a porous medium, described in detail in Slattery (1967), Whitaker (1969, 1999) and Gray and Lee (1977) makes use of the concept of a *representative elementary volume* (REV) over which local equations are integrated. After integration, detailed information within the volume is lost and, instead, overall properties referring to an REV are considered. In a similar fashion, statistical analysis of turbulent flow leads to time mean properties. Transport equations for statistical values are considered in lieu of instantaneous information on the flow.

Before undertaking the task of developing macroscopic equations, it is convenient to recall the definition of volume average and time average operators.

## 2.2. THE AVERAGING OPERATORS

### 2.2.1. Local Volume Averaging

The macroscopic governing equations for flow through a porous substratum can be obtained by volume averaging the corresponding microscopic equations over a Representative Elementary Volume of size $\Delta V$ (Bear, 1972; Figure 2.1). For a general fluid property $\varphi$, the volumetric average taken over an REV can be written as (Slattery, 1967):

$$\langle\varphi\rangle^v = \frac{1}{\Delta V}\int_{\Delta V}\varphi\,dV \tag{2.17}$$

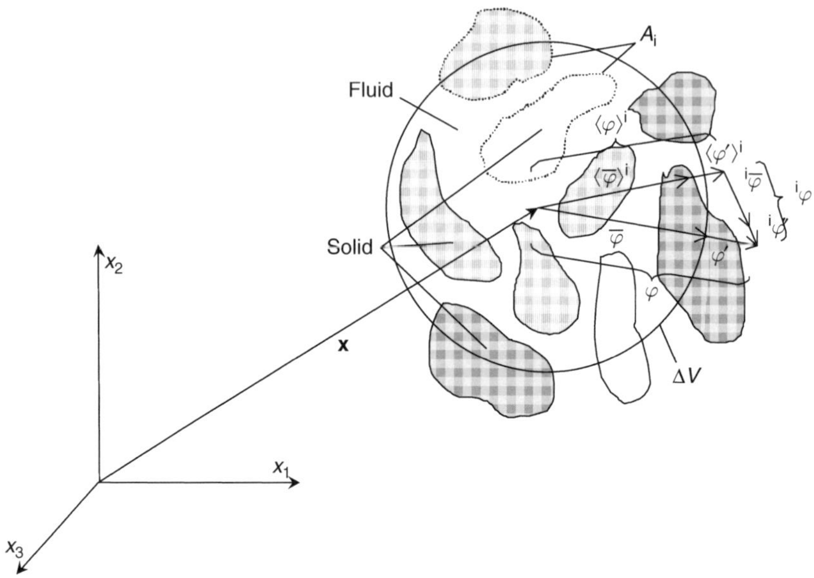


**Figure 2.1.** Representative elementary volume (REV), intrinsic average; space and time fluctuations (from Pedras and de Lemos, 2001a, with permission).

The value $\langle\varphi\rangle^{\mathrm{v}}$ is defined for any point **x** surrounded by an REV of size $\Delta V$. This average is related to the intrinsic average for the fluid phase as follows:

$$\langle\varphi_{\mathrm{f}}\rangle^{\mathrm{v}} = \phi\langle\varphi_{\mathrm{f}}\rangle^{\mathrm{i}} \tag{2.18}$$

where $\phi = \Delta V_{\mathrm{f}}/\Delta V$ is the local medium porosity and $\Delta V_{\mathrm{f}}$ is the volume occupied by the fluid in an REV.

The property $\varphi$ can then be defined as the sum of $\langle\varphi\rangle^{\mathrm{i}}$ and a term related to its distribution within the REV, ${}^{\mathrm{i}}\varphi$ (Whitaker, 1969):

$$\varphi = \langle\varphi\rangle^{\mathrm{i}} + {}^{\mathrm{i}}\varphi \tag{2.19}$$

In Equation (2.19), ${}^{\mathrm{i}}\varphi$ is the spatial deviation of $\varphi$ with respect to the intrinsic average $\langle\varphi\rangle^{\mathrm{i}}$. From (2.17) and (2.19), one derives $\langle{}^{\mathrm{i}}\varphi\rangle^{\mathrm{i}} = 0$. Figure 2.1 illustrates the idea underlined by Equation (2.19) for the value of a property of vectorial nature (e.g. velocity) in a position **x**. The spatial deviation is the difference between the local value (microscopic) and its intrinsic (fluid-based) average.

For deriving the flow-governing equations, it is necessary to know the relationship between the volumetric average of derivatives and the derivatives of the volumetric average. These relationships are presented in a number of works (Slattery (1967),

Whitaker (1969, 1999), Gray and Lee (1977), and others). They are known as the "Theorem of Local Volumetric Average", and are written as follows:

$$\langle \nabla \varphi \rangle^{\mathrm{v}} = \nabla(\phi \langle \varphi \rangle^{\mathrm{i}}) + \frac{1}{\Delta V} \int_{A_{\mathrm{i}}} \mathbf{n} \varphi \, \mathrm{d}S \tag{2.20}$$

$$\langle \nabla \cdot \varphi \rangle^{\mathrm{v}} = \nabla \cdot (\phi \langle \varphi \rangle^{\mathrm{i}}) + \frac{1}{\Delta V} \int_{A_{\mathrm{i}}} \mathbf{n} \cdot \varphi \, \mathrm{d}S \tag{2.21}$$

$$\left\langle \frac{\partial \varphi}{\partial t} \right\rangle^{\mathrm{v}} = \frac{\partial}{\partial t}(\phi \langle \varphi \rangle^{\mathrm{i}}) - \frac{1}{\Delta V} \int_{A_{\mathrm{i}}} \mathbf{n} \cdot (\mathbf{u}_{\mathrm{i}} \varphi) \, \mathrm{d}S \tag{2.22}$$

where $A_{\mathrm{i}}$, $\mathbf{u}_{\mathrm{i}}$ and $\mathbf{n}$ are the interfacial area, the velocity of phase $f$ and the unit vector normal to $A_{\mathrm{i}}$, respectively. The area $A_{\mathrm{i}}$ should not be confused with the surface area surrounding volume $\Delta V$ in Figure 2.1.

For single-phase flow, phase $f$ is the fluid itself and $\mathbf{u}_{\mathrm{i}} = 0$ if the porous substrate is assumed to be fixed. If the medium is further assumed to be rigid, then $\Delta V_{\mathrm{f}}$ may be dependent on space but is independent of time (Gray and Lee, 1977). To the interested reader, details on the Theorem of Local Volumetric Average can be found in Slattery (1967), Whitaker (1969, 1999) and Gray and Lee (1977).

### 2.2.2. Instantaneous Time Averaging

The need for considering time fluctuations occurs when turbulence effects are of concern. Traditional analyses of turbulence are based on statistical quantities. By applying statistical tools to the instantaneous flow-governing equations, local time-averaged equations are obtained. For that, the time average value of the general property, $\varphi$, associated with the fluid is given as (see Figure 2.2; Tennekes and Lumley, 1972; Hinze, 1975; Libby, 1996; Davidson, 2004):

$$\overline{\varphi} = \frac{1}{\Delta t} \int_{t}^{t+\Delta t} \varphi \, \mathrm{d}t \tag{2.23}$$

where the time interval $\Delta t$ is small compared to the fluctuations of the average value $\overline{\varphi}$, but large enough to capture turbulent fluctuations of $\varphi$. Time decomposition can, then, be written such that the instantaneous property $\varphi$ can be defined as the sum of the time average $\overline{\varphi}$ and the fluctuating component $\varphi'$ as follows:

$$\varphi = \overline{\varphi} + \varphi' \tag{2.24}$$

with $\overline{\varphi'} = 0$. Here, $\varphi'$ is the *time fluctuation* of $\varphi$ around its average $\overline{\varphi}$.

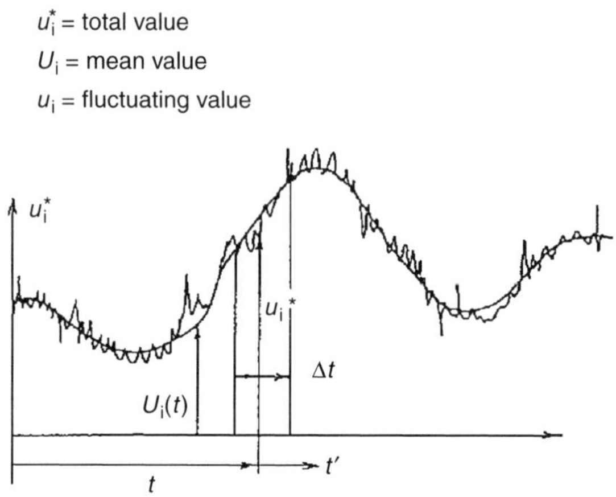


**Figure 2.2.** Time averaging over a length of time $\Delta t$.

### 2.2.3. Commutative Properties

From the definition of volume average (2.17) and time average (2.23), one can conclude that the time average of the volume average of property $\varphi$ is given by:

$$\overline{\langle \varphi \rangle^{\mathrm{v}}} = \frac{1}{\Delta t} \int_{t}^{t+\Delta t} \left[ \frac{1}{\Delta V} \int_{\Delta V_{\mathrm{f}}} \varphi \, \mathrm{d}V \right] \mathrm{d}t \tag{2.25}$$

The volume average of the time average is:

$$\langle \overline{\varphi} \rangle^{\mathrm{v}} = \frac{1}{\Delta V} \int_{\Delta V_{\mathrm{f}}} \left[ \frac{1}{\Delta t} \int_{t}^{t+\Delta t} \varphi \, \mathrm{d}t \right] \mathrm{d}V \tag{2.26}$$

As mentioned, for a rigid medium, the volume of fluid $\Delta V_{\mathrm{f}}$ might be dependent on space but not on time. If the time interval chosen for temporal averaging, $\Delta t$, is the same for all REV, then the volumetric average commutes with time average because both integration domains in (2.25) and (2.26) are independent of each other. In this case, the order of application of average operators is immaterial so that Equations (2.25) and (2.26) will lead to:

$$\overline{\langle \varphi \rangle^{\mathrm{v}}} = \langle \overline{\varphi} \rangle^{\mathrm{v}} \quad \text{or} \quad \overline{\langle \varphi \rangle^{\mathrm{i}}} = \langle \overline{\varphi} \rangle^{\mathrm{i}} \tag{2.27}$$

## 2.3. TIME AVERAGED TRANSPORT EQUATIONS

In order to apply the time average operator to the relevant transport equations, we consider:

$$\mathbf{u} = \overline{\mathbf{u}} + \mathbf{u}'; \quad T = \overline{T} + T'; \quad C = \overline{C} + C'; \quad p = \overline{p} + p' \tag{2.28}$$

Substituting expression (2.28) into Equations (2.1), (2.3), (2.9) and (2.13), and considering constant flow properties, we obtain, respectively

$$\nabla \cdot \overline{\mathbf{u}} = 0 \tag{2.29}$$

$$(\rho c_p)\nabla \cdot (\overline{\mathbf{u}}\overline{T}) = \nabla \cdot (\lambda \nabla \overline{T}) + \nabla \cdot (-\rho c_p \overline{\mathbf{u}'T'}) \tag{2.30}$$

$$\nabla \cdot (\mathbf{u}C) = -\nabla \cdot (D\nabla C) + \nabla \cdot (-\overline{\mathbf{u}'C'}) \tag{2.31}$$

$$\rho \nabla \cdot (\overline{\mathbf{u}}\overline{\mathbf{u}}) = -(\nabla \overline{p})^* + \mu \nabla^2 \overline{\mathbf{u}} + \nabla \cdot (-\rho \overline{\mathbf{u}'\mathbf{u}'}) - \rho\, \mathbf{g}[\beta(\overline{T} - T_{ref}) + \beta_C(\overline{C} - C_{ref})] \tag{2.32}$$

A transient form of (2.32), not using Bousinesq approximation for the buoyancy term, comes from time averaging (2.2) with $\mathbf{u} = \overline{\mathbf{u}} + \mathbf{u}'$:

$$\rho \left[ \frac{\partial \overline{\mathbf{u}}}{\partial t} + \nabla \cdot (\overline{\mathbf{u}}\overline{\mathbf{u}}) \right] = -\nabla \overline{p} + \mu \nabla^2 \overline{\mathbf{u}} + \nabla \cdot (-\rho \overline{\mathbf{u}'\mathbf{u}'}) + \rho \mathbf{g} \tag{2.33}$$

where the apparent stresses $-\rho\overline{\mathbf{u}'\mathbf{u}'}$, known as Reynolds stresses, appear after the averaging process.

For a clear fluid, the use of the eddy-diffusivity concept for expressing the *stress–rate of strain* relationship for the Reynolds stress appearing in Equations (2.32) or (2.33) gives:

$$-\rho \overline{\mathbf{u}'\mathbf{u}'} = \mu_t 2\overline{\mathbf{D}} - \frac{2}{3}\rho k \mathbf{I} \tag{2.34}$$

where $\overline{\mathbf{D}} = [\nabla \overline{\mathbf{u}} + (\nabla \overline{\mathbf{u}})^T]/2$ is the mean deformation tensor, $k = \overline{\mathbf{u}' \cdot \mathbf{u}'}/2$ is the turbulent kinetic energy per unit mass, $\mu_t$ is the turbulent viscosity and $\mathbf{I}$ is the unity tensor.

Similarly, for the turbulent fluxes on the rhs of (2.30) and (2.31), the eddy-diffusivity concept reads:

$$-\rho c_p \overline{\mathbf{u}'T'} = c_p \frac{\mu_t}{\sigma_t} \nabla \overline{T}; \quad -\rho \overline{\mathbf{u}'C'} = \frac{\mu_t}{Sc_t} \nabla \overline{C} \tag{2.35}$$

respectively, where $\sigma_t$ and $Sc_t$ are known as the turbulent Prandtl and Schmidt numbers, respectively.

The transport equation for the turbulent kinetic energy is obtained by multiplying first the difference between the instantaneous and the time averaged momentum equations by $\mathbf{u}'$. Then applying further the time average operator to the resulting product, we obtain,

$$\rho\nabla\cdot(\overline{\mathbf{u}}k) = -\rho\nabla\cdot\overline{\left[\mathbf{u}'\left(\frac{p'}{\rho}+q\right)\right]} + \mu\nabla^2 k + P + G_\mathrm{T} + G_\mathrm{C} - \rho\varepsilon \tag{2.36}$$

where $P_\mathrm{k} = -\rho\overline{\mathbf{u}'\mathbf{u}'} : \nabla\overline{\mathbf{u}}$ is the generation rate of $k$ due to gradients of the mean velocity and

$$G_\mathrm{T} = -\rho\beta\mathbf{g}\cdot\overline{\mathbf{u}'T'} \tag{2.37}$$

$$G_\mathrm{C} = -\rho\beta_\mathrm{C}\mathbf{g}\cdot\overline{\mathbf{u}'C'} \tag{2.38}$$

are the thermal and concentration generation rates of $k$ due to temperature and concentration fluctuations, respectively. Also, $q = \frac{\mathbf{u}'\cdot\mathbf{u}'}{2}$.

## 2.4. VOLUME AVERAGED TRANSPORT EQUATIONS

The volumetric average of Equation (2.4) using the Theorem of Local Volumetric Average (Equations (2.20) (2.22)) results in the following:

$$\rho\left[\frac{\partial}{\partial t}(\phi\langle\mathbf{u}\rangle^\mathrm{i}) + \nabla\cdot[\phi\langle\mathbf{u}\mathbf{u}\rangle^\mathrm{i}]\right] = -\nabla(\phi\langle p\rangle^\mathrm{i}) + \mu\nabla^2(\phi\langle\mathbf{u}\rangle^\mathrm{i}) + \phi\rho\mathbf{g} + \mathbf{R} \tag{2.39}$$

where

$$\mathbf{R} = \frac{\mu}{\Delta V}\int_{A_\mathrm{i}}\mathbf{n}\cdot(\nabla\mathbf{u})\,\mathrm{d}S - \frac{1}{\Delta V}\int_{A_\mathrm{i}}\mathbf{n}p\,\mathrm{d}S \tag{2.40}$$

represents the total drag force per unit volume due to the presence of the porous matrix, being composed by both viscous drag and form (pressure) drag. Further, using spatial decomposition to write $\mathbf{u} = \langle\mathbf{u}\rangle^\mathrm{i} + {}^\mathrm{i}\mathbf{u}$ in the inertia term, we obtain the following:

$$\rho\left[\frac{\partial}{\partial t}(\phi\langle\mathbf{u}\rangle^\mathrm{i}) + \nabla\cdot[\phi\langle\mathbf{u}\rangle^\mathrm{i}\langle\mathbf{u}\rangle^\mathrm{i}]\right]$$
$$= -\nabla(\phi\langle p\rangle^\mathrm{i}) + \mu\nabla^2(\phi\langle\mathbf{u}\rangle^\mathrm{i}) - \nabla\cdot[\phi\langle{}^\mathrm{i}\mathbf{u}{}^\mathrm{i}\mathbf{u}\rangle^\mathrm{i}] + \phi\rho\mathbf{g} + \mathbf{R} \tag{2.41}$$

Hsu and Cheng (1990) pointed out that the third term on the rhs represents the hydrodynamic dispersion due to spatial deviations. Note that Equation (2.41) models typical porous media flow for Reynolds number based on the pore size, $Re_\mathrm{p}$, up to the range

150–300. In fact, the literature recognizes distinct flow regimes, namely: (a) Darcy or creeping flow regime ($Re_p < 1$), (b) Forchheimer flow regime ($1 \sim 10 < Re_p < 150$), (c) Post-Forchheimer flow regime (unsteady laminar flow, $150 < Re_p < 300$) and (d) Fully turbulent flow ($Re_p > 300$). The mathematical description of the last regime has given rise to interesting discussions in the literature and remains a controversial issue. This is so because when extending the analysis to turbulent flow, time varying quantities have to be considered.

Chapter 3

# The Double-Decomposition Concept

Properties of averaging procedures when the two mathematical operators are applied simultaneously over the instantaneous equations are presented in this chapter. Fundamentals of the double-decomposition concept are clarified and discussed. A comparison of different approaches published in the literature is presented, followed by a critical review of the distinct analytical tools used up to the present date.

The double-decomposition idea, herein used for obtaining macroscopic equations, has been detailed in Pedras and de Lemos (1999, 2000a,b, 2001a–c, 2003). In this chapter, a general overview is presented. Further, the resulting equations using this concept for the flow (Pedras and de Lemos, 2000a,b) and for non-buoyant thermal fields (Rocamora and de Lemos, 2000a; de Lemos and Rocamora, 2002) are already available in the literature, and hence they are not reviewed here in great detail. As mentioned, extensions of the double-decomposition methodology to buoyant flows (de Lemos and Braga, 2003), mass transport (de Lemos and Mesquita, 2003) and double-diffusive convection (de Lemos and Tofaneli, 2004) have also been presented in the open literature.

In addition to the aforementioned research papers, the double-decomposition concept has been briefly overviewed in some recent books (Lage *et al.*, 2002; de Lemos, 2005b,c). Basically, for porous media analysis, a macroscopic form of governing equations is obtained by taking the volumetric average of the entire equation set. In that development, the porous medium is considered rigid and saturated by an incompressible fluid.

## 3.1. BASIC RELATIONSHIPS

From the work in Pedras and de Lemos (2000a) and Rocamora and de Lemos (2000a), one can write for any property $\varphi$ after combining decompositions (2.19) and (2.24):

$$\langle\varphi\rangle^{\mathrm{i}} = \langle\overline{\varphi}\rangle^{\mathrm{i}} + \langle\varphi'\rangle^{\mathrm{i}} \tag{3.1}$$

$$\overline{\varphi} = \langle\overline{\varphi}\rangle^{\mathrm{i}} + {}^{\mathrm{i}}\overline{\varphi} \tag{3.2}$$

$${}^{\mathrm{i}}\varphi = {}^{\mathrm{i}}\overline{\varphi} + {}^{\mathrm{i}}(\varphi') \tag{3.3}$$

$$\varphi' = \langle\varphi'\rangle^{\mathrm{i}} + {}^{\mathrm{i}}(\varphi') \tag{3.4}$$

or further

$$\varphi' = \langle\varphi\rangle^{\mathrm{i}'} + ({}^{\mathrm{i}}\varphi)' = \langle\varphi'\rangle^{\mathrm{i}} + {}^{\mathrm{i}}(\varphi') \tag{3.5}$$

$$\varphi = \overline{\langle\varphi\rangle^{\mathrm{i}}} + \langle\varphi\rangle^{\mathrm{i}} + \overline{{}^{\mathrm{i}}\varphi} + ({}^{\mathrm{i}}\varphi)' \tag{3.6}$$

or

$$\varphi = \langle\overline{\varphi}\rangle^{\mathrm{i}} + {}^{\mathrm{i}}\overline{\varphi} + \langle\varphi'\rangle^{\mathrm{i}} + {}^{\mathrm{i}}(\varphi') \tag{3.7}$$

where ${}^{\mathrm{i}}(\varphi')$ can be understood as the *time fluctuation of the spatial deviation* and $({}^{\mathrm{i}}\varphi)'$ is the *spatial deviation of the time fluctuation*. The relationship between these two terms will be shown below.

Invoking the commutative property mentioned in Chapter 2(see p. 24), we can write:

$$\overline{\langle\varphi\rangle^{\mathrm{v}}} = \langle\overline{\varphi}\rangle^{\mathrm{v}} \tag{3.8}$$

and

$$\overline{\langle\varphi\rangle^{\mathrm{i}}} = \langle\overline{\varphi}\rangle^{\mathrm{i}} \tag{3.9}$$

i.e. the time and volume averages commute. Also, we can prove that, Pedras and de Lemos (2000 a,b):

$${}^{\mathrm{i}}\overline{\varphi} = \overline{{}^{\mathrm{i}}\varphi}; \quad \langle\varphi'\rangle^{\mathrm{i}} = \langle\varphi\rangle^{\mathrm{i}'} \tag{3.10}$$

or

$$\langle\varphi\rangle^{\mathrm{i}} = \frac{1}{\Delta V_{\mathrm{f}}} \int_{\Delta V_{\mathrm{f}}} \varphi \, \mathrm{d}V = \frac{1}{\Delta V_{\mathrm{f}}} \int_{\Delta V_{\mathrm{f}}} (\overline{\varphi} + \varphi') \, \mathrm{d}V = \langle\overline{\varphi}\rangle^{\mathrm{i}} + \langle\varphi'\rangle^{\mathrm{i}} \tag{3.11}$$

$${}^{\mathrm{i}}\varphi = {}^{\mathrm{i}}\overline{\varphi} + {}^{\mathrm{i}}(\varphi') = \overline{{}^{\mathrm{i}}\varphi} + ({}^{\mathrm{i}}\varphi)' \tag{3.12}$$

so that

$$\varphi' = \langle\varphi'\rangle^{\mathrm{i}} + {}^{\mathrm{i}}(\varphi'); \, {}^{\mathrm{i}}\varphi = \overline{{}^{\mathrm{i}}\varphi} + ({}^{\mathrm{i}}\varphi)' \quad \text{where} \quad {}^{\mathrm{i}}(\varphi') = \varphi' - \langle\varphi'\rangle^{\mathrm{i}}; \; ({}^{\mathrm{i}}\varphi)' = {}^{\mathrm{i}}\varphi - \overline{{}^{\mathrm{i}}\varphi} \tag{3.13}$$

Finally, we can have a full variable decomposition as follows:

$$\varphi = \langle\overline{\varphi}\rangle^{\mathrm{i}} + \langle\varphi'\rangle^{\mathrm{i}} + {}^{\mathrm{i}}\overline{\varphi} + {}^{\mathrm{i}}(\varphi') = \overline{\langle\varphi\rangle^{\mathrm{i}}} + \langle\varphi\rangle^{\mathrm{i}'} + \overline{{}^{\mathrm{i}}\varphi} + ({}^{\mathrm{i}}\varphi)' \tag{3.14}$$

or further

$$\varphi = \underbrace{\langle\overline{\varphi}\rangle^{\mathrm{i}} + \langle\varphi'\rangle^{\mathrm{i}}}_{\langle\varphi\rangle^{\mathrm{i}}} + \overbrace{{}^{\mathrm{i}}\overline{\varphi} + {}^{\mathrm{i}}(\varphi')}^{{}^{\mathrm{i}}\varphi} = \underbrace{\overline{\langle\varphi\rangle^{\mathrm{i}}} + \overline{{}^{\mathrm{i}}\varphi}}_{\overline{\varphi}} + \overbrace{\langle\varphi\rangle^{\mathrm{i}'} + ({}^{\mathrm{i}}\varphi)'}^{\varphi'} \tag{3.15}$$

Equation (3.14) comprises the *double-decomposition* concept. The significance of the four terms in Equation (3.15) can be reviewed as:

(a) $\langle \overline{\varphi} \rangle^{\rm i}$ is the intrinsic average of the time mean value of $\varphi$, i.e. we compute first the time-averaged values of all points composing the REV, and then we find their volumetric mean to obtain $\langle \overline{\varphi} \rangle^{\rm i}$. Instead, we could also consider a certain point $\mathbf{x}$ surrounded by the REV, according to Equations (2.17) and (2.18), and take the volumetric average at different time steps. Thus, we calculate the average over such different values of $\langle \varphi \rangle^{\rm i}$ in time. We, then, get $\overline{\langle \varphi \rangle^{\rm i}}$ and, according to expression (3.9), $\langle \overline{\varphi} \rangle^{\rm i} = \overline{\langle \varphi \rangle^{\rm i}}$, i.e. the volumetric and time averages commute.

(b) If we now take the volume average of all fluctuating components of $\varphi$, which compose the REV, we end up with $\langle \varphi' \rangle^{\rm i}$. Instead, with the volumetric average around the point $\mathbf{x}$ taken at different time steps, we can determine the difference between the instantaneous and a time averaged value. This will be $\langle \varphi \rangle^{\rm i'}$ that, according to Equation (3.10), equals $\langle \varphi' \rangle^{\rm i}$.

(c) Further, on performing first a time averaging operation over all the points that contribute with their local values to the REV, we get a distribution of $\overline{\varphi}$ within this volume. If we calculate the intrinsic average of this distribution of $\overline{\varphi}$, we get $\langle \overline{\varphi} \rangle^{\rm i}$. The difference or deviation between these two values is ${}^{\rm i}\overline{\varphi}$. Now, using the same space decomposition approach, we can find, for any instant of time $t$, the deviation ${}^{\rm i}\varphi$. This value also fluctuates with time, and as such a time mean $\overline{{}^{\rm i}\varphi}$ can be calculated. Again the use of Equation (3.10) gives ${}^{\rm i}\overline{\varphi} = \overline{{}^{\rm i}\varphi}$.

(d) Finally, it is interesting to note the meaning of the last terms of the two expressions for $\varphi$ of Equation (3.15). That of the first expression ${}^{\rm i}(\varphi')$ is the time fluctuation of the spatial component whereas that of the second expression $({}^{\rm i}\varphi)'$ means the spatial component of the time-varying term. If, however, one makes use of relationships (3.8) and (3.10) to simplify Equation (3.15), we finally conclude,

$$ {}^{\rm i}(\varphi') = ({}^{\rm i}\varphi)' \tag{3.16} $$

and, for simplicity of notation, we can drop the parentheses and write both superscripts at the same level in the format: ${}^{\rm i}\varphi'$. Also, $\langle {}^{\rm i}\varphi' \rangle^{\rm i} = \overline{{}^{\rm i}\varphi'} = 0$.

With the help of Figure 3.1 taken from Rocamora and de Lemos (2000a), the concept of *double-decomposition* can be better understood. The figure shows a three-dimensional diagram for a general vector variable $\varphi$. For a scalar, all the quantities shown would be drawn on a single line. Also, notice that points B, C, D and E fall in the same plane, with segments BC and BE parallel to ED and CD, respectively. Line ACF represents standard space decomposition given by (2.19) and the line AEF represents that given by Equation (2.24). Further, Equation (3.1) is represented by the triangle ABC and (3.2) by ABE. Triangles CDF and EDF are associated with Equations (3.3) and (3.4),

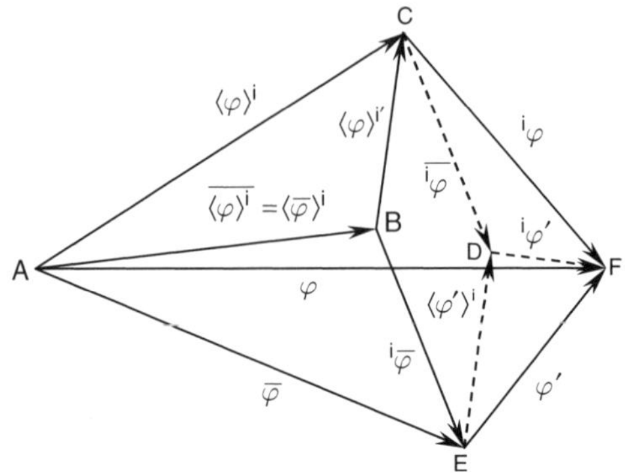


**Figure 3.1.** General three-dimensional vector diagram for a quantity $\varphi$ (see Rocamora and de Lemos, 2000a).

respectively. Equality (3.9) is represented by AB, and the two Equations (3.10) by the equivalence between the parallel segments BE and CD and between BC and ED, respectively. Finally, Equation (3.14) follows the sequence ABCDF or the path ABEDF, both of them decomposing the same general variable $\varphi$.

The basic advantage of the double-decomposition concept is to serve as a mathematical framework for analysis of flows where within the fluid phase there is enough room for turbulence to be established. As such, the double-decomposition methodology would be useful in situations where a solid phase is present in the domain under analysis so that a macroscopic view is appropriate. At the same time, properties in the fluid phase are subjected to the turbulent regime, and a statistical approach is appropriate. Examples of possible applications of such methodology can be found in engineering systems such as heat exchangers, porous combustors, nuclear reactor cores, etc. Natural systems include atmospheric boundary layer over forests and crops.

## 3.2. CLASSIFICATION OF MACROSCOPIC TURBULENCE MODELS

Based on the derivations above, one can establish a general classification of the models presented so far in the literature. Table 3.1 classifies all proposals into four major categories. These classes are based on the sequence of application of averaging operators, on the handling of surface integrals and on the applications reported so far.

The A-L models make use of transport equations for $k_m$ instead of $\langle k\rangle^{i}$. Consequently, this methodology applies only time averaging procedure to already established macroscopic equations (see for example Hsu and Cheng, 1990, for macroscopic equations). In this sense, the sequence *space–time* integration is employed and surface integrals are

**Table 3.1.** Classification of turbulence models for porous media (from de Lemos and Pedras, 2001, with permission).

| Model class | Authors | General characteristics and treatment of surface integrals | Sequence of integration | Applications |
|---|---|---|---|---|
| A-L | Lee and Howell (1987), Wang and Takle (1995), Antohe and Lage (1997), Getachewa *et al.* (2000). | Surface integrals are not applied since models are based on macroscopic quantities subjected to time averaging only. | Space–time | Only theory presented. Numerical results using this model are found in Chan *et al.* (2000). |
| N-K | Masuoka and Takatsu (1996), Kuwahara and Nakayama (1998), Takatsu and Masuoka (1998), Nakayama and Kuwahara (1999). | Masuoka and Takatsu (1996) assumed a non-null value in their Equation (11) for the turbulent shear stress $\mathbf{S}_t = -\rho\overline{\mathbf{u}'\mathbf{u}'}$ along the interfacial area $A_i$. Takatsu and Masuoka (1998) assume for their volume integral in Equation (14) a different from zero for $\mathbf{d} = (\mu/\rho + \mu_t/\sigma_k\rho)\nabla k$ at the interface $A_i$. | Time–space | Microscopic computations on periodic cell of square rods. Macroscopic model computations presented. |
| T-C | Gratton *et al.* (1994), Travkin and Catton (1992, 1995, 1998), Travkin *et al.* (1993, 1999). | Morphology-based theory. Surface integrals and volume average operators depend on media morphology. | Time–space | Only theory presented. |
| P-D | Pedras and de Lemos (2000a,b, 2001b), Rocamora and de Lemos (2000a). | Double-decomposition theory. Surface integrals involving null quantities at surfaces are neglected. The connection between *space–time* and *time–space* theories is unveiled. | Time–space | Microscopic computation on periodic cell of circular rods. Macroscopic computations for porous media presented. Results for hybrid domains are found in de Lemos and Pedras (2000b) and Rocamora and de Lemos (2000b,c,d). |

not manipulated since macroscopic quantities are the sole independent variables used. Application of this theory is found in Chan *et al.* (2000).

The N-K models constitute the second class of models compiled here. It is interesting to mention that Masuoka and Takatsu (1996) assumed a non-null value for the turbulent shear stress, $\mathbf{S}_t = -\rho\overline{\mathbf{u}'\mathbf{u}'}$, along the interfacial area A-L in their Equation (11). With that, their surface integral $\int_{A_i} \mathbf{S}_t \cdot \mathbf{n}\,dA$ was associated with the Darcy flow resistance term. Yet, using the Boussinesq approximation as in their Equation (7), $\mathbf{S}_t = 2\mu_t\mathbf{D} - \frac{2}{3}k\,\mathbf{I}$, one can also see that both $\mu_t$ and $k$ vanish at the surface $A_i$, ultimately indicating that the surface integral in question is actually equal to zero. Similarly, Takatsu and Masuoka (1998) assumed for their surface integral in Equation (14), $\int_{A_i} \mathbf{d} \cdot \mathbf{n}\,dA$, a non-null value where $\mathbf{d} = (\mu/\rho + \mu_t/\sigma_k\rho)\nabla k$. Here, also it is worth noting that $\nabla k = \overline{\mathbf{u}' \cdot (\nabla\mathbf{u}')^{\mathrm{T}}}$ and that, at the interface $A_i$, $\nabla k = 0$ due to the non-slip condition. Consequently, in this case also the surface integral of $\mathbf{d}$ over $A_i$ is of zero value. In regard to the average operators used, N-K models follow the *time–space* integration sequence. Calibration of the model required microscopic computations on a periodic cell of square rods. Macroscopic results in a channel filled with a porous material was also a test case run by Nakayama and Kuwahara (1999).

The work developed in a series of papers using a morphology-based theory is here grouped in the T-C model category shown in Table 3.1. In this morphology-based theory, surface integrals resulting after application of volume average operators depend on the media morphology. Governing equations set up for turbulent flow, although seem to be complicated at first sight, just follow the usual volume-integration technique applied to standard $k$–$\varepsilon$ and $k$–$L$ turbulence models. In this model, time–space integration sequence is followed. No closure is proposed for the unknown surface integrals (and morphology parameters) so that practical applications of such development in solving *real-world* engineering flows are still a challenge to be overcome. Nevertheless, the developed theory seems to be mathematically correct even though additional *ad-hoc* information is still necessary to fully model the remaining unknowns and medium-dependent parameters.

Lastly, the model group named P-D uses the recently developed *double-decomposition* theory just reviewed above. In this development, all surface integrals involving null quantities at interface $A_i$ are neglected. The connection between *space–time* and *time–space* theories is made possible due to the splitting of the dependent variables into four (rather that two) components, as expressed by Equation (3.14) above. For the momentum and energy equations, the double-decomposition approach has proved that the order of application of averaging operators (time–space or space–time) is immaterial. For the turbulence kinetic energy equation, however, the order of application of such mathematical operators will define different quantities being transported (Pedras and de Lemos, 2000a; Rocamora and de Lemos, 2000a). Further results for hybrid domains (porous medium – clear fluid) are found in de Lemos and Pedras (2000b) and Rocamora and de Lemos (2000b,c,d).

Chapter 4

# Turbulent Momentum Transport

The problem of how to characterize the value of turbulent kinetic energy in the flow is reviewed in this chapter. The methodology for numerically adjusting a proposed macroscopic turbulence model is presented.

## 4.1. MOMENTUM EQUATION

### 4.1.1. Mean Flow

The development to follow assumes single-phase flow in a saturated, rigid, porous medium ($\Delta V_f$ independent of time) for which, in accordance with Equation (3.8), time average operation on the variable $\varphi$ commutes with the space-average operation. Application of the double-decomposition idea in Equation (3.15) to the inertia term in the momentum equation leads to four different terms. Not all of these terms are considered in the same analysis in the literature.

#### *4.1.1.1. Continuity*

The microscopic continuity equation for an incompressible fluid flowing in a clear (non-porous) domain was given by Equation (2.1). Using the double-decomposition idea of Equation (3.15) gives:

$$\nabla \cdot \mathbf{u} = \nabla \cdot (\langle \overline{\mathbf{u}} \rangle^{\mathrm{i}} + \langle \mathbf{u}' \rangle^{\mathrm{i}} + {}^{\mathrm{i}}\overline{\mathbf{u}} + {}^{\mathrm{i}}\mathbf{u}') = 0 \tag{4.1}$$

On applying both volume and time average in either order:

$$\nabla \cdot (\phi \langle \overline{\mathbf{u}} \rangle^{\mathrm{i}}) = 0 \tag{4.2}$$

For the continuity equation, the averaging order is immaterial in regard to obtaining the final result shown in Equation (4.2).

#### *4.1.1.2. Momentum equation – two average operators*

Equations (2.33) and (2.41) are used when treating turbulent flow in clear fluid and low $Re_p$ porous media flow, respectively. In each one of those equations, only one averaging operator was applied, time average in Equation (2.33) and volume average in Equation (2.41). In this chapter, the use of both operators is discussed with the objective of modeling turbulent flow in porous media.

The volume average of Equation (2.33) becomes:

$$\rho\left[\frac{\partial}{\partial t}(\phi\langle\overline{\mathbf{u}}\rangle^{\mathrm{i}})+\nabla\cdot(\phi\langle\overline{\mathbf{u}}\,\overline{\mathbf{u}}\rangle^{\mathrm{i}})\right]=-\nabla(\phi\langle\overline{p}\rangle^{\mathrm{i}})+\mu\nabla^{2}(\phi\langle\overline{\mathbf{u}}\rangle^{\mathrm{i}})$$
$$+\nabla\cdot(-\rho\phi\langle\overline{\mathbf{u}'\mathbf{u}'}\rangle^{\mathrm{i}})+\phi\rho\mathbf{g}+\overline{\mathbf{R}} \tag{4.3}$$

with

$$\overline{\mathbf{R}}=\frac{\mu}{\Delta V}\int_{A_{\mathrm{i}}}\mathbf{n}\cdot(\nabla\overline{\mathbf{u}})\mathrm{d}S-\frac{1}{\Delta V}\int_{A_{\mathrm{i}}}\mathbf{n}\overline{p}\mathrm{d}S \tag{4.4}$$

where $\overline{\mathbf{R}}$ is the time averaged total drag per unit volume due to the solid matrix, which is composed by both viscous and form (pressure) drags.

Likewise, applying now the time averaging operation to Equation (2.39), we obtain:

$$\rho\left[\frac{\partial}{\partial t}\overline{(\phi\langle\overline{\mathbf{u}}+\mathbf{u}'\rangle^{\mathrm{i}})}+\nabla\cdot\overline{(\phi\langle(\overline{\mathbf{u}}+\mathbf{u}')(\overline{\mathbf{u}}+\mathbf{u}')\rangle^{\mathrm{i}})}\right]$$
$$=-\nabla\overline{(\phi\langle\overline{p}+p'\rangle^{\mathrm{i}})}+\mu\nabla^{2}\overline{(\phi\langle\overline{\mathbf{u}}+\mathbf{u}'\rangle^{\mathrm{i}})}+\phi\rho\mathbf{g}+\overline{\mathbf{R}} \tag{4.5}$$

Dropping terms containing only one fluctuating quantity results in:

$$\rho\left[\frac{\partial}{\partial t}(\phi\langle\overline{\mathbf{u}}\rangle^{\mathrm{i}})+\nabla\cdot(\phi\langle\overline{\mathbf{u}}\,\overline{\mathbf{u}}\rangle^{\mathrm{i}})\right]=-\nabla(\phi\langle\overline{p}\rangle^{\mathrm{i}})+\mu\nabla^{2}(\phi\langle\overline{\mathbf{u}}\rangle^{\mathrm{i}})$$
$$+\nabla\cdot(-\rho\phi\langle\overline{\mathbf{u}'\mathbf{u}'}\rangle^{\mathrm{i}})+\phi\rho\mathbf{g}+\overline{\mathbf{R}} \tag{4.6}$$

where

$$\overline{\mathbf{R}}=\frac{\mu}{\Delta V}\int_{A_{\mathrm{i}}}\mathbf{n}\cdot[\nabla\overline{(\overline{\mathbf{u}}+\mathbf{u}')}]\,\mathrm{d}S-\frac{1}{\Delta V}\int_{A_{\mathrm{i}}}\mathbf{n}\overline{(\overline{p}+p')}\,\mathrm{d}S$$
$$=\frac{\mu}{\Delta V}\int_{A_{\mathrm{i}}}\mathbf{n}\cdot(\nabla\overline{\mathbf{u}})\mathrm{d}S-\frac{1}{\Delta V}\int_{A_{\mathrm{i}}}\mathbf{n}\overline{p}\,\mathrm{d}S \tag{4.7}$$

Comparing Equations (4.3) and (4.6), we can see that for the momentum equation, too, the order of the application of the two averaging operators is immaterial.

It is interesting to emphasize that both views in the literature use the same final form for the additional drag forces. The term $\overline{\mathbf{R}}$ is modeled by the Darcy–Forcheimer (Dupuit) expression after either order of application of the average operators giving:

$$\rho\nabla\cdot\left(\frac{\overline{\mathbf{u}}_{\mathrm{D}}\overline{\mathbf{u}}_{\mathrm{D}}}{\phi}\right)=-\nabla(\phi\langle\overline{p}\rangle^{\mathrm{i}})+\mu\nabla^{2}\overline{\mathbf{u}}_{\mathrm{D}}+\nabla\cdot(-\rho\phi\langle\overline{\mathbf{u}'\mathbf{u}'}\rangle^{\mathrm{i}})$$
$$-\left[\frac{\mu\phi}{K}\overline{\mathbf{u}}_{\mathrm{D}}+\frac{c_{\mathrm{F}}\phi\rho|\overline{\mathbf{u}}_{\mathrm{D}}|\overline{\mathbf{u}}_{\mathrm{D}}}{\sqrt{K}}\right] \tag{4.8}$$

Since both orders of integration lead to the same equation, namely Equation (4.4) or (4.7), there would be no reason for modeling them in different forms. Had the outcome of both integration processes been distinct, the use of a different model for each case would have been consistent. In fact, it has been pointed out by Pedras and de Lemos (2000a) that the major difference between those two processes lies in the definition of a suitable turbulent kinetic energy for the flow. Accordingly, the source of controversies comes from the inertia term, as seen below.

Applying the double decomposition idea (Equation (3.15)) for velocity in the inertia term of Equation (2.4) will lead to different sets of terms. In the literature, not all of them are used in the same analysis.

Starting with time decomposition, applying both average operators (see Equation (4.3)) gives:

$$\nabla\cdot\overline{(\phi\langle\mathbf{u}\mathbf{u}\rangle^{i})}=\nabla\cdot\overline{(\phi\langle(\overline{\mathbf{u}}+\mathbf{u}')(\overline{\mathbf{u}}+\mathbf{u}')\rangle^{i})}=\nabla\cdot[\phi(\langle\overline{\mathbf{u}}\,\overline{\mathbf{u}}\rangle^{i}+\langle\overline{\mathbf{u}'\mathbf{u}'}\rangle^{i})] \tag{4.9}$$

using spatial decomposition to write $\overline{\mathbf{u}}=\langle\overline{\mathbf{u}}\rangle^{i}+{}^{i}\overline{\mathbf{u}}$ we obtain,

$$\begin{aligned}\nabla\cdot[\phi(\langle\overline{\mathbf{u}}\,\overline{\mathbf{u}}\rangle^{i}+\langle\overline{\mathbf{u}'\mathbf{u}'}\rangle^{i})]&=\nabla\cdot\{\phi[\langle(\langle\overline{\mathbf{u}}\rangle^{i}+{}^{i}\overline{\mathbf{u}})(\langle\overline{\mathbf{u}}\rangle^{i}+{}^{i}\overline{\mathbf{u}})\rangle^{i}+\langle\overline{\mathbf{u}'\mathbf{u}'}\rangle^{i}]\}\\&=\nabla\cdot\{\phi[\langle\overline{\mathbf{u}}\rangle^{i}\langle\overline{\mathbf{u}}\rangle^{i}+\langle{}^{i}\overline{\mathbf{u}}\,{}^{i}\overline{\mathbf{u}}\rangle^{i}+\langle\overline{\mathbf{u}'\mathbf{u}'}\rangle^{i}]\}\end{aligned} \tag{4.10}$$

Now, applying Equation (3.5) to write $\mathbf{u}'=\langle\mathbf{u}'\rangle^{i}+{}^{i}\mathbf{u}'$, and substituting into Equation (4.10), we get:

$$\begin{aligned}&\nabla\cdot\{\phi[\langle\overline{\mathbf{u}}\rangle^{i}\langle\overline{\mathbf{u}}\rangle^{i}+\langle{}^{i}\overline{\mathbf{u}}{}^{i}\overline{\mathbf{u}}\rangle^{i}+\langle\overline{\mathbf{u}'\mathbf{u}'}\rangle^{i}]\}\\&\quad=\nabla\cdot\{\phi[\langle\overline{\mathbf{u}}\rangle^{i}\langle\overline{\mathbf{u}}\rangle^{i}+\langle{}^{i}\overline{\mathbf{u}}\,{}^{i}\overline{\mathbf{u}}\rangle^{i}+\langle\overline{(\langle\mathbf{u}'\rangle^{i}+{}^{i}\mathbf{u}')(\langle\mathbf{u}'\rangle^{i}+{}^{i}\mathbf{u}')}\rangle^{i}]\}\\&\quad=\nabla\cdot\{\phi[\langle\overline{\mathbf{u}}\rangle^{i}\langle\overline{\mathbf{u}}\rangle^{i}+\langle{}^{i}\overline{\mathbf{u}}\,{}^{i}\overline{\mathbf{u}}\rangle^{i}+\langle\overline{(\langle\mathbf{u}'\rangle^{i}\langle\mathbf{u}'\rangle^{i}+\langle\mathbf{u}'\rangle^{i}\,{}^{i}\mathbf{u}'+{}^{i}\mathbf{u}'\langle\mathbf{u}'\rangle^{i}+{}^{i}\mathbf{u}'\,{}^{i}\mathbf{u}')}\rangle^{i}]\}\\&\quad=\nabla\cdot\{\phi[\langle\overline{\mathbf{u}}\rangle^{i}\langle\overline{\mathbf{u}}\rangle^{i}+\langle{}^{i}\overline{\mathbf{u}}\,{}^{i}\overline{\mathbf{u}}\rangle^{i}+\overline{\langle\mathbf{u}'\rangle^{i}\langle\mathbf{u}'\rangle^{i}}+\overline{\langle\langle\mathbf{u}'\rangle^{i}\,{}^{i}\mathbf{u}'\rangle^{i}}+\overline{\langle{}^{i}\mathbf{u}'\langle\mathbf{u}'\rangle^{i}\rangle^{i}}+\langle\overline{{}^{i}\mathbf{u}'\,{}^{i}\mathbf{u}'}\rangle^{i}]\}\end{aligned} \tag{4.11}$$

The fourth and fifth terms on the final expression contain only one space-varying quantity and will vanish under the application of volume integration. Equation (4.11) will then be reduced to:

$$\nabla\cdot\overline{(\phi\langle\mathbf{u}\mathbf{u}\rangle^{i})}=\nabla\cdot\{\phi[\langle\overline{\mathbf{u}}\rangle^{i}\langle\overline{\mathbf{u}}\rangle^{i}+\overline{\langle\mathbf{u}'\rangle^{i}\langle\mathbf{u}'\rangle^{i}}+\langle{}^{i}\overline{\mathbf{u}}\,{}^{i}\overline{\mathbf{u}}\rangle^{i}+\langle\overline{{}^{i}\mathbf{u}'\,{}^{i}\mathbf{u}'}\rangle^{i}]\} \tag{4.12}$$

Using the equivalence expressions (3.9) and (3.10), Equation (4.12) can be further rewritten as follows:

$$\nabla\cdot\overline{(\phi\langle\mathbf{u}\mathbf{u}\rangle^{\mathrm{i}})}=\nabla\cdot\{\phi[\overline{\langle\mathbf{u}\rangle^{\mathrm{i}}}\;\overline{\langle\mathbf{u}\rangle^{\mathrm{i}}}+\overline{\langle\mathbf{u}\rangle^{\mathrm{i}'}\langle\mathbf{u}\rangle^{\mathrm{i}'}}+\langle{}^{\mathrm{i}}\overline{\mathbf{u}}\,{}^{\mathrm{i}}\overline{\mathbf{u}}\rangle^{\mathrm{i}}+\langle\overline{{}^{\mathrm{i}}\mathbf{u}'\,{}^{\mathrm{i}}\mathbf{u}'}\rangle^{\mathrm{i}}]\}\tag{4.13}$$

with an interpretation of the terms in Equation (4.12) given later.

Another method to get the same results is to start out with the application of the space decomposition in the inertia term, as usually done in classical mathematical treatment of flow in porous media. Then we obtain:

$$\nabla\cdot\overline{(\phi\langle\mathbf{u}\mathbf{u}\rangle^{\mathrm{i}})}=\nabla\cdot\overline{(\phi\langle(\langle\mathbf{u}\rangle^{\mathrm{i}}+{}^{\mathrm{i}}\mathbf{u})(\langle\mathbf{u}\rangle^{\mathrm{i}}+{}^{\mathrm{i}}\mathbf{u})\rangle^{\mathrm{i}})}=\nabla\cdot\overline{[\phi(\langle\mathbf{u}\rangle^{\mathrm{i}}\langle\mathbf{u}\rangle^{\mathrm{i}}+\langle{}^{\mathrm{i}}\mathbf{u}{}^{\mathrm{i}}\mathbf{u}\rangle^{\mathrm{i}})]}\tag{4.14}$$

and on using Equation (3.11) to express $\langle\mathbf{u}\rangle^{\mathrm{i}}=\langle\overline{\mathbf{u}}\rangle^{\mathrm{i}}+\langle\mathbf{u}'\rangle^{\mathrm{i}}$, we get:

$$\begin{aligned}\nabla\cdot\overline{[\phi(\langle\mathbf{u}\rangle^{\mathrm{i}}\langle\mathbf{u}\rangle^{\mathrm{i}}+\langle{}^{\mathrm{i}}\mathbf{u}{}^{\mathrm{i}}\mathbf{u}\rangle^{\mathrm{i}})]}&=\nabla\cdot\{\overline{\phi[(\langle\overline{\mathbf{u}}\rangle^{\mathrm{i}}+\langle\mathbf{u}'\rangle^{\mathrm{i}})(\langle\overline{\mathbf{u}}\rangle^{\mathrm{i}}+\langle\mathbf{u}'\rangle^{\mathrm{i}})+\langle{}^{\mathrm{i}}\mathbf{u}{}^{\mathrm{i}}\mathbf{u}\rangle^{\mathrm{i}}]}\}\\&=\nabla\cdot\{\phi[\langle\overline{\mathbf{u}}\rangle^{\mathrm{i}}\langle\overline{\mathbf{u}}\rangle^{\mathrm{i}}+\overline{\langle\mathbf{u}'\rangle^{\mathrm{i}}\langle\mathbf{u}'\rangle^{\mathrm{i}}}+\overline{\langle{}^{\mathrm{i}}\mathbf{u}{}^{\mathrm{i}}\mathbf{u}\rangle^{\mathrm{i}}}]\}\end{aligned}\tag{4.15}$$

With the help of Equation (3.12) one can write ${}^{\mathrm{i}}\mathbf{u}={}^{\mathrm{i}}\overline{\mathbf{u}}+{}^{\mathrm{i}}\mathbf{u}'$, which inserted into Equation (4.15) gives:

$$\begin{aligned}&\nabla\cdot\{\phi[\langle\overline{\mathbf{u}}\rangle^{\mathrm{i}}\langle\overline{\mathbf{u}}\rangle^{\mathrm{i}}+\overline{\langle\mathbf{u}'\rangle^{\mathrm{i}}\langle\mathbf{u}'\rangle^{\mathrm{i}}}+\overline{\langle{}^{\mathrm{i}}\mathbf{u}{}^{\mathrm{i}}\mathbf{u}\rangle^{\mathrm{i}}}]\}\\&\quad=\nabla\cdot\{\phi[\langle\overline{\mathbf{u}}\rangle^{\mathrm{i}}\langle\overline{\mathbf{u}}\rangle^{\mathrm{i}}+\overline{\langle\mathbf{u}'\rangle^{\mathrm{i}}\langle\mathbf{u}'\rangle^{\mathrm{i}}}+\overline{\langle({}^{\mathrm{i}}\overline{\mathbf{u}}+{}^{\mathrm{i}}\mathbf{u}')({}^{\mathrm{i}}\overline{\mathbf{u}}+{}^{\mathrm{i}}\mathbf{u}')\rangle^{\mathrm{i}}}]\}\\&\quad=\nabla\cdot\{\phi[\langle\overline{\mathbf{u}}\rangle^{\mathrm{i}}\langle\overline{\mathbf{u}}\rangle^{\mathrm{i}}+\overline{\langle\mathbf{u}'\rangle^{\mathrm{i}}\langle\mathbf{u}'\rangle^{\mathrm{i}}}+\langle\overline{{}^{\mathrm{i}}\overline{\mathbf{u}}{}^{\mathrm{i}}\overline{\mathbf{u}}}+\overline{{}^{\mathrm{i}}\overline{\mathbf{u}}{}^{\mathrm{i}}\mathbf{u}'}+\overline{{}^{\mathrm{i}}\mathbf{u}'{}^{\mathrm{i}}\overline{\mathbf{u}}}+\overline{{}^{\mathrm{i}}\mathbf{u}'{}^{\mathrm{i}}\mathbf{u}'}\rangle^{\mathrm{i}}]\}\end{aligned}\tag{4.16}$$

The only one fluctuating component in the fourth and fifth terms of the last expression of Equation (4.16) vanishes on applying the time average operator to these terms. In addition, remembering that with Equation (3.10) the equivalences ${}^{\mathrm{i}}\overline{\mathbf{u}}=\overline{{}^{\mathrm{i}}\mathbf{u}}$ and $\langle\mathbf{u}'\rangle^{\mathrm{i}}=\langle\mathbf{u}\rangle^{\mathrm{i}'}$ are valid, and that with Equation (3.9) we can write $\overline{\langle\mathbf{u}\rangle^{\mathrm{i}}}=\langle\overline{\mathbf{u}}\rangle^{\mathrm{i}}$, we obtain the following alternative form for Equation (4.16):

$$\begin{aligned}&\nabla\cdot\overline{[\phi(\langle\mathbf{u}\rangle^{\mathrm{i}}\langle\mathbf{u}\rangle^{\mathrm{i}}+\langle{}^{\mathrm{i}}\mathbf{u}\,{}^{\mathrm{i}}\mathbf{u}\rangle^{\mathrm{i}})]}\\&\quad=\nabla\cdot\{\phi[\overline{\langle\mathbf{u}\rangle^{\mathrm{i}}}\;\overline{\langle\mathbf{u}\rangle^{\mathrm{i}}}+\overline{\langle\mathbf{u}\rangle^{\mathrm{i}'}\langle\mathbf{u}\rangle^{\mathrm{i}'}}+\langle{}^{\mathrm{i}}\overline{\mathbf{u}}\,{}^{\mathrm{i}}\overline{\mathbf{u}}\rangle^{\mathrm{i}}+\langle\overline{{}^{\mathrm{i}}\mathbf{u}'\,{}^{\mathrm{i}}\mathbf{u}'}\rangle^{\mathrm{i}}]\}\\&\quad=\nabla\cdot\{\phi[\underbrace{\langle\overline{\mathbf{u}}\rangle^{\mathrm{i}}\langle\overline{\mathbf{u}}\rangle^{\mathrm{i}}}_{1}+\underbrace{\overline{\langle\mathbf{u}'\rangle^{\mathrm{i}}\langle\mathbf{u}'\rangle^{\mathrm{i}}}}_{2}+\underbrace{\langle{}^{\mathrm{i}}\overline{\mathbf{u}}\,{}^{\mathrm{i}}\overline{\mathbf{u}}\rangle^{\mathrm{i}}}_{3}+\underbrace{\langle\overline{{}^{\mathrm{i}}\mathbf{u}'\,{}^{\mathrm{i}}\mathbf{u}'}\rangle^{\mathrm{i}}}_{4}]\}\end{aligned}\tag{4.17}$$

whose rhs is the same as that of Equation (4.12).

The physical significance of all the four terms on the right hand side of (4.17) are as follows:

1 represents the *convective* term of macroscopic mean velocity,
2 signifies the *turbulent (Reynolds) stresses* divided by density $\rho$ due to the fluctuating component of the macroscopic velocity,

3 is the *dispersion* associated with spatial deviations of microscopic time mean velocity. Note that this term is also present in laminar flow (say, when $Re_p < 150$).
4 is the *turbulent dispersion* in a porous medium due to both time and spatial fluctuations of the microscopic velocity.

Further, the macroscopic Reynolds stress tensor (MRST) is given in Antohe and Lage (1997) based on Equation (2.34) as follows:

$$-\rho\phi\langle\overline{\mathbf{u}'\mathbf{u}'}\rangle^{\mathrm{i}} = \mu_{\mathrm{t}_\phi} 2\langle\overline{\mathbf{D}}\rangle^{\mathrm{v}} - \frac{2}{3}\phi\rho\langle k\rangle^{\mathrm{i}}\mathbf{I} \tag{4.18}$$

where

$$\langle\overline{\mathbf{D}}\rangle^{\mathrm{v}} = \frac{1}{2}\{\nabla(\phi\langle\overline{\mathbf{u}}\rangle^{\mathrm{i}}) + [\nabla(\phi\langle\overline{\mathbf{u}}\rangle^{\mathrm{i}})]^{\mathrm{T}}\} \tag{4.19}$$

is the macroscopic deformation tensor, $\langle k\rangle^{\mathrm{i}}$ is the intrinsic average for $k$ and $\mu_{\mathrm{t}_\phi}$ is the macroscopic turbulent viscosity assumed to be in Lee and Howell (1987) as follows:

$$\mu_{\mathrm{t}_\phi} = \rho c_\mu \frac{\langle k\rangle^{\mathrm{i}^2}}{\langle\varepsilon\rangle^{\mathrm{i}}} \tag{4.20}$$

### 4.1.2. Fluctuating Velocity

The starting point for an equation for the turbulent kinetic energy of the flow is an equation for the microscopic velocity fluctuation $\mathbf{u}'$. Such a relationship can be written, after subtracting the equation for the mean velocity $\overline{\mathbf{u}}$ from the instantaneous momentum equation, as follows:

$$\rho\left\{\frac{\partial\mathbf{u}'}{\partial t} + \nabla\cdot[\overline{\mathbf{u}}\mathbf{u}' + \mathbf{u}'\overline{\mathbf{u}} + \mathbf{u}'\mathbf{u}' - \overline{\mathbf{u}'\mathbf{u}'}]\right\} = -\nabla p' + \mu\nabla^2\mathbf{u}' \tag{4.21}$$

Now, the volumetric average of Equation (4.21) using the theorem of local volumetric average gives:

$$\begin{aligned}&\rho\frac{\partial}{\partial t}(\phi\langle\mathbf{u}'\rangle^{\mathrm{i}}) + \rho\nabla\cdot\{\phi[\langle\overline{\mathbf{u}}\mathbf{u}'\rangle^{\mathrm{i}} + \langle\mathbf{u}'\overline{\mathbf{u}}\rangle^{\mathrm{i}} + \langle\mathbf{u}'\mathbf{u}'\rangle^{\mathrm{i}} - \langle\overline{\mathbf{u}'\mathbf{u}'}\rangle^{\mathrm{i}}]\}\\ &= -\nabla(\phi\langle p'\rangle^{\mathrm{i}}) + \mu\nabla^2(\phi\langle\mathbf{u}'\rangle^{\mathrm{i}}) + \mathbf{R}'\end{aligned} \tag{4.22}$$

where

$$\mathbf{R}' = \frac{\mu}{\Delta V}\int_{A_{\mathrm{i}}}\mathbf{n}\cdot(\nabla\mathbf{u}')\,\mathrm{d}S - \frac{1}{\Delta V}\int_{A_{\mathrm{i}}}\mathbf{n}p'\,\mathrm{d}S \tag{4.23}$$

is the fluctuating part of the total drag due to the porous structure.

Expanding further the divergent operator in Equation (4.22) by means of Equation (3.15), one ends up with an equation for $\langle \mathbf{u}' \rangle^{\mathrm{i}}$ as follows:

$$\rho \frac{\partial}{\partial t}(\phi \langle \mathbf{u}' \rangle^{\mathrm{i}}) + \rho \nabla \cdot \{\phi[\langle \overline{\mathbf{u}} \rangle^{\mathrm{i}} \langle \mathbf{u}' \rangle^{\mathrm{i}} + \langle \mathbf{u}' \rangle^{\mathrm{i}} \langle \overline{\mathbf{u}} \rangle^{\mathrm{i}} + \langle \mathbf{u}' \rangle^{\mathrm{i}} \langle \mathbf{u}' \rangle^{\mathrm{i}} + \langle {}^{\mathrm{i}}\overline{\mathbf{u}}\, {}^{\mathrm{i}}\mathbf{u}' \rangle^{\mathrm{i}} + \langle {}^{\mathrm{i}}\mathbf{u}'\, {}^{\mathrm{i}}\overline{\mathbf{u}} \rangle^{\mathrm{i}}$$
$$+ \langle {}^{\mathrm{i}}\mathbf{u}'\, {}^{\mathrm{i}}\mathbf{u}' \rangle^{\mathrm{i}} - \overline{\langle \mathbf{u}' \rangle^{\mathrm{i}} \langle \mathbf{u}' \rangle^{\mathrm{i}}} - \langle \overline{{}^{\mathrm{i}}\mathbf{u}'\, {}^{\mathrm{i}}\mathbf{u}'} \rangle^{\mathrm{i}}]\} = -\nabla(\phi \langle p' \rangle^{\mathrm{i}}) + \mu \nabla^2 (\phi \langle \mathbf{u}' \rangle^{\mathrm{i}}) + \mathbf{R}' \tag{4.24}$$

Another route to follow in order to obtain Equation (4.24) is to start out with the macroscopic instantaneous momentum equation for an incompressible fluid given by Hsu and Cheng (1990):

$$\rho \left\{ \frac{\partial}{\partial t}(\phi \langle \mathbf{u} \rangle^{\mathrm{i}} + \nabla \cdot (\phi \langle \mathbf{u} \rangle^{\mathrm{i}} \langle \mathbf{u} \rangle^{\mathrm{i}}) \right\} = -\nabla[\phi(\langle p \rangle^{\mathrm{i}})] + \mu \nabla^2 (\phi \langle \mathbf{u} \rangle^{\mathrm{i}})$$
$$+ \phi \rho \mathbf{g} - \nabla \cdot (\phi \langle {}^{\mathrm{i}}\mathbf{u}\, {}^{\mathrm{i}}\mathbf{u} \rangle^{\mathrm{i}}) + \mathbf{R} \tag{4.25}$$

where $\mathbf{R}$ was given earlier by Equation (2.40) and the term $\langle {}^{\mathrm{i}}\mathbf{u}\, {}^{\mathrm{i}}\mathbf{u} \rangle^{\mathrm{i}}$ is known as *dispersion*. The mathematical meaning of dispersion can be seen as a correlation between spatial deviations of velocity components.

Making use of the double-decomposition concept given by Equation (3.14), Equation (4.25) can be expanded as:

$$\rho \left\{ \frac{\partial}{\partial t}[\phi(\langle \overline{\mathbf{u}} \rangle^{\mathrm{i}} + \langle \mathbf{u}' \rangle^{\mathrm{i}})] + \nabla \cdot [\phi \langle [\langle \overline{\mathbf{u}} \rangle^{\mathrm{i}} + \langle \mathbf{u}' \rangle^{\mathrm{i}} + {}^{\mathrm{i}}\overline{\mathbf{u}} + {}^{\mathrm{i}}\mathbf{u}'][\langle \overline{\mathbf{u}} \rangle^{\mathrm{i}} + \langle \mathbf{u}' \rangle^{\mathrm{i}} + {}^{\mathrm{i}}\overline{\mathbf{u}} + {}^{\mathrm{i}}\mathbf{u}'] \rangle^{\mathrm{i}}] \right\}$$
$$= -\nabla[\phi(\langle \overline{p} \rangle^{\mathrm{i}} + \langle p' \rangle^{\mathrm{i}})] + \mu \nabla^2 [\phi(\langle \overline{\mathbf{u}} \rangle^{\mathrm{i}} + \langle \mathbf{u}' \rangle^{\mathrm{i}})] + \phi \rho \mathbf{g} + \mathbf{R} \tag{4.26}$$

which results, after some manipulation, in the following:

$$\rho \left\{ \frac{\partial}{\partial t}[\phi(\langle \overline{\mathbf{u}} \rangle^{\mathrm{i}} + \langle \mathbf{u}' \rangle^{\mathrm{i}})] + \nabla \cdot [\phi[\langle \overline{\mathbf{u}} \rangle^{\mathrm{i}} \langle \overline{\mathbf{u}} \rangle^{\mathrm{i}} + \langle \overline{\mathbf{u}} \rangle^{\mathrm{i}} \langle \mathbf{u}' \rangle^{\mathrm{i}} + \langle \mathbf{u}' \rangle^{\mathrm{i}} \langle \overline{\mathbf{u}} \rangle^{\mathrm{i}} + \langle \mathbf{u}' \rangle^{\mathrm{i}} \langle \mathbf{u}' \rangle^{\mathrm{i}} \right.$$
$$\left. + \langle {}^{\mathrm{i}}\overline{\mathbf{u}}\, {}^{\mathrm{i}}\overline{\mathbf{u}} \rangle^{\mathrm{i}} + \langle {}^{\mathrm{i}}\overline{\mathbf{u}}\, \mathbf{u}' \rangle^{\mathrm{i}} + \langle {}^{\mathrm{i}}\mathbf{u}'\, {}^{\mathrm{i}}\overline{\mathbf{u}} \rangle^{\mathrm{i}} + \langle {}^{\mathrm{i}}\mathbf{u}'\, {}^{\mathrm{i}}\mathbf{u}' \rangle^{\mathrm{i}}]] \right\}$$
$$= -\nabla[\phi(\langle \overline{p} \rangle^{\mathrm{i}} + \langle p' \rangle^{\mathrm{i}})] + \mu \nabla^2 [\phi(\langle \overline{\mathbf{u}} \rangle^{\mathrm{i}} + \langle \mathbf{u}' \rangle^{\mathrm{i}})] + \phi \rho \mathbf{g} + \mathbf{R} \tag{4.27}$$

Taking the time average of Equation (4.27) gives further:

$$\rho \left\{ \frac{\partial}{\partial t}(\phi \langle \overline{\mathbf{u}} \rangle^{\mathrm{i}}) + \nabla \cdot \{\phi[\langle \overline{\mathbf{u}} \rangle^{\mathrm{i}} \langle \overline{\mathbf{u}} \rangle^{\mathrm{i}} + \overline{\langle \mathbf{u}' \rangle^{\mathrm{i}} \langle \mathbf{u}' \rangle^{\mathrm{i}}} + \langle {}^{\mathrm{i}}\overline{\mathbf{u}}\, {}^{\mathrm{i}}\overline{\mathbf{u}} \rangle^{\mathrm{i}} + \langle \overline{{}^{\mathrm{i}}\mathbf{u}'\, {}^{\mathrm{i}}\mathbf{u}'} \rangle^{\mathrm{i}}]\} \right\}$$
$$= -\nabla(\phi \langle \overline{p} \rangle^{\mathrm{i}}) + \mu \nabla^2 (\phi \langle \overline{\mathbf{u}} \rangle^{\mathrm{i}}) + \phi \rho \mathbf{g} + \overline{\mathbf{R}} \tag{4.28}$$

where

$$\overline{\mathbf{R}} = \frac{\mu}{\Delta V} \int_{A_{\mathrm{i}}} \mathbf{n} \cdot (\nabla \overline{\mathbf{u}})\, \mathrm{d}S - \frac{1}{\Delta V} \int_{A_{\mathrm{i}}} \mathbf{n} \overline{p}\, \mathrm{d}S \tag{4.29}$$

represents the time averaged value of the instantaneous total drag given by Equation (2.40).

An expression for the fluctuating macroscopic velocity is obtained by subtracting Equation (4.28) from (4.27), which results in the following:

$$\rho\frac{\partial}{\partial t}(\phi\langle\mathbf{u}'\rangle^{\mathrm{i}})+\rho\nabla\cdot\{\phi[\langle\overline{\mathbf{u}}\rangle^{\mathrm{i}}\langle\mathbf{u}'\rangle^{\mathrm{i}}+\langle\mathbf{u}'\rangle^{\mathrm{i}}\langle\overline{\mathbf{u}}\rangle^{\mathrm{i}}+\langle\mathbf{u}'\rangle^{\mathrm{i}}\langle\mathbf{u}'\rangle^{\mathrm{i}}+\langle{}^{\mathrm{i}}\overline{\mathbf{u}}\,{}^{\mathrm{i}}\mathbf{u}'\rangle^{\mathrm{i}}+\langle{}^{\mathrm{i}}\mathbf{u}'\,{}^{\mathrm{i}}\overline{\mathbf{u}}\rangle^{\mathrm{i}}$$
$$+\langle{}^{\mathrm{i}}\mathbf{u}'\,{}^{\mathrm{i}}\mathbf{u}'\rangle^{\mathrm{i}}-\overline{\langle\mathbf{u}'\rangle^{\mathrm{i}}\langle\mathbf{u}'\rangle^{\mathrm{i}}}-\langle\overline{{}^{\mathrm{i}}\mathbf{u}'\,{}^{\mathrm{i}}\mathbf{u}'}\rangle^{\mathrm{i}}]\}=-\nabla(\phi\langle p'\rangle^{\mathrm{i}})+\mu\nabla^2(\phi\langle\mathbf{u}'\rangle^{\mathrm{i}})+\mathbf{R}' \quad (4.30)$$

where $\mathbf{R}'$ is given by Equation (4.23) such that Equation (4.30) is the same as Equation (4.24).

## 4.2. TURBULENT KINETIC ENERGY

As mentioned, the determination of the turbulent kinetic energy of macroscopic flow follows two different paths in the literature. In the models of Lee and Howell (1987), Wang and Takle (1995), Antohe and Lage (1997) and Getachewa *et al.* (2000), the turbulence kinetic energy was based on $k_{\mathrm{m}}=\overline{\langle\mathbf{u}'\rangle^{\mathrm{i}}\cdot\langle\mathbf{u}'\rangle^{\mathrm{i}}}/2$. They started with a simplified form of Equation (4.24) neglecting the 5th, 6th, 7th and 9th terms. Then they took the scalar product of this simplified form with $\langle\mathbf{u}'\rangle^{\mathrm{i}}$ and applied the time average operator. On the other hand, if one starts with Equation (4.21) and takes the scalar product of it with $\mathbf{u}'$ followed by application of time averaging first, one ends up, after volume averaging, with an equation for $\langle k\rangle^{\mathrm{i}}=\langle\overline{\mathbf{u}'\cdot\mathbf{u}'}\rangle^{\mathrm{i}}/2$. This was the path followed by Masuoka and Takatsu (1996), Takatsu and Masuoka (1998) and Nakayama and Kuwahara (1999).

The objective of this section is to derive both transport equations for $k_{\mathrm{m}}$ and $\langle k\rangle^{\mathrm{i}}$ in order to compare similar terms.

### 4.2.1. Equation for $k_{\mathrm{m}}=\overline{\langle\mathbf{u}'\rangle^{\mathrm{i}}\cdot\langle\mathbf{u}'\rangle^{\mathrm{i}}}/2$

From the instantaneous microscopic continuity equation for a constant-property fluid one obtains:

$$\nabla\cdot(\phi\langle\mathbf{u}\rangle^{\mathrm{i}})=0 \quad\Rightarrow\quad \nabla\cdot[\phi(\langle\overline{\mathbf{u}}\rangle^{\mathrm{i}}+\langle\mathbf{u}'\rangle^{\mathrm{i}})]=0 \quad (4.31)$$

with time average:

$$\nabla\cdot(\phi\langle\overline{\mathbf{u}}\rangle^{\mathrm{i}})=0 \quad (4.32)$$

From Equations (4.31) and (4.32) one obtains,

$$\nabla\cdot(\phi\langle\mathbf{u}'\rangle^{\mathrm{i}})=0 \quad (4.33)$$

Taking the scalar product of Equation (4.22) with $\langle \mathbf{u}' \rangle^{\mathrm{i}}$, making use of Equations (4.31)–(4.33) and time averaging it, an equation for $k_{\mathrm{m}}$ will have for each of its terms (note that $\phi$ is here considered as independent of time):

$$\rho \overline{\langle \mathbf{u}' \rangle^{\mathrm{i}} \cdot \frac{\partial}{\partial t}(\phi \langle \mathbf{u}' \rangle^{\mathrm{i}})} = \rho \frac{\partial (\phi k_{\mathrm{m}})}{\partial t} \tag{4.34}$$

$$\begin{aligned} \rho \overline{\langle \mathbf{u}' \rangle^{\mathrm{i}} \cdot \{\nabla \cdot (\phi \langle \overline{\mathbf{u}}\, \mathbf{u}' \rangle^{\mathrm{i}})\}} &= \rho \overline{\langle \mathbf{u}' \rangle^{\mathrm{i}} \cdot \{\nabla \cdot [\phi \langle \overline{\mathbf{u}} \rangle^{\mathrm{i}} \langle \mathbf{u}' \rangle^{\mathrm{i}} + \phi \langle {}^{\mathrm{i}}\overline{\mathbf{u}}\, {}^{\mathrm{i}}\mathbf{u}' \rangle^{\mathrm{i}}]\}} \\ &= \rho \nabla \cdot [\phi \langle \overline{\mathbf{u}} \rangle^{\mathrm{i}} k_{\mathrm{m}}] + \rho \overline{\langle \mathbf{u}' \rangle^{\mathrm{i}} \cdot \{\nabla \cdot [\phi \langle {}^{\mathrm{i}}\overline{\mathbf{u}}\, {}^{\mathrm{i}}\mathbf{u}' \rangle^{\mathrm{i}}]\}} \end{aligned} \tag{4.35}$$

$$\begin{aligned} \rho \overline{\langle \mathbf{u}' \rangle^{\mathrm{i}} \cdot \{\nabla \cdot (\phi \langle \mathbf{u}'\, \overline{\mathbf{u}} \rangle^{\mathrm{i}})\}} &= \rho \overline{\langle \mathbf{u}' \rangle^{\mathrm{i}} \cdot \{\nabla \cdot [\phi \langle \mathbf{u}' \rangle^{\mathrm{i}} \langle \overline{\mathbf{u}} \rangle^{\mathrm{i}} + \phi \langle {}^{\mathrm{i}}\mathbf{u}'\, {}^{\mathrm{i}}\overline{\mathbf{u}} \rangle^{\mathrm{i}}]\}} \\ &= \rho \phi \overline{\langle \mathbf{u}' \rangle^{\mathrm{i}} \langle \mathbf{u}' \rangle^{\mathrm{i}}} : \nabla \langle \overline{\mathbf{u}} \rangle^{\mathrm{i}} + \rho \overline{\langle \mathbf{u}' \rangle^{\mathrm{i}} \cdot \{\nabla \cdot [\phi \langle {}^{\mathrm{i}}\mathbf{u}'\, {}^{\mathrm{i}}\overline{\mathbf{u}} \rangle^{\mathrm{i}}]\}} \end{aligned} \tag{4.36}$$

$$\begin{aligned} \rho \overline{\langle \mathbf{u}' \rangle^{\mathrm{i}} \cdot \{\nabla \cdot (\phi \langle \mathbf{u}'\, \mathbf{u}' \rangle^{\mathrm{i}})\}} &= \rho \overline{\langle \mathbf{u}' \rangle^{\mathrm{i}} \cdot \{\nabla \cdot [\phi \langle \mathbf{u}' \rangle^{\mathrm{i}} \langle \mathbf{u}' \rangle^{\mathrm{i}} + \phi \langle {}^{\mathrm{i}}\mathbf{u}'\, {}^{\mathrm{i}}\mathbf{u}' \rangle^{\mathrm{i}}]\}} \\ &= \rho \nabla \cdot [\phi \overline{\langle \mathbf{u}' \rangle^{\mathrm{i}} \frac{\langle \mathbf{u}' \rangle^{\mathrm{i}} \cdot \langle \mathbf{u}' \rangle^{\mathrm{i}}}{2}}] + \rho \overline{\langle \mathbf{u}' \rangle^{\mathrm{i}} \cdot \{\nabla \cdot [\phi \langle {}^{\mathrm{i}}\mathbf{u}'\, {}^{\mathrm{i}}\mathbf{u}' \rangle^{\mathrm{i}}]\}} \end{aligned} \tag{4.37}$$

$$\rho \overline{\langle \mathbf{u}' \rangle^{\mathrm{i}} \cdot \{\nabla \cdot (-\phi \langle \overline{\mathbf{u}'\, \mathbf{u}'} \rangle^{\mathrm{i}})\}} = 0 \tag{4.38}$$

$$-\overline{\langle \mathbf{u}' \rangle^{\mathrm{i}} \cdot \nabla(\phi \langle p' \rangle^{\mathrm{i}})} = -\nabla \cdot \overline{[\phi \langle \mathbf{u}' \rangle^{\mathrm{i}} \langle p' \rangle^{\mathrm{i}}]} \tag{4.39}$$

$$\mu \overline{\langle \mathbf{u}' \rangle^{\mathrm{i}} \cdot \nabla^2(\phi \langle \mathbf{u}' \rangle^{\mathrm{i}})} = \mu \nabla^2(\phi k_{\mathrm{m}}) - \rho \phi \varepsilon_{\mathrm{m}} \tag{4.40}$$

$$\overline{\langle \mathbf{u}' \rangle^{\mathrm{i}} \cdot \mathbf{R}'} \equiv 0 \tag{4.41}$$

where $\varepsilon_{\mathrm{m}} = \nu \overline{\nabla \langle \mathbf{u}' \rangle^{\mathrm{i}} : (\nabla \langle \mathbf{u}' \rangle^{\mathrm{i}})^{\mathrm{T}}}$. In handling Equation (4.39), the porosity $\phi$ was assumed to be constant only for simplifying the manipulation to be shown next. However, this assumption does not represent a limitation in deriving a general transport equation for $k_{\mathrm{m}}$.

Another important point is the treatment of the scalar product shown in Equation (4.41). Here, a view different from the work of Lee and Howell (1987), Wang and Takle (1995), Antohe and Lage (1997) and Getachewa *et al.* (2000) is considered. The fluctuating drag form $\mathbf{R}'$ acts through the solid–fluid interfacial area and, as such, on fluid particles at rest. The fluctuating mechanical energy represented by the operation in Equation (4.41) is not associated with any fluid particle movement and, as a result, is here considered to be of null value. This point is further discussed later in this chapter.

The final equation for $k_{\mathrm{m}}$ gives:

$$\begin{aligned} \rho \frac{\partial (\phi k_{\mathrm{m}})}{\partial t} + \rho \nabla \cdot [\phi \langle \overline{\mathbf{u}} \rangle^{\mathrm{i}} k_{\mathrm{m}}] = &- \rho \nabla \cdot \left\{ \overline{\phi \langle \mathbf{u}' \rangle^{\mathrm{i}} \left[ \frac{\langle p' \rangle^{\mathrm{i}}}{\rho} + \frac{\langle \mathbf{u}' \rangle^{\mathrm{i}} \cdot \langle \mathbf{u}' \rangle^{\mathrm{i}}}{2} \right]} \right\} + \mu \nabla^2(\phi k_{\mathrm{m}}) \\ &- \rho \phi \overline{\langle \mathbf{u}' \rangle^{\mathrm{i}} \langle \mathbf{u}' \rangle^{\mathrm{i}}} : \nabla \langle \overline{\mathbf{u}} \rangle^{\mathrm{i}} - \rho \phi \varepsilon_{\mathrm{m}} - D_{\mathrm{m}} \end{aligned} \tag{4.42}$$

where

$$D_{\mathrm{m}} = \rho \overline{\langle \mathbf{u}' \rangle^{\mathrm{i}} \cdot \{\nabla \cdot [\phi (\langle {}^{\mathrm{i}}\overline{\mathbf{u}}\, {}^{\mathrm{i}}\mathbf{u}' \rangle^{\mathrm{i}} + \langle {}^{\mathrm{i}}\mathbf{u}'\, {}^{\mathrm{i}}\overline{\mathbf{u}} \rangle^{\mathrm{i}} + \langle {}^{\mathrm{i}}\mathbf{u}'\, {}^{\mathrm{i}}\mathbf{u}' \rangle^{\mathrm{i}})]\}} \tag{4.43}$$

represents the dispersion of $k_m$ and the three terms on the rhs of Equation 4.43 are the last terms on the rhs of Equations (4.35)–(4.37), respectively. It is interesting to note that this term can be either negative or positive.

The first term on the rhs of Equation (4.42) represents the turbulent diffusion of $k_m$ and is normally modeled via a diffusion-like expression resulting in the transport equation for $k_m$ (Antohe and Lage, 1997; Getachewa *et al.*, 2000):

$$\rho\frac{\partial(\phi k_m)}{\partial t}+\rho\nabla\cdot[\phi\langle\overline{\mathbf{u}}\rangle^i k_m]=\nabla\cdot\left[\mu+\frac{\mu_{t_m}}{\sigma_{k_m}}\nabla(\phi k_m)\right]+P_m-\rho\phi\varepsilon_m-D_m \tag{4.44}$$

where

$$P_m=-\rho\phi\overline{\langle\mathbf{u}'\rangle^i\langle\mathbf{u}'\rangle^i}:\nabla\langle\overline{\mathbf{u}}\rangle^i \tag{4.45}$$

is the production rate of $k_m$ due to the gradient of the macroscopic time mean velocity $\langle\overline{\mathbf{u}}\rangle^i$.

Lee and Howell (1987), Wang and Takle (1995), Antohe and Lage (1997) and Getachewa *et al.* (2000) made use of the above equation for $k_m$ considering for $\mathbf{R}'$ the Darcy–Forchheimer extended model with macroscopic time-fluctuation velocities $\langle\mathbf{u}'\rangle^i$. They have also neglected all dispersion terms that were grouped into $D_m$ in Equation (4.43). Note also that the order of application of both volume and time average operators in this case cannot be changed. The quantity $k_m$ is defined by applying first the volume operator to the fluctuating velocity field.

### 4.2.2. Equation for $\langle k\rangle^i=\langle\overline{\mathbf{u}'\cdot\mathbf{u}'}\rangle^i/2$

The other procedure for composing the flow turbulent kinetic energy is to take the scalar product of Equation (4.21) and the microscopic fluctuating velocity $\mathbf{u}'$. Then apply both time and volume average operators for obtaining an equation for $\langle k\rangle^i=\langle\overline{\mathbf{u}'\cdot\mathbf{u}'}\rangle^i/2$. It is worth noting that in this case the order of application of both operations is immaterial since no additional mathematical operation (the scalar product) is conducted between the averaging processes. Therefore, this is the same as applying the volume operator to a transport equation for $k$.

The volumetric average of a transport equation for $k$ has been carried out in detail by de Lemos and Pedras (2000a) and Pedras and de Lemos (2001a). Nevertheless, for the sake of completeness, a few steps of that derivation are reproduced here. Application of the volume average theorem to the transport equation for the turbulence kinetic energy $k$ gives:

$$\rho\left[\frac{\partial}{\partial t}(\phi\langle k\rangle^i)+\nabla\cdot(\phi\langle\overline{\mathbf{u}k}\rangle^i)\right]=-\rho\nabla\cdot\left\{\phi\left\langle\overline{\mathbf{u}'\left(\frac{p'}{\rho}+k\right)}\right\rangle^i\right\}+\mu\nabla^2(\phi\langle k\rangle^i)$$
$$-\rho\phi\langle\overline{\mathbf{u}'\mathbf{u}'}:\nabla\overline{\mathbf{u}}\rangle^i-\rho\phi\langle\varepsilon\rangle^i \tag{4.46}$$

The divergence of the rhs can be expanded as:

$$\nabla \cdot (\phi \langle \overline{\mathbf{u}} k \rangle^{\mathrm{i}}) = \nabla \cdot [\phi(\langle \overline{\mathbf{u}} \rangle^{\mathrm{i}} \langle k \rangle^{\mathrm{i}} + \langle {}^{\mathrm{i}}\overline{\mathbf{u}}^{\mathrm{i}} k \rangle^{\mathrm{i}})] \tag{4.47}$$

where the first term on the rhs is the convection of $\langle k \rangle^{\mathrm{i}}$ due to the macroscopic velocity whereas the second is the dispersive transport due to spatial deviations of both $k$ and $u$. Likewise, the production term on the rhs of Equation (4.46) can be expanded as:

$$-\rho\phi\langle \overline{\mathbf{u}'\mathbf{u}'} : \nabla \overline{\mathbf{u}} \rangle^{\mathrm{i}} = -\rho\phi[\langle \overline{\mathbf{u}'\mathbf{u}'} \rangle^{\mathrm{i}} : \langle \nabla \overline{\mathbf{u}} \rangle^{\mathrm{i}} + \langle {}^{\mathrm{i}}(\overline{\mathbf{u}'\mathbf{u}'}) : {}^{\mathrm{i}}(\nabla \overline{\mathbf{u}}) \rangle^{\mathrm{i}}] \tag{4.48}$$

Similarly, the first term on the rhs of Equation (4.48) is the production of $\langle k \rangle^{\mathrm{i}}$ due to the mean macroscopic flow whereas the second term is the generation of $\langle k \rangle^{\mathrm{i}}$ associated with spatial deviations of the Reynolds stresses (divided by $\rho$) and gradients of the mean velocity.

The other terms appearing in Equations (4.47) and (4.48) represent, respectively, extra transport/production of $\langle k \rangle^{\mathrm{i}}$ due to the presence of solid material inside the integration volume. They should be null for the limiting case of clear fluid flow, or, say, when $\phi \to 1 \Rightarrow K \to \infty$. Also, they should be proportional to the macroscopic velocity and to $\langle k \rangle^{\mathrm{i}}$.

In Pedras and de Lemos (2001a), a proposal for those two extra transport/production rates of $\langle k \rangle^{\mathrm{i}}$ was made as follows:

$$\nabla \cdot (\phi \langle {}^{\mathrm{i}}\overline{\mathbf{u}}^{\mathrm{i}} k \rangle^{\mathrm{i}}) - \rho\phi \langle {}^{\mathrm{i}}(\overline{\mathbf{u}'\mathbf{u}'}) : {}^{\mathrm{i}}(\nabla \overline{\mathbf{u}}) \rangle^{\mathrm{i}} = G_{\mathrm{i}} = c_k \rho \phi \frac{\langle k \rangle^{\mathrm{i}} |\overline{\mathbf{u}}_{\mathrm{D}}|}{\sqrt{K}} \tag{4.49}$$

where $K$ is the medium permeability and the constant $c_k$ was numerically determined by fine flow computations considering the medium to be formed by circular rods (Pedras and de Lemos, 2001b), as well as longitudinal (Pedras and de Lemos, 2001c) and transversal rods (Pedras and de Lemos, 2003). In spite of the variation in the medium morphology and the use of a wide range of porosity and Reynolds number, a value of 0.28 was found to be suitable for most calculations. Details on the determination of the numerical value of $c_K$ will be shown below.

The final resulting equation for $\langle k \rangle^{\mathrm{i}}$ proposed be Pedras and de Lemos (2001a) is,

$$\rho \left[ \frac{\partial}{\partial t}(\phi \langle k \rangle^{\mathrm{i}}) + \nabla \cdot (\overline{\mathbf{u}}_{\mathrm{D}} \langle k \rangle^{\mathrm{i}}) \right] = \nabla \cdot \left[ \left( \mu + \frac{\mu_{t_\phi}}{\sigma_k} \right) \nabla(\phi \langle k \rangle^{\mathrm{i}}) \right] + P_{\mathrm{i}} + G_{\mathrm{i}} - \rho\phi\langle \varepsilon \rangle^{\mathrm{i}} \tag{4.50}$$

where

$$P_{\mathrm{i}} = -\rho \langle \overline{\mathbf{u}'\mathbf{u}'} \rangle^{\mathrm{i}} : \nabla \overline{\mathbf{u}}_{\mathrm{D}}, \quad G_{\mathrm{i}} = c_k \rho \phi \frac{\langle k \rangle^{\mathrm{i}} |\overline{\mathbf{u}}_{\mathrm{D}}|}{\sqrt{K}} \tag{4.51}$$

### 4.2.3. Comparison of Macroscopic Transport Equations

A comparison between the terms in the transport equation for $k_m$ and $\langle k\rangle^i$ can now be made. Pedras and de Lemos (2000a) have already showed the connection between these two quantities as being:

$$\langle k\rangle^i = \langle\overline{\mathbf{u}'\cdot\mathbf{u}'}\rangle^i/2 = \overline{\langle\mathbf{u}'\rangle^i\cdot\langle\mathbf{u}'\rangle^i}/2 + \langle\overline{{}^i\mathbf{u}'\cdot{}^i\mathbf{u}'}\rangle^i/2 = k_m + \langle\overline{{}^i\mathbf{u}'\cdot{}^i\mathbf{u}'}\rangle^i/2 \tag{4.52}$$

Expanding the correlation, forming the production term $P_i$ by means of Equation (2.19), a connection between the two generation rates can also be written as follows:

$$\begin{aligned} P_i = -\rho\langle\overline{\mathbf{u}'\mathbf{u}'}\rangle^i : \nabla\bar{\mathbf{u}}_D &= -\rho\left(\overline{\langle\mathbf{u}'\rangle^i\langle\mathbf{u}'\rangle^i} : \nabla\bar{\mathbf{u}}_D + \overline{\langle{}^i\mathbf{u}'{}^i\mathbf{u}'\rangle^i} : \nabla\bar{\mathbf{u}}_D\right) \\ &= P_m - \rho\overline{\langle{}^i\mathbf{u}'{}^i\mathbf{u}'\rangle^i} : \nabla\bar{\mathbf{u}}_D \end{aligned} \tag{4.53}$$

We note that all the production rate of $k_m$, due to the mean flow, constitutes only part of the general production rate responsible for maintaining the overall level of $\langle k\rangle^i$.

The dissipation rates also carry a correspondence if we expand

$$\begin{aligned} \langle\varepsilon\rangle^i = \nu\langle\overline{\nabla\mathbf{u}' : (\nabla\mathbf{u}')^T}\rangle^i &= \nu\overline{\langle\nabla\mathbf{u}'\rangle^i : [\langle\nabla\mathbf{u}'\rangle^i]^T} + \nu\langle\overline{{}^i(\nabla\mathbf{u}') : {}^i(\nabla\mathbf{u}')^T}\rangle^i \\ &= \frac{\nu}{\phi^2}\overline{\nabla(\phi\langle\mathbf{u}'\rangle^i) : [\nabla(\phi\langle\mathbf{u}'\rangle^i)]^T} + \nu\langle\overline{{}^i(\nabla\mathbf{u}') : {}^i(\nabla\mathbf{u}')^T}\rangle^i \end{aligned} \tag{4.54}$$

If the porosity $\phi$ is considered to be constant:

$$\langle\varepsilon\rangle^i = \varepsilon_m + \nu\langle\overline{{}^i(\nabla\mathbf{u}') : {}^i(\nabla\mathbf{u}')^T}\rangle^i \tag{4.55}$$

and this indicates that an additional dissipation rate is necessary to fully account for the energy decay process inside the REV.

## 4.3. MACROSCOPIC TURBULENCE MODEL

In the work presented by de Lemos and Pedras (2000a) and Pedras and de Lemos (2001a), the authors have applied the volume average operator to the microscopic $k$–$\varepsilon$ equations and proposed the following macroscopic model:

$$\begin{aligned} \rho\left[\frac{\partial}{\partial t}(\phi\langle k\rangle^i) + \nabla\cdot(\bar{\mathbf{u}}_D\langle k\rangle^i)\right] &= \nabla\cdot\left[\left(\mu + \frac{\mu_{t_\phi}}{\sigma_k}\right)\nabla(\phi\langle k\rangle^i)\right] \\ &\quad - \rho\langle\overline{\mathbf{u}'\mathbf{u}'}\rangle^i : \nabla\bar{\mathbf{u}}_D + c_k\rho\phi\frac{\langle k\rangle^i|\bar{\mathbf{u}}_D|}{\sqrt{K}} - \rho\phi\langle\varepsilon\rangle^i \end{aligned} \tag{4.56}$$

$$\rho\left[\frac{\partial}{\partial t}(\phi\langle\varepsilon\rangle^{\mathrm{i}})+\nabla\cdot(\overline{\mathbf{u}}_{\mathrm{D}}\langle\varepsilon\rangle^{\mathrm{i}})\right]=\nabla\cdot\left[\left((\mu+\frac{\mu_{t_\phi}}{\sigma_\varepsilon})\nabla(\phi\langle\varepsilon\rangle^{\mathrm{i}}\right)\right]+c_{1\varepsilon}(-\rho\langle\overline{\mathbf{u}'\mathbf{u}'}\rangle^{\mathrm{i}}:\nabla\overline{\mathbf{u}}_{\mathrm{D}})\frac{\langle\varepsilon\rangle^{\mathrm{i}}}{\langle k\rangle^{\mathrm{i}}}$$
$$+c_{2\varepsilon}\rho\phi\left\{c_k\frac{\langle\varepsilon\rangle^{\mathrm{i}}|\overline{\mathbf{u}}_{\mathrm{D}}|}{\sqrt{K}}-\frac{\langle\varepsilon\rangle^{\mathrm{i}2}}{\langle k\rangle^{\mathrm{i}}}\right\} \tag{4.57}$$

with,

$$-\rho\phi\langle\overline{\mathbf{u}'\mathbf{u}'}\rangle^{\mathrm{i}}=\mu_{t_\phi}2\langle\overline{\mathbf{D}}\rangle^{\mathrm{v}}-\frac{2}{3}\phi\rho\langle k\rangle^{\mathrm{i}}\mathbf{I} \tag{4.58}$$

$$\mu_{t_\phi}=\rho c_\mu\frac{\langle k\rangle^{\mathrm{i}2}}{\langle\varepsilon\rangle^{\mathrm{i}}} \tag{4.59}$$

where $c_{1\varepsilon}$, $c_{2\varepsilon}$ and $c_\mu$ are non-dimensional constants. The additional constant $c_k$ in the extra production term $G_i$ in Equation (4.51) needs to be determined for completeness of the turbulence model. The methodology followed for determining such a constant is presented below:

### 4.3.1. Numerical Determination of Constant $c_k$

For fully developed, uni-dimensional, macroscopic flow in isotropic and homogeneous media, the limiting values for $\langle k\rangle^{\mathrm{i}}$ and $\langle\varepsilon\rangle^{\mathrm{i}}$ are given by $k_\phi$ and $\varepsilon_\phi$, respectively. In this limiting condition, Equations (4.56) and (4.57) reduce to:

$$\left.\begin{aligned}\langle\varepsilon\rangle^{\mathrm{i}}=\varepsilon_\phi=c_k\frac{k_\phi|\overline{\mathbf{u}}_{\mathrm{D}}|}{\sqrt{K}}\\ \frac{\langle\varepsilon\rangle^{\mathrm{i}2}}{\langle k\rangle^{\mathrm{i}}}=c_k\frac{\varepsilon_\phi|\overline{\mathbf{u}}_{\mathrm{D}}|}{\sqrt{K}}\end{aligned}\right\}\Rightarrow\langle k\rangle^{\mathrm{i}}=k_\phi \tag{4.60}$$

or in the following dimensionless form:

$$\frac{\varepsilon_\phi\sqrt{K}}{|\overline{\mathbf{u}}_{\mathrm{D}}|^3}=c_k\frac{k_\phi}{|\overline{\mathbf{u}}_{\mathrm{D}}|^2} \tag{4.61}$$

The coefficient $c_k$ was adjusted in this limiting condition for the spatially periodic cells shown in Figure 4.1. The figure represents different solid phase shapes and is a step toward mapping a number of different morphological description of distinct media. Ultimately, one intends to gather information on a variety of structures in order to validate the macroscopic two-equation model using Equations (4.56)–(4.57). In the first geometry shown in Figure 4.1a, the ratio of ellipse axes is $a/b=5/3$ and the flow is from left to right along the longer axis of the ellipse (longitudinal case). Both the longitudinal rods of Figure 4.1a and the cylindrical case shown in Figure 4.1b were also investigated in Pedras and de Lemos (2001a,c), respectively, and are here included for the sake of comparison. In the third periodic cell (Figure 4.1c), one has the transversal positioning

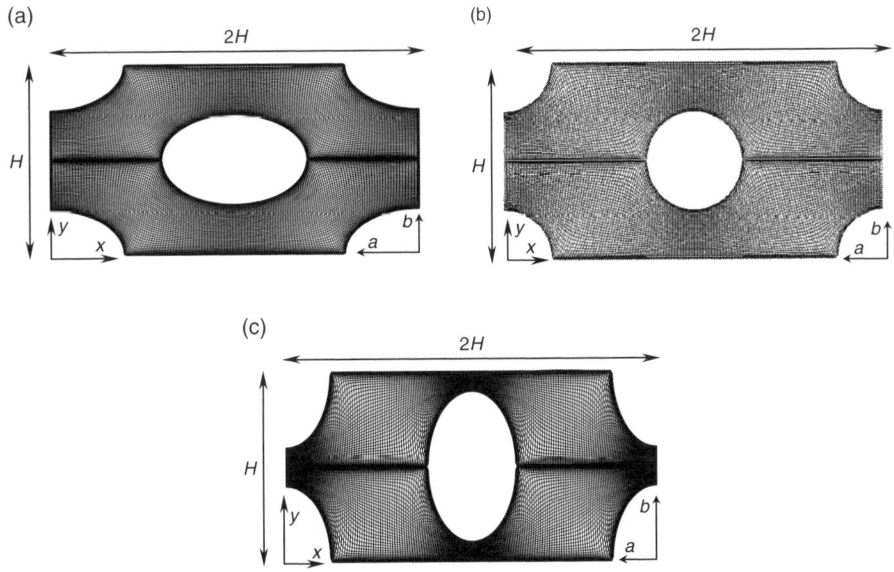


**Figure 4.1.** Model of REV – periodic cell and elliptically generated grids: (a) longitudinal elliptic rods, $a/b = 5/3$ (Pedras and de Lemos, 2001c); (b) cylindrical rods, $a/b = 1$ (Pedras and de Lemos, 2001a); (c) transverse elliptic rods, $a/b = 3/5$ (Pedras and de Lemos, 2003).

of the same elliptical rod shown in Figure 4.1a. Also important to note is that both the cylindrical and elliptical arrangements had nearly the same value for porosity so that all comparisons shown below, resulting from microscopic computations, reflect changes due to other medium properties such as permeability and morphology of the different geometric models. It is also important to emphasize the influence of medium morphology on macroscopic models that, in principle, do not explicitly account for any effect of turbulence. In fact, recent literature results by Bhattacharya *et al.* (2002) propose correlations for the inertia coefficient as a function of medium and flow properties. In the path followed here, however, one unique value for the inertia coefficient will be used when presenting macroscopic results later. Here, the explicit accounting for turbulent transport for high *Re* numbers, while keeping a unique macroscopic inertia coefficient, can be seen as an alternative path on adjusting the Forchheimer coefficient for large values of *Re*. Also important to remember is that a distinction between laminar, non-linear and fully turbulent flow in porous media is not as evident as in unobstructed flow and that adequate models covering a wide range of medium ($K$, $\phi$) and flow properties ($Re_H$) are still to be developed.

In all cases computed, the flow was assumed to enter through the left aperture so that symmetry along the $y$-direction and periodic boundary conditions along the $x$-coordinate

**Table 4.1.** Parameters for microscopic computations, velocities in m/s (Pedras and de Lemos, 2001b).

| | $\phi = 0.4$ | | $\phi = 0.6$ | | $\phi = 0.8$ | | |
|---|---|---|---|---|---|---|---|
| $Re_H$ | $\bar{\mathbf{u}}_D$ | $\langle \bar{\mathbf{u}} \rangle^i$ | $\bar{\mathbf{u}}_D$ | $\langle \bar{\mathbf{u}} \rangle^i$ | $\bar{\mathbf{u}}_D$ | $\langle \bar{\mathbf{u}} \rangle^i$ | Turbulence model |
| 1.20E+01 | 1.80E−04 | 4.50E−04 | 1.79E−04 | 2.99E−04 | 1.79E−04 | 2.24E−04 | Laminar |
| 1.20E+04 | 1.80E−01 | 4.50E−01 | 1.79E−01 | 2.99E−01 | 1.79E−01 | 2.24E−01 | Low Re |
| 1.20E+05 | 1.80E+00 | 4.50E+00 | 1.79E+00 | 2.99E+00 | 1.79E+00 | 2.24E+00 | Low Re |
| 1.20E+05 | 1.80E+00 | 4.50E+00 | 1.79E+00 | 2.99E+00 | 1.79E+00 | 2.24E+00 | High Re |
| 1.20E+06 | 1.80E+01 | 4.50E+01 | 1.79E+01 | 2.99E+01 | 1.79E+01 | 2.24E+01 | High Re |

were applied. Values of $k_\phi$ and $\varepsilon_\phi$ were obtained by integrating the microscopic flow field for Reynolds number, $Re_H = |\langle \bar{\mathbf{u}} \rangle^v| H/\nu$, ranging from $10^4$ to $10^6$. The porosity, given by $\phi = 1 - ab\pi/H^2$, was varied from 0.53 to 0.85 for longitudinal ellipses and from 0.70 to 0.90 for the transversal case.

The numerical method SIMPLE was employed for relaxing the mean and turbulence equations within the domain. The dimensions of the periodic cell for circular rods, considered in Pedras and de Lemos (2001b) were $H = 0.1$ m, $S = 2H$, and $D = 0.03$ m ($\phi = 0.8$), 0.05 m ($\phi = 0.6$) and 0.06 m ($\phi = 0.4$). The solutions were grid independent and all normalized residuals were brought down to $10^{-5}$. Also, relaxation parameters for all the variables $\mathbf{u}$, $p$, $k$ and $\varepsilon$ were kept equal to 0.8. A summary of all relevant parameters is presented in Table 4.1.

### 4.3.2. Microscopic Results and Integrated Values

In this section, numerical results from Pedras and de Lemos (2001b,c, 2003) are reviewed. Eighteen runs were carried out for each case (longitudinal and transversal ellipses), being six for laminar flow, six with the low *Re* model and six using the high *Re* theory. The main objective in Pedras and de Lemos (2001b,c, 2003) was the numerical determination of the introduced constant $c_k$ appearing in Equation (4.47).

Some of the results for the longitudinal ellipses were presented in Pedras and de Lemos (2001c) and are here referred to for the sake of completeness and comparison. Table 4.2 summarizes the integrated values for the longitudinal ellipses (volumetric averaging over the periodic cell obtained for turbulent flow), whereas Table 4.3 compiles the integrated quantities for the transversal cases. In all runs, the medium permeability was calculated using the procedure adopted by Kuwahara and Nakayama (1998).

Figure 4.2 presents velocity, pressure, $k$ and $\varepsilon$ fields for the *longitudinal* ellipses with $Re_H = 1.67 \times 10^5$ (low *Re* model) and $\phi = 0.70$, whereas Figure 4.3 presents the same variables, at the same conditions, for *transversal* ellipses. It is observed that the flow accelerates in the upper and lower passages around the ellipse and separates at the back. As expected, transversal ellipses present a larger wake region that will contribute for larger pressure drop for the same mass flow rate through the bed.

**Table 4.2.** Summary of the integrated results for the longitudinal ellipses, permeability in $m^2$, velocities in m/s, $k$ in $m^2/s^2$ and $\varepsilon$ in $m^2/s^3$ (Pedras and de Lemos, 2001c).

| | Medium permeability | $Re_H$ | $k$–$\varepsilon$ Model | $\lvert\langle\overline{\mathbf{u}}\rangle^v\rvert$ | $\langle k\rangle^i$ | $\langle\varepsilon\rangle^i$ |
|---|---|---|---|---|---|---|
| $\phi = 0.53$ | $K = 4.12E-05$ | 1.67E+04 | Low | 2.51E−01 | 1.36E−02 | 1.26E−01 |
| | | 1.67E+05 | Low | 2.51E+00 | 1.09E+00 | 1.17E+02 |
| | | 1.67E+05 | High | 2.51E+00 | 1.40E+00 | 1.21E+02 |
| | | 1.67E+06 | High | 2.51E+01 | 1.62E+02 | 1.34E+05 |
| $\phi = 0.70$ | $K = 1.29E-04$ | 1.67E+04 | Low | 2.51E−01 | 1.06E−02 | 5.72E−02 |
| | | 1.67E+05 | Low | 2.51E+00 | 8.16E−01 | 4.71E+01 |
| | | 1.67E+05 | High | 2.51E+00 | 8.71E−01 | 4.45E+01 |
| | | 1.67E+06 | High | 2.51E+01 | 9.99E+01 | 5.00E+04 |
| $\phi = 0.85$ | $K = 3.25E-04$ | 1.67E+04 | Low | 2.51E−01 | 7.52E−03 | 2.83E−02 |
| | | 1.67E+05 | Low | 2.51E+00 | 5.48E−01 | 2.14E+01 |
| | | 1.67E+05 | High | 2.51E+00 | 5.17E−01 | 1.79E+01 |
| | | 1.67E+06 | High | 2.51E+01 | 7.52E+01 | 2.70E+04 |

**Table 4.3.** Summary of the integrated results for the transversal ellipses, permeability in $m^2$, velocities in m/s, $k$ in $m^2/s^2$ and $\varepsilon$ in $m^2/s^3$ (Pedras and de Lemos, 2003).

| | Medium permeability | $Re_H$ | $k$–$\varepsilon$ Model | $\lvert\langle\overline{\mathbf{u}}\rangle^v\rvert$ | $\langle k\rangle^i$ | $\langle\varepsilon\rangle^i$ |
|---|---|---|---|---|---|---|
| $\phi = 0.70$ | $K = 2.31E-05$ | 1.67E+04 | Low | 2.51E−01 | 1.22E−01 | 1.67E+00 |
| | | 1.67E+05 | Low | 2.51E+00 | 1.10E+01 | 1.53E+03 |
| | | 1.67E+05 | High | 2.51E+00 | 1.12E+01 | 1.58E+03 |
| | | 1.67E+06 | High | 2.51E+01 | 1.16E+03 | 1.58E+06 |
| $\phi = 0.80$ | $K = 8.69E-05$ | 1.67E+04 | Low | 2.51E−01 | 6.10E−02 | 4.68E−01 |
| | | 1.67E+05 | Low | 2.51E+00 | 4.60E+00 | 3.73E+02 |
| | | 1.67E+05 | High | 2.51E+00 | 5.40E+00 | 4.23E+02 |
| | | 1.67E+06 | High | 2.51E+01 | 5.61E+02 | 4.27E+05 |
| $\phi = 0.90$ | $K = 2.32E-04$ | 1.67E+04 | Low | 2.51E−01 | 3.10E−02 | 1.80E−01 |
| | | 1.67E+05 | Low | 2.51E+00 | 2.36E+00 | 1.57E+02 |
| | | 1.67E+05 | High | 2.51E+00 | 2.24E+00 | 1.29E+02 |
| | | 1.67E+06 | High | 2.51E+01 | 2.75E+02 | 1.78E+05 |

In the remainder fields, it is verified that the pressure increases at the front of the ellipse and decreases at the upper and lower faces. The turbulence kinetic energy is high at the front, on the top and below the ellipse. The dissipation rate of $k$ presents a behavior similar to the turbulence kinetic energy.

Figure 4.4 shows the overall pressure drop as a function of $Re_H$ obtained for elliptic, cylindrical (Pedras and de Lemos, 2001b) and square rods (Nakayama and Kuwahara, 1999). The pressure drop across the cell is defined as:

$$\frac{\mathrm{d}\langle\overline{p}\rangle^i}{\mathrm{d}S} = \frac{1}{2H(\frac{H}{2} - \frac{D}{2})}\int_{D/2}^{H/2} (p|_{x=2H} - p|_{x=0})\mathrm{d}y \tag{4.62}$$

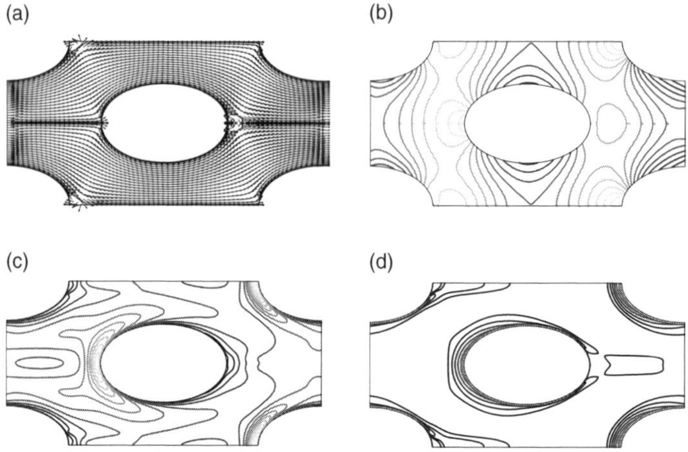


**Figure 4.2.** Microscopic results at $Re_H = 1.67 \times 10^5$ and $\phi = 0.70$ for longitudinal ellipses: (a) velocity, (b) pressure, (c) $k$ and (d) $\varepsilon$.

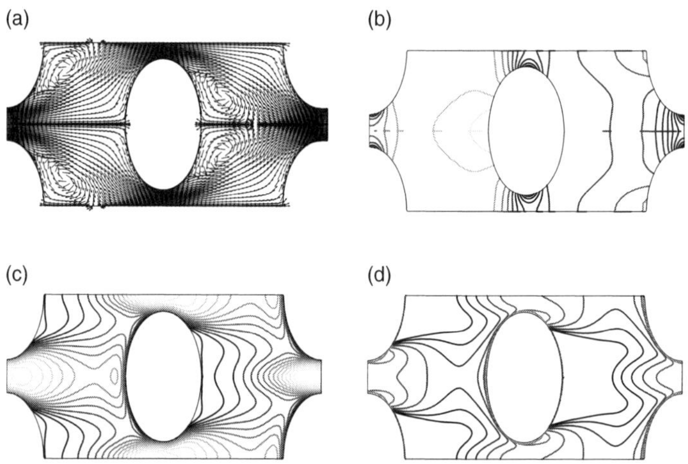


**Figure 4.3.** Microscopic results at $Re_H = 1.67 \times 10^5$ and $\phi = 0.70$ for transversal ellipses: (a) velocity, (b) pressure, (c) $k$ and (d) $\varepsilon$.

Due to the periodic boundary conditions applied (the inlet and outlet momentum are the same), the overall pressure drop can be interpreted as the total drag, including form and friction forces, inside the periodic cell. As expected, for the same porosity and for transversal ellipses one gets the greater drag, followed by square, cylindrical and longitudinal elliptic rods. Although not shown here, one could speculate that, for the same rod type, the higher the porosity the lower the pressure drop, since smaller rods,

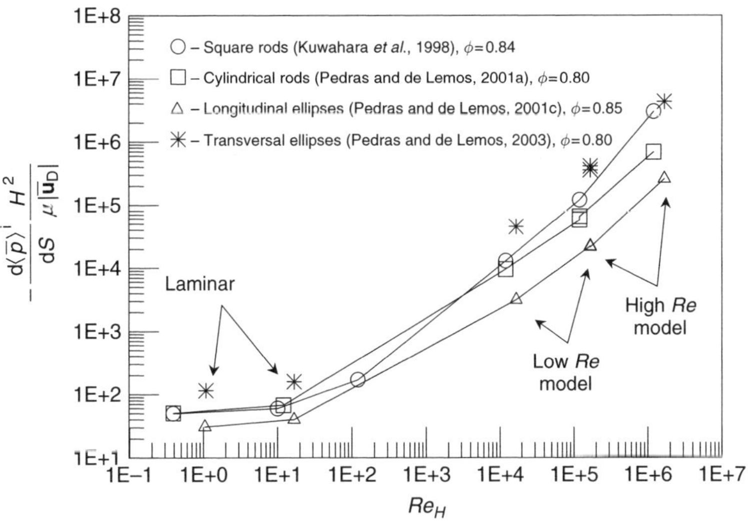


**Figure 4.4.** Overall pressure drop as a function of $Re_H$ and medium morphology.

spaced wider apart, would not only provoke a lower frictional drag (due to smaller interfacial area) but also yield smaller wakes (and then smaller pressure drag) behind the obstacles.

Macroscopic turbulent kinetic energy as a function of medium morphology is presented in Figure 4.5. As porosity decreases maintaining $Re_H$ constant, or, say, reducing the flow passage and increasing the local fluid speed, the integrated turbulence kinetic energy, $\langle k \rangle^i$, increases (see also Tables 4.2 and 4.3). In others words, for a fixed mass flow rate through a certain bed, a decrease in porosity implies accentuated velocity gradients which, in turn, result in larger production rates of $k$ within the fluid. Also, the effect of the medium morphology when comparing the two rod dispositions is clearly indicated in the figure. For the same porosity and Reynolds number, a larger frontal area of the transverse case forces the fluid to a much more irregular path and induces large wake regions. Velocity gradients are everywhere of larger values than those for the longitudinal set-up, ultimately increasing the production rates of $k$ within the entire cell.

Accordingly, it is also interesting to point out that for the same $\phi$ and $Re_H$ the integrated values shown in Table 4.2 for $\langle k \rangle^i$ (longitudinal ellipses) are lower than those obtained for square (Nakayama and Kuwahara, 1999) and cylindrical rods (Pedras and de Lemos, 2001b), whereas for transversal ellipses $\langle k \rangle^i$ values are greater among other cases compared. Figure 4.6 further plots values for the non-dimensional turbulent kinetic energy for all cases computed. It is interesting to note that the results in non-dimensional form are nearly independent of $Re_H$. Also important to note in Figure 4.6 is the inappropriateness of the wall function approach (high $Re$ computations) when a large recirculation bubble

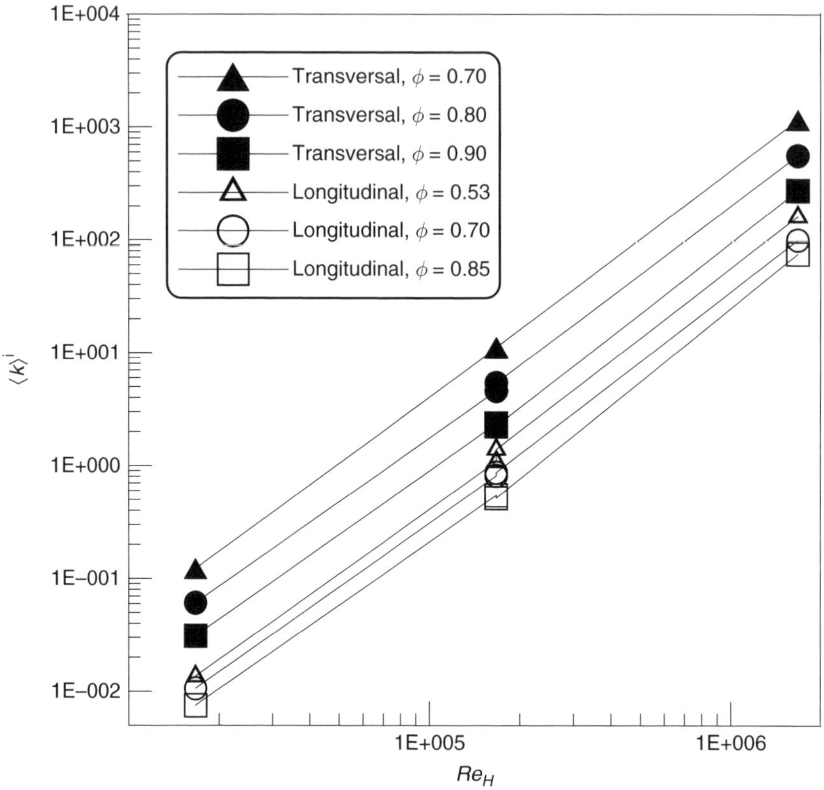


**Figure 4.5.** Effect of porosity and medium morphology on the overall level of turbulent kinetic energy.

covers most of the area of the solid (transversal ellipses). For $Re_H = 1.67 \times 10^5$, the low *Re* model was also computed and, for the transversal ellipses, the discrepancy between the two wall treatments is large due to the large wake region behind the solid. From Figures 4.5–4.6 one can finally conclude that smoother passages in between longitudinal elliptic rods contribute for reducing sudden flow acceleration within the flow, reducing the overall velocity gradients and, consequently, lowering production rates and levels of $\langle k \rangle^i$. Additional results for the permeability $K$ are presented in Figure 4.7 for the different cells identified in Figure 4.1. Their values were calculated using the method described by Kuwahara and Nakayama (1998). In this method, the flow through the cell was computed with a very small inlet mass flow rate and results were compared with the Darcy formula. Here also one can have the behavior of medium properties as a function of medium morphology. As the porosity $\phi$ increases, all values for values of $K$ also increase as expected, and results for the cases of cylindrical rod lie in between the two other cells, in a consistent way.

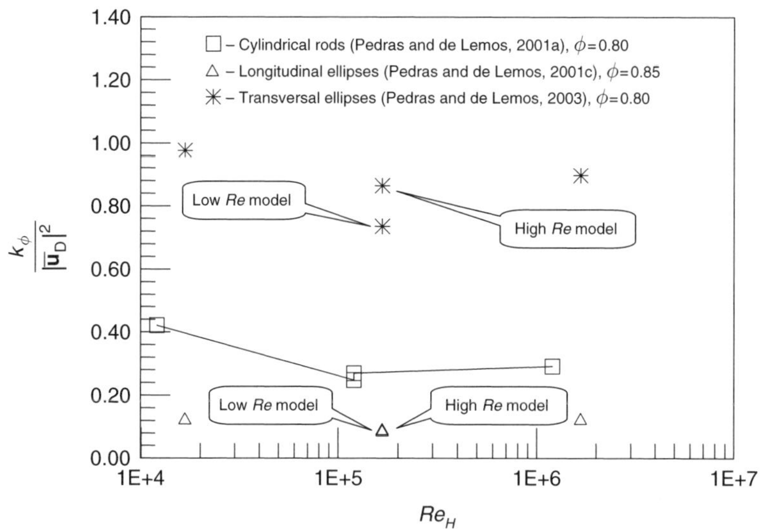


**Figure 4.6.** Macroscopic turbulent kinetic energy as a function of medium morphology and $Re_H$.

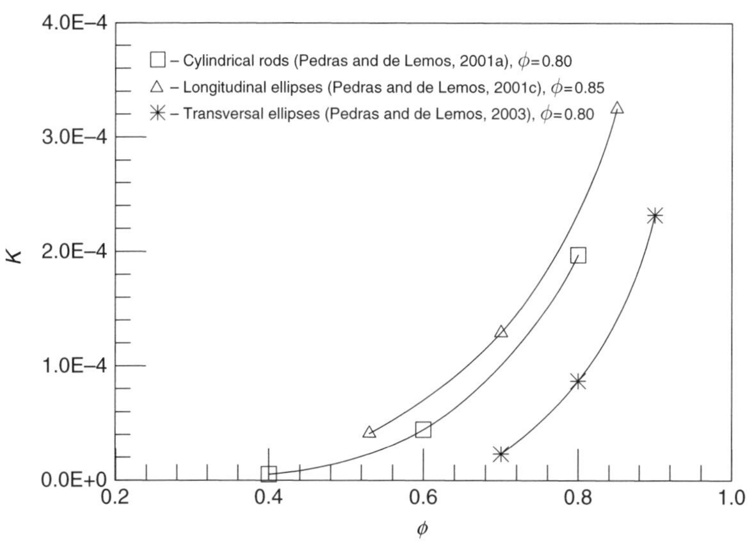


**Figure 4.7.** Numerically obtained permeability $K$ ($m^2$) as a function of medium morphology.

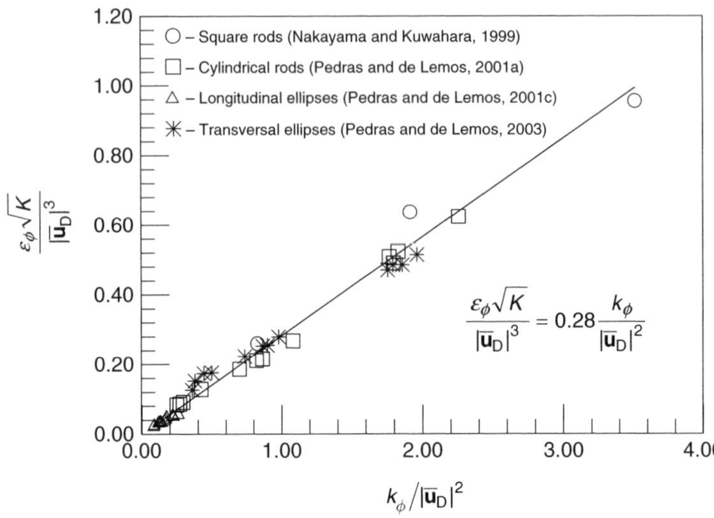


**Figure 4.8.** Determination of value for $c_k$ using data for different medium morphology.

Once the intrinsic values of $k_\phi$ and $\varepsilon_\phi$ were obtained, they were plugged into Equation (4.61). The value of $c_k$ equal to 0.28 was determined in Pedras and de Lemos (2001b) for cylindrical rods by noting the collapse of all the data into the straight line. Here, Figure 4.8 compiles results for cylindrical rods, square bars (Nakayama and Kuwahara, 1999), longitudinal ellipses (Pedras and de Lemos, 2001c) and transversal ellipses. The figure indicates that in spite of having a number of different shapes for representing the solid phase, the turbulence closure for porous media proposed by Pedras and de Lemos (2001a) seems to be of a reasonable degree of universality. Spanning from a streamlined, low-drag longitudinal-ellipse case to a high-pressure-loss, large-wake flow past transversal ellipses, a unique value for the introduced constant $c_k$ stimulates further development on such model and indicates the appropriateness of the macroscopic treatment followed so far.

Chapter 5

# Turbulent Heat Transport

## 5.1. MACROSCOPIC ENERGY EQUATION

In this chapter, the macroscopic energy equation is obtained for a porous medium starting from the local energy equations (for the fluid and solid phases). Then, time averaging is applied followed by volume averaging (or vice versa) using the local thermal equilibrium hypothesis. This procedure leads to the so-called *one-energy equation model.*

### 5.1.1. Time Average Followed by Volume Average

In order to apply the time average operator to (2.14) and (2.15), one substitutes (2.28) into the energy equations obtaining:

$$(\rho c_{\mathrm{p}})_{\mathrm{f}} \nabla \cdot (\overline{\mathbf{u}}\overline{T_{\mathrm{f}}} + \overline{\mathbf{u}}T_{\mathrm{f}}' + \mathbf{u}'\overline{T_{\mathrm{f}}} + \mathbf{u}'T_{\mathrm{f}}') = \nabla \cdot (\lambda_{\mathrm{f}} \nabla(\overline{T_{\mathrm{f}}} + T_{\mathrm{f}}')) \tag{5.1}$$

$$0 = \nabla \cdot (\lambda_{\mathrm{s}} \nabla(\overline{T}_{\mathrm{s}} + T_{\mathrm{s}}')) \tag{5.2}$$

Applying time average to (5.1) and (5.2), one obtains:

$$(\rho c_{\mathrm{p}})_{\mathrm{f}} \nabla \cdot (\overline{\mathbf{u}}\overline{T_{\mathrm{f}}} + \overline{\mathbf{u}'T_{\mathrm{f}}'}) = \nabla \cdot (\lambda_{\mathrm{f}} \nabla \overline{T_{\mathrm{f}}}) \tag{5.3}$$

$$0 = \nabla \cdot (\lambda_{\mathrm{s}} \nabla \overline{T_{\mathrm{s}}}) \tag{5.4}$$

The second term on the lhs of Equation (5.3) is known as *turbulent heat flux*. It requires a model for closure of the mathematical problem. Also, in order to apply the volume average to (5.3) and (5.4), one must first define the spatial deviations with respect to the time averages, given by:

$$\overline{T} = \langle \overline{T} \rangle^{\mathrm{i}} + {}^{\mathrm{i}}\overline{T} \tag{5.5}$$

$$\overline{\mathbf{u}} = \langle \overline{\mathbf{u}} \rangle^{\mathrm{i}} + {}^{\mathrm{i}}\overline{\mathbf{u}} \tag{5.6}$$

Substituting now (5.5) and (5.6) into (5.1) and (5.2), respectively, and performing the volume average operation in those equaitons, one has:

$$(\rho c_{\mathrm{p}})_{\mathrm{f}} \left[ \nabla \cdot \left\{ \phi \left( \langle \overline{\mathbf{u}} \rangle^{\mathrm{i}} \langle \overline{T_{\mathrm{f}}} \rangle^{\mathrm{i}} + \langle {}^{\mathrm{i}}\overline{\mathbf{u}}\, {}^{\mathrm{i}}\overline{T_{\mathrm{f}}} \rangle^{\mathrm{i}} + \langle \overline{\mathbf{u}'T_{\mathrm{f}}'} \rangle^{\mathrm{i}} \right) \right\} \right]$$
$$= \nabla \cdot \left[ \lambda_{\mathrm{f}} \nabla \left( \phi \langle \overline{T_{\mathrm{f}}} \rangle^{\mathrm{i}} \right) \right] + \nabla \cdot \left[ \frac{1}{\Delta V} \int_{A_{\mathrm{i}}} \mathbf{n} \lambda_{\mathrm{f}} \overline{T_{\mathrm{f}}} \, \mathrm{d}S \right] + \frac{1}{\Delta V} \int_{A_{\mathrm{i}}} \mathbf{n} \cdot \lambda_{\mathrm{f}} \nabla \overline{T_{\mathrm{f}}} \, \mathrm{d}S \tag{5.7}$$

$$0 = \nabla \cdot \{\lambda_s \nabla [(1-\phi)\langle \overline{T_s} \rangle^i]\} - \nabla \cdot \left[ \frac{1}{\Delta V} \int_{A_i} \mathbf{n} \lambda_s \overline{T_s} \, \mathrm{d}S \right]$$

$$- \frac{1}{\Delta V} \int_{A_i} \mathbf{n} \cdot \lambda_s \nabla \overline{T_s} \, \mathrm{d}S \tag{5.8}$$

where $A_i$ is the interface area between the fluid and solid phases, within the representative elementary volume REV of size $\Delta V$, and $\mathbf{n}$ is the unit vector normal to the fluid–solid interface. Equations (5.7) and (5.8) are the macroscopic energy equations for the fluid and the porous matrix (solid), respectively, taking first the time average followed by the volume average operator.

### 5.1.2. Volume Average Followed by Time Average

Applying the volume average to (2.14) and (2.15) one has:

$$T = \langle T \rangle^i + {}^iT \tag{5.9}$$

$$\mathbf{u} = \langle \mathbf{u} \rangle^i + {}^i\mathbf{u} \tag{5.10}$$

in addition, considering that the intrinsic average can be applied to either the fluid ($\langle\,\rangle^f$) or the solid phase ($\langle\,\rangle^s$), we have:

$$\begin{aligned} \langle T \rangle^v &= \phi \langle T \rangle^f + (1-\phi)\langle T \rangle^s \\ \langle \mathbf{u} \rangle^v &= \phi \langle \mathbf{u} \rangle^f \end{aligned} \tag{5.11}$$

Substituting (5.9) and (5.10) into (2.14) and (2.15), one obtains:

$$(\rho c_p)_f \nabla \cdot (\langle \mathbf{u} \rangle^i \langle T_f \rangle^i + \langle \mathbf{u} \rangle^{ii} T_f + {}^i\mathbf{u} \langle T_f \rangle^i + {}^i\mathbf{u}^i T_f) = \nabla \cdot [\lambda_f \nabla (\langle T_f \rangle^i + {}^iT_f)] \tag{5.12}$$

$$0 = \nabla \cdot [\lambda_s \nabla (\langle T_s \rangle^i + {}^iT_s)] \tag{5.13}$$

Taking the volume average of (5.12) and (5.13), one has:

$$(\rho c_p)_f \nabla \cdot [\phi(\langle \mathbf{u} \rangle^i \langle T_f \rangle^i + \langle {}^i\mathbf{u}^i T_f \rangle^i)] = \nabla \cdot [\lambda_f \nabla (\phi \langle T_f \rangle^i)] + \nabla \cdot \left[ \frac{1}{\Delta V} \int_{A_i} \mathbf{n} \lambda_f T_f \, \mathrm{d}S \right]$$

$$+ \frac{1}{\Delta V} \int_{A_i} \mathbf{n} \cdot \lambda_f \nabla T_f \, \mathrm{d}S \tag{5.14}$$

$$\nabla \cdot \{\lambda_s \nabla [(1-\phi)\langle T_s \rangle^i]\} - \nabla \cdot \left[ \frac{1}{\Delta V} \int_{A_i} \mathbf{n} \lambda_s T_s \, \mathrm{d}S \right] - \frac{1}{\Delta V} \int_{A_i} \mathbf{n} \cdot \lambda_s \nabla T_s \, \mathrm{d}S = 0 \tag{5.15}$$

The second term on the lhs of Equation (5.14) appears in the classical analysis of convection in porous media (e.g. Hsu and Cheng, 1990) and is known as *thermal dispersion.*

In order to apply thc timc avcrage to (5.14) and (5.15), one defines the intrinsic volume average as:

$$\langle T\rangle^{\mathrm{i}} = \overline{\langle T\rangle^{\mathrm{i}}} + \langle T\rangle^{\mathrm{i}'} \tag{5.16}$$

$$\langle \mathbf{u}\rangle^{\mathrm{i}} = \overline{\langle \mathbf{u}\rangle^{\mathrm{i}}} + \langle \mathbf{u}\rangle^{\mathrm{i}'} \tag{5.17}$$

Substituting (5.16) and (5.17) in (5.14) and (5.15) and taking the time average, we obtain:

$$\begin{aligned}
&(\rho c_{\mathrm{p}})_{\mathrm{f}}\nabla\cdot\left[\phi\left(\overline{\langle \mathbf{u}\rangle^{\mathrm{i}}}\;\overline{\langle T_{\mathrm{f}}\rangle^{\mathrm{i}}} + \overline{\langle \mathbf{u}\rangle^{\mathrm{i}'}\langle T_{\mathrm{f}}\rangle^{\mathrm{i}'}} + \overline{\langle {}^{\mathrm{i}}\mathbf{u}^{\mathrm{i}}T_{\mathrm{f}}\rangle^{\mathrm{i}}}\right)\right] \\
&\quad = \nabla\cdot[\lambda_{\mathrm{f}}\nabla(\phi\overline{\langle T_{\mathrm{f}}\rangle^{\mathrm{i}}})] + \nabla\cdot\left[\frac{1}{\Delta V}\int_{A_{\mathrm{i}}} \mathbf{n}\lambda_{\mathrm{f}}\overline{T_{\mathrm{f}}}\,\mathrm{d}S\right] + \frac{1}{\Delta V}\int_{A_{\mathrm{i}}} \mathbf{n}\cdot\lambda_{\mathrm{f}}\nabla\overline{T_{\mathrm{f}}}\,\mathrm{d}S
\end{aligned} \tag{5.18}$$

$$\nabla\cdot\{\lambda_{\mathrm{s}}\nabla[(1-\phi)\overline{\langle T_{\mathrm{s}}\rangle^{\mathrm{i}}}]\} - \nabla\cdot\left[\frac{1}{\Delta V}\int_{A_{\mathrm{i}}} \mathbf{n}\lambda_{\mathrm{s}}\overline{T_{\mathrm{s}}}\,\mathrm{d}S\right] - \frac{1}{\Delta V}\int_{A_{\mathrm{i}}} \mathbf{n}\cdot\lambda_{\mathrm{s}}\nabla\overline{T_{\mathrm{s}}}\,\mathrm{d}S = 0 \tag{5.19}$$

Equations (5.18) and (5.19) are the macroscopic energy equations for the fluid and the porous matrix (solid), respectively, taking first the volume average followed by the time average.

It is interesting to observe that Equations (5.7) and (5.8), obtained through the first procedure (time–volume average), are equivalent to (5.18) and (5.19), respectively, obtained through the second method (volume–time average). In order to prove that, our next step is to demonstrate that the sum of the 2nd and 3rd terms on the lhs of either Equation (5.7) or Equation (5.18) are identical.

### 5.1.3. Turbulent Thermal Dispersion

Using now (3.3) and (3.4), the third terms on the lhs of Equations (5.7) and (5.18), namely $\langle\overline{\mathbf{u}'T_{\mathrm{f}}'}\rangle^{\mathrm{i}}$ and $\langle\overline{{}^{\mathrm{i}}\mathbf{u}^{\mathrm{i}}T_{\mathrm{f}}}\rangle^{\mathrm{i}}$, respectively, can be expanded as:

$$\langle\overline{\mathbf{u}'T_{\mathrm{f}}'}\rangle^{\mathrm{i}} = \langle\overline{((\langle\mathbf{u}'\rangle^{\mathrm{i}} + {}^{\mathrm{i}}\mathbf{u}')((\langle T_{\mathrm{f}}'\rangle^{\mathrm{i}} + {}^{\mathrm{i}}T')}\rangle^{\mathrm{i}} = \overline{\langle\mathbf{u}'\rangle^{\mathrm{i}}\langle T_{\mathrm{f}}'\rangle^{\mathrm{i}}} + \langle\overline{{}^{\mathrm{i}}\mathbf{u}'{}^{\mathrm{i}}T_{\mathrm{f}}'}\rangle^{\mathrm{i}} \tag{5.20}$$

$$\overline{\langle{}^{\mathrm{i}}\mathbf{u}^{\mathrm{i}}T_{\mathrm{f}}\rangle^{\mathrm{i}}} = \overline{\langle({}^{\mathrm{i}}\overline{\mathbf{u}} + {}^{\mathrm{i}}\mathbf{u}')({}^{\mathrm{i}}\overline{T_{\mathrm{f}}} + {}^{\mathrm{i}}T'))\rangle^{\mathrm{i}}} = \langle{}^{\mathrm{i}}\overline{\mathbf{u}}{}^{\mathrm{i}}\overline{T_{\mathrm{f}}}\rangle^{\mathrm{i}} + \overline{\langle{}^{\mathrm{i}}\mathbf{u}'{}^{\mathrm{i}}T_{\mathrm{f}}'\rangle^{\mathrm{i}}} \tag{5.21}$$

Substituting (5.20) into (5.7), the convection term will read:

$$(\rho c_{\mathrm{p}})_{\mathrm{f}}\nabla\cdot(\overline{\phi\langle\mathbf{u}T\rangle^{\mathrm{i}}}) = (\rho c_{\mathrm{p}})_{\mathrm{f}}\nabla\cdot\{\phi(\langle\overline{\mathbf{u}}\rangle^{\mathrm{i}}\langle\overline{T_{\mathrm{f}}}\rangle^{\mathrm{i}} + \langle{}^{\mathrm{i}}\overline{\mathbf{u}}{}^{\mathrm{i}}\overline{T_{\mathrm{f}}}\rangle^{\mathrm{i}} + \overline{\langle\mathbf{u}'\rangle^{\mathrm{i}}\langle T_{\mathrm{f}}'\rangle^{\mathrm{i}}} + \langle\overline{{}^{\mathrm{i}}\mathbf{u}'{}^{\mathrm{i}}T_{\mathrm{f}}'}\rangle^{\mathrm{i}})\} \tag{5.22}$$

Also, plugging (5.21) into (5.18) will give, for the same convection term:

$$(\rho c_p)_f \nabla \cdot (\overline{\phi \langle \mathbf{u} T \rangle^i}) = (\rho c_p)_f \nabla \cdot \{\phi(\underset{\mathbf{1}}{\underset{\uparrow}{\overline{\langle \mathbf{u} \rangle^i} \, \overline{\langle T_f \rangle^i}}} + \underset{\mathbf{2}}{\underset{\uparrow}{\overline{\langle \mathbf{u} \rangle^{i'}} \, \overline{\langle T_f \rangle^{i'}}}} + \underset{\mathbf{3}}{\underset{\uparrow}{\langle {}^i\overline{\mathbf{u}}{}^i\overline{T_f} \rangle^i}} + \underset{\mathbf{4}}{\underset{\uparrow}{\langle \overline{{}^i\mathbf{u}'} \, \overline{{}^iT_f'} \rangle^i}})\} \tag{5.23}$$

Comparing Equation (5.22) to Equation (5.23), in light of Equations (3.9) and (3.10), one can conclude that Equations (5.7) and (5.8) are, in fact, equal to Equations (5.18) and (5.19), respectively. This demonstrates that the final expanded form of the macroscopic energy equation for a rigid, homogeneous porous medium saturated with an incompressible fluid does not depend on the averaging order, i.e., both procedures lead to the same results.

Further, the four terms on the rhs of Equation (5.23) could be given by the following physical significance:

1 is *convective heat flux* based on macroscopic time mean velocity and temperature.
2 is *turbulent heat flux* due to the fluctuating components of macroscopic velocity and temperature.
3 is *thermal dispersion* associated with deviations of microscopic time mean velocity and temperature. Note that this term is also present when analyzing laminar convective heat transfer in porous media.
4 is *turbulent thermal dispersion* in a porous medium due to both time fluctuations and spatial deviations of both microscopic velocity and temperature.

## 5.2. THERMAL EQUILIBRIUM MODEL

Assuming local thermal equilibrium between the fluid and solid phases, i.e., $\langle \overline{T_f} \rangle^i = \langle \overline{T_s} \rangle^i = \langle \overline{T} \rangle^i$, and adding the two Equations (5.18) and (5.19) in their transient form, one has:

$$\begin{aligned}
&\left\{(\rho c_p)_f \phi + (\rho c_p)_s (1-\phi)\right\} \frac{\partial \langle \overline{T} \rangle^i}{\partial t} + (\rho c_p)_f \nabla \cdot (\overline{\mathbf{u}}_D \langle \overline{T} \rangle^i) \\
&= \nabla \cdot \left\{[\lambda_f \phi + \lambda_s (1-\phi)] \nabla \langle \overline{T} \rangle^i\right\} + \nabla \cdot \left[\frac{1}{\Delta V} \int_{A_i} \mathbf{n} \left(\lambda_f \overline{T_f} - \lambda_s \overline{T_s}\right) \mathrm{d}S\right] \\
&\quad - (\rho c_p)_f \nabla \cdot \left[\phi \left(\langle {}^i\overline{\mathbf{u}}\,{}^i\overline{T_f} \rangle^i + \langle \overline{\mathbf{u}' T_f'} \rangle^i\right)\right]
\end{aligned} \tag{5.24}$$

where use has been made of the time averaged Dupuit–Forchheimer relationship, $\overline{\mathbf{u}}_D = \langle \overline{\mathbf{u}} \rangle^v = \phi \langle \overline{\mathbf{u}} \rangle^i$. The interfacial conditions at $A_i$ are further given by:

$$\left.\begin{aligned} T_f &= T_s \\ \mathbf{n} \cdot (\lambda_f \nabla T_f) &= \mathbf{n} \cdot (\lambda_s \nabla T_s) \end{aligned}\right\} \quad \text{in} \quad A_i \tag{5.25}$$

Equation (5.24) express the one-energy equation model for heat transport in porous media.

Also, in view of Equation (5.20), Equation (5.24) can be rewritten as:

$$\begin{aligned}
&\left\{(\rho c_p)_f\,\phi+(\rho c_p)_s\,(1-\phi)\right\}\frac{\partial\langle\overline{T}\rangle^i}{\partial t}+(\rho c_p)_f\,\nabla\cdot\left(\overline{\mathbf{u}}_D\langle\overline{T}\rangle^i\right)\\
&=\nabla\cdot\left\{\left[\lambda_f\phi+\lambda_s\,(1-\phi)\right]\nabla\langle\overline{T}\rangle^i\right\}+\nabla\cdot\left[\frac{1}{\Delta V}\int_{A_i}\mathbf{n}\left(\lambda_f\overline{T_f}-\lambda_s\overline{T_s}\right)\mathrm{d}S\right]\\
&-(\rho c_p)_f\,\nabla\cdot\left[\phi\left(\overline{\langle\mathbf{u}'\rangle^i\langle T_f'\rangle^i}+\langle{}^i\overline{\mathbf{u}}{}^i\overline{T_f}\rangle^i+\langle\overline{{}^i\mathbf{u}'{}^iT_f'}\rangle^i\right)\right]
\end{aligned}\tag{5.26}$$

where the second term on the rhs is the tortuosity, which depends on local time average temperatures of the solid and fluid and their respective thermal conductivities on the interfacial area $A_i$.

### 5.2.1. Effective Conductivity Tensor

In order to apply Equation 5.26 to obtain the temperature field for turbulent flow in porous media, the last four terms in the rhs have to be modeled in some way as a function of the surface average temperature, $\langle\overline{T}\rangle^i$. To accomplish this, a gradient-type diffusion model is used so that we can write:

$$\text{Tortuosity:}\left[\frac{1}{\Delta V}\int_{A_i}\mathbf{n}\left(\lambda_f\overline{T_f}-\lambda_s\overline{T_s}\right)\mathrm{d}S\right]=\mathbf{K}_{tor}\cdot\nabla\langle\overline{T}\rangle^i\tag{5.27}$$

$$\text{Turbulent heat flux:}-(\rho c_p)_f\,(\phi\overline{\langle\mathbf{u}'\rangle^i\langle T_f'\rangle^i})=\mathbf{K}_t\cdot\nabla\langle\overline{T}\rangle^i\tag{5.28}$$

$$\text{Thermal dispersion:}-(\rho c_p)_f\,(\phi\langle{}^i\overline{\mathbf{u}}{}^i\overline{T_f}\rangle^i)=\mathbf{K}_{disp}\cdot\nabla\langle\overline{T}\rangle^i\tag{5.29}$$

$$\text{Turbulent thermal dispersion:}-(\rho c_p)_f\,(\phi\langle\overline{{}^i\mathbf{u}'{}^iT_f'}\rangle^i)=\mathbf{K}_{disp,t}\cdot\nabla\langle\overline{T}\rangle^i\tag{5.30}$$

For the above shown Equations 5.26 can be rewritten as:

$$\left\{(\rho c_p)_f\,\phi+(\rho c_p)_s\,(1-\phi)\right\}\frac{\partial\langle\overline{T}\rangle^i}{\partial t}+(\rho c_p)_f\,\nabla\cdot\left(\overline{\mathbf{u}}_D\langle\overline{T}\rangle^i\right)=\nabla\cdot\left\{\mathbf{K}_{eff}\cdot\nabla\langle\overline{T}\rangle^i\right\}\tag{5.31}$$

where $\mathbf{K}_{eff}$ is given by:

$$\mathbf{K}_{eff}=\lfloor\phi\lambda_f+(1-\phi)\lambda_s\rfloor\,\mathbf{I}+\mathbf{K}_{tor}+\mathbf{K}_t+\mathbf{K}_{disp}+\mathbf{K}_{disp,t}\tag{5.32}$$

and is the effective conductivity tensor. A steady state form of 5.31 reads,

$$(\rho c_p)_f\,\nabla\cdot\left(\overline{\mathbf{u}}_D\langle T\rangle^i\right)=\nabla\cdot\left\{\mathbf{K}_{eff}\cdot\nabla\langle\overline{T}\rangle^i\right\}\tag{5.33}$$

In order to be able to apply Equation 5.33, it is necessary to determine the conductivity tensors in Equation (5.32), i.e., $\mathbf{K}_{\rm tor}$, $\mathbf{K}_{\rm t}$, $\mathbf{K}_{\rm disp}$ and $\mathbf{K}_{\rm disp,t}$. Following Kuwahara and Nakayama (1998) and Pedras and de Lemos (2001b), this can be accomplished for the tortuosity and thermal dispersion conductivity tensors, $\mathbf{K}_{\rm tor}$ and $\mathbf{K}_{\rm disp}$, by making use of a unit cell subjected to periodic boundary conditions for the flow and a linear temperature gradient imposed over the domain. The conductivity tensors are then obtained directly from the microscopic results for the unit cell, using Equations (5.27) and (5.29).

The turbulent heat flux and turbulent thermal dispersion terms, $\mathbf{K}_{\rm t}$ and $\mathbf{K}_{\rm disp,t}$, which cannot be determined from such a microscopic calculation, are modeled through the eddy diffusivity concept, similar to Nakayama and Kuwahara (1999). It should be noticed that these terms arise only if the flow is turbulent, whereas the tortuosity and the thermal dispersion terms exist for both laminar and turbulent flow regimes. Starting out from the time averaged energy equation coupled with the modeling for the "turbulent heat flux" using the eddy diffusivity concept, $\alpha_{\rm t} = \mu_{\rm t}/\sigma_{\rm t}$, one can write:

$$-\left(\rho c_{\rm p}\right)_{\rm f}\overline{\mathbf{u}'T'_{\rm f}} = \left(\rho c_{\rm p}\right)_{\rm f}\frac{\nu_{\rm t}}{\sigma_{\rm t}}\nabla\overline{T}_{\rm f} \tag{5.34}$$

In Equation (5.34), $\mu_{\rm t}$ is the eddy or turbulent viscosity given by:

$$\mu_{\rm t} = \rho_{\rm f} c_\mu \frac{k^2}{\varepsilon} \tag{5.35}$$

and $\sigma_{\rm t}$ is the turbulent Prandtl number, which is taken here as a constant.

Applying the volume average to the resulting equation, one obtains the macroscopic form of the "turbulent heat flux", modeled as:

$$-\left(\rho c_{\rm p}\right)_{\rm f}\langle\overline{\mathbf{u}'T'_{\rm f}}\rangle^{\rm i} = \left(\rho c_{\rm p}\right)_{\rm f}\frac{\nu_{{\rm t}_\phi}}{\sigma_{{\rm t}_\phi}}\nabla\langle\overline{T}_{\rm f}\rangle^{\rm i} \tag{5.36}$$

where we have adopted the symbol $\nu_{{\rm t}_\phi}$ to express the macroscopic eddy viscosity, $\mu_{{\rm t}_\phi} = \rho_{\rm f}\nu_{{\rm t}_\phi}$, given by:

$$\mu_{{\rm t}_\phi} = \rho_{\rm f} c_\mu \frac{\langle k\rangle^{{\rm i}^2}}{\langle\varepsilon\rangle^{\rm i}} \tag{5.37}$$

According to Equations (5.20) (5.28) and (5.30), the macroscopic heat flux due to turbulence is taken as the sum of the turbulent heat flux and the turbulent thermal dispersion found by Rocamora and de Lemos (2000a). In view of such argument, the tensors, $\mathbf{K}_{\rm t}$ and $\mathbf{K}_{\rm disp,t,}$ will be combined as:

$$\mathbf{K}_{\rm t} + \mathbf{K}_{\rm disp,t} = \phi(\rho c_{\rm p})_{\rm f}\frac{\nu_{{\rm t}_\phi}}{\sigma_{{\rm t}_\phi}}\mathbf{I} \tag{5.38}$$

### 5.2.2. Determination of Dispersion Tensor $\mathrm{K}_{\mathbf{disp}}$

The thermal dispersion modeling utilized in Pedras and de Lemos (2005) follows the same procedure of Pedras *et al.* (2003a,b) and Pedras and de Lemos (2004). The macroscopic energy equation is obtained by volume averaging the microscopic energy equations

$$\rho_f c_{pf}\left[\frac{\partial \overline{T}_f}{\partial t}+\nabla\cdot(\overline{\mathbf{u}}\overline{T}_f)\right]=\nabla\cdot\left[\lambda_f\nabla\overline{T}_f-\rho_f c_{pf}\overline{\mathbf{u}'T_f'}\right] \tag{5.39}$$

and

$$\rho_s c_{ps}\frac{\partial \overline{T}_s}{\partial t}=\nabla\cdot[\lambda_s\nabla\overline{T}_s] \tag{5.40}$$

over the REV of Figure 2.1 (p. 22) assuming local thermal equilibrium assumption, i.e., $\langle\overline{T}_f\rangle^f=\langle\overline{T}_s\rangle^s=\langle\overline{T}\rangle$. The result is again(see Equation 5.31 with $\langle\overline{T}_s\rangle^i, =\langle\overline{T}\rangle, \overline{\mathbf{u}}_D=\langle\overline{\mathbf{u}}\rangle^v$)

$$[\phi_f\rho_f c_{pf}+\phi_s\rho_s c_{ps}]\frac{\partial\langle\overline{T}\rangle}{\partial t}+\rho_f c_{pf}\nabla\cdot(\langle\overline{\mathbf{u}}\rangle\langle\overline{T}\rangle)=\nabla\cdot\mathbf{K}_{eff}\cdot\nabla\langle\overline{T}\rangle \tag{5.41}$$

Combining Equations (5.32) and (5.38), the effective conductivity becomes:

$$\mathbf{K}_{eff}=\left(\phi_f\frac{\rho_f c_{pf}\nu_{t_\phi}}{\sigma_t}+\phi_f\lambda_f+\phi_s\lambda_s\right)\mathbf{I}+\mathbf{K}_{tor}+\mathbf{K}_{disp} \tag{5.42}$$

Assuming that on the interfacial area $A_i$ the equality $\overline{T}_f=\overline{T}_s$ prevails, Equation (5.27) gives:

$$\mathbf{K}_{tor}\cdot\nabla\langle\overline{T}\rangle=\frac{(\lambda_f-\lambda_s)}{\Delta V}\int_{A_i}\mathbf{n}_f\,{}^i\overline{T}_f\,dS \tag{5.43}$$

In addition, the dispersion tensor, $\mathbf{K}_{disp}$, is defined such that:

$$\mathbf{K}_{disp}\cdot\nabla\langle\overline{T}\rangle=-\rho_f c_{pf}\phi_f\langle{}^i\overline{\mathbf{u}}\,{}^i\overline{T}_f\rangle^f=-\frac{\rho_f c_{pf}}{\Delta V}\int_{\Delta V_f}{}^i\overline{\mathbf{u}}\,{}^i\overline{T}_f\,dV \tag{5.44}$$

where ${}^i\varphi$ is the space deviation of $\varphi$.

According to Figure 5.1a, the macroscopic velocity and temperature fields are given by

$$\langle\overline{\mathbf{u}}\rangle^v=|\langle\overline{\mathbf{u}}\rangle^v|(\cos\theta\vec{i}+\sin\theta\vec{j}) \tag{5.45}$$

$$\nabla\langle\overline{T}\rangle=\frac{\Delta T}{H}(-\sin\theta\vec{i}+\cos\theta\vec{j})\quad\text{Transversal component} \tag{5.46}$$

$$\nabla\langle\overline{T}\rangle=\frac{\Delta T}{H}(\cos\theta\vec{i}+\sin\theta\vec{j})\quad\text{Longitudinal component} \tag{5.47}$$

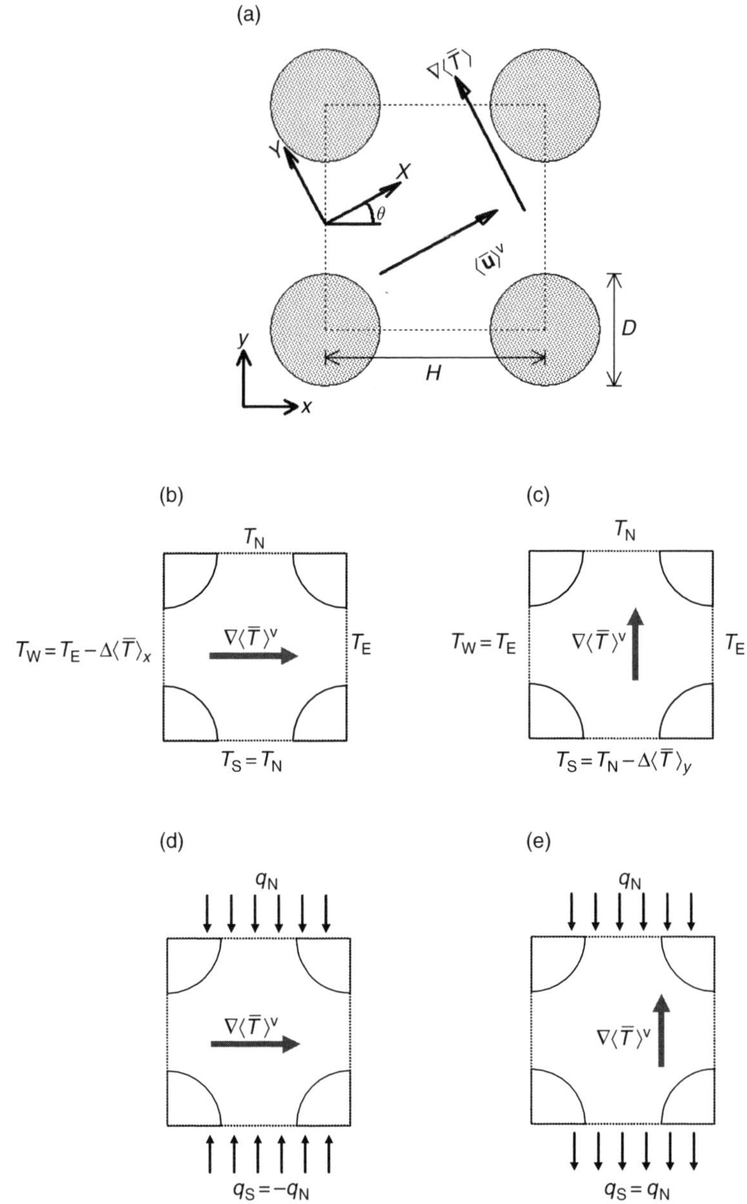


**Figure 5.1.** Unit-cell and boundary conditions: (a) macroscopic velocity and temperature gradients; given temperature difference at East–West boundaries: (b) longitudinal gradient, Equation (5.52), (c) transversal gradient, Equation (5.53); given heat fluxes at North–South boundaries, (d) longitudinal gradient, (e) transversal gradient.

If the gradient of the average temperature is in the same direction of the macroscopic flow or transverse to it, only diagonal components of $\mathbf{K}_{\text{disp}}$ remain non-zero. In these conditions, Equation (5.44) renders, respectively, for the diagonal components of $\mathbf{K}_{\text{disp}}$:

$$(\mathbf{K}_{\text{disp}})_{XX} \approx -\frac{\dfrac{\rho_{\mathrm{f}} c_{\mathrm{pf}}}{\Delta V}}{\dfrac{\Delta \langle \overline{T} \rangle_x}{H}} \int_{\Delta V_{\mathrm{f}}} {}^{\mathrm{i}}\overline{u}\,{}^{\mathrm{i}}\overline{T}_{\mathrm{f}}\, \mathrm{d}V \tag{5.48}$$

or

$$(\mathbf{K}_{\text{disp}})_{XX} \approx -\frac{\dfrac{\rho_{\mathrm{f}} c_{\mathrm{pf}}}{\Delta V}}{\dfrac{\Delta \langle \overline{T} \rangle_x}{H}} \int_0^H \int_0^H \left(\overline{\mathbf{u}} - \langle \overline{\mathbf{u}} \rangle^{\mathrm{i}}\right) \left(\overline{T} - \langle \overline{T} \rangle\right) \mathrm{d}x\, \mathrm{d}y \cdot (\cos\theta \vec{i} + \sin\theta \vec{j}) \tag{5.49}$$

and

$$(\mathbf{K}_{\text{disp}})_{YY} \approx -\frac{\dfrac{\rho_{\mathrm{f}} c_{\mathrm{pf}}}{\Delta V}}{\dfrac{\Delta \langle \overline{T} \rangle_y}{H}} \int_{\Delta V_{\mathrm{f}}} {}^{\mathrm{i}}\overline{v}\,{}^{\mathrm{i}}\overline{T}_{\mathrm{f}}\, \mathrm{d}V \tag{5.50}$$

or

$$(\mathbf{K}_{\text{disp}})_{YY} \approx -\frac{\dfrac{\rho_{\mathrm{f}} c_{\mathrm{pf}}}{\Delta V}}{\dfrac{\Delta \langle \overline{T} \rangle_y}{H}} \int_0^H \int_0^H \left(\overline{\mathbf{u}} - \langle \overline{\mathbf{u}} \rangle^{\mathrm{i}}\right) \left(\overline{T} - \langle \overline{T} \rangle\right) \mathrm{d}x\, \mathrm{d}y \cdot (-\sin\theta \vec{i} + \cos\theta \vec{j}) \tag{5.51}$$

Solution of the flow and energy equations inside the unit cell provides the velocity and temperature distributions necessary for the integrands of Equations (5.48) and (5.50). These values are needed in order to calculate the dispersion components. Further, in Equations (5.48) and (5.50), the gradients $\Delta \langle \overline{T} \rangle_x$ and $\Delta \langle \overline{T} \rangle_y$ can be calculated in two ways, as presented next.

### 5.2.3. Imposed Boundary Temperature Difference

In the first method, a temperature difference or Dirichlet boundary conditions are imposed for the energy equation at the faces of the computational cell (Kuwahara and Nakayama, 1998). Accordingly, two distinct macroscopic temperature gradients are considered to obtain the transverse and longitudinal dispersion coefficients given in Equations (5.48) and (5.50), respectively.

For obtaining $(\mathbf{K}_{\text{disp}})_{XX}$ we have (see Figure 5.1b):

$$\overline{T}_{x=0} = \overline{T}_{x=H} - \Delta \langle \overline{T} \rangle_x \quad \text{and} \quad \overline{T}_{y=0} = \overline{T}_{y=H} \tag{5.52}$$

and for the $(\mathbf{K}_{\text{disp}})_{YY}$ calculation, we have (Figure 5.1c):

$$\overline{T}_{x=0} = \overline{T}_{x=H} \quad \text{and} \quad \overline{T}_{y=0} = \overline{T}_{y=H} - \Delta\langle\overline{T}\rangle_y \tag{5.53}$$

In both Equations (5.52) and (5.53), $\Delta\langle\overline{T}\rangle_x$ and $\Delta\langle\overline{T}\rangle_y$ are given constants.

### 5.2.4. Imposed Boundary Heat Flux

The other possibility for getting a macroscopic temperature difference across the cell, either in the longitudinal direction for having $\Delta\langle\overline{T}\rangle_x$ to be used in Equation (5.48) or for calculating $\Delta\langle\overline{T}\rangle_y$ for applying it in Equation (5.50), is to impose heat fluxes at the north and south boundaries of the unit cell shown in Figure 5.1a. When the two fluxes "enter" the cell (Figure 5.1d), $\Delta\langle\overline{T}\rangle_x$ is obtained. For heat entering from above and leaving at the south surface, the situation is analogous of having a uniform transverse temperature difference $\Delta\langle\overline{T}\rangle_y$ across the $y$-direction (Figure 5.1e).

In those cases $\Delta\langle\overline{T}\rangle_x$ and $\Delta\langle\overline{T}\rangle_y$ are no longer given values, but rather a consequence of the imposed heat fluxes (Neumann conditions) at the north and south boundaries. Their values are then calculated as:

$$\Delta\langle\overline{T}\rangle_x = \frac{1}{H}\int_{y=0}^{y=H} [\overline{T}_{x=H} - \overline{T}_{x=0}]\mathrm{d}y \tag{5.54}$$

$$\Delta\langle\overline{T}\rangle_y = \frac{1}{H}\int_{x=0}^{x=H} [\overline{T}_{y=H} - \overline{T}_{y=0}]\mathrm{d}x \tag{5.55}$$

### 5.2.5. Numerical Results

The transport equations at the pore-scale were numerically solved using the SIMPLE method on a non-orthogonal boundary-fitted coordinate system. The equations were discretized using the finite-volume procedure of Patankar (1980). The relaxation process starts with the solution of the two momentum equations, and the velocity field is adjusted in order to satisfy the continuity principle. This adjustment is attained by solving the pressure correction equation. The turbulence model and the energy equations are relaxed to update the the $k$, $\varepsilon$ and temperature fields. Details on the numerical discretization can be found in Pedras and de Lemos (2001b).

In Pedras and de Lemos (2005) just one unit cell, together with periodic boundary conditions for mass, momentum and Neumann and Dirichlet conditions for the energy equation, was used to represent the porous medium. In all runs, flow was always in the horizontal direction and from left to right. For given $\Delta\langle\overline{T}\rangle_x$ and $\Delta\langle\overline{T}\rangle_y$ in Figure 5.1b,c, respectively, all boundary temperatures were varied during the relaxation process until all equations converged. For the Neumann temperature conditions (Figure 5.1d,e), the thermal dispersion tensors were calculated after a sequence of converged loops on the same run. This sequence of loops followed the same procedure detailed in Saito and de Lemos (2005). After convergence with initial profiles at the west faces, outlet profiles at $x = H$ were plugged back at the inlet in $x = 0$. Although the volume average temperature $\langle\overline{T}\rangle^{\mathrm{v}}$ changed after increasing the inlet temperature profile in each run, the spatial deviation temperature field ${}^{\mathrm{v}}T = T - \langle\overline{T}\rangle^{\mathrm{v}}$ within the cell becomes established as the flow develops along the $x$-direction. In this situation, the flow is considered to have a thermal field macroscopically developed.

Also, in the low *Re* model, the node adjacent to the wall requires that $u_\tau n/\nu \leq 1$. To accomplish this requirement, the grid needs points close to the wall leading to computational meshes of $40 \times 54$ nodes. A highly non-uniform grid arrangement was employed with concentration of nodes close to the wall. Values for $(\mathbf{K}_{\mathrm{disp}})_{XX}$ and $(\mathbf{K}_{\mathrm{disp}})_{YY}$ were obtained varying $Pe_H = |\langle\overline{\mathbf{u}}\rangle|H/\alpha_{\mathrm{f}}$ from 1 to $4 \times 10^3$ and the $\phi = 1 - ab\pi/H^2$, from 0.60 to 0.90.

A total of 27 runs were carried out, 23 for laminar flow and 4 for turbulent flow with the low *Re* model theory. In all runs, for the fluid phase a Prandtl number of 0.72 and a thermal conductivity ratio between the solid and fluid phase, $\lambda_{\mathrm{s}}/\lambda_{\mathrm{f}}$, of 10 were used.

Temperature fields calculated with given boundary temperatures conditions sketched in Figure 5.1b (see Equation (5.52)) are presented in the Figure 5.2a and Figure 5.2b, for $Pe_H = 10$ and $4 \times 10^3$, respectively. On the other hand, results for the same cases but using the given flux boundary type of Figure 5.1d are shown in Figure 5.2c,d for the same $Pe_H$ numbers. In Figure 5.2, the macroscopic temperature gradient $\nabla\langle\overline{T}\rangle^{\mathrm{v}}$ is in the same horizontal direction as the macroscopic flow $\langle\overline{\mathbf{u}}\rangle^{\mathrm{v}}$. As will be shown further, in spite of the differences in temperature fields, occuring mainly in the solid phase, the values of the longitudinal component of $\mathbf{K}_{\mathrm{disp}}$ were very similar regardless of the boundary condition applied. This behavior can be explained by recalling the definition of $\mathbf{K}_{\mathrm{disp}}$ (Equation (5.44)), i.e., the determination of $\mathbf{K}_{\mathrm{disp}}$ is dependent on the deviation fields of velocity and temperature within the fluid phase. As such, inspecting Figure 5.2a,c for small Peclet numbers and Figure 5.2b,d for $Pe_H = 4 \times 10^3$, one can see that velocity and temperature fields within the fluid phase, regardless of the boundary type used, resemble fairly well each other.

Temperature fields calculated with boundary conditions sketched in Figure 5.1c (Equation (5.53)) and in Figure 5.1d are presented, respectively, in the Figure 5.3a,c for $Pe_H = 10$ and in Figure 5.3b,d for $Pe_H = 4 \times 10^3$. In Figure 5.3, $\nabla\langle\overline{T}\rangle^{\mathrm{v}}$ is transversal to $\langle\overline{\mathbf{u}}\rangle^{\mathrm{v}}$. Also here the temperature fields obtained with the two methodologies, namely given boundary $T$ values and heat fluxes, are close to each other. As before, here also the transverse component of $\mathbf{K}_{\mathrm{disp}}$ calculated with these two boundary conditions will be very close to one another, as will be seen below. Furthermore, as $Pe_H$ increases, Figure (5.3a,b) and (5.3c,d) show the same behavior of Figure (5.2a,b) and (5.2c,d), i.e., as the flow rate increases, the fluid temperature becomes more homogeneous due to enhancement of the convection strength.

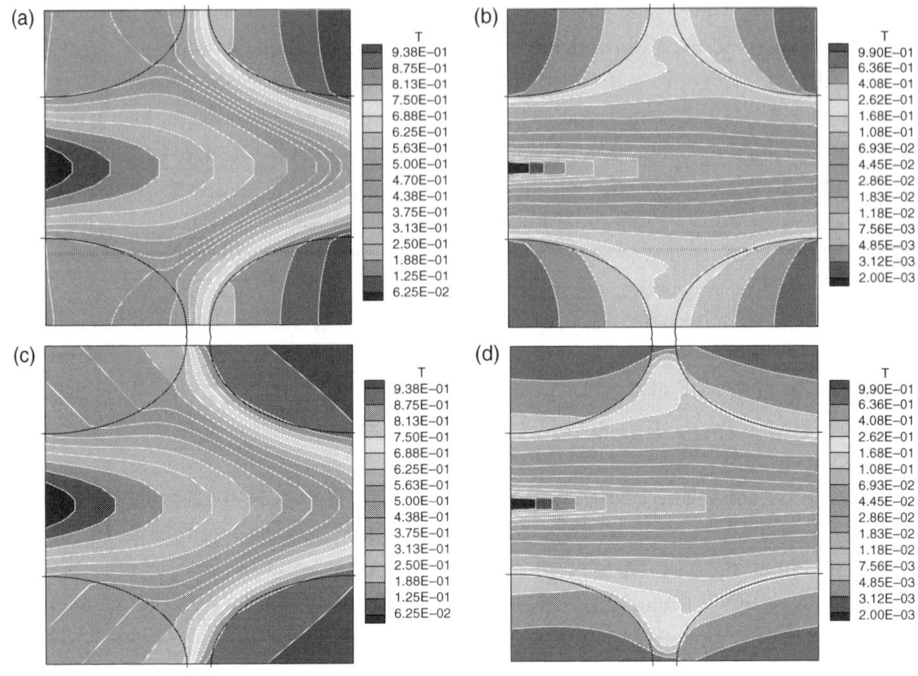


**Figure 5.2.** Temperature field with imposed longitudinal temperature gradient, $\phi = 0.60$: given temperature difference at East–West boundaries (see Equation (5.52)): (a) $Pe_H = 10$, (b) $Pe_H = 4 \times 10^3$; given heat fluxes at North–South boundaries (see Figure (4.1d)): (c) $Pe_H = 10$, (d) $Pe_H = 4 \times 10^3$.

Figure 5.4 shows the longitudinal component of the thermal dispersion tensor as a function of the Peclet number (Figure 5.4a) and for different porosities (Figure 5.4b). For simplicity of notation in the figures and text to follow, the thermal conductivity is given the symbol $k$ and, also due to the fact that flow here considered is always in the horizontal direction ($\theta = 0$ in Figure 5.1a), the longitudinal component of tensor $\mathbf{K}_{\text{disp}}$ is named either by $(\mathbf{K}_{\text{disp}})_{XX}$ or $(\mathbf{K}_{\text{disp}})_{xx}$. The same applies for the transversal direction. Results in Figure 5.4a show good agreement when compared with the data of Kuwahara and Nakayama (1998), for square and cylindrical rods, respectively, and for the same porosity. Also, as mentioned before, the use of different boundary conditions (Figure 5.1b,d) yields very little differences in the longitudinal component of $\left(\mathbf{K}_{\text{disp}}\right)_{XX}$. Figure 5.4a also shows that for different medium morphologies (longitudinally displaced elliptic, square and cylindrical rods), $\left(\mathbf{K}_{\text{disp}}\right)_{XX}$ is little sensitive. For the same fluid, $Pe_H$ and porosity, the mass flow rate through the bed will be the same. Thus, the overall convection strength and temperatures along the $x$-direction will vary little, which, in turn, will yield similar values for $\left(\mathbf{K}_{\text{disp}}\right)_{XX}$.

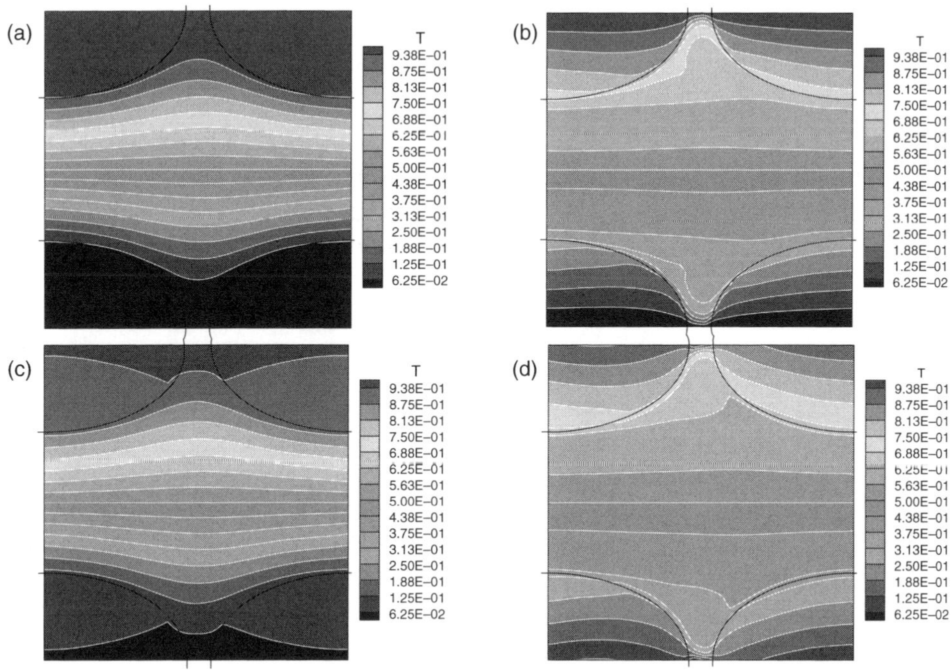


**Figure 5.3.** Temperature field with imposed transversal temperature gradient, $\phi = 0.60$: given temperature difference at North–South boundaries (see Equation (5.53)): (a) $Pe_H = 10$, (b) $Pe_H = 4 \times 10^3$; given heat fluxes at North–South boundaries (see Figure 4.1e): (c) $Pe_H = 10$, (d) $Pe_H = 4 \times 10^3$.

Figure 5.4b presents longitudinal dispersion coefficients for porosities covering the range 0.6–0.9. For lower Peclet numbers, application of both boundary conditions results in nearly the same values for $(\mathbf{K}_{\text{disp}})_{XX}$. Also for higher $Pe_H$, dependency of $(\mathbf{K}_{\text{disp}})_{XX}$ with $\phi$ is small. On the overall, for all porosities here considered, a general equation which is dependent on Peclet number, can be inferred as:

$$(\mathbf{K}_{\text{disp}})_{XX}/\mathbf{k}_{\text{f}} = 3.45 \times 10^{-2} {Pe_H}^{1.65} \tag{5.56}$$

showing, as expected, the usual behavior of $(\mathbf{K}_{\text{disp}})_{XX}/\mathbf{k}_{\text{f}} \sim {Pe_H}^{\text{n}}$.

The longitudinal components calculated with $\mathbf{K}_{\text{s}}/\mathbf{k}_{\text{f}} = 2$ was given in Pedras *et al.* (2003b) as,

$$(\mathbf{K}_{\text{disp}})_{XX}/\mathbf{k}_{\text{f}} = 3.52 \times 10^{-2} {Pe_H}^{1.65} \tag{5.57}$$

These results are plotted in Figure 5.5a with those calculated with $k_{\text{s}}/k_{\text{f}} = 10$. One can see that the conductivity ratio $k_{\text{s}}/k_{\text{f}}$ has little influence on the behavior of the

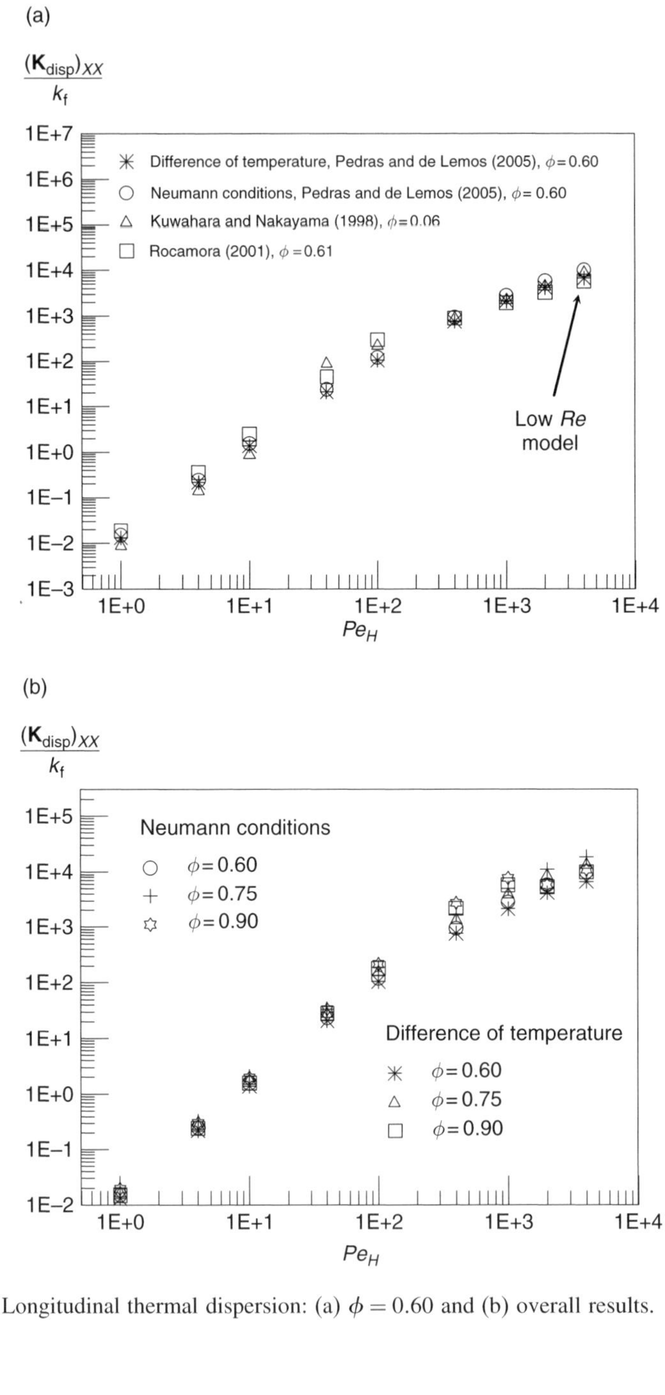


**Figure 5.4.** Longitudinal thermal dispersion: (a) $\phi = 0.60$ and (b) overall results.

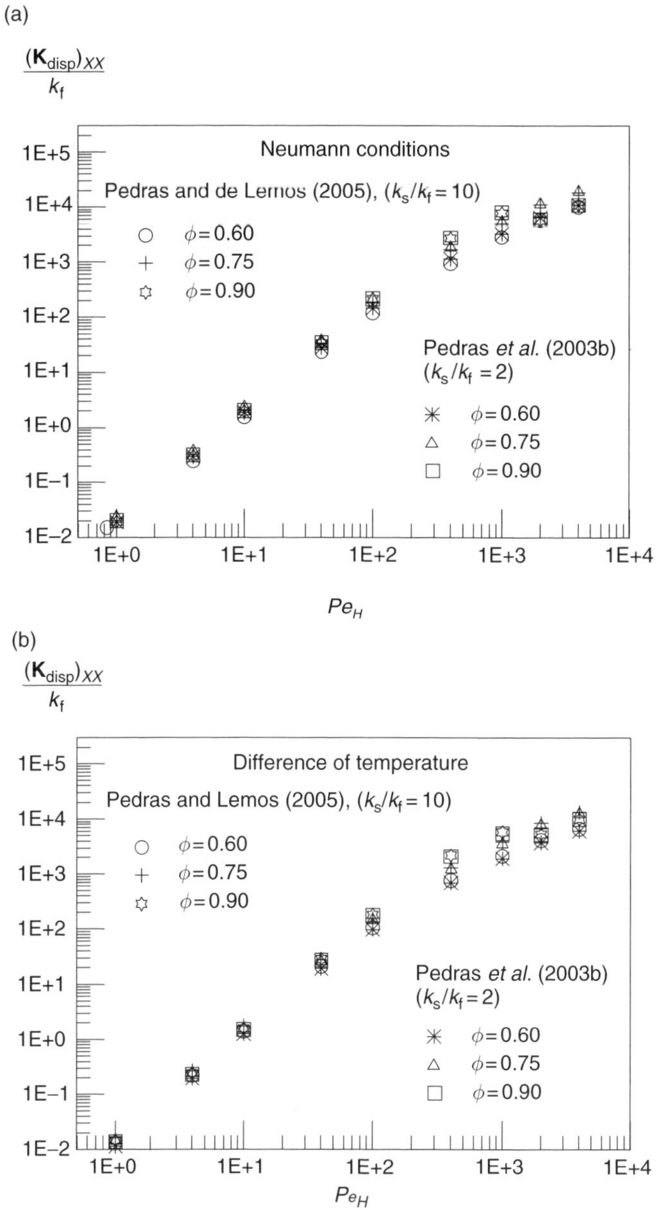


**Figure 5.5.** Longitudinal thermal dispersion compared with $K_s/K_f = 2$ and $K_s/K_f = 10$: (a) Neumann boundary conditions (Figure 5.1d), (b) temperature boundary conditions (Figure 5.1b, Equation (5.52)).

$(\mathbf{K}_{\text{disp}})_{XX} \times {Pe_H}^{1.65}$ curve. Temperatures inside the solid will be mostly affected by increasing the solid thermal conductivity and, as seen before, $(\mathbf{K}_{\text{disp}})_{XX}$ values are basically dependent on the fluid phase temperature (see Equation (5.48)).

The transverse component of the thermal dispersion, $(\mathbf{K}_{\text{disp}})_{YY}$, is shown next in Figure 5.6. As already mentioned, the use of different boundary conditions (Figure 5.1c,e) yields very little differences in the transverse component of the dispersion tensor $\mathbf{K}_{\text{disp}}$. However, in Figure 5.6a one can also see that different medium morphologies, such as longitudinally displaced elliptic rods (Pedras and de Lemos, 2005), as well as square (Kuwahara and Nakayama, 1998) and cylindrical rods (Rocamora, 2001), yield substantially different values for $(\mathbf{K}_{\text{disp}})_{YY}$. Results for square rods by Kuwahara and Nakayama 1998) were greater than for circular rods (Rocamora, 2001), which were further greater than computations by Pedras and de Lemos (2005) for longitudinally displaced rods. If one recalls that $(\mathbf{K}_{\text{disp}})_{YY}$ is associated with dispersive transport in the $y$-direction, one can infer that the easier the fluid flows in the longitudinal $x$-direction, due to a streamwise optimized geometric shape (for example (longitudinally displaced elliptic rods) less exchange in the transversal direction will take place. Also, for the same porosity or void-to-cell volume ratio, square, cylindrical and elliptical rods will have a reducing opening area in the north and south faces of the unit cell (see Figure 5.1a), reducing then the exchange of energy in the transverse $y$-direction. It is also interesting to point out that $(\mathbf{K}_{\text{disp}})_{YY}$ is several orders of magnitude smaller than $(\mathbf{K}_{\text{disp}})_{XX}$ because of the fact that for a macroscopically horizontal flow most dispersive transport will be along the main flow direction. The overall dependence of the transverse component on the Peclet number was found to be:

$$\frac{(\mathbf{K}_{\text{disp}})_{YY}}{k_{\text{f}}} = 1.55 \times 10^{-4} {Pe_H}^{0.94} \tag{5.58}$$

Figure 5.6b further shows the dependence of $(\mathbf{K}_{\text{disp}})_{YY}$ on the porosity $\phi$ for the case here investigated, namely the longitudinally displaced elliptic rods. Results confirm the already observed insensitivity of the results on the type of boundary condition used. Dependency on the porosity of the cell is more difficult to access due to the spread of the results. Nevertheless, a general observation can be made that by reducing $\phi$, via increasing the size of the ellipses, lower values for $(\mathbf{K}_{\text{disp}})_{YY}$ are obtained, at least for low $Pe_H$. This observation is in line with an argument already used that the more obstructed the passages are in between cells along $y$, the lower the values of $(\mathbf{K}_{\text{disp}})_{YY}$ will be.

Figure 5.7 finally shows comparisons between the transverse components calculated with $k_{\text{s}}/k_{\text{f}} = 2$ (Pedras *et al.*, 2003b), where

$$(\mathbf{K}_{\text{disp}})_{YY}/k_{\text{f}} = 2.29 \times 10^{-4} {Pe_H}^{0.88} \tag{5.59}$$

with the cases having $k_{\text{s}}/k_{\text{f}} = 10$. The comparison shows that the transverse component is more sensitive to the variation of the conductivity ratio than the longitudinal component.

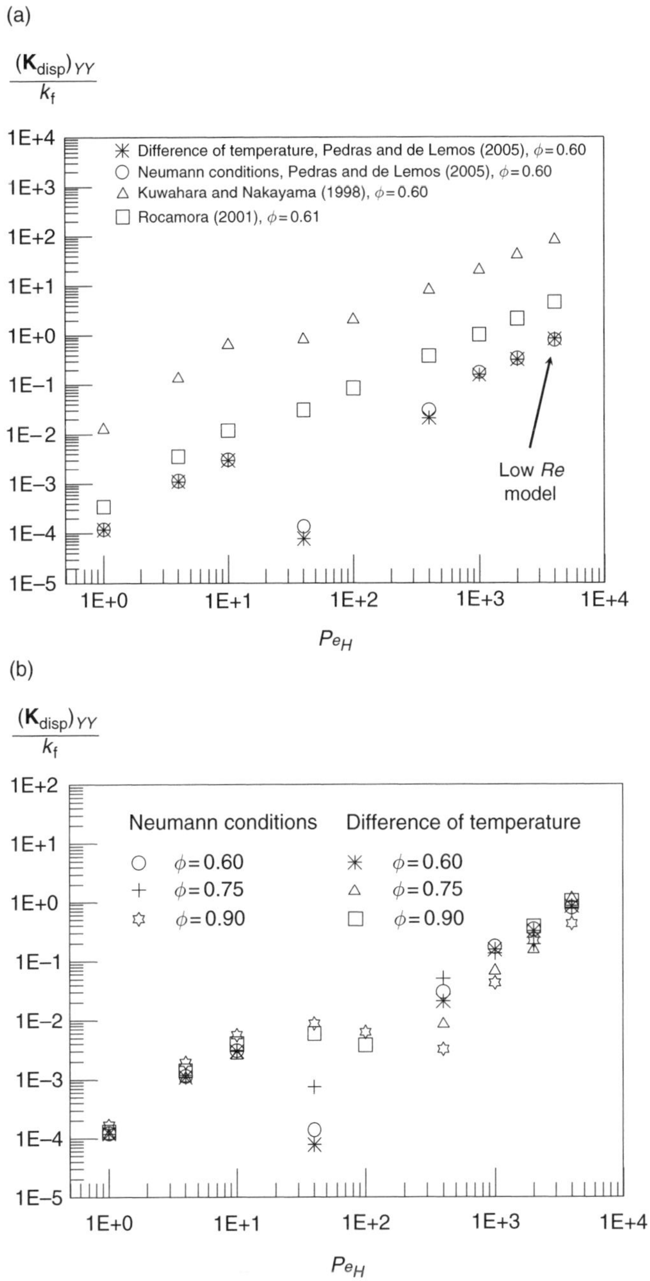


**Figure 5.6.** Transverse thermal dispersion: (a) $\phi = 0.60$ and (b) overall results.

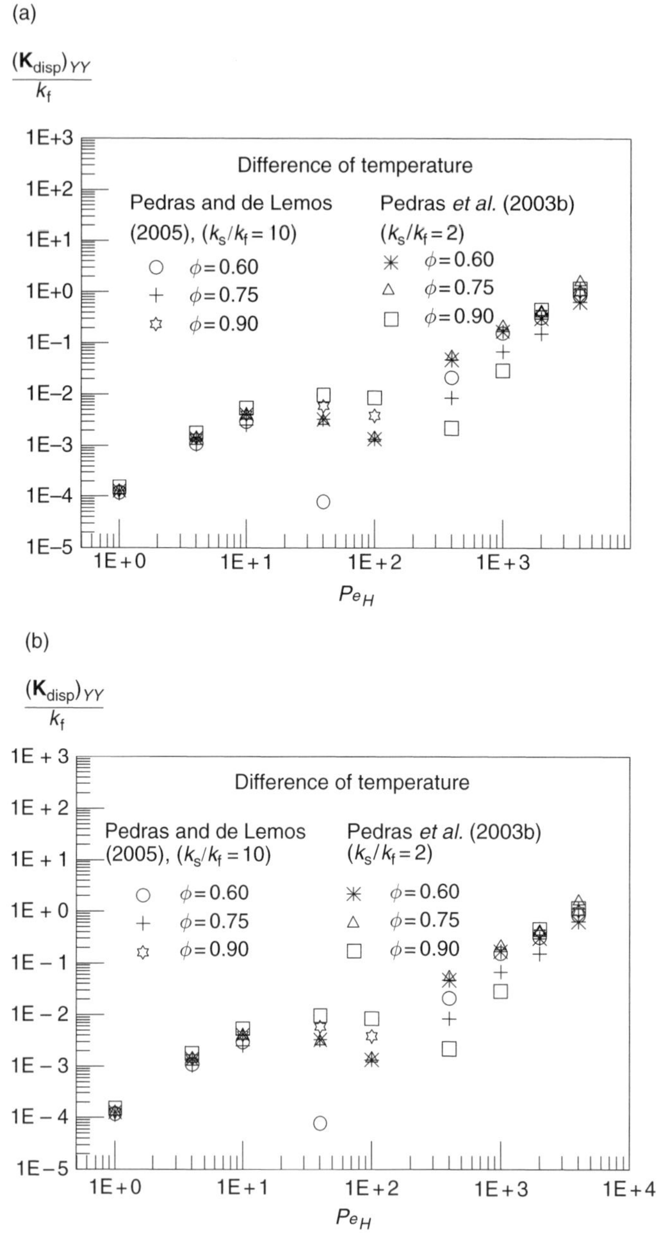


**Figure 5.7.** Transverse thermal dispersion comparing $K_s/K_f = 2$ and $K_s/K_f = 10$: (a) Neumann boundary conditions (Figure 5.1e), (b) temperature boundary conditions (Figure 5.1c, Equation (5.33)).

In general, for higher $\mathbf{K}_s/\mathbf{K}_f$ ratios, lower $(\mathbf{K}_{disp})_{YY}/\mathbf{K}_f$ coefficients were obtained. Such difference mainly occurs for Peclet numbers of about $10^2$. In this range of $Pe_H$, little recirculation was observed behind the rods (not shown here). A possible explanation for this behavior might be associated with the temperature deviation values ${}^iT$ existing in the fluid, in each case. For $\mathbf{K}_s/\mathbf{K}_f = 2$, temperature gradients in the recirculating zones (behind the rods in the $x$-direction) were larger that those for $\mathbf{K}_s/\mathbf{K}_f = 10$, which, in turn, were almost negligible. This almost-zero temperature gradient along with the recirculating zone for $\mathbf{K}_s/\mathbf{K}_f = 10$ reduced the temperature deviations ${}^iT$, which, from Equation (5.50), reduced the transverse dispersion.

## 5.3. THERMAL NON-EQUILIBRIUM MODELS

In many industrial applications, turbulent flow through a packed bed represents an important configuration for efficient heat and mass transfer. A common model used for analyzing such system is the so-called "local thermal equilibrium assumption" where both solid and fluid phase temperatures are represented by a unique value (see Section 5.2, p. 58). This model simplifies theoretical and numerical research, but the assumption of local thermal equilibrium between the fluid and the solid is inadequate for a number of problems (Kaviany, 1995; Quintard, 1998). Consequently, in many instances it is important to take into account distinct temperatures for the porous material and for the working fluid. In transient heat conduction processes, for example, the assumption of local thermal equilibrium must be discarded, according to Kaviany (1995) and Hsu (1999). Also, when there is significant heat generation in any one of the two phases, namely solid and fluid, average temperatures are no longer identical so the hypothesis of local thermal equilibrium must be reevaluated. This suggests the use of equations governing thermal non-equilibrium involving distinct energy balances for both the solid and fluid phases. In recent years more attention has been paid to the local thermal non-equilibrium model and its use has increased in theoretical and numerical research for convection heat transfer processes in porous media (Ochoa-Tapia and Whitaker, 1997; Quintard *et al.*, 1997). Accordingly, the use of such two-energy equation model requires an extra parameter to be determined, namely the heat transfer coefficient between the fluid and the solid material (Kuznetsov, 1998).

Quintard (1998) argues that assessing the validity of the assumption of local thermal equilibrium is not a simple task since the temperature difference between the two phases cannot be easily measured. He suggests that the use of a two-energy equation model is a possible approach to solving the problem.

Kuwahara *et al.* (2001) proposed a numerical procedure to determine macroscopic transport coefficients from a theoretical basis without any empiricism. They used a single unit cell and determined the interfacial heat transfer coefficient for the asymptotic case of infinite conductivity of the solid phase. Nakayama *et al.* (2001) extended the conduction model of Hsu (1999) for treating convection also in porous media. Having established the macroscopic energy equations for both phases, useful exact solutions were obtained for two fundamental heat transfer processes associated with porous media – namely, steady conduction in a porous slab with internal heat generation within the solid, and thermally developing flow through a semi-infinite porous medium.

Based on the double-decomposition concept introduced in Chapter 3, de Lemos and Rocamora (2002) developed a macroscopic energy equation considering local thermal equilibrium between the fluid and the solid matrix. This section reviews the heat transfer analysis of Saito and de Lemos (2005), who extended the transport model of de Lemos and Rocamora (2002) by considering on additional energy equation for the solid phase. The contribution therein consisted in computing the heat transfer coefficient at the interface between the two phases, which considered laminar flow through a bed formed by square rods. In Saito and de Lemos (2006), a new correlation for the interfacial heat transfer coefficient was proposed for Reynolds numbers, up to $10^7$. Some numerical results by Saito and de Lemos (2005, 2006) are here reviewed.

### 5.3.1. Laminar Flow through Packed Bed

#### *5.3.1.1. Interfacial Heat Transfer Coefficient*

In Equations (5.18) and (5.19) the heat transferred between the two phases can be modeled by means of a film coefficient $h_i$ such that:

$$h_i a_i \left( \langle \overline{T_s} \rangle^i - \langle \overline{T_f} \rangle^i \right) = \frac{1}{\Delta V} \int_{A_i} \mathbf{n}_i \cdot \lambda_f \nabla \overline{T_f} \, dA = \frac{1}{\Delta V} \int_{A_i} \mathbf{n}_i \cdot \lambda_s \nabla \overline{T_s} \, dA \qquad (5.60)$$

where, $h_i$ is known as the interfacial convective heat transfer coefficient, $A_i$ is the interfacial heat transfer area and $a_i = A_i/\Delta V$ is the surface area per unit volume.

For determining $h_i$, Kuwahara *et al.* (2001) modeled a porous medium by considering an infinite number of solid square rods of size $D$, arranged in a regular triangular pattern (Figure 5.8). They numerically solved the governing equations in the void region, exploiting to advantage the fact that for an infinite and geometrically ordered medium a repetitive cell can be identified. Periodic boundary conditions were then applied for obtaining the temperature distribution under fully developed flow conditions. A numerical correlation for the interfacial convective heat transfer coefficient was proposed by Kuwahara *et al.* (2001) as:

$$\frac{h_i D}{\lambda_f} = \left( 1 + \frac{4(1-\phi)}{\phi} \right) + \frac{1}{2}(1-\phi)^{1/2} Re_D^{0.6} Pr^{1/3}, \quad \text{valid for } 0.2 < \phi < 0.9 \qquad (5.61)$$

Equation (5.61) is based on porosity dependency and is valid for packed beds of particle diameter $D$.

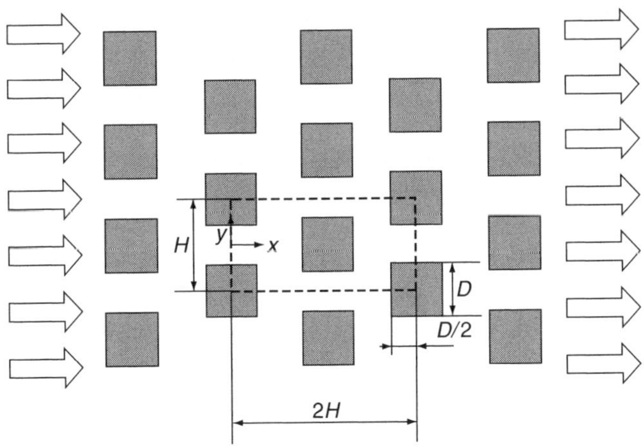


**Figure 5.8.** Physical model and coordinate system.

This same physical model will be used here for obtaining the interfacial heat transfer coefficient $h_i$ for macroscopic flows.

### *5.3.1.2. Periodic Cell and Boundary Conditions*

In order to evaluate the numerical tool to be used in the determination of the film coefficient given by Equation (5.60), a test case was run for obtaining the flow field in a periodic cell, which is here assumed to represent the porous medium. Consider a macroscopically uniform flow through an infinite number of square rods of lateral size $D$, placed in a staggered fashion and maintained at constant temperature $T_w$. The periodic cell or representative elementary volume, $\Delta V$, is schematically shown in Figure 5.8 and has dimensions $2H \times H$. Computations within this cell were carried out using a non-uniform grid of size $90 \times 70$ nodes, as shown in Figure 5.9, to ensure that the results were

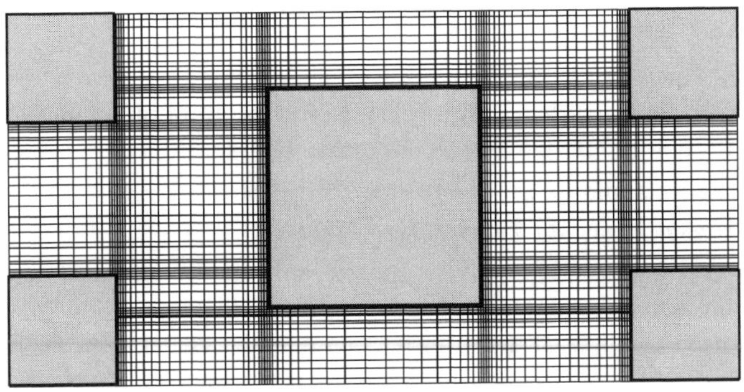

**Figure 5.9.** Non-uniform computational grid.

grid independent. The Reynolds number $Re_D = \rho \mathbf{u}_D D/\mu$ was varied from 4 to $4 \times 10^2$. Further, porosity was calculated to be in the range $0.44 < \phi < 0.65$ using the formula $\phi = 1 - (D/H)^2$.

The numerical method utilized to discretize the microscopic flow and energy equations in the unit cell is the Control Volume. The SIMPLE method of Patankar (1980) was used for the velocity–pressure coupling. Convergence was monitored in terms of the normalized residue for each variable. The maximum residue allowed for convergence check was set to $10^{-9}$, being the variables normalized by appropriate reference values.

For fully developed flow in the cell shown in Figure 5.8, the velocity at exit $(x/H = 2)$ must be identical to that at the inlet $(x/H = 0)$. Temperature profiles, however, are only identical at both cell exit and inlet if presented in terms of an appropriate non-dimensional variable. The situation is analogous to the case of forced convection in a channel with isothermal walls. Thus, boundary conditions and periodic constraints are given by:

On the solid walls:

$$\mathbf{u} = 0, \quad T = T_w \tag{5.62}$$

On the periodic boundaries:

$$\mathbf{u}|_{\text{inlet}} = \mathbf{u}|_{\text{outlet}} \tag{5.63}$$

$$\left.\int_0^H u\,\mathrm{d}y\right|_{\text{inlet}} = \left.\int_0^H u\,\mathrm{d}y\right|_{\text{outlet}} = H\,|\mathbf{u}_D| \tag{5.64}$$

$$\theta\,|_{\text{inlet}} = \theta\,|_{\text{outlet}} \Leftrightarrow \left.\frac{T - T_w}{T_b(x) - T_w}\right|_{\text{inlet}} = \left.\frac{T - T_w}{T_b(x) - T_w}\right|_{\text{outlet}} \tag{5.65}$$

where the bulk mean temperature of the fluid is given by:

$$T_b(x) = \frac{\int uT\,\mathrm{d}y}{\int u\,\mathrm{d}y} \tag{5.66}$$

Computations are based on the Darcy velocity, the length of structural unit $H$ and the temperature difference $(T_b(x) - T_w)$, as references scales.

### *5.3.1.3. Developed Flow and Temperature Fields*

Macroscopically developed flow field for $Pr = 1$ and $Re_D = 100$ is presented in Figure 5.10, corresponding to $x/D = 6$ at the cell inlet. The expression "macroscopically developed" is used herein to account for the fact that periodic flow has been achieved at that axial position. Figure 5.10 indicates that the flow impinges on the left face of the obstacles and surrounds the rod faces, forming a weak recirculation bubble past the rod.

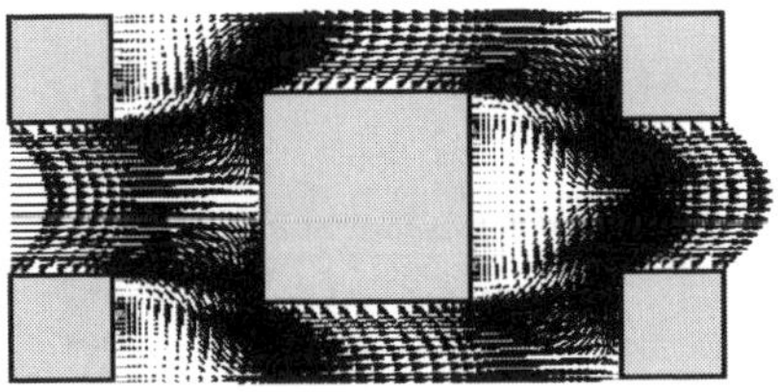

**Figure 5.10.** Velocity field for $Pr = 1$ and $Re_D = 100$.

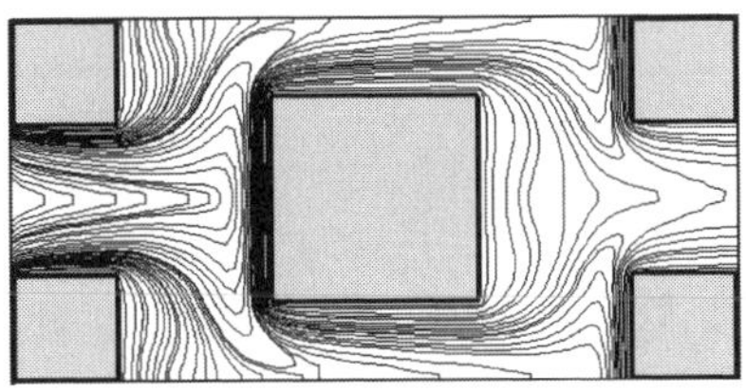

**Figure 5.11.** Isotherms for $Pr = 1$ and $Re_D = 100$.

When $Re_D$ is low (not shown here), the horizontal velocity field in between two rods appears to be very similar to what we observe in a channel, namely the parabolic profile, particularly at inlet and outlet of unit cell. As $Re_D$ increases, stronger recirculation bubbles appear further behind the rods. Temperature distribution pattern is shown in Figure 5.11, which is also for $Re_D = 100$. Colder fluid impinges on the left surface yielding strong temperature gradients on that face. Downstream the obstacle, fluid recirculation smoothes temperature gradients and deforms isotherms within the mixing region. When $Re_D$ is sufficiently high (not shown here), the thermal boundary layers covering the rod surfaces indicate that convective heat transfer overwhelms thermal diffusion.

### 5.3.1.4. *Film Coefficient $h_i$*

For the unit cell of Figure 5.8, determination of $h_i$ is given by:

$$h_i = \frac{Q_{total}}{A_i \Delta T_{ml}} \tag{5.67}$$

where $A_i = 8D \times 1$. The overall heat transferred in the cell, $Q_{total}$, is given by,

$$Q_{total} = (H - D)\rho \mathbf{u}_b c_p \left( T_b|_{outlet} - T_b|_{inlet} \right) \tag{5.68}$$

where $\mathbf{u}_B$ is the bulk mean velocity of the fluid, and $\Delta T_{ml}$, the logarithm mean temperature difference, is given by:

$$\Delta T_{ml} = \frac{(T_w - T_b|_{outlet}) - (T_w - T_b|_{inlet})}{\ln[(T_w - T_b|_{outlet})/(T_w - T_b|_{inlet})]} \tag{5.69}$$

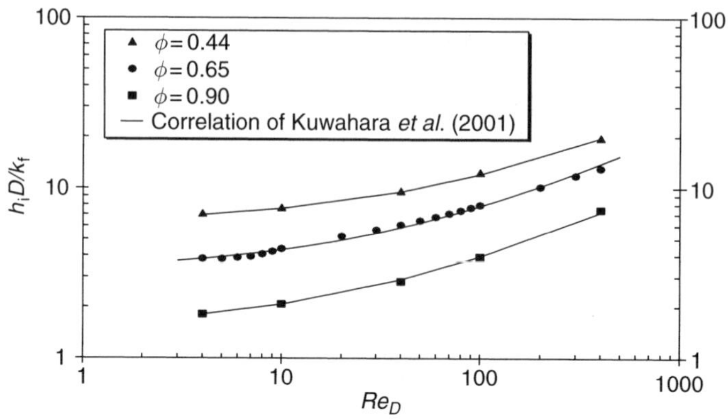


**Figure 5.12.** Effect of $Re_D$ on $h_i$ for $Pr = 1$; solid symbols denote present results; solid lines denote the results by Kuwahara *et al.* (2001).

Equation (5.68) represents an overall heat balance on the entire cell and associates the heat transferred to the fluid to a suitable temperature difference $\Delta T_{ml}$. As mentioned earlier, Equations (2.1)–(2.3) were numerically solved in the unit cell until conditions Equations (5.63)–(5.65) were satisfied.

Once fully developed flow and temperature fields were achieved, for the condition $x > 6H$, bulk temperatures were calculated according to Equation (5.66), at both inlet and outlet positions. They were then used to calculate $h_i$ using Equations (5.68) and (5.69). Results for $h_i$ are plotted in Figure 5.12 for $Re_D$ up to 400. Also plotted in this figure are results computed with correlation (5.61) using different porosity values. The figure seems to indicate that both computations show a reasonable agreement.

### 5.3.2. Turbulent Flow through Packed Bed

#### *5.3.2.1. Interfacial Heat Transfer Coefficient*

Wakao *et al.* (1979) obtained a heuristic correlation for closely packed bed of particle diameter $D$, and compared their results with experimental data. This correlation for the interfacial heat transfer coefficient is given by:

$$\frac{h_i D}{\lambda_f} = 2 + 1.1 Re_D^{0.6} Pr^{1/3} \tag{5.70}$$

Saito and de Lemos (2005) obtained the interfacial heat transfer coefficient for laminar flows through an infinite, staggered array of square rods; this same physical model will be used here for obtaining the interfacial heat transfer coefficient $h_i$ for turbulent flows.

For the staggered configuration, Zhukauskas (1972) has proposed a correlation of the form:

$$\frac{h_{\mathrm{i}}D}{\lambda_{\mathrm{f}}} = 0.022 Re_D^{0.84} Pr^{0.36} \tag{5.71}$$

where the values 0.022 and 0.84 are constants for tube bank in cross flow and for this particular case $2 \times 10^5 < Re_D < 2 \times 10^6$

### 5.3.2.2. *Turbulence Models and Boundary Conditions*

For turbulent flows the time averaged transport equations can be written as:

*Continuity*:

$$\nabla \cdot \bar{\mathbf{u}} = 0 \tag{5.72}$$

*Momentum*:

$$\rho_{\mathrm{f}}[\nabla \cdot (\bar{\mathbf{u}}\bar{\mathbf{u}})] = -\nabla \bar{p} + \nabla \cdot \{\mu [\nabla \bar{\mathbf{u}} + (\nabla \bar{\mathbf{u}})^{\mathrm{T}}] - \rho \overline{\mathbf{u}'\mathbf{u}'}\} \tag{5.73}$$

where the low and high *Re* $k - \varepsilon$ model is used to obtain the eddy viscosity $\mu_{\mathrm{t}}$, whose equations for the turbulent kinetic energy per unit mass and for its dissipation rate read:

Turbulent kinetic energy per unit mass:

$$\rho_{\mathrm{f}}[\nabla \cdot (\bar{\mathbf{u}}k)] = \nabla \cdot \left[\left(\mu + \frac{\mu_{\mathrm{t}}}{\sigma_k}\right) \nabla k\right] - \rho \overline{\mathbf{u}'\mathbf{u}'} : \nabla \bar{\mathbf{u}} - \rho \varepsilon \tag{5.74}$$

Turbulent kinetic energy per unit mass dissipation rate:

$$\rho_{\mathrm{f}}[\nabla \cdot (\bar{\mathbf{u}}\varepsilon)] = \nabla \cdot \left[\left(\mu + \frac{\mu_{\mathrm{t}}}{\sigma_\varepsilon}\right) \nabla \varepsilon\right] + \left[c_1 \left(-\rho \overline{\mathbf{u}'\mathbf{u}'} : \nabla \bar{\mathbf{u}}\right) - c_2 f_2 \rho \varepsilon\right] \frac{\varepsilon}{k} \tag{5.75}$$

Reynolds stresses and the eddy viscosity are given by, respectively:

$$-\rho \overline{\mathbf{u}'\mathbf{u}'} = \mu_{\mathrm{t}} [\nabla \bar{\mathbf{u}} + (\nabla \bar{\mathbf{u}})^{\mathrm{T}}] - \frac{2}{3} \rho k \mathbf{I} \tag{5.76}$$

$$\mu_{\mathrm{t}} = \rho c_\mu f_\mu \frac{k^2}{\varepsilon} \tag{5.77}$$

where $\rho$ is the fluid density, $p$ is the pressure and $\mu$ represents the fluid viscosity.

In the above equation set, $\sigma_k$, $\sigma_\varepsilon$, $c_1$, $c_2$ and $c_\mu$ are dimensionless constants whereas $f_2$ and $f_\mu$ are damping functions of the low *Re* $k-\varepsilon$ turbulence model, which read:

$$f_\mu = \left\{1-\exp\left[-\frac{(\nu\varepsilon)^{0.25}y}{14\nu}\right]\right\}^2 \left\{1+\frac{5}{(k^2/\nu\varepsilon)^{0.75}}\exp\left[-\left(\frac{(k^2/\nu\varepsilon)}{200}\right)^2\right]\right\} \tag{5.78}$$

$$f_2 = \left\{1-\exp\left[-\frac{(\nu\varepsilon)^{0.25}y}{3.1\nu}\right]\right\}^2 \left\{1-0.3\exp\left[-\left(\frac{(k^2/\nu\varepsilon)}{6.5}\right)^2\right]\right\} \tag{5.79}$$

where $y$ is the coordinate normal to the wall. The model constants are given as follows:

$$c_\mu = 0.09, \quad c_1 = 1.5, \quad c_2 = 1.9, \quad \sigma_k = 1.4, \quad \sigma_\varepsilon = 1.3$$

For the high *Re* model the standard constants of Launder and Spalding (1974) were used. Also, the time averaged energy equations become:

*Energy – Fluid phase*:

$$(\rho c_{\mathrm{p}})_{\mathrm{f}}\left[\nabla\cdot(\overline{\mathbf{u}}\overline{T}_{\mathrm{f}})\right] = \nabla\cdot(\lambda_{\mathrm{f}}\nabla\overline{T}_{\mathrm{f}}) - (\rho c_{\mathrm{p}})_{\mathrm{f}}\nabla\cdot\left(\overline{\mathbf{u}'T'_{\mathrm{f}}}\right) \tag{5.80}$$

*Energy – Solid phase (porous matrix)*:

$$\nabla\cdot(\lambda_{\mathrm{s}}\nabla\overline{T}_{\mathrm{s}}) + S_{\mathrm{s}} = 0 \tag{5.81}$$

The boundary conditions and periodic constraints are given by:

On the solid walls (Low *Re*):

$$\overline{u} = 0, \quad k = 0, \quad \varepsilon = \nu\frac{\partial^2 k}{\partial y^2}, \quad \overline{T} = \overline{T}_{\mathrm{w}} \tag{5.82}$$

On the solid walls (high *Re*):

$$\frac{\overline{u}}{u_\tau} = \frac{1}{\kappa}\ln\left(y^+E\right), \quad k = \frac{u_\tau^2}{c_\mu{}^{1/2}}, \quad \varepsilon = \frac{c_\mu{}^{3/4}\lambda_{\mathrm{w}}{}^{3/2}}{\kappa y_{\mathrm{w}}},$$

$$q_{\mathrm{w}} = \frac{(\rho c_{\mathrm{p}})_{\mathrm{f}}\, c_\mu{}^{1/4}\lambda_{\mathrm{w}}{}^{1/2}\left(\overline{T}-T_{\mathrm{w}}\right)}{\left(\frac{\sigma_{\mathrm{t}}}{\kappa}\ln(y_{\mathrm{w}}^+) + c_{\mathrm{Q}}(Pr)\right)} \tag{5.83}$$

where,

$$u_\tau = \left(\frac{\tau_{\mathrm{w}}}{\rho}\right)^{1/2}, \quad y_{\mathrm{w}}^+ = \frac{y_{\mathrm{w}}u_\tau}{\nu}, \quad c_{\mathrm{Q}} = 12.5Pr^{2/3} + 2.12\ln(Pr) - 5.3 \quad \text{for } Pr > 0.5$$

where, $Pr$ and $\sigma_t$ are Prandtl and turbulent Prandtl numbers, respectively, $q_w$ is wall heat flux, $u_\tau$ is wall-friction velocity, $y_w$ is the coordinate normal to wall, $\kappa$ is constant for turbulent flow past smooth impermeable walls, or von Kármán's constant, and $E$ is an integration constant that depends on the roughness of the wall. For smooth walls $E = 9$.

On the symmetry planes:

$$\frac{\partial \overline{u}}{\partial y} = \frac{\partial \overline{v}}{\partial y} = \frac{\partial k}{\partial y} = \frac{\partial \varepsilon}{\partial y} = 0 \tag{5.84}$$

where $\overline{u}$ and $\overline{v}$ are components of $\overline{\mathbf{u}}$.

On the periodic boundaries:

$$\overline{u}\big|_{\text{inlet}} = \overline{u}\big|_{\text{outlet}}, \quad \overline{v}\big|_{\text{inlet}} = \overline{v}\big|_{\text{outlet}}, \quad k\big|_{\text{inlet}} = k\big|_{\text{outlet}}, \quad \varepsilon\big|_{\text{inlet}} = \varepsilon\big|_{\text{outlet}} \tag{5.85}$$

$$\theta\Bigg|_{\text{inlet}} = \frac{\overline{T} - \overline{T}_w}{\overline{T}_b(x) - \overline{T}_w}\Bigg|_{\text{inlet}} = \theta\Bigg|_{\text{outlet}} = \frac{\overline{T} - \overline{T}_w}{\overline{T}_b(x) - \overline{T}_w}\Bigg|_{\text{outlet}} \tag{5.86}$$

The bulk mean temperature of the fluid is given by:

$$\overline{T}_b(x) = \frac{\int \overline{u}\,\overline{T}\,\mathrm{d}y}{\int \overline{u}\,\mathrm{d}y} \tag{5.87}$$

Computations are based on the Darcy velocity, the length of structural unit $H$ and the temperature difference $\left(\overline{T}_b(x) - \overline{T}_w\right)$, as references scales.

### 5.3.2.3. Turbulent Results

*Periodic Flow* Results for velocity and temperature fields were obtained for different Reynolds numbers. In order to assure that the flow was hydrodynamically and thermally developed in the periodic cell of Figure 5.8, the governing equations were solved repetitively in the cell, taking the outlet profiles for $\overline{\mathbf{u}}$ and $\theta$ at exit and plugging them back at inlet. In the first run, uniform velocity and temperature profiles were set at the cell entrance for $Pr = 1$ giving $\theta = 1$ at $x/H = 0$. Then, after convergence of the flow and temperature fields, $\overline{\mathbf{u}}$ and $\theta$ at $x/H = 2$ were used as inlet profiles for a second run, corresponding to solving again the flow for a similar cell beginning in $x/H = 2$. Similarly, a third run was carried out and again outlet results, this time corresponding to an axial position $x/H = 4$, were recorded. This procedure was repeated several times until $\overline{\mathbf{u}}$ and $\theta$ did not differ substantially at both inlet and outlet positions. Resulting non-dimensional velocity and temperature profiles are shown in Figures 5.13 and 5.14, respectively, showing that the periodicity constraints imposed by Equations (5.63)–(5.65) were satisfied for $x/H > 4$. For the entrance region $(0 < x/H < 4)$, $\theta$ profiles change

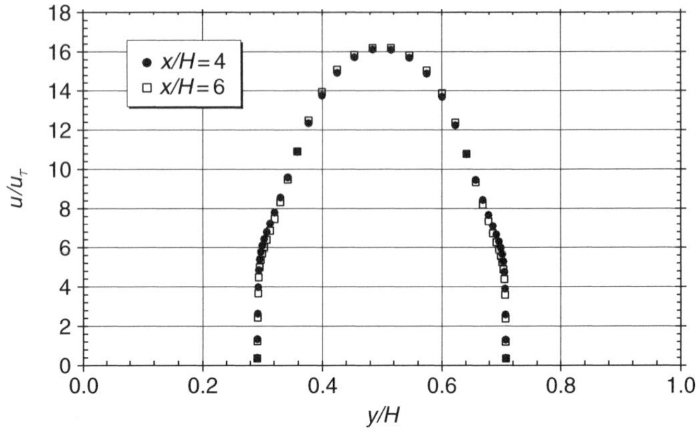


**Figure 5.13.** Dimensionless velocity profile for $Pr = 1$ and $Re_D = 5 \times 10^4$.

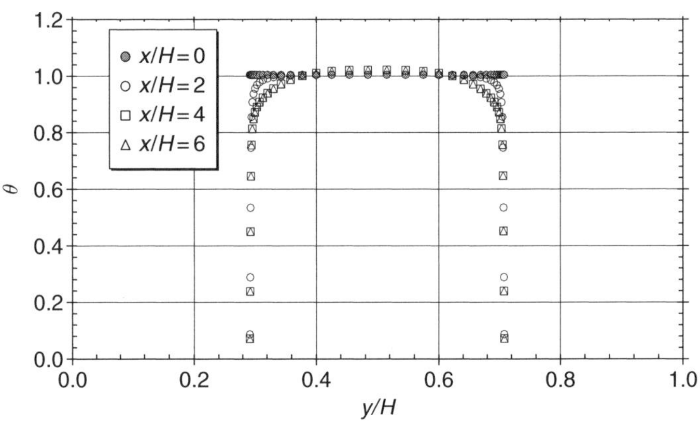


**Figure 5.14.** Dimensionless temperature profile for $Pr = 1$ and $Re_D = 5 \times 10^4$.

with length $x/H$ being essentially invariable after this distance. Under this condition of constant-$\theta$ profile, the flow was considered to be macroscopically developed for $Re_D$ up to $10^7$.

For the low *Re* model, the first node adjacent to the wall requires that the non-dimensional wall distance be such that $y^+ \; (= u_\tau y \rho / \mu) \leq 1$. To accomplish this requirement, the grid needs a great number of points close to the wall leading to computational meshes of large sizes.

*Developed Flow and Temperature Fields* As mentioned, the macroscopically developed flow field for $Pr = 1$ and $Re_D = 5 \times 10^4$ was presented in Figures 5.13 corresponding to $x/D = 4$ at the cell inlet. For a cell beginning at $x/D = 6$, Figures (5.15) through (5.17)

**Figure 5.15.** Non-dimensional pressure field for $Re_D = 10^5$ and $\phi = 0.65$.

**Figure 5.16.** Isotherms for $Pr = 1$, $Re_D = 10^5$ and $\phi = 0.65$.

show distributions of pressure, isotherms and turbulence kinetic energy in a microscopic porous structure, obtained at $Re_D = 10^5$ for $\phi = 0.65$. Pressure increases at the front face of the square rod and drastically decreases around the corner, as can be seen from the pressure contours shown in Figure 5.15. Temperature distribution is shown in Figure 5.16. Colder fluid impinges on the left surfaces of the rod yielding strong temperature gradients on that face. Downstream the obstacles, fluid recirculation smoothes temperature gradients and deforms isotherms within the mixing region. When the Reynolds number is sufficiently high (not shown here), thermal boundary layers cover the rod surfaces indicating that convective heat transfer overwhelms thermal diffusion. Figure 5.17 presents levels of

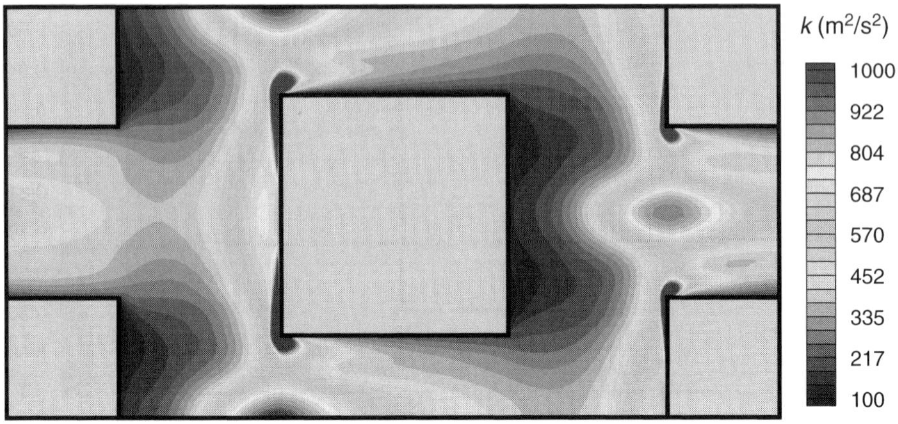


**Figure 5.17.** Turbulence kinetic energy for $Re_D = 10^5$ and $\phi = 0.65$.

turbulence kinetic energy, which are higher around the corners of the rod where a strong shear layer is formed. Further downstream the rods, in the wake region, steep velocity gradients appear due to flow deceleration, also increasing there the local level of $k$.

Once fully developed flow and temperature fields are achieved, for the fully developed condition ($x > 6H$), bulk temperatures were calculated according to Equation (5.66), at both inlet and outlet positions. They were then used to calculate $h_i$ using Equations. (5.67)–(5.69). Results for $h_i$ are plotted in Figure (5.18) for $\mathrm{Re}_D$ up to $10^7$. Also plotted in this figure are results computed with correlation (5.61) given by Kuwahara *et al.* (2001) for $\phi = 0.65$. The figure seems to indicate that both computations show a reasonable agreement for laminar results.

Figure (5.19) shows numerical results of the interfacial convective heat transfer coefficient for various porosities ($\phi = 0.44$, 0.65 and 0.90). Results for $h_i$ are plotted for $Re_D$ up to $10^7$.

In order to obtain a correlation for $h_i$ in the turbulent regime, all curves were first collapsed after plotting them in terms of $Re_D/\phi$, as showed in Figure (5.20). Furthermore, the least square technique was applied in order to determine the best correlation, which leads to a minimum overall error. Thus, the following expression was proposed in Saito and de Lemos (2006):

$$\frac{h_i D}{\lambda_f} = 0.08\left(\frac{Re_D}{\phi}\right)^{0.8} Pr^{1/3}; \quad \text{for} \quad 1.0 \times 10^4 < \frac{Re_D}{\phi} < 2.0 \times 10^7, \quad \text{valid for} \quad 0.2 < \phi < 0.9 \tag{5.88}$$

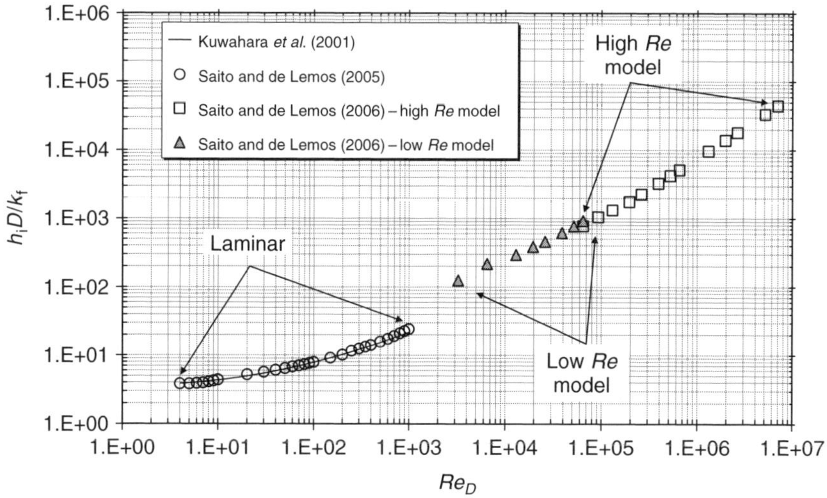


**Figure 5.18.** Effect of $Re_D$ on $h_i$ for $Pr = 1$ and $\phi = 0.65$.

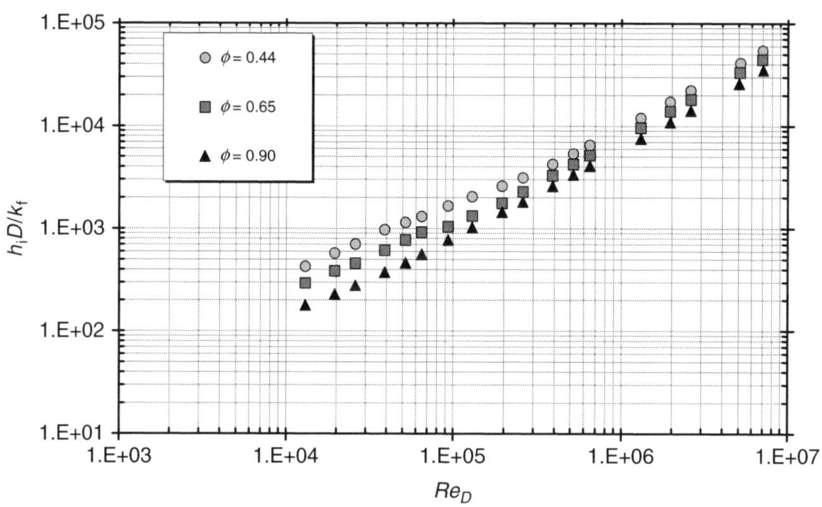


**Figure 5.19.** Effect of porosity on $h_i$ for $Pr = 1$.

Equation (5.88), which gives the heat transfer coefficient for turbulent flow, is compared with numerical results obtained with low and high *Re* models. Such comparison is presented in Figure (5.21), which also shows computations using correlations given by Equation (5.70) and Equation (5.71) by Wakao *et al.* (1979) and Zhukauskas (1972),

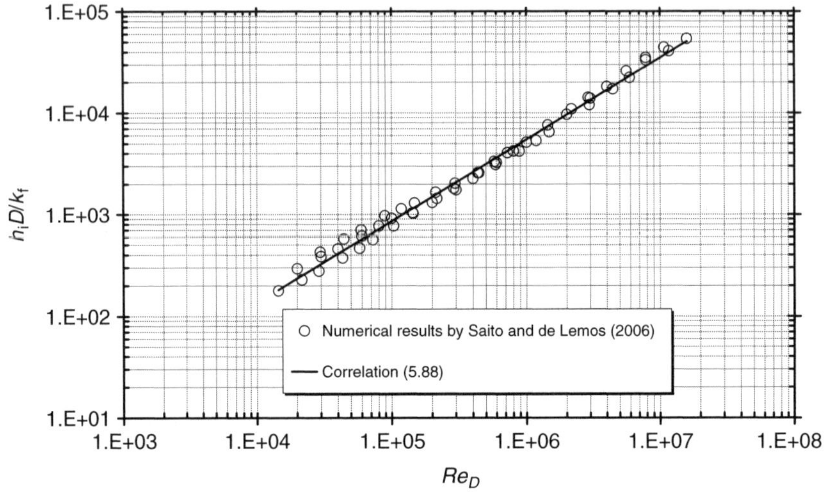


**Figure 5.20.** Comparison of the numerical results and proposed correlation.

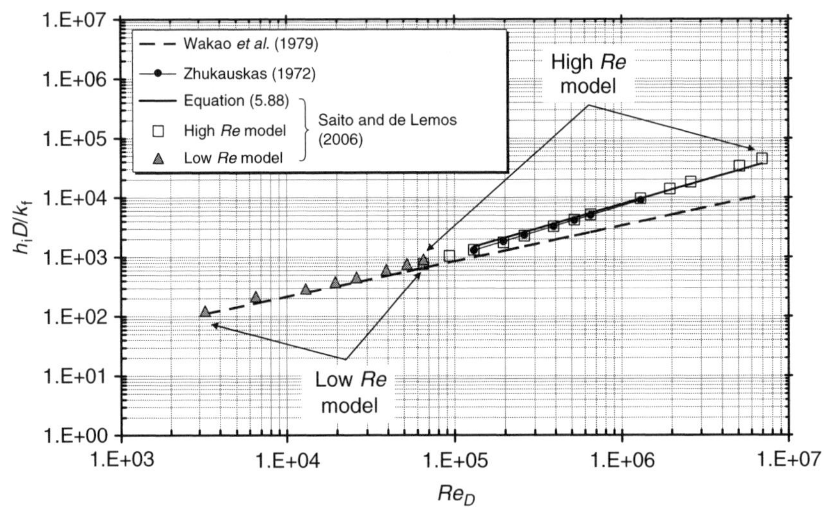


**Figure 5.21.** Comparison of the numerical results and various correlations for $\phi = 0.65$.

respectively. The agreement between the proposal in Saito and de Lemos (2006), other correlations in the literature and the numerical simulations stimulates further investigation on this subject, contributing towards the building of a more general expression for the interfacial heat transfer coefficient for porous media.

## 5.4. MACROSCOPIC BUOYANCY EFFECTS

### 5.4.1. Mean Flow

Now focusing attention only to buoyancy effects, application of the volume average procedure to the last term of (2.32) leads to:

$$\langle \rho \mathbf{g} \beta (\overline{T} - T_{\text{ref}}) \rangle^{\text{v}} = \frac{\Delta V_{\text{f}}}{\Delta V} \frac{1}{\Delta V_{\text{f}}} \int_{\Delta V_{\text{f}}} \rho \mathbf{g} \beta (\overline{T} - T_{\text{ref}}) \, \mathrm{d}V \tag{5.89}$$

Expanding the lhs of (5.89) in light of (2.19), the buoyancy term becomes:

$$\langle \rho \mathbf{g} \beta (\overline{T} - T_{\text{ref}}) \rangle^{\text{v}} = \rho \beta_{\phi} \mathbf{g} \phi (\langle \overline{T} \rangle^{\text{i}} - T_{\text{ref}}) + \underbrace{\rho \mathbf{g} \beta \phi \langle {}^{\text{i}}\overline{T} \rangle^{\text{i}}}_{=0} \tag{5.90}$$

where the second term on the rhs is null since $\langle {}^{\text{i}}\varphi \rangle^{\text{i}} = 0$. Here, the coefficient $\beta_{\phi}$ is the macroscopic thermal expansion coefficient. Assuming that gravity is constant over the REV, an expression for it based on Equation (5.90) is given as:

$$\beta_{\phi} = \frac{\langle \rho \beta (\overline{T} - T_{\text{ref}}) \rangle^{\text{v}}}{\rho \phi (\langle \overline{T} \rangle^{\text{i}} - T_{\text{ref}})} \tag{5.91}$$

Including Equation (5.90) into Equation (4.8), the macroscopic time-mean Navier-Stokes (NS) equation for an incompressible fluid with constant properties is given as:

$$\begin{aligned} \rho \nabla \cdot \left( \frac{\overline{\mathbf{u}}_{\text{D}} \overline{\mathbf{u}}_{\text{D}}}{\phi} \right) = & - \nabla (\phi \langle \overline{p} \rangle^{\text{i}}) + \mu \nabla^2 \overline{\mathbf{u}}_{\text{D}} + \nabla \cdot (-\rho \phi \langle \overline{\mathbf{u}' \mathbf{u}'} \rangle^{\text{i}}) \\ & + \rho \beta_{\phi} \mathbf{g} \phi (\langle \overline{T} \rangle^{\text{i}} - T_{\text{ref}}) - \left[ \frac{\mu \phi}{K} \overline{\mathbf{u}}_{\text{D}} + \frac{c_{\text{F}} \phi \rho |\overline{\mathbf{u}}_{\text{D}}| \overline{\mathbf{u}}_{\text{D}}}{\sqrt{K}} \right] \end{aligned} \tag{5.92}$$

### 5.4.2. Turbulent Field

In de Lemos and Braga (2003), the development in Pedras and de Lemos (2001a) was extended to include the buoyancy production rate in the turbulence model equations. For clear flows, the buoyancy contribution to the $k$ equation is given by Equation (2.37). Applying the volume average operator to that term, one has:

$$\langle G_{\text{T}} \rangle^{\text{v}} = G_{\beta}^{\text{i}} = \langle -\rho \beta \mathbf{g} \cdot \overline{\mathbf{u}' T'} \rangle^{\text{v}} = -\rho \beta_{\phi}^{\text{k}} \phi \mathbf{g} \cdot \langle \overline{\mathbf{u}' T_{\text{f}}'} \rangle^{\text{i}} \tag{5.93}$$

where the coefficient $\beta_{\phi}^{\text{k}}$, for a constant value of $\mathbf{g}$ within the REV, is given by $\beta_{\phi}^{\text{k}} = \dfrac{\langle \rho \beta \overline{\mathbf{u}' T'} \rangle^{\text{v}}}{\rho \phi \langle \overline{\mathbf{u}' T_{\text{f}}'} \rangle^{\text{i}}}$, which, in turn, is not necessarily equal to $\beta_{\phi}$ given by Equation (5.91).

However, for the sake of simplicity and in the absence of better information, one can make use of the assumption $\beta_\phi^{\mathrm{k}} = \beta_\phi = \beta$. Further, expanding the rhs of Equation (5.93) in light of Equations (3.4), one has:

$$
\begin{aligned}
&-\rho\beta_\phi^{\mathrm{k}}\phi\mathbf{g}\cdot\langle\overline{\mathbf{u}'T_{\mathrm{f}}'}\rangle^{\mathrm{i}} \\
&= -\rho\beta_\phi^{\mathrm{k}}\phi\mathbf{g}\cdot\langle\overline{(\langle\mathbf{u}'\rangle^{\mathrm{i}}+{}^{\mathrm{i}}\mathbf{u}')(\langle T_{\mathrm{f}}'\rangle^{\mathrm{i}}+{}^{\mathrm{i}}T_{\mathrm{f}}')}\rangle^{\mathrm{i}} \\
&= -\rho\beta_\phi^{\mathrm{k}}\phi\mathbf{g}\cdot\left(\langle\overline{\langle\mathbf{u}'\rangle^{\mathrm{i}}\langle T_{\mathrm{f}}'\rangle^{\mathrm{i}}}\rangle^{\mathrm{i}} + \langle\overline{{}^{\mathrm{i}}\mathbf{u}'{}^{\mathrm{i}}T_{\mathrm{f}}'}\rangle^{\mathrm{i}} + \langle\overline{\langle\mathbf{u}'\rangle^{\mathrm{i}}\,{}^{\mathrm{i}}T_{\mathrm{f}}'}\rangle^{\mathrm{i}} + \langle\overline{{}^{\mathrm{i}}\mathbf{u}'\langle T_{\mathrm{f}}'\rangle^{\mathrm{i}}}\rangle^{\mathrm{i}}\right) \\
&= -\rho\beta_\phi^{\mathrm{k}}\phi\mathbf{g}\cdot\left(\underbrace{\overline{\langle\mathbf{u}\rangle^{\mathrm{i}'}\langle T_{\mathrm{f}}\rangle^{\mathrm{i}'}}}_{\mathbf{1}} + \underbrace{\langle\overline{{}^{\mathrm{i}}\mathbf{u}'{}^{\mathrm{i}}T_{\mathrm{f}}'}\rangle^{\mathrm{i}}}_{\mathbf{2}} + \underbrace{\overline{\langle\mathbf{u}'\rangle^{\mathrm{i}}\langle{}^{\mathrm{i}}T_{\mathrm{f}}'\rangle^{\mathrm{i}}}}_{=0} + \underbrace{\overline{\langle{}^{\mathrm{i}}\mathbf{u}'\rangle^{\mathrm{i}}\langle T_{\mathrm{f}}'\rangle^{\mathrm{i}}}}_{=0}\right)
\end{aligned}
\tag{5.94}
$$

The last two terms on last expression of the rhs of (5.94) are null since $\langle{}^{\mathrm{i}}T_{\mathrm{f}}'\rangle^{\mathrm{i}} = 0$ and $\langle{}^{\mathrm{i}}\mathbf{u}'\rangle^{\mathrm{i}} = 0$. In addition, the following physical significance can be inferred to the two remaining:

1 *Buoyancy generation/destruction rate of $\langle k\rangle^{\mathrm{i}}$ due to macroscopic time fluctuations of u and T*: This term is also present in turbulent flow in clear (non-obstructed) domains and represents an exchange between the energy associated with the macroscopic turbulent motion and potential energy. In stable stratification, this term damps turbulence by being of negative value whereas the potential energy of the system is increased. On the other hand, in unstable stratification, it enhances $\langle k\rangle^{\mathrm{i}}$ at the expense of potential energy.
2 *Buoyancy generation/destruction rate of $\langle k\rangle^{\mathrm{i}}$ due to turbulent buoyant dispersion*: Extra generation/destruction rate due to time fluctuations and spatial deviations of both **u** and T. This term might be interpreted as an additional source/sink of turbulence kinetic energy due to the fact that time fluctuations of local velocities and temperatures present a spatial deviation in relation to their macroscopic value. Then, additional exchange between turbulent kinetic energy and potential energy in systems may occur due to the presence of a porous matrix.

A model for Equation (5.94) is still needed in order to solve an equation for $\langle k\rangle^{\mathrm{i}}$, which is a necessary information when computing $\mu_{t_\phi}$ using Equation (4.20). Consequently, terms 1 and 2 above have to be modeled as a function of average temperature, $\langle\overline{T}\rangle^{\mathrm{i}}$. To accomplish this, a gradient-type diffusion model is used, in the following form:

- Buoyancy generation of $\langle k\rangle^{\mathrm{i}}$ due to turbulent fluctuations:

$$
-\rho\beta_\phi^{\mathrm{k}}\phi\mathbf{g}\cdot\overline{\langle\mathbf{u}\rangle^{\mathrm{i}'}\langle T_{\mathrm{f}}\rangle^{\mathrm{i}'}} = \mathbf{B}_{\mathrm{t}}\cdot\nabla\langle\overline{T}\rangle^{\mathrm{i}} \tag{5.95}
$$

- Buoyancy generation of $\langle k\rangle^{\mathrm{i}}$ due to turbulent buoyant dispersion:

$$-\rho\beta_{\phi}^{\mathrm{k}}\phi\mathbf{g}\cdot\langle\overline{{}^{\mathrm{i}}\mathbf{u}'^{\mathrm{i}}T_{\mathrm{f}}'}\rangle^{\mathrm{i}} = \mathbf{B}_{\mathrm{disp,t}}\cdot\nabla\langle\overline{T}\rangle^{\mathrm{i}} \tag{5.96}$$

The buoyancy coefficients seem above, namely $\mathbf{B}_{\mathrm{t}}$ and $\mathbf{B}_{\mathrm{disp,t}}$, are modeled here through the eddy-diffusivity concept, similar to the work in Nakayama and Kuwahara (1999). It should be noticed that these terms arise only if the flow is turbulent and if buoyancy is of importance.

Then using an expression similar to Equation (5.36), the macroscopic buoyancy generation of $k$ can be modeled as,

$$G_{\beta}^{\mathrm{i}} = -\rho\beta_{\phi}^{\mathrm{k}}\phi\mathbf{g}\cdot\langle\overline{\mathbf{u}'T_{\mathrm{f}}'}\rangle^{\mathrm{i}} = \beta_{\phi}^{\mathrm{k}}\phi\frac{\mu_{\mathrm{t}_{\phi}}}{\sigma_{\mathrm{t}_{\phi}}}\mathbf{g}\cdot\nabla\langle\overline{T}\rangle^{\mathrm{i}} = \mathbf{B}_{\mathrm{eff}}\cdot\nabla\langle\overline{T}\rangle^{\mathrm{i}} \tag{5.97}$$

where $\mu_{\mathrm{t}_{\phi}}$ and $\sigma_{\mathrm{t}_{\phi}}$ have been defined before and the two coefficients $\mathbf{B}_{\mathrm{t}}$ and $\mathbf{B}_{\mathrm{disp,t}}$ are expressed as:

$$\mathbf{B}_{\mathrm{t}} + \mathbf{B}_{\mathrm{disp,t}} = \mathbf{B}_{\mathrm{eff}} = \beta_{\phi}^{\mathrm{k}}\phi\frac{\mu_{\mathrm{t}_{\phi}}}{\sigma_{\mathrm{t}_{\phi}}}\mathbf{g} \tag{5.98}$$

Final transport equations for $\langle k\rangle^{\mathrm{i}} = \langle\overline{\mathbf{u}'\cdot\mathbf{u}'}\rangle^{\mathrm{i}}/2$ and $\langle\varepsilon\rangle^{\mathrm{i}} = \mu\langle\overline{\nabla\mathbf{u}':(\nabla\mathbf{u}')^{\mathrm{T}}}\rangle^{\mathrm{i}}/\rho$, in their so-called high Reynolds number form, as proposed in Pedras and de Lemos (2001a), can now include the buoyancy generation terms seen above as:

$$\rho\nabla\cdot(\overline{\mathbf{u}}_{\mathrm{D}}\langle k\rangle^{\mathrm{i}}) = \nabla\cdot\left[\left(\mu+\frac{\mu_{\mathrm{t}_{\phi}}}{\sigma_k}\right)\nabla(\phi\langle k\rangle^{\mathrm{i}})\right] + P^{\mathrm{i}} + G^{\mathrm{i}} + G_{\beta}^{\mathrm{i}} - \rho\phi\langle\varepsilon\rangle^{\mathrm{i}} \tag{5.99}$$

$$\rho\nabla\cdot(\overline{\mathbf{u}}_{\mathrm{D}}\langle\varepsilon\rangle^{\mathrm{i}}) = \nabla\cdot\left[\left(\mu+\frac{\mu_{\mathrm{t}_{\phi}}}{\sigma_{\varepsilon}}\right)\nabla\left(\phi\langle\varepsilon\rangle^{\mathrm{i}}\right)\right] + \frac{\langle\varepsilon\rangle^{\mathrm{i}}}{\langle k\rangle^{\mathrm{i}}}\left[c_1P^{\mathrm{i}} + c_2G^{\mathrm{i}} + c_1c_3G_{\beta}^{\mathrm{i}} - c_2\rho\phi\langle\varepsilon\rangle^{\mathrm{i}}\right] \tag{5.100}$$

where $c_1$, $c_2$, $c_3$ and $c_k$ are constants, $P^{\mathrm{i}} = -\rho\langle\overline{\mathbf{u}'\mathbf{u}'}\rangle^{\mathrm{i}}:\nabla\overline{\mathbf{u}}_{\mathrm{D}}$ is the production rate of $\langle k\rangle^{\mathrm{i}}$ due to gradients of $\overline{\mathbf{u}}_{\mathrm{D}}$, $G^{\mathrm{i}} = c_k\rho\dfrac{\phi\langle k\rangle^{\mathrm{i}}|\overline{\mathbf{u}}_{\mathrm{D}}|}{\sqrt{K}}$ is the generation rate of the intrinsic average of $\langle k\rangle^{\mathrm{i}}$ due to the action of the porous matrix and $G_{\beta}^{\mathrm{i}} = \mathbf{B}_{\mathrm{eff}}\cdot\nabla\langle\overline{T}\rangle^{\mathrm{i}}$ is the generation of $\langle k\rangle^{\mathrm{i}}$ due to buoyancy.

Chapter 6

# Turbulent Mass Transport

An extension to mass transport follows the well-established analogy between heat and mass transfer existing also for clear flow.

## 6.1. MEAN FIELD

Mass transfer analysis using the double-decomposition concept follows similar steps applied to heat transfer as seen in Chapter 5. To apply the volume average to the mass transport Equation (2.5), we have,

$$m_\ell = \langle m_\ell \rangle^{\mathrm{i}} + {}^{\mathrm{i}}m_\ell; \quad \mathbf{u} = \langle \mathbf{u} \rangle^{\mathrm{i}} + {}^{\mathrm{i}}\mathbf{u} \tag{6.1}$$

which on substituting into Equation (2.5) and constitutive relation (2.6), we obtain:

$$\frac{\partial(\langle m_\ell \rangle^{\mathrm{i}} + {}^{\mathrm{i}}m_\ell)}{\partial t} + \nabla \cdot [(\langle \mathbf{u} \rangle^{\mathrm{i}} + {}^{\mathrm{i}}\mathbf{u})(\langle m_\ell \rangle^{\mathrm{i}} + {}^{\mathrm{i}}m_\ell)] = \langle R_\ell \rangle^{\mathrm{i}} + {}^{\mathrm{i}}R_\ell + D_\ell \nabla^2(\langle m_\ell \rangle^{\mathrm{i}} + {}^{\mathrm{i}}m_\ell) \tag{6.2}$$

where the mixture density $\rho$ and the coefficient $D_\ell$ in Equation (6.2) have been assumed to be constant. Expanding the convection term and taking the volume average, with the help of Equations (2.20), (2.21) and (2.22), we have,

$$\begin{aligned} &\frac{\partial \phi \langle \langle m_\ell \rangle^{\mathrm{i}} + {}^{\mathrm{i}}m_\ell \rangle^{\mathrm{i}}}{\partial t} + \nabla \cdot [\phi \langle (\langle \mathbf{u} \rangle^{\mathrm{i}} \langle m_\ell \rangle^{\mathrm{i}} + {}^{\mathrm{i}}\mathbf{u} \langle m_\ell \rangle^{\mathrm{i}} + \langle \mathbf{u} \rangle^{\mathrm{i}}\, {}^{\mathrm{i}}m_\ell + {}^{\mathrm{i}}\mathbf{u}\, {}^{\mathrm{i}}m_\ell) \rangle^{\mathrm{i}}] \\ &= \phi \langle \langle R_\ell \rangle^{\mathrm{i}} + {}^{\mathrm{i}}R_\ell \rangle^{\mathrm{i}} + D_\ell \langle \nabla^2 \phi (\langle m_\ell \rangle^{\mathrm{i}} + {}^{\mathrm{i}}m_\ell) \rangle^{\mathrm{i}} \end{aligned} \tag{6.3}$$

or

$$\frac{\partial \phi \langle m_\ell \rangle^{\mathrm{i}}}{\partial t} + \nabla \cdot [\phi(\langle \mathbf{u} \rangle^{\mathrm{i}} \langle m_\ell \rangle^{\mathrm{i}} + \langle {}^{\mathrm{i}}\mathbf{u}\, {}^{\mathrm{i}}m_\ell \rangle^{\mathrm{i}})] = \phi \langle R_\ell \rangle^{\mathrm{i}} + D_\ell \nabla^2(\phi \langle m_\ell \rangle^{\mathrm{i}}) \tag{6.4}$$

where the third term on the lhs of Equation (6.4) appears in classical analysis of mass transport in porous media, e.g. Bear and Bachmat (1967), Bear (1972), and is known as the *mass dispersion*.

In order to apply the time average to Equation (6.4), we define the intrinsic volume average as follows:

$$\langle m_\ell \rangle^{\mathrm{i}} = \overline{\langle m_\ell \rangle^{\mathrm{i}}} + \langle m_\ell \rangle^{\mathrm{i}\prime}; \quad \langle \mathbf{u} \rangle^{\mathrm{i}} = \overline{\langle \mathbf{u} \rangle^{\mathrm{i}}} + \langle \mathbf{u} \rangle^{\mathrm{i}\prime} \tag{6.5}$$

which on substituting into Equation (6.4) and taking the time average, gives:

$$\frac{\partial \phi \overline{\langle m_\ell \rangle^{\mathrm{i}}}}{\partial t} + \nabla \cdot \phi (\overline{\langle \mathbf{u} \rangle^{\mathrm{i}}} \, \overline{\langle m_\ell \rangle^{\mathrm{i}}} + \overline{\langle \mathbf{u} \rangle^{\mathrm{i}'} \langle m_\ell \rangle^{\mathrm{i}'}} + \overline{\langle {}^{\mathrm{i}}\mathbf{u} \, {}^{\mathrm{i}}m_\ell \rangle^{\mathrm{i}}}) = \phi \overline{\langle R_\ell \rangle^{\mathrm{i}}} + D_\ell \nabla^2 (\phi \overline{\langle m_\ell \rangle^{\mathrm{i}}}) \quad (6.6)$$

Equation (6.6) is the macroscopic mass transfer equation for the species $\ell$ in the porous matrix taking first the volume average followed by the time average.

Another route to reach a macroscopic transport equation for turbulent flow is to invert the order of application of the same average operators applied over Equation (2.5). Therefore, starting now with the time average, one needs to consider time decomposition first:

$$m_\ell = \overline{m}_\ell + m'_\ell \quad (6.7)$$

$$\mathbf{u} = \overline{\mathbf{u}} + \mathbf{u}' \quad (6.8)$$

Substituting Equations (6.7) and (6.8) into Equation (2.5) one has:

$$\frac{\partial (\overline{m}_\ell + m'_\ell)}{\partial t} + \nabla \cdot [(\overline{\mathbf{u}} + \mathbf{u}')(\overline{m}_\ell + m'_\ell)] = \overline{R}_\ell + R'_\ell + D_\ell \nabla^2 (\overline{m}_\ell + m'_\ell) \quad (6.9)$$

where again the mixture density $\rho$ and the diffusion coefficient $D_\ell$ were kept constants. Applying time average to Equation (6.9), one obtains:

$$\overline{\frac{\partial (\overline{m}_\ell + m'_\ell)}{\partial t}} + \overline{\nabla \cdot (\overline{\mathbf{u}}\,\overline{m}_\ell + \overline{\mathbf{u}} m'_\ell + \mathbf{u}' \overline{m}_\ell + \mathbf{u}' m'_\ell)} = \overline{\overline{R}_\ell + R'_\ell} + D_\ell \overline{\nabla^2 (\overline{m}_\ell + m'_\ell)} \quad (6.10)$$

or

$$\frac{\partial \overline{m}_\ell}{\partial t} + \nabla \cdot (\overline{\mathbf{u}}\,\overline{m}_\ell + \overline{\mathbf{u}' m'_\ell}) = \overline{R}_\ell + D_\ell \nabla^2 \overline{m}_\ell \quad (6.11)$$

The second term on the lhs of Equation (6.11) is known as *turbulent mass flux* (divided by $\rho$). It requires a model for closure of the mathematical problem.

Further, in order to apply the volume average to Equation (6.11), one must first define the spatial deviations with respect to the time averages, given by:

$$\overline{m}_\ell = \langle \overline{m}_\ell \rangle^{\mathrm{i}} + {}^{\mathrm{i}}\overline{m}_\ell \quad (6.12)$$

$$\overline{\mathbf{u}} = \langle \overline{\mathbf{u}} \rangle^{\mathrm{i}} + {}^{\mathrm{i}}\overline{\mathbf{u}} \quad (6.13)$$

Substituting now Equations (6.12) and (6.13) into Equation (6.11) and performing the volume average operation, one has:

$$\frac{\partial \phi \langle \overline{m}_\ell \rangle^{\mathrm{i}}}{\partial t} + \nabla \cdot \phi (\langle \overline{\mathbf{u}} \rangle^{\mathrm{i}} \langle \overline{m}_\ell \rangle^{\mathrm{i}} + \langle {}^{\mathrm{i}}\overline{\mathbf{u}}\, {}^{\mathrm{i}}\overline{m}_\ell \rangle^{\mathrm{i}} + \langle \overline{\mathbf{u}' m'_\ell} \rangle^{\mathrm{i}}) = \phi \langle \overline{R}_\ell \rangle^{\mathrm{i}} + D_\ell \nabla^2 \phi \langle \overline{m}_\ell \rangle^{\mathrm{i}} \quad (6.14)$$

Equation (6.14) is the macroscopic mass diffusion equation taking first the time average followed by the volume average operator.

It is interesting to observe that Equation (6.14), obtained through the second procedure (time–volume average), is equivalent to Equation (6.6) as will be shown below.

## 6.2. TURBULENT MASS DISPERSION

Using now Equation (3.13), the fourth term on the lhs of Equation (6.6) can be expanded as follows:

$$\overline{\langle {}^{\mathrm{i}}\mathbf{u}{}^{\mathrm{i}}m_\ell \rangle^{\mathrm{i}}} = \overline{\langle ({}^{\mathrm{i}}\overline{\mathbf{u}} + {}^{\mathrm{i}}\mathbf{u}')({}^{\mathrm{i}}\overline{m}_\ell + {}^{\mathrm{i}}m'_\ell)\rangle^{\mathrm{i}}} = \langle {}^{\mathrm{i}}\overline{\mathbf{u}}\,{}^{\mathrm{i}}\overline{m}_\ell \rangle^{\mathrm{i}} + \overline{\langle {}^{\mathrm{i}}\mathbf{u}'{}^{\mathrm{i}}m'_\ell \rangle^{\mathrm{i}}} \tag{6.15}$$

If we substitute this into Equation (6.6), the convection term becomes:

$$\nabla\cdot(\overline{\phi\langle \mathbf{u}m_\ell\rangle^{\mathrm{i}}}) = \nabla\cdot\{\phi(\langle\overline{\mathbf{u}}\rangle^{\mathrm{i}}\langle\overline{m}_\ell\rangle^{\mathrm{i}} + \langle {}^{\mathrm{i}}\overline{\mathbf{u}}\,{}^{\mathrm{i}}\overline{m}_\ell\rangle^{\mathrm{i}} + \overline{\langle\mathbf{u}\rangle^{\mathrm{i}'}\langle m_\ell\rangle^{\mathrm{i}'}} + \langle\overline{{}^{\mathrm{i}}\mathbf{u}'\,{}^{\mathrm{i}}m'_\ell}\rangle^{\mathrm{i}})\} \tag{6.16}$$

Likewise, applying again Equation (3.13) to the fourth term on the lhs of Equation (6.14), we obtain:

$$\langle\overline{\mathbf{u}'m'_\ell}\rangle^{\mathrm{i}} = \langle\overline{((\langle\mathbf{u}'\rangle^{\mathrm{i}} + {}^{\mathrm{i}}\mathbf{u}')(\langle m'_\ell\rangle^{\mathrm{i}} + {}^{\mathrm{i}}m'_\ell))}\rangle^{\mathrm{i}} = \overline{\langle\mathbf{u}'\rangle^{\mathrm{i}}\langle m'_\ell\rangle^{\mathrm{i}}} + \langle\overline{{}^{\mathrm{i}}\mathbf{u}'{}^{\mathrm{i}}m'_\ell}\rangle^{\mathrm{i}} \tag{6.17}$$

which on substituting back into Equation (6.14) gives for the same convection term:

$$\nabla\cdot(\overline{\phi\langle \mathbf{u}m_\ell\rangle^{\mathrm{i}}}) = \nabla\cdot\{\phi(\underset{\substack{\uparrow\\ \mathbf{1}}}{\langle\overline{\mathbf{u}}\rangle^{\mathrm{i}}\langle\overline{m}_\ell\rangle^{\mathrm{i}}} + \underset{\substack{\uparrow\\ \mathbf{2}}}{\langle {}^{\mathrm{i}}\overline{\mathbf{u}}{}^{\mathrm{i}}\overline{m}_\ell\rangle^{\mathrm{i}}} + \underset{\substack{\uparrow\\ \mathbf{3}}}{\overline{\langle\mathbf{u}'\rangle^{\mathrm{i}}\langle m'_\ell\rangle^{\mathrm{i}}}} + \underset{\substack{\uparrow\\ \mathbf{4}}}{\langle\overline{{}^{\mathrm{i}}\mathbf{u}'{}^{\mathrm{i}}m'_\ell}\rangle^{\mathrm{i}}})\} \tag{6.18}$$

Here also, the order of application of both the averaging operators to the entire equation is immaterial and the proof is left to the interested reader. Further, the four terms on the right of Equation (6.18) could be given the following physical significance (multiplied by $\rho$):

1 *Convective mass flux* based on macroscopic time mean velocity and mass fraction.
2 *Mass dispersion* associated with deviations of the microscopic time mean velocity and mass fraction. Note that this term is also present when analyzing the laminar mass transfer in porous media, but it does not exist if a volume average is not performed.
3 *Turbulent mass flux* due to the fluctuating components of both macroscopic velocity and mass fraction. This term is also present in turbulent flow in clear (non-porous)

domains. It is not defined for laminar flow in porous media where time fluctuations are not considered.

4 *Turbulent mass dispersion* in a porous medium due to both time fluctuations and spatial deviations of both microscopic velocity and mass fraction.

Thus, the macroscopic mass transport equation for an incompressible flow in a rigid, homogeneous and saturated porous medium can be written as follows:

$$\frac{\partial \phi \langle \overline{m}_\ell \rangle^{\mathrm{i}}}{\partial t} + \nabla \cdot \phi (\langle \overline{\mathbf{u}} \rangle^{\mathrm{i}} \langle \overline{m}_\ell \rangle^{\mathrm{i}} + \langle {}^{\mathrm{i}}\overline{\mathbf{u}}\, {}^{\mathrm{i}}\overline{m}_\ell \rangle^{\mathrm{i}} + \overline{\langle \mathbf{u}' \rangle^{\mathrm{i}} \langle m'_\ell \rangle} + \langle \overline{{}^{\mathrm{i}}\mathbf{u}'\, {}^{\mathrm{i}}m'_\ell} \rangle^{\mathrm{i}})$$
$$= \phi \langle \overline{R}_\ell \rangle^{\mathrm{i}} + D_\ell \nabla^2 (\phi \langle \overline{m}_\ell \rangle^{\mathrm{i}}) \tag{6.19}$$

or in its equivalent form (see Equation (3.10)):

$$\frac{\partial \phi \overline{\langle m_\ell \rangle^{\mathrm{i}}}}{\partial t} + \nabla \cdot \phi (\langle \overline{\mathbf{u}} \rangle^{\mathrm{i}} \langle \overline{m}_\ell \rangle^{\mathrm{i}} + \langle {}^{\mathrm{i}}\overline{\mathbf{u}}\, {}^{\mathrm{i}}\overline{m}_\ell \rangle^{\mathrm{i}} + \overline{\langle \mathbf{u} \rangle^{\mathrm{i}\prime} \langle m_\ell \rangle^{\mathrm{i}\prime}} + \overline{\langle {}^{\mathrm{i}}\mathbf{u}'\, {}^{\mathrm{i}}m'_\ell \rangle^{\mathrm{i}}})$$
$$= \phi \overline{\langle R_\ell \rangle^{\mathrm{i}}} + D_\ell \nabla^2 (\phi \overline{\langle m_\ell \rangle^{\mathrm{i}}}) \tag{6.20}$$

## 6.3. MACROSCOPIC TRANSPORT MODELS

All terms in Equation (6.19) or (6.20) need to be represented or modeled as functions of the macroscopic mass fraction $\langle \overline{m}_\ell \rangle^{\mathrm{i}}$. Using gradient-type diffusion models, the proposed forms for the different mechanisms are as follows:

*Mass dispersion*: Following the literature (Whitaker, 1966–1967; Bear and Bachmat, 1967; Bear, 1972), a time-mean version for a dispersion model is given as:

$$-\rho \langle {}^{\mathrm{i}}\overline{\mathbf{u}}\, {}^{\mathrm{i}}\overline{m}_\ell \rangle^{\mathrm{i}} = \rho \mathbf{D}_{\mathrm{disp}} \cdot \nabla \langle \overline{m}_\ell \rangle^{\mathrm{i}} \tag{6.21}$$

where the dispersion coefficient $\mathbf{D}_{\mathrm{disp}}$ is a second order tensor.

*Turbulent mass flux*:

$$-\rho \overline{\langle \mathbf{u}' \rangle^{\mathrm{i}} \langle m'_\ell \rangle^{\mathrm{i}}} = -\rho \overline{\langle \mathbf{u} \rangle^{\mathrm{i}\prime} \langle m_\ell \rangle^{\mathrm{i}\prime}} = \rho \mathbf{D}_{\mathrm{t}} \cdot \nabla \langle \overline{m}_\ell \rangle^{\mathrm{i}} \tag{6.22}$$

*Turbulent mass dispersion*:

$$-\rho \langle \overline{{}^{\mathrm{i}}\mathbf{u}'\, {}^{\mathrm{i}}m'_\ell} \rangle^{\mathrm{i}} = \rho \mathbf{D}_{\mathrm{disp,t}} \cdot \nabla \langle \overline{m}_\ell \rangle^{\mathrm{i}} \tag{6.23}$$

The coefficients $\mathbf{D}_t$ and $\mathbf{D}_{disp,t}$ in Equations (6.22) and (6.23), respectively, will be combined as suggested by de Lemos and Mesquita (2003). Therefore, the two additional transport mechanisms, namely the turbulent mass flux and turbulent mass dispersion, can be added up so that a model for $\langle \overline{\mathbf{u}'m'_\ell} \rangle^i$ will be necessary for the closure of the mathematical problem.

Starting out from the time averaged mass transfer equation coupled with modeling for the turbulent mass flux through the eddy diffusivity concept, $D_{\ell,t} = \mu_t / \rho Sc_{\ell,t}$, one can write:

$$-\rho \overline{\mathbf{u}'m'_\ell} = \rho D_{\ell,t} \nabla \overline{m}_\ell = \frac{\mu_t}{Sc_{\ell,t}} \nabla \overline{m}_\ell \tag{6.24}$$

where the microscopic eddy viscosity in Equation (6.24), $\mu_t$, is given again by:

$$\mu_t = \rho c_\mu \frac{k^2}{\varepsilon} \tag{6.25}$$

and $Sc_{\ell,t}$ is the turbulent Schmidt number for the species $\ell$, which is taken here as a constant.

Applying volume average to the resulting equation, one obtains the macroscopic turbulent mass flux, given by:

$$-\rho \langle \overline{\mathbf{u}'m'_\ell} \rangle^i = \rho \langle D_{\ell,t} \rangle^i \nabla \langle \overline{m}_\ell \rangle^i = \rho (\mathbf{D}_t + \mathbf{D}_{disp,t}) \cdot \nabla \langle \overline{m}_\ell \rangle^i = \frac{\mu_{t_\phi}}{Sc_{\ell,t}} \nabla \langle \overline{m}_\ell \rangle^i \tag{6.26}$$

where the symbol $\mu_{t_\phi}$ expresses the macroscopic version of the eddy viscosity, given by:

$$\mu_{t_\phi} = \rho c_\mu \frac{\langle k \rangle^{i^2}}{\langle \varepsilon \rangle^i} \tag{6.27}$$

As mentioned above, following the same idea embodied in Equation (5.38), the overall macroscopic mass flux due to turbulence is taken as the sum of the turbulent mass flux and the turbulent mass dispersion appearing in either Equation (6.19) or (6.20). Further, the isotropic nature of Equation (6.27) suggests the equality:

$$\mathbf{D}_t + \mathbf{D}_{disp,t} = \frac{1}{\rho} \frac{\mu_{t_\phi}}{Sc_{\ell,t}} \mathbf{I} \tag{6.28}$$

Likewise, a constant value for the molecular diffusion coefficient in Equation (2.6) leads to a macroscopic diffusion coefficient, $\mathbf{D}_{diff}$, of the form:

$$\mathbf{D}_{diff} = \langle D_\ell \rangle^i \mathbf{I} = \frac{1}{\rho} \frac{\mu_\phi}{Sc_\ell} \mathbf{I} \tag{6.29}$$

Finally, using the Dupuit–Forchheimer relationship, $\mathbf{u}_{\mathrm{D}} = \langle \overline{\mathbf{u}} \rangle^{v} = \phi \langle \overline{\mathbf{u}} \rangle^{\mathrm{i}}$, in combination with Equations (6.21), (6.22) and (6.23), the final modeled form for a transport equation can be written as:

$$\frac{\partial \phi \langle \overline{m}_\ell \rangle^{\mathrm{i}}}{\partial t} + \nabla \cdot (\overline{\mathbf{u}}_{\mathrm{D}} \langle \overline{m}_\ell \rangle^{\mathrm{i}}) = \nabla \cdot \mathbf{D}_{\mathrm{eff}} \cdot \nabla(\phi \langle \overline{m}_\ell \rangle^{\mathrm{i}}) + \phi \langle \overline{R}_\ell \rangle^{\mathrm{i}}$$

where

$$\begin{aligned} \mathbf{D}_{\mathrm{eff}} = \mathbf{D}_{\mathrm{disp}} + \mathbf{D}_{\mathrm{diff}} + \mathbf{D}_{\mathrm{t}} + \mathbf{D}_{\mathrm{disp,t}} &= \mathbf{D}_{\mathrm{disp}} + \frac{1}{\rho}\left(\frac{\mu_\phi}{Sc_\ell} + \frac{\mu_{\mathrm{t}_\phi}}{Sc_{\ell,\mathrm{t}}}\right)\mathbf{I} \\ &= \mathbf{D}_{\mathrm{disp}} + \frac{1}{\rho}\left(\frac{\mu_{\phi,\mathrm{ef}}}{Sc_{\ell,\mathrm{ef}}}\right)\mathbf{I} \end{aligned} \tag{6.30}$$

In Equation (6.30), $Sc_{\ell,\mathrm{ef}}$ is the effective macroscopic turbulent Schmidt number given by:

$$Sc_{\ell,\mathrm{ef}} = \frac{\mu_{\mathrm{ef}_\phi}}{\frac{\mu_\phi}{Sc_\ell} + \frac{\mu_{\mathrm{t}_\phi}}{Sc_{\ell,\mathrm{t}}}} \tag{6.31}$$

## 6.4. MASS DISPERSION COEFFICIENTS

In order to calculate the dispersion coefficients in Equation (6.21), the methodology described in Mesquita and de Lemos (2004) is here reviewed. For steady laminar and turbulent flow regimes, we consider a macroscopically uniform flow, making an angle $\theta$ with the horizontal direction and meandering through an infinite number of square rods placed in a regular fashion, as illustrated in Figure 6.1. The macroscopic velocity and mass fraction fields are given by:

$$\langle \overline{\mathbf{u}} \rangle^{v} = |\langle \overline{\mathbf{u}} \rangle^{v}| (\cos\theta \vec{i} + \sin\theta \vec{j}) \tag{6.32}$$

$$\nabla \langle \overline{m}_\ell \rangle_x = \frac{\Delta m_\ell}{H} (\cos\theta \vec{i} + \sin\theta \vec{j}) \quad \text{Longitudinal component} \tag{6.33}$$

$$\nabla \langle \overline{m}_\ell \rangle_y = \frac{\Delta m_\ell}{H} (-\sin\theta \vec{i} + \cos\theta \vec{j}) \quad \text{Transversal component} \tag{6.34}$$

Following Kuwahara and Nakayama (1998), we integrate the microscopic Equation (6.21) for an incompressible fluid over an REV and obtain:

$$-\rho \left[ \frac{1}{V} \int_{A_{\mathrm{i}}} (\overline{\mathbf{u}} - \langle \overline{\mathbf{u}} \rangle^{\mathrm{i}}) (\overline{m}_\ell - \langle \overline{m}_\ell \rangle^{\mathrm{i}}) \, \mathrm{d}\mathbf{A} \right] = \rho \mathbf{D}_{\mathrm{disp}} \cdot \frac{1}{V} \int_{A_{\mathrm{i}}} \nabla \langle \overline{m}_\ell \rangle^{\mathrm{i}} \, \mathrm{d}\mathbf{A} \tag{6.35}$$

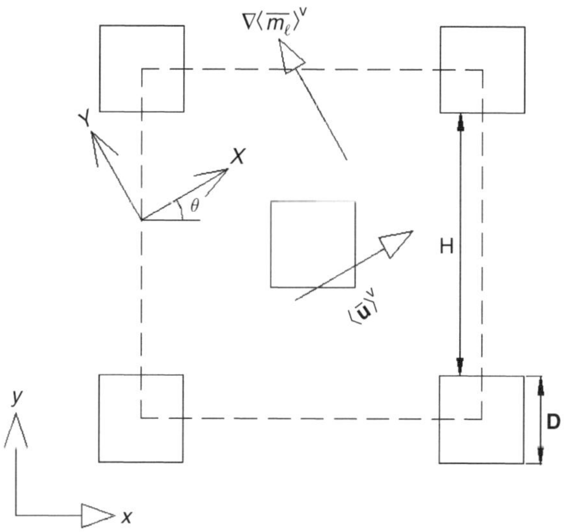


**Figure 6.1.** Physical model and its coordinate system.

where $A_{\mathrm{i}}$ is the total area in the fluid phase within a control volume $V$, while d$\mathbf{A}$ is its vector element pointing outward from the fluid side to solid side.

If the gradient of the average mass fraction is in the same direction of the macroscopic flow or transverse to it, only the diagonal components of $\mathbf{D}_{\mathrm{disp}}$ remain non-zero. In these conditions, Equation (6.35) renders, respectively, for the diagonal components of $\mathbf{D}_{\mathrm{disp}}$:

$$\left(\mathbf{D}_{\mathrm{disp}}\right)_{XX} = \frac{-\frac{1}{H^2}\int_0^H\int_0^H \left(\overline{m_\ell} - \langle\overline{m_\ell}\rangle^{\mathrm{i}}\right)\left(\overline{\mathbf{u}} - \langle\overline{\mathbf{u}}\rangle^{\mathrm{i}}\right)\cdot\left(\cos\theta\vec{i} + \sin\theta\vec{j}\right)\mathrm{d}x\,\mathrm{d}y}{\left(\dfrac{\Delta\langle\overline{m_\ell}\rangle_x}{H}\right)} \tag{6.36}$$

and

$$\left(\mathbf{D}_{\mathrm{disp}}\right)_{YY} = \frac{-\frac{1}{H^2}\int_0^H\int_0^H \left(\overline{m_\ell} - \langle\overline{m_\ell}\rangle^{\mathrm{i}}\right)\left(\overline{\mathbf{u}} - \langle\overline{\mathbf{u}}\rangle^{\mathrm{i}}\right)\cdot\left(-\sin\theta\vec{i} + \cos\theta\vec{j}\right)\mathrm{d}x\,\mathrm{d}y}{\left(\dfrac{\Delta\langle\overline{m_\ell}\rangle_y}{H}\right)} \tag{6.37}$$

Solutions of the flow and mass fraction equations inside the unit cell provide the velocity and mass fraction distributions necessary for the integrands in Equations (6.36) and (6.37).

These values are needed in order to calculate the dispersion components. Further, in (6.36) and (6.37) the gradients $\Delta\langle\overline{m_\ell}\rangle_x$ and $\Delta\langle\overline{m_\ell}\rangle_y$ can be calculated as presented in the next section. Also, values of $(\mathbf{D}_{\mathrm{disp}})_{XX}$ and $(\mathbf{D}_{\mathrm{disp}})_{YY}$ obtained with Equations (6.36) and (6.37), are usually adjusted according to the formulas:

$$\frac{(\mathbf{D}_{\mathrm{disp}})_{XX}}{D_\ell} = C \cdot Pe^{\mathrm{m}} \tag{6.38}$$

$$\frac{(\mathbf{D}_{\mathrm{disp}})_{YY}}{D_\ell} = C \cdot Pe^{\mathrm{m}} \tag{6.39}$$

and proposed as correlations in the literature with constants $C$ and $m$ depending on the case analyzed.

### 6.4.1. Imposed Mass Fraction Flux at Boundaries

One possibility to obtain a macroscopic mass fraction difference across the cell of Figure 6.1, either in the longitudinal direction for having $\Delta\langle\overline{m_\ell}\rangle_x$ to be used in Equation (6.36) or for calculating $\Delta\langle\overline{m_\ell}\rangle_y$ for applying in Equation (6.37), is to impose mass fraction fluxes at the north and south boundaries of the unit cell schematically shown in Figure 6.2. This same strategy has been used a by Pedras and de Lemos (2005) for heat transfer modeling (see Chapter 5, p. 55).

Figure 6.2a,b illustrates two types of periodic cell arrays commonly found in the literature (Brenner, 1980; Carbonel and Whitaker, 1984). The first model arrangement for an REV is named a *square* array (or in-line arrangement) and the second one is known as *staggered* array (or triangular arrangement). In Mesquita and de Lemos (2004) only the staggered array was considered. Their unit cell is shown in Figure 6.2c.

When the two fluxes "enter" into the cell (Figure 6.3a), $\Delta\langle\overline{m_\ell}\rangle_x$ is obtained. For the mass fraction flux entering from above (north boundary) and leaving at the south surface, the situation is analogous of having a uniform transverse mass fraction difference, $\Delta\langle\overline{m_\ell}\rangle_y$ across the $y$-direction (see Figure 6.3c).

In those cases $\Delta\langle\overline{m_\ell}\rangle_x$ and $\Delta\langle\overline{m_\ell}\rangle_y$ are no longer given values but rather a consequence of the imposed mass fraction fluxes (Neumann conditions) at the north and south boundaries. Their values are then calculated as,

$$\Delta\langle\overline{m_\ell}\rangle_x = \frac{1}{H}\int_{y=0}^{y=H} \left(\overline{m_\ell}_{x=H} - \overline{m_\ell}_{x=0}\right) \mathrm{d}y \tag{6.40}$$

for determining the longitudinal mass fraction dispersion, and

$$\Delta\langle\overline{m_\ell}\rangle_y = \frac{1}{H}\int_{x=0}^{x=H} \left(\overline{m_\ell}_{y=H} - \overline{m_\ell}_{y=0}\right) \mathrm{d}x \tag{6.41}$$

for obtaining the transversal mass fraction dispersion.

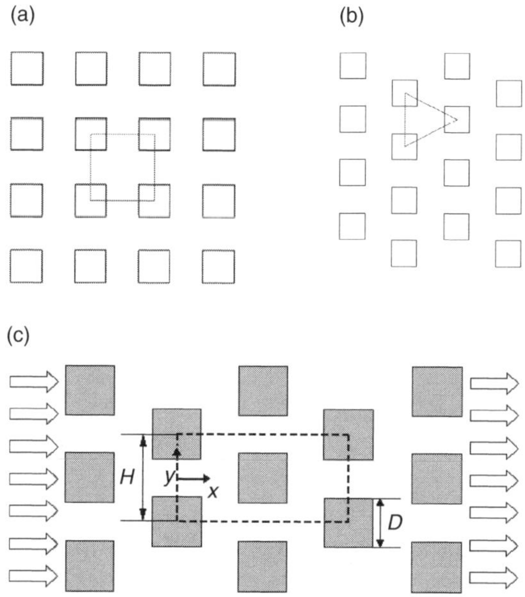


**Figure 6.2.** Porous media modeling, REV: (a) in-line array of square rods; (b) triangular array of square rods; (c) unit cell for determining $\mathbf{D}_{disp}$.

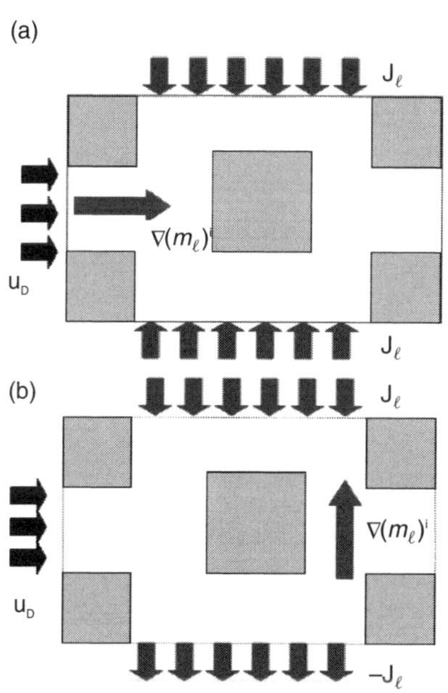


**Figure 6.3.** Neumann boundary conditions for mass fraction: (a) longitudinal gradient, (b) transverse gradient.

### 6.4.2. Numerical Results

In Mesquita and de Lemos (2004) a total of 48 runs were carried out, 30 for laminar flow and 18 for turbulent flow (6 with the low *Re* model and 12 with the high *Re* model). In all runs, a constant Schmidt number of 1.0 was assumed and the fluid was air ($\rho = 1.23\,\mathrm{kg/m^3}$ and $\mu = 1.79 \times 10^{-5}\,\mathrm{kg/m\,s}$).

Figure 6.4 shows computational grids corresponding to different values of cell porosity ($\phi = 0.65$, 0.75 and 0.90). For numerical accuracy, grid nodes are concentrated around the square obstacles representing the solid phase.

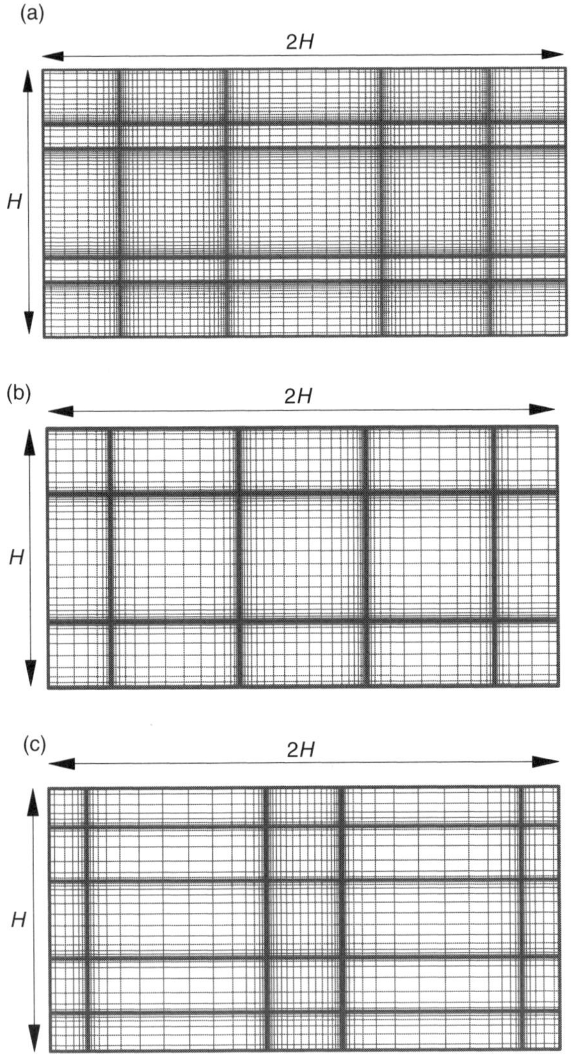


**Figure 6.4.** Computational grids: (a) $\phi = 0.65$, (b) $\phi = 0.75$ and (c) $\phi = 0.90$.

Mass concentration fields calculated with boundary conditions sketched in Figure 6.3a for $(\mathbf{D}_{\text{disp}})_{XX}$, namely imposed longitudinal mass concentration gradients (see Equation 6.36), are presented in Figure 6.5 for $Re_H = 1.0E+01$ (laminar regime), and in Figurc 6.6 for $Re_H = 1.0E \mid 06$ (turbulent regime). In Figures 6.5 and 6.6 the macroscopic mass concentration gradient $(\nabla\langle\overline{m_\ell}\rangle^{\text{v}})$ is in the same horizontal direction as the macroscopic flow $\langle\overline{\mathbf{u}}\rangle^{\text{v}}$. As will be shown below, the value of $(\mathbf{D}_{\text{disp}})_{XX}$ is dependent on the deviation fields of velocity and mass concentration within the fluid phase, i.e., it depends on the differences $(\overline{\mathbf{u}} - \langle\overline{\mathbf{u}}\rangle^{\text{i}})$ and $(\overline{m_\ell} - \langle\overline{m_\ell}\rangle^{\text{i}})$ within the flow. Consequently, it depends on the Reynolds number, $Re_H$. This behavior can be explained by observing that Equation 6.36 uses the deviation of both fields to calculate $(\mathbf{D}_{\text{disp}})_{XX}$. As such, by inspecting Figures 6.5 and 6.6, we can conclude that as the flow rate increases, the mass concentration distribution becomes more homogeneous due to enhancement of the convection strength and turbulence effects.

The boundary conditions sketched in Figure 6.3b are employed when calculating $(\mathbf{D}_{\text{disp}})_{YY}$. An imposed mass flux enters the cell at the north boundary, leaving the REV from below. Likewise here, once the velocity and mass concentration fields are solved, they are applied to the numerical determination of $(\mathbf{D}_{\text{disp}})_{YY}$ by means of Equation 6.37. Results in Figures 6.7 (laminar flow) and 6.8 (turbulent flow) assume $\nabla\langle\overline{m_\ell}\rangle^{\text{v}}$ as transversal to $\langle\overline{\mathbf{u}}\rangle^{\text{v}}$. As $Re_H$ increases, Figures 6.7 and 6.8 show the same behavior of Figures 6.5 and 6.6, i.e, as the flow increase, the mass concentration distribution becomes more homogeneous due to enhancement of convection and appearance of turbulence.

Integrated results in Mesquita and de Lemos (2004) are summarized in Tables 6.1, 6.2 and 6.3 for different values of the porosity $\phi$. In all runs, the medium permeability $K$ was calculated using the procedure adopted in Kuwahara and Nakayama (1998).

Figure 6.9 shows computed values for $(\mathbf{D}_{\text{disp}})_{XX}$ given by Mesquita and de Lemos (2004) compared with experimental data compiled by Han *et al.* (1985), which are also shown in Table 6.4. Values are plotted as a function of Peclet number based on the particle diameter, $Pe_{\text{p}}$, covering the range $0.65 < \phi < 0.90$. Results in Figure 6.9 show good agreement when compared with the data summarized in Table 6.4. For higher $Pe_{\text{p}}$, different values of $\phi$ yield results for $(\mathbf{D}_{\text{disp}})_{YY}$ which are nearly the same, indicating that for higher Peclet numbers the dependency of $(\mathbf{D}_{\text{disp}})_{XX}$ on $\phi$ is small.

For lower values of Peclet, however, $\left(\mathbf{D}_{\text{disp}}\right)_{XX}$ increases as porosity is reduced. This behavior is better visualized in Figure 6.10, which shows the effect of porosity on longitudinal dispersion.

Transverse component of the mass dispersion tensor $(\mathbf{D}_{\text{disp}})_{YY}$ obtained in Mesquita and de Lemos (2004) is shown in Figure 6.11, also plotted along with experimental data compiled by Han *et al.* (1985). The numerical results in Figure 6.11 are not as good as the ones presented for the case of the longitudinal component (Figure 6.9). Difference on the values of the $(\mathbf{D}_{\text{disp}})_{YY}$ between the results in Mesquita and de Lemos (2004) and

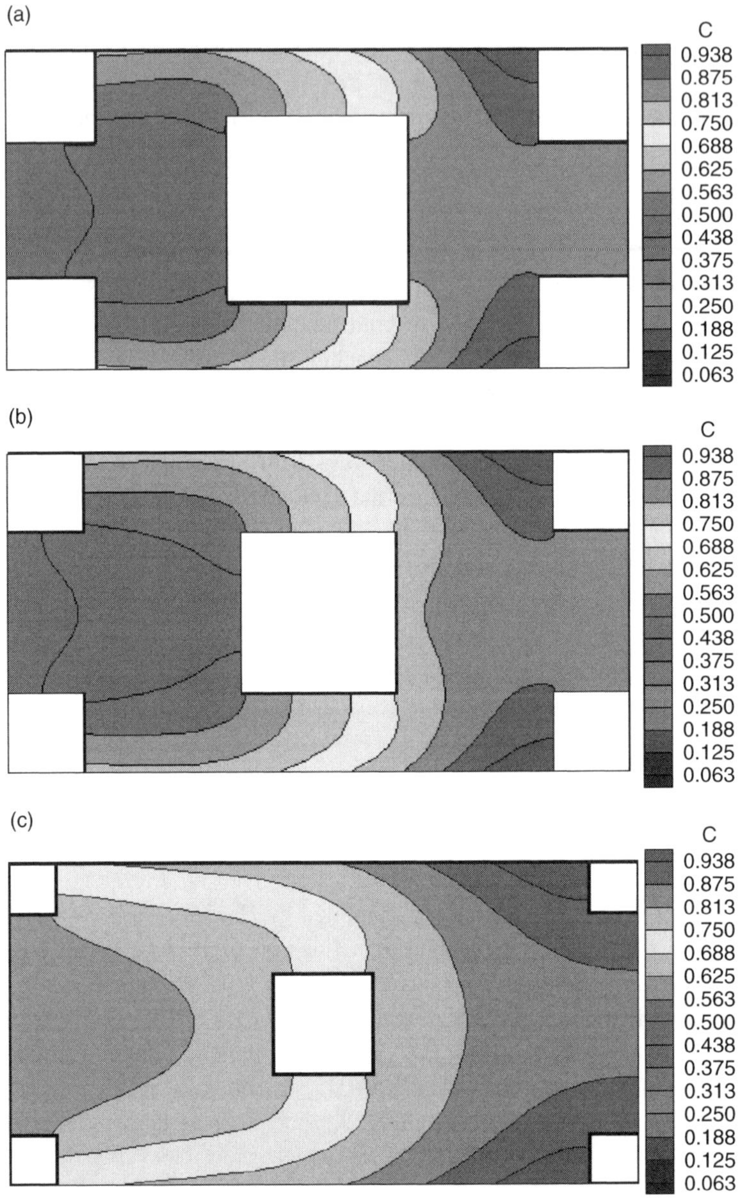


**Figure 6.5.** Mass concentration fields for unit cell calculated with longitudinal mass concentration gradients – (laminar regime – $Re_H = 1.0E+01$): (a) $\phi = 0.65$, (b) $\phi = 0.75$ and (c) $\phi = 0.90$.

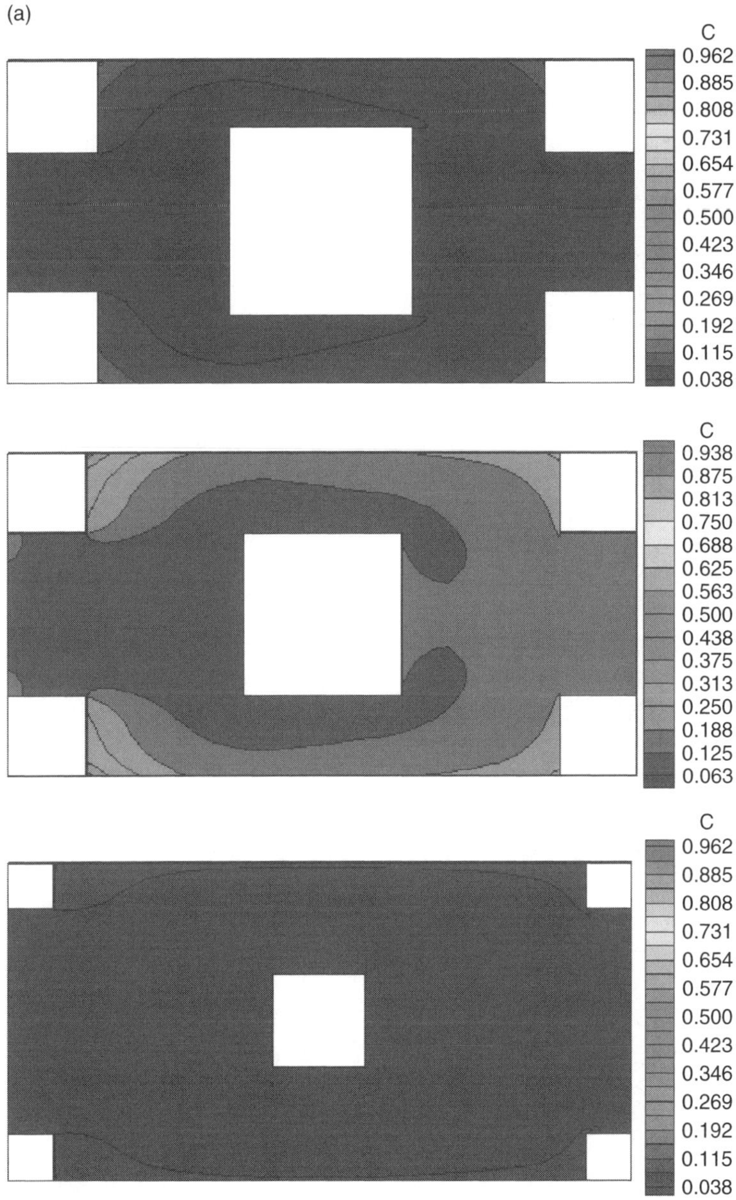


**Figure 6.6.** Mass concentration fields for unit cell calculated with longitudinal mass concentration gradients (turbulent regime – high Reynolds model – $Re_H = 1.0E+06$): (a) $\phi = 0.65$, (b) $\phi = 0.75$ and (c) $\phi = 0.90$.

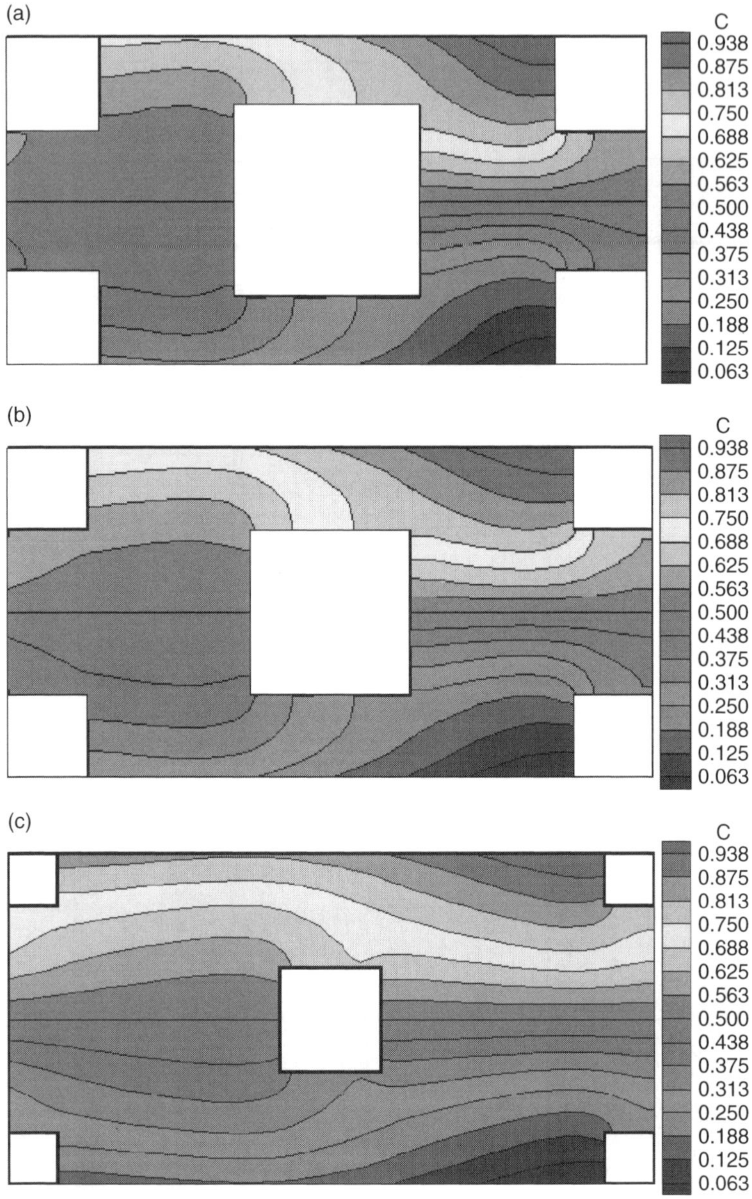


**Figure 6.7.** Mass concentration fields for unit cell calculated with transverse mass concentration gradients (laminar regime – $Re_H = 1.0E+01$): (a) $\phi = 0.65$, (b) $\phi = 0.75$ and (c) $\phi = 0.90$.

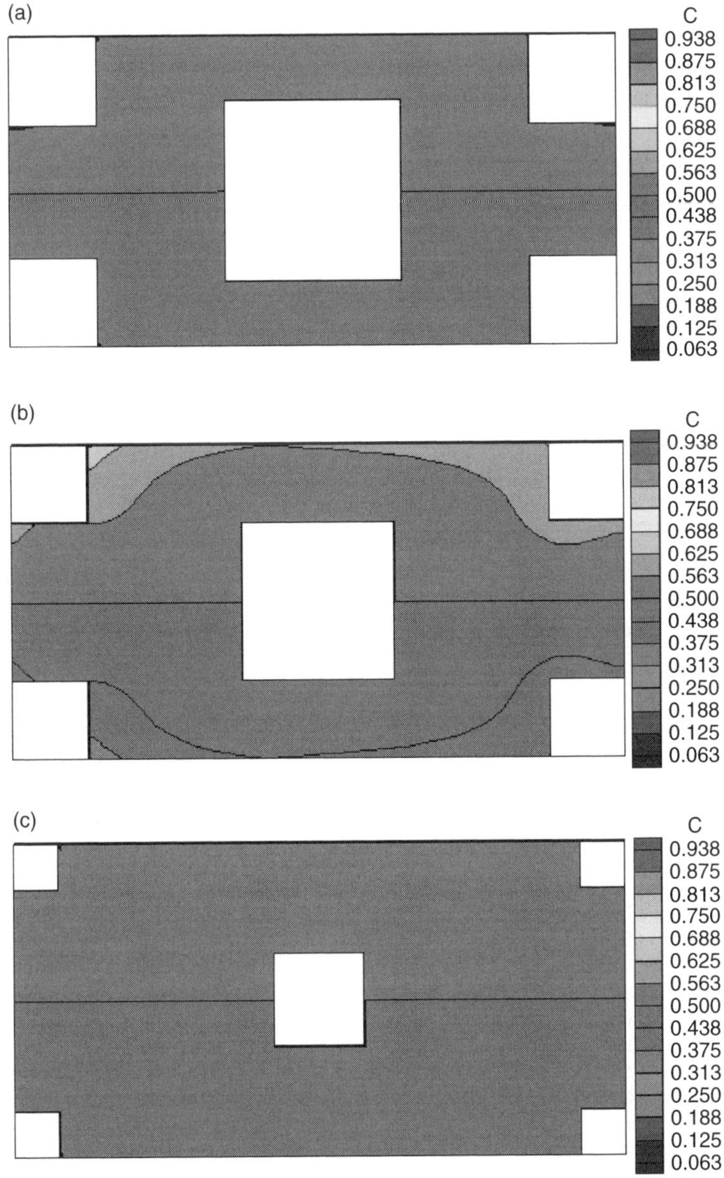


**Figure 6.8.** Mass concentration fields for unit cell calculated with transverse mass concentration gradients (turbulent regime – high Reynolds model – $Re_H = 1.0E+06$): (a) $\phi = 0.65$, (b) $\phi = 0.75$ and (c) $\phi = 0.90$.

**Table 6.1.** Summary of the integrated results for the square rods, $\phi = 0.65$.

| $\phi$ | $Re_H$ | $Pe_p$ | $k$–$\varepsilon$ model | $\frac{(\mathbf{D}_{disp})_{XX}}{D_\ell}$ | $\frac{(\mathbf{D}_{disp})_{YY}}{D_\ell}$ |
|---|---|---|---|---|---|
| | 1.0 | 0.549 | Laminar | 2.356 | 1.009 |
| | 2.0 | 1.098 | Laminar | 3.658 | 1.033 |
| | 3.0 | 1.648 | Laminar | 4.923 | 1.068 |
| | 4.0 | 2.197 | Laminar | 6.167 | 1.112 |
| | 5.0 | 2.746 | Laminar | 7.398 | 1.162 |
| | 1.0E+01 | 5.494 | Laminar | 13.481 | 1.438 |
| | 5.0E+01 | 27.469 | Laminar | 60.384 | 2.609 |
| | 1.0E+02 | 54.943 | Laminar | 109.984 | 2.901 |
| 0.65 | 5.0E+02 | 274.779 | Laminar | 378.016 | 6.878 |
| | 1.0E+03 | 549.627 | Laminar | 1540.016 | 11.864 |
| | 2.0E+03 | 1098.954 | Low *Re* | 4041.941 | 68.238 |
| | 4.0E+03 | 2197.862 | Low *Re* | 11333.724 | 132.726 |
| | 4.0E+03 | 2197.946 | High *Re* | 10881.373 | 100.375 |
| | 1.0E+04 | 5494.905 | High *Re* | 19119.294 | 251.692 |
| | 1.0E+05 | 54949.124 | High *Re* | 109437.226 | 2330.466 |

**Table 6.2.** Summary of the integrated results for the square rods, $\phi = 0.75$.

| $\phi$ | $Re_H$ | $Pe_p$ | $k$–$\varepsilon$ model | $\frac{(\mathbf{D}_{disp})_{XX}}{D_\ell}$ | $\frac{(\mathbf{D}_{disp})_{YY}}{D_\ell}$ |
|---|---|---|---|---|---|
| | 1.0 | 0.750 | Laminar | 1.828 | 1.006 |
| | 2.0 | 1.500 | Laminar | 2.626 | 1.021 |
| | 3.0 | 2.250 | Laminar | 3.402 | 1.044 |
| | 4.0 | 3.000 | Laminar | 4.163 | 1.074 |
| | 5.0 | 3.750 | Laminar | 4.914 | 1.107 |
| | 1.0E+01 | 7.500 | Laminar | 8.597 | 1.289 |
| | 5.0E+01 | 37.525 | Laminar | 36.205 | 2.023 |
| | 1.0E+02 | 75.081 | Laminar | 113.375 | 2.261 |
| 0.75 | 5.0E+02 | 375.772 | Laminar | 646.003 | 5.136 |
| | 1.0E+03 | 751.899 | Laminar | 1229.766 | 7.474 |
| | 2.0E+03 | 1501.608 | Low *Re* | 2404.648 | 40.629 |
| | 4.0E+03 | 3002.248 | Low *Re* | 6837.993 | 83.104 |
| | 4.0E+03 | 3003.233 | High *Re* | 6304.745 | 63.170 |
| | 1.0E+04 | 7506.793 | High *Re* | 10847.580 | 135.440 |
| | 1.0E+05 | 75078.424 | High *Re* | 160051.706 | 1116.993 |

the data in the literature, are more evident for $Pe_p$ greater than 10 and porosity greater than 0.75. One should point out that real consolidated porous media used in the reported experiments are by far different from the physical model of Figure 6.2, particularly when compared with cases with large void spaces, which happens when porosity is high.

**Table 6.3.** Summary of the integrated results for the square rods, $\phi = 0.90$.

| $\phi$ | $Re_H$ | $Pe_p$ | $k$–$\varepsilon$ model | $\frac{(\mathbf{D}_{disp})_{XX}}{D_\ell}$ | $\frac{(\mathbf{D}_{disp})_{YY}}{D_\ell}$ |
|---|---|---|---|---|---|
| | 1.0 | 1.993 | Laminar | 1.292 | 1.003 |
| | 2.0 | 3.985 | Laminar | 1.571 | 1.012 |
| | 3.0 | 5.977 | Laminar | 1.841 | 1.024 |
| | 4.0 | 7.970 | Laminar | 2.104 | 1.039 |
| | 5.0 | 9.962 | Laminar | 2.362 | 1.055 |
| | 1.0E+01 | 19.925 | Laminar | 5.408 | 1.136 |
| | 5.0E+01 | 99.661 | Laminar | 38.964 | 1.404 |
| | 1.0E+02 | 199.365 | Laminar | 106.028 | 1.524 |
| 0.90 | 5.0E+02 | 997.601 | Laminar | 814.465 | 2.837 |
| | 1.0E+03 | 1995.678 | Laminar | 1392.311 | 3.634 |
| | 2.0E+03 | 3989.089 | Low *Re* | 3197.117 | 10.575 |
| | 4.0E+03 | 7973.195 | Low *Re* | 6924.148 | 22.045 |
| | 4.0E+03 | 7976.414 | High *Re* | 6816.244 | 18.418 |
| | 1.0E+04 | 19936.028 | High *Re* | 13674.875 | 33.217 |
| | 1.0E+05 | 199284.067 | High *Re* | 125572.335 | 330.346 |

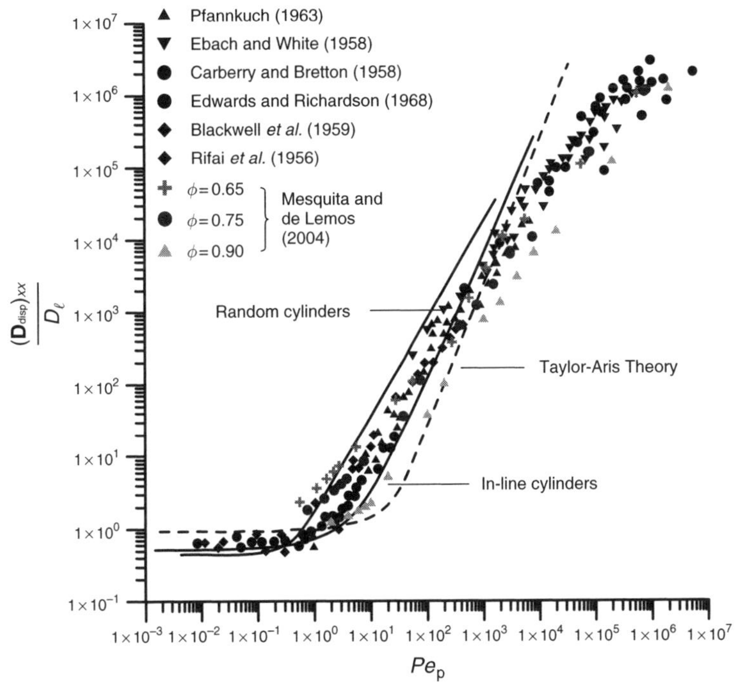


**Figure 6.9.** Longitudinal mass dispersion coefficient for $\phi = 0.65$, 0.75 and 0.90, experimental data compiled by Han *et al.* (1985).

**Table 6.4.** Experimental conditions in the literature, complied by Han *et al.* (1985), for determination of longitudinal dispersion coefficients.

| Authors | Solvent | Solute | $d_p$ (cm) | $\phi$ | $Re_p$ | $Sc$ |
|---|---|---|---|---|---|---|
| Harleman and Rumer (1963) | Water | NaCl | 0.096 | 0.36 | 0.035–21.4 | 560 |
| Edwards and Richardson (1968) | Air | Argon | 0.0377–0.607 | 0.361–0.420 | 0.0133–96.1 | 0.72 |
| Rifai *et al.* (1956) | Water | NaCl | 0.025–0.045 | 0.375–0.395 | 0.000807–0.506 | 1,858 |
| Carberry and Bretton (1958) | Water | Dye | 0.05–0.6 | 0.365–0.645 | 0.258–2144 | 1,858 |
| Ebach and White (1958) | Water | Dye | 0.021–0.673 | 0.34–0.632 | 0.0275–1023 | 1,858 |
| Pfannkuch (1963) | Water | NaCl | 0.0355–0.21 | 0.34–0.388 | 0.00069–9.59 | 560 |
| Blackwell *et al.* (1959) | Argon | Helium | 0.021 | 0.339 | 0.00298–0.200 | 1.82 |
| Gunn and Pryce (1969) | Air | Argon | 0.037–0.6 | 0.37 | 0.234–246.6 | 0.88 |

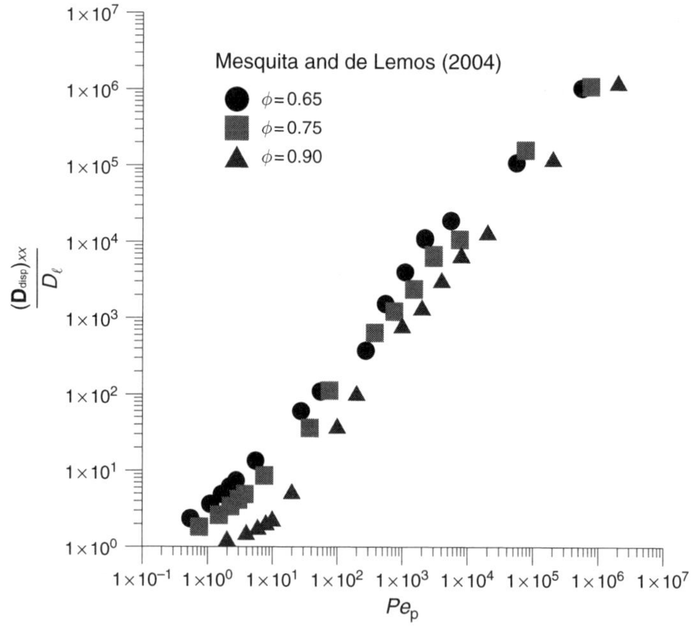


**Figure 6.10.** Effect of porosity ($\phi$) on longitudinal dispersion coefficient $(\mathbf{D}_{disp})_{XX}$.

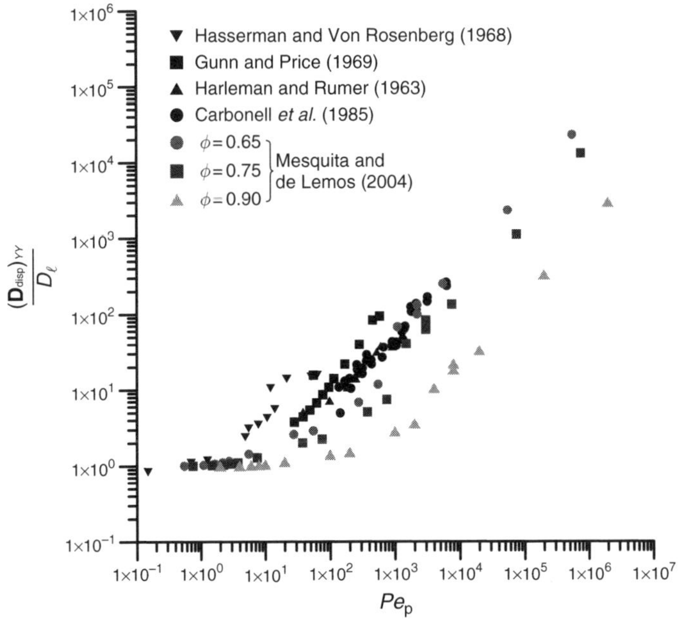


**Figure 6.11.** Transverse mass dispersion for $\phi = 0.65$, 0.75 and 0.90.

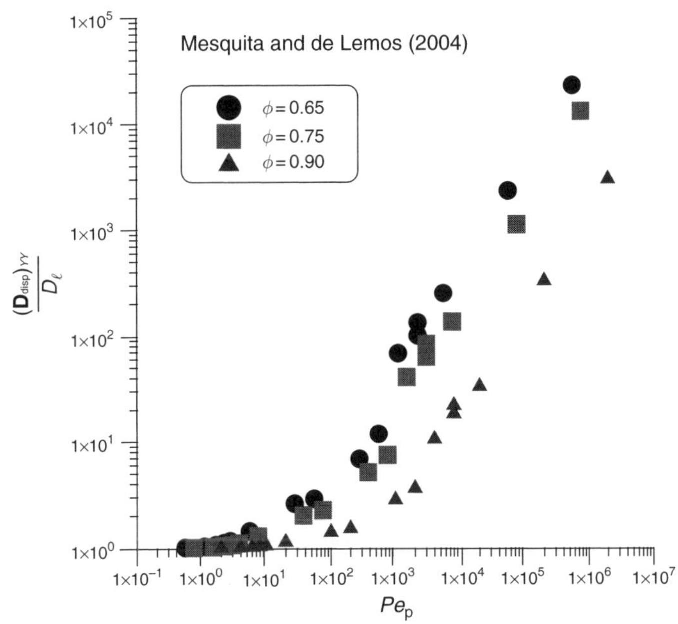


**Figure 6.12.** Effect of porosity ($\phi$) on transversal dispersion coefficient $(\mathbf{D}_{disp})_{YY}$.

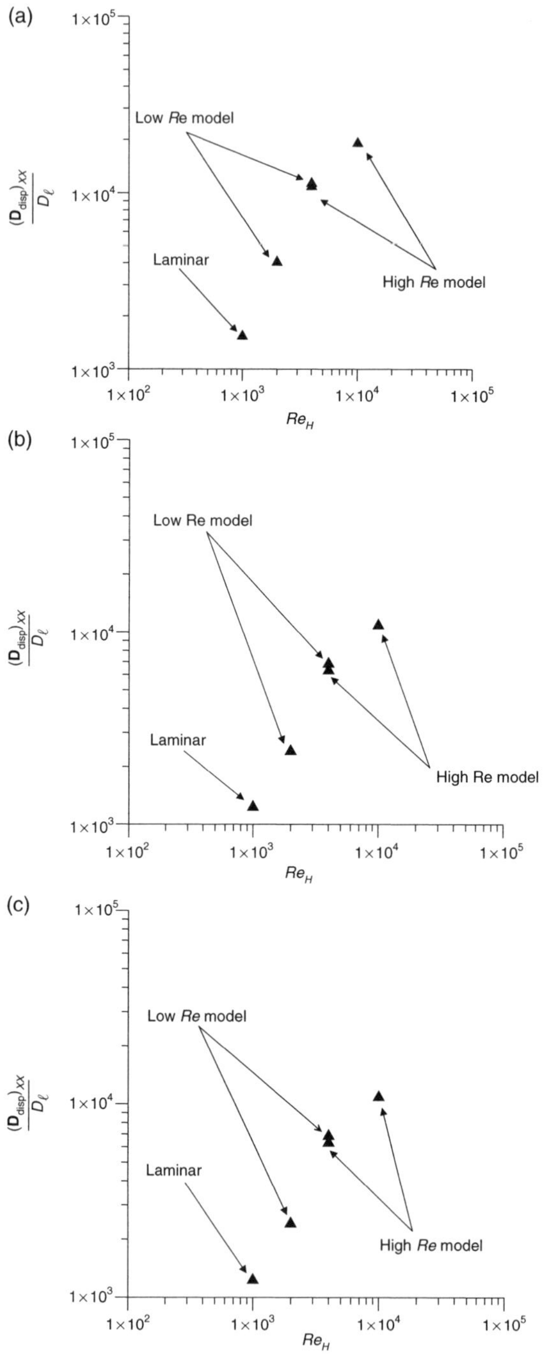


**Figure 6.13.** Effect of turbulence model on longitudinal mass dispersion: (a) $\phi = 0.65$, (b) $\phi = 0.75$ and (c) $\phi = 0.90$.

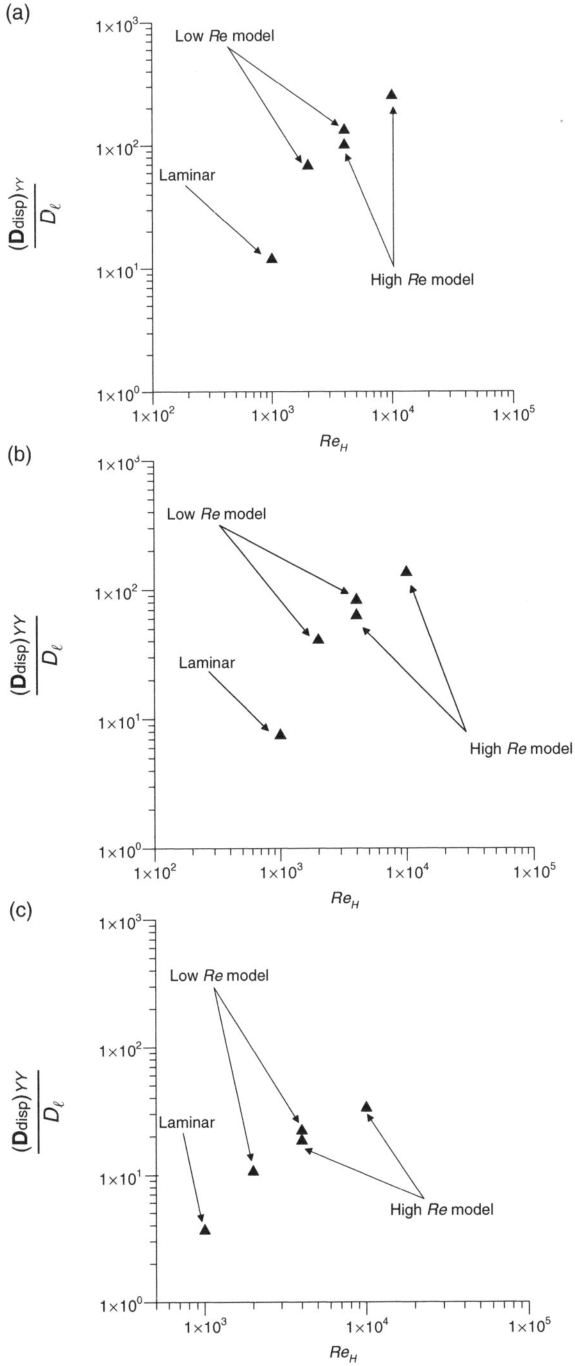


**Figure 6.14.** Effect of turbulence model on transversal mass dispersion: (a) $\phi = 0.65$, (b) $\phi = 0.75$ and (c) $\phi = 0.90$.

Low and high Reynolds turbulence models were employed for calculating $(\mathbf{D}_{\text{disp}})_{XX}$ and $(\mathbf{D}_{\text{disp}})_{YY}$. The higher the Reynolds number through the bed, the thinner the boundary layers over the rod surfaces making the use of High *Re* models more economical and suitable to apply. Figures 6.13 and 6.14 shows the effect of turbulence model on the calculated dispersion tensor components. The figures show that appropriate models were applied according to the range of $Re_H$ investigated. For $(\mathbf{D}_{\text{disp}})_{XX}$ (Figure 6.13), results are nearly independent of the two turbulence models used for $Re_H = 4 \times 10^3$ (High and low *Re* models). Superposition of results for $(\mathbf{D}_{\text{disp}})_{YY}$ (Figure 6.14, $Re_H = 4 \times 10^3$) is less evident, showing that for the transversal dispersion the choice of an adequate turbulence model might affect the quality of the predictions.

Chapter 7

# Turbulent Double Diffusion

## 7.1. INTRODUCTION

The study of double-diffusive natural convection in porous media has many environmental and industrial applications, including grain storage and drying, petrochemical processes, oil and gas extraction, contaminant dispersion in underground water reservoirs, electrochemical processes, etc. (Mamou *et al.*, 1995; Goyeau *et al.*, 1996; Nithiarasu *et al.*, 1997; Mamou *et al.*, 1998; Bennacer *et al.*, 2001; Mohamad and Bennacer, 2002; Bennacer *et al.*, 2003). In some specific applications, the fluid mixture may become turbulent and difficulties arise in the proper mathematical modeling of the transport processes under both temperature and concentration gradients.

As mentioned already in this book, modeling of macroscopic transport for incompressible flows in rigid porous media has been based on the volume-average methodology for either heat transfer (Hsu and Cheng, 1990) or mass transfer (Whitaker, 1966, 1967; Bear and Bachmat, 1967; Bear, 1972). If time fluctuations of the flow properties are considered, in addition to spatial deviations, there are two possible methodologies to follow to obtain macroscopic equations: (a) application of time average operator followed by volume averaging (Kuwahara *et al.*, 1996; Masuoka and Takatsu, 1996; Kuwahara and Nakayama, 1998; Nakayama and Kuwahara, 1999), and (b) use of volume average operator before time averaging is applied (Lee and Howell, 1987; Wang and Takle, 1995; Antohe and Lage, 1997; Getachewa *et al.*, 2000). This book intends to present a set of macroscopic mass transport equations derived under the recently established double-decomposition concept, which was discussed in Chapter 3 and through which the connection between the two paths (a) and (b) above is unveiled (Pedras and de Lemos, 2000a–2003). This methodology, initially developed for the flow variables, has been extended to heat transfer in porous media where both time fluctuations and spatial deviations were considered for velocity and temperature (Rocamora and de Lemos, 2000a). In this chapter, double-diffusive turbulent natural convection is considered.

### 7.1.1. Macroscopic Equations for Buoyancy-Free flows

For non-buoyant flows, macroscopic equations considering turbulence have been already presented before for momentum (Chapter 4 – Turbulent Momentum Transport), heat (Chapter 5 – Turbulent Heat Transport) and mass transfer (Chapter 6 – Turbulent Mass Transport) and for this reason their derivation need not to be repeated here. They are summarized below.

#### *7.1.1.1. Momentum Transport*

$$\rho\nabla\cdot\left(\frac{\bar{\mathbf{u}}_{\mathrm{D}}\bar{\mathbf{u}}_{\mathrm{D}}}{\phi}\right) = -\nabla(\phi\langle\bar{p}\rangle^{\mathrm{i}}) + \mu\nabla^2\bar{\mathbf{u}}_{\mathrm{D}} + \nabla\cdot(-\rho\phi\langle\overline{\mathbf{u}'\mathbf{u}'}\rangle^{\mathrm{i}}) - \left[\frac{\mu\phi}{K}\bar{\mathbf{u}}_{\mathrm{D}} + \frac{c_{\mathrm{F}}\phi\rho|\bar{\mathbf{u}}_{\mathrm{D}}|\bar{\mathbf{u}}_{\mathrm{D}}}{\sqrt{K}}\right] \tag{7.1}$$

$$-\rho\phi\langle\overline{\mathbf{u}'\mathbf{u}'}\rangle^{\mathrm{i}} = \mu_{\mathrm{t}_\phi}2\langle\bar{\mathbf{D}}\rangle^{\mathrm{v}} - \frac{2}{3}\phi\rho\langle k\rangle^{\mathrm{i}}\mathbf{I} \tag{7.2}$$

$$\langle\bar{\mathbf{D}}\rangle^{\mathrm{v}} = \frac{1}{2}\{\nabla(\phi\langle\bar{\mathbf{u}}\rangle^{\mathrm{i}}) + [\nabla(\phi\langle\bar{\mathbf{u}}\rangle^{\mathrm{i}})]^{\mathrm{T}}\} \tag{7.3}$$

$$\mu_{\mathrm{t}_\phi} = \rho c_\mu \langle k\rangle^{\mathrm{i}^2}/\langle\varepsilon\rangle^{\mathrm{i}} \tag{7.4}$$

#### *7.1.1.2. Heat Transport*

$$(\rho c_{\mathrm{p}})_{\mathrm{f}}\nabla\cdot(\bar{\mathbf{u}}_{\mathrm{D}}\langle\bar{T}\rangle^{\mathrm{i}}) = \nabla\cdot\{\mathbf{K}_{\mathrm{eff}}\cdot\nabla\langle\bar{T}\rangle^{\mathrm{i}}\} \tag{7.5}$$

$$\mathbf{K}_{\mathrm{eff}} = [\phi\lambda_{\mathrm{f}} + (1-\phi)\lambda_{\mathrm{s}}]\mathbf{I} + \mathbf{K}_{\mathrm{tor}} + \mathbf{K}_{\mathrm{t}} + \mathbf{K}_{\mathrm{disp}} + \mathbf{K}_{\mathrm{disp,t}} \tag{7.6}$$

The subscripts f and s refer to fluid and solid phases, respectively, and the components of $\mathbf{K}_{\mathrm{eff}}$ in Equation (7.6) come from the modeling of the following mechanisms:

- Tortuosity: $\left[\frac{1}{\Delta V}\int_{A_{\mathrm{i}}}\mathbf{n}\left(\lambda_{\mathrm{f}}\bar{T}_{\mathrm{f}} - \lambda_{\mathrm{s}}\bar{T}_{\mathrm{s}}\right)\mathrm{d}S\right] = \mathbf{K}_{\mathrm{tor}}\cdot\nabla\langle\bar{T}\rangle^{\mathrm{i}}$ (7.7)
- Thermal dispersion: $-(\rho c_{\mathrm{p}})_{\mathrm{f}}\phi\langle{}^{\mathrm{i}}\bar{\mathbf{u}}\,{}^{\mathrm{i}}\bar{T}_{\mathrm{f}}\rangle^{\mathrm{i}} = \mathbf{K}_{\mathrm{disp}}\cdot\nabla\langle\bar{T}\rangle^{\mathrm{i}}$ (7.8)
- Turbulent heat flux: $-(\rho c_{\mathrm{p}})_{\mathrm{f}}\phi\overline{\langle\mathbf{u}'\rangle^{\mathrm{i}}\langle T_{\mathrm{f}}'\rangle^{\mathrm{i}}} = \mathbf{K}_{\mathrm{t}}\cdot\nabla\langle\bar{T}\rangle^{\mathrm{i}}$ (7.9)
- Turbulent thermal dispersion: $-(\rho c_{\mathrm{p}})_{\mathrm{f}}\phi\langle\overline{{}^{\mathrm{i}}\mathbf{u}'\,{}^{\mathrm{i}}T_{\mathrm{f}}'}\rangle^{\mathrm{i}} = \mathbf{K}_{\mathrm{disp,t}}\cdot\nabla\langle\bar{T}\rangle^{\mathrm{i}}$ (7.10)

Mechanisms (7.9) and (7.10) were modeled together in de Lemos and Braga (2003) and de Lemos and Rocamora (2002) by assuming:

$$-(\rho c_{\mathrm{p}})_{\mathrm{f}}\langle\overline{\mathbf{u}'T_{\mathrm{f}}'}\rangle^{\mathrm{i}} = c_{\mathrm{p_f}}\frac{\mu_{\mathrm{t}_\phi}}{\sigma_{\mathrm{t}_\phi}}\nabla\langle\bar{T}_{\mathrm{f}}\rangle^{\mathrm{i}} \tag{7.11}$$

or

$$\mathbf{K}_{\mathrm{t}} + \mathbf{K}_{\mathrm{disp,t}} = \phi c_{\mathrm{p_f}}\frac{\mu_{\mathrm{t}_\phi}}{\sigma_{\mathrm{t}_\phi}}\mathbf{I} \tag{7.12}$$

#### *7.1.1.3. Mass Transport*

$$\nabla \cdot (\bar{\mathbf{u}}_{\mathrm{D}} \langle \overline{C} \rangle^{\mathrm{i}}) = \nabla \cdot \mathbf{D}_{\mathrm{eff}} \cdot \nabla(\phi \langle \overline{C} \rangle^{\mathrm{i}}) \tag{7.13}$$

$$\mathbf{D}_{\mathrm{eff}} = \mathbf{D}_{\mathrm{disp}} + \mathbf{D}_{\mathrm{diff}} + \mathbf{D}_{\mathrm{t}} + \mathbf{D}_{\mathrm{disp,t}} \tag{7.14}$$

$$\mathbf{D}_{\mathrm{diff}} = \langle D \rangle^{\mathrm{i}} \mathbf{I} = \frac{1}{\rho} \frac{\mu_{\phi}}{Sc} \mathbf{I} \tag{7.15}$$

$$\mathbf{D}_{\mathrm{t}} + \mathbf{D}_{\mathrm{disp,t}} = \frac{1}{\rho} \frac{\mu_{\mathrm{t}_{\phi}}}{Sc_{\mathrm{t}}} \mathbf{I} \tag{7.16}$$

Coefficients $\mathbf{D}_{\mathrm{disp}}$, $\mathbf{D}_{\mathrm{t}}$ and $\mathbf{D}_{\mathrm{disp,t}}$ in Equation (7.14) appear due to the non-linearity of the convection term. They come from the modeling of the following mechanisms:

- Mass dispersion: $-\langle {}^{\mathrm{i}}\bar{\mathbf{u}}\, {}^{\mathrm{i}}\overline{C} \rangle^{\mathrm{i}} - \mathbf{D}_{\mathrm{disp}} \cdot \nabla \langle \overline{C} \rangle^{\mathrm{i}}$ (7.17)
- Turbulent mass flux: $-\overline{\langle \mathbf{u}' \rangle^{\mathrm{i}} \langle C' \rangle^{\mathrm{i}}} = -\overline{\langle \mathbf{u} \rangle^{\mathrm{i}'} \langle C \rangle^{\mathrm{i}'}} = \mathbf{D}_{\mathrm{t}} \cdot \nabla \langle \overline{C} \rangle^{\mathrm{i}}$ (7.18)
- Turbulent mass dispersion: $-\langle \overline{{}^{\mathrm{i}}\mathbf{u}'\, {}^{\mathrm{i}}C'} \rangle^{\mathrm{i}} = \mathbf{D}_{\mathrm{disp,t}} \cdot \nabla \langle \overline{C} \rangle^{\mathrm{i}}$ (7.19)

Here also mechanisms (7.18) and (7.19) are added up as:

$$-\langle \overline{\mathbf{u}' C'} \rangle^{\mathrm{i}} = \frac{1}{\rho} \frac{\mu_{\mathrm{t}_{\phi}}}{Sc_{\mathrm{t}\phi}} \nabla \langle \overline{C} \rangle^{\mathrm{i}} = \langle D_{\mathrm{t}} \rangle^{\mathrm{i}} \nabla \langle \overline{C} \rangle^{\mathrm{i}} = (\mathbf{D}_{\mathrm{t}} + \mathbf{D}_{\mathrm{disp,t}}) \cdot \nabla \langle \overline{C} \rangle^{\mathrm{i}} \tag{7.20}$$

## 7.2. MACROSCOPIC DOUBLE-DIFFUSION EFFECTS

### 7.2.1. Mean Flow

Focusing now attention to buoyancy effects only, application of the volume average procedure to the last term of Equation (2.32) leads to:

$$\langle \rho \mathbf{g}[\beta(\overline{T} - T_{\mathrm{ref}}) + \beta_{\mathrm{C}}(\overline{C} - C_{\mathrm{ref}})] \rangle^{\mathrm{v}} = \frac{\Delta V_{\mathrm{f}}}{\Delta V} \frac{1}{\Delta V_{\mathrm{f}}} \int_{\Delta V_{\mathrm{f}}} \rho \mathbf{g}[\beta(\overline{T} - T_{\mathrm{ref}}) + \beta_{\mathrm{C}}(\overline{C} - C_{\mathrm{ref}})] \mathrm{d}V \tag{7.21}$$

Expanding the lhs of the above equation in light of Equation (2.19), the buoyancy term becomes:

$$\begin{aligned} \langle \rho \mathbf{g}[\beta(\overline{T} - T_{\mathrm{ref}}) + \beta_{\mathrm{C}}(\overline{C} - C_{\mathrm{ref}})] \rangle^{\mathrm{v}} =& \rho \mathbf{g} \phi [\beta_{\phi}(\langle \overline{T} \rangle^{\mathrm{i}} - T_{\mathrm{ref}}) + \beta_{\mathrm{C}_{\phi}}(\langle \overline{C} \rangle^{\mathrm{i}} - C_{\mathrm{ref}})] \\ &+ \underbrace{\rho \mathbf{g} \beta \phi \overline{\langle {}^{\mathrm{i}}T \rangle^{\mathrm{i}}}}_{=0} + \underbrace{\rho \mathbf{g} \beta_{\mathrm{C}} \phi \overline{\langle {}^{\mathrm{i}}C \rangle^{\mathrm{i}}}}_{=0} \end{aligned} \tag{7.22}$$

where the third and forth terms on the rhs are null since $\langle {}^{i}\varphi \rangle^{i} = 0$. Here, coefficients $\beta_\phi$ and $\beta_{C_\phi}$ are the *macroscopic* thermal and salute expansion coefficients, respectively. Assuming that gravity is constant over the REV, expressions for them based on Equation (7.22) are given as:

$$\beta_\phi = \frac{\langle \rho\beta(\overline{T} - T_{\mathrm{ref}})\rangle^{\mathrm{v}}}{\rho\phi(\langle \overline{T}\rangle^{\mathrm{i}} - T_{\mathrm{ref}})}; \quad \beta_{C_\phi} = \frac{\langle \rho\beta_{\mathrm{C}}(\overline{C} - C_{\mathrm{ref}})\rangle^{\mathrm{v}}}{\rho\phi(\langle \overline{C}\rangle^{\mathrm{i}} - C_{\mathrm{ref}})} \tag{7.23}$$

Adding Equation (7.22) to the rhs of Equation (4.8), the macroscopic time-mean Navier-Stokes (NS) equation for an incompressible fluid is given as:

$$\begin{aligned} \rho\nabla\cdot\left(\frac{\overline{\mathbf{u}}_{\mathrm{D}}\overline{\mathbf{u}}_{\mathrm{D}}}{\phi}\right) = &- \nabla(\phi\langle\overline{p}\rangle^{\mathrm{i}}) + \mu\nabla^2\overline{\mathbf{u}}_{\mathrm{D}} + \nabla\cdot(-\rho\phi\langle\overline{\mathbf{u}'\mathbf{u}'}\rangle^{\mathrm{i}}) + \rho\mathbf{g}\phi[\beta_\phi(\langle\overline{T}\rangle^{\mathrm{i}} - T_{\mathrm{ref}}) \\ &+ \beta_{C_\phi}(\langle\overline{C}\rangle^{\mathrm{i}} - C_{\mathrm{ref}})] - \left[\frac{\mu\phi}{K}\overline{\mathbf{u}}_{\mathrm{D}} + \frac{c_{\mathrm{F}}\phi\rho|\overline{\mathbf{u}}_{\mathrm{D}}|\overline{\mathbf{u}}_{\mathrm{D}}}{\sqrt{K}}\right] \end{aligned} \tag{7.24}$$

Before proceeding, it is interesting to comment on the role of coefficients $\beta_\phi$ and $\beta_{C_\phi}$ on the overall mixture density value. Figure 7.1 presents the variation of $\rho$ as a function of temperature or concentration gradients. Here, only fluids that became less dense with increasing temperature are considered (Figure 7.1a). However, two situations might occur when increasing $\langle\overline{C}\rangle^{\mathrm{i}}$, namely the mixture might became less dense with the addition of a lighter solute (Figure 7.1b) or else, a denser fluid may result by mixing a heavier component to it (Figure 7.1c). Implications of this on the stability of the entire fluid system will be discussed below.

### 7.2.2. Turbulent Field

For clear fluid, the buoyancy contribution to the $k$ equation is given by Equations (2.37) and (2.38).

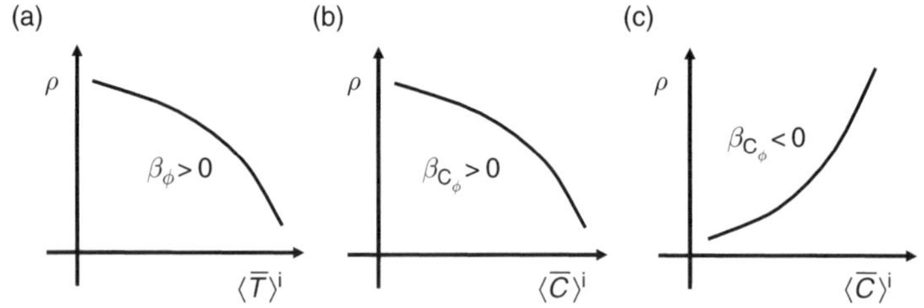


**Figure 7.1.** Behaviour of mixture density: (a) lighter mixture with increasing $\langle\overline{T}\rangle^{\mathrm{i}}$, (b) lighter mixture with increasing $\langle\overline{C}\rangle^{\mathrm{i}}$, (c) heavier mixture with increasing $\langle\overline{C}\rangle^{\mathrm{i}}$.

For thermally driven flows, volume averaging of Equation (2.38) in de Lemos and Braga (2003) has resulted in the term:

$$G^{\mathrm{i}}_{\beta} = \beta^{\mathrm{k}}_{\phi}\phi\frac{\mu_{\mathrm{t}_\phi}}{\sigma_{\mathrm{t}_\phi}}\mathbf{g}\cdot\nabla\langle\overline{T}\rangle^{\mathrm{i}} \tag{7.25}$$

as an additional macroscopic generation/destruction rate of $\langle k\rangle^{\mathrm{i}}$ due to temperature variation in porous media, where $\beta^{\mathrm{k}}_{\phi}$ is a macroscopic coefficient. In de Lemos and Braga (2003), coefficients $\beta$ (Equation (2.11)), $\beta_{\phi}$ (Equation (5.91)) and $\beta^{\mathrm{k}}_{\phi}$ (Equation (5.93)) are all assumed to be equal, for simplicity.

In order to add the effect of concentration variation within the fluid, one applies the volume average operator to Equation (2.38) such that,

$$\langle G_{\mathrm{C}}\rangle^{\mathrm{v}} = G^{\mathrm{i}}_{\beta_{\mathrm{C}}} = \langle -\rho\beta_{\mathrm{C}}\mathbf{g}\cdot\overline{\mathbf{u}'C'}\rangle^{\mathrm{v}} = -\rho\beta^{\mathrm{k}}_{\mathrm{C}_\phi}\phi\mathbf{g}\cdot\langle\overline{\mathbf{u}'C'}\rangle^{\mathrm{i}} \tag{7.26}$$

where the coefficient $\beta^{\mathrm{k}}_{\mathrm{C}_\phi}$, for a constant value of $\mathbf{g}$ within the REV, is given by $\beta^{\mathrm{k}}_{\mathrm{C}_\phi} = \dfrac{\langle\rho\beta_{\mathrm{C}}\overline{\mathbf{u}'C'}\rangle^{\mathrm{v}}}{\rho\phi\langle\overline{\mathbf{u}'C'}\rangle^{\mathrm{i}}}$, which, in turn, is not necessarily equal to $\beta_{\mathrm{C}_\phi}$ given by Equation (7.23). However, for the sake of simplicity and in the absence of better information, one can use a similar argument as in de Lemos and Braga (2003) and make use of the assumption $\beta^{\mathrm{k}}_{\mathrm{C}_\phi} = \beta_{\mathrm{C}_\phi} = \beta_{\mathrm{C}}$. Further, expanding the rhs of Equation (7.26) in light of Equations (2.19) and (3.10), one has:

$$\begin{aligned}
&-\rho\beta^{\mathrm{k}}_{\mathrm{C}_\phi}\phi\mathbf{g}\cdot\langle\overline{\mathbf{u}'C'}\rangle^{\mathrm{i}}\\
&= -\rho\beta^{\mathrm{k}}_{\mathrm{C}_\phi}\phi\mathbf{g}\cdot\langle\overline{(\langle\mathbf{u}'\rangle^{\mathrm{i}}+{}^{\mathrm{i}}\mathbf{u}')(\langle C'\rangle^{\mathrm{i}}+{}^{\mathrm{i}}C')}\rangle^{\mathrm{i}}\\
&= -\rho\beta^{\mathrm{k}}_{\mathrm{C}_\phi}\phi\mathbf{g}\cdot\left(\langle\overline{\langle\mathbf{u}'\rangle^{\mathrm{i}}\langle C'\rangle^{\mathrm{i}}}\rangle^{\mathrm{i}} + \langle\overline{{}^{\mathrm{i}}\mathbf{u}'\,{}^{\mathrm{i}}C'}\rangle^{\mathrm{i}} + \langle\overline{\langle\mathbf{u}'\rangle^{\mathrm{i}}\,{}^{\mathrm{i}}C'}\rangle^{\mathrm{i}} + \langle\overline{{}^{\mathrm{i}}\mathbf{u}'\langle C'\rangle^{\mathrm{i}}}\rangle^{\mathrm{i}}\right)\\
&= -\rho\beta^{\mathrm{k}}_{\mathrm{C}_\phi}\phi\mathbf{g}\cdot\left(\underbrace{\overline{\langle\mathbf{u}\rangle^{\mathrm{i}'}\langle C\rangle^{\mathrm{i}'}}}_{1} + \underbrace{\langle\overline{{}^{\mathrm{i}}\mathbf{u}'\,{}^{\mathrm{i}}C'}\rangle^{\mathrm{i}}}_{2} + \underbrace{\overline{\langle\mathbf{u}'\rangle^{\mathrm{i}}\langle{}^{\mathrm{i}}C'\rangle^{\mathrm{i}}}}_{=0} + \underbrace{\overline{\langle{}^{\mathrm{i}}\mathbf{u}'\rangle^{\mathrm{i}}\langle{}^{\mathrm{i}}C'\rangle^{\mathrm{i}}}}_{=0}\right)
\end{aligned} \tag{7.27}$$

The last two terms on the right of Equation (7.27) are null since $\langle{}^{\mathrm{i}}C'\rangle^{\mathrm{i}} = 0$ and $\langle{}^{\mathrm{i}}\mathbf{u}'\rangle^{\mathrm{i}} = 0$. In addition, the following physical significance can be attached to the two remaining terms:

1 *Buoyancy generation/destruction rate of turbulent kinetic energy due to macroscopic velocity concentration fluctuations*: This term is also present in turbulent flow in clear (non-obstructed) domains and represents an exchange between the energy associated with the macroscopic turbulent motion and potential energy. In stable stratification, within regions of high concentration of heavier solutes ($\beta^{\mathrm{k}}_{\mathrm{C}_\phi} < 0$), this term damps

turbulence by being of negative value whereas the potential energy of the system is increased. On the other hand, in unstable stratification, for high concentration regions of lighter mixtures ($\beta^{\mathrm{k}}_{\mathrm{C}_\phi} > 0$), it enhances $\langle k \rangle^{\mathrm{i}}$ at the expense of potential energy. A more detailed analysis on the stability of mixture systems is presented below.

2 *Buoyancy generation/destruction rate of* $\langle k \rangle^{\mathrm{i}}$ *due to turbulent mass dispersion*: Extra generation/destruction rate due to time fluctuations and spatial deviations of both local velocity and concentration. This term might be interpreted as an additional source/sink of turbulence kinetic energy due to the fact that time fluctuations of local velocities and concentration present a spatial deviation in relation to their macroscopic value. Then, additional exchange between turbulent kinetic energy and potential energy in systems may occur due to the presence of a porous matrix.

A model for Equation (7.27) is still needed to solve an equation for $\langle k \rangle^{\mathrm{i}}$, which is a necessary information when computing $\mu_{\mathrm{t}_\phi}$ using Equation (4.20). Consequently, terms 1 and 2 above have to be modeled as a function of average concentration $\langle \overline{C} \rangle^{\mathrm{i}}$. To accomplish this, a gradient-type diffusion model is used, in the form:

- Buoyancy generation of $\langle k \rangle^{\mathrm{i}}$ due to *turbulent salute fluctuations*:

$$-\rho \beta^{\mathrm{k}}_{\mathrm{C}_\phi} \phi \mathbf{g} \cdot \overline{\langle \mathbf{u} \rangle^{\mathrm{i}\prime} \langle C \rangle^{\mathrm{i}\prime}} = \rho \beta^{\mathrm{k}}_{\mathrm{C}_\phi} \phi \mathbf{g} \cdot (\mathbf{D}_{\mathrm{t}} \cdot \nabla \langle \overline{C} \rangle^{\mathrm{i}}) \tag{7.28}$$

- Buoyancy generation of $\langle k \rangle^{\mathrm{i}}$ due to *turbulent salute dispersion*:

$$-\rho \beta^{\mathrm{k}}_{\mathrm{C}_\phi} \phi \mathbf{g} \cdot \langle \overline{{}^{\mathrm{i}}\mathbf{u}' \, {}^{\mathrm{i}}C'} \rangle^{\mathrm{i}} = \rho \beta^{\mathrm{k}}_{\mathrm{C}_\phi} \phi \mathbf{g} \cdot (\mathbf{D}_{\mathrm{disp,t}} \cdot \nabla \langle \overline{C} \rangle^{\mathrm{i}}) \tag{7.29}$$

The buoyancy concentration coefficients above, namely $\mathbf{D}_{\mathrm{t}}$ and $\mathbf{D}_{\mathrm{disp,t}}$, were used before in Equations (7.18) and (7.19), respectively. It should be noticed that these terms arise only if the flow is turbulent and if buoyancy is of importance.

Using then Equation (7.20), the macroscopic buoyancy generation of $k$ due to concentration fluctuations can be modeled as,

$$\begin{aligned} G^{\mathrm{i}}_{\beta_{\mathrm{C}}} &= -\rho \beta^{\mathrm{k}}_{\mathrm{C}_\phi} \phi \mathbf{g} \cdot \langle \overline{\mathbf{u}' C'} \rangle^{\mathrm{i}} \\ &= \rho \beta^{\mathrm{k}}_{\mathrm{C}_\phi} \phi \mathbf{g} \cdot [(\mathbf{D}_{\mathrm{t}} + \mathbf{D}_{\mathrm{disp,t}}) \cdot \nabla \langle \overline{C} \rangle^{\mathrm{i}}] \\ &= \beta^{\mathrm{k}}_{\mathrm{C}_\phi} \phi \frac{\mu_{\mathrm{t}_\phi}}{Sc_{\mathrm{t}_\phi}} \mathbf{g} \cdot \nabla \langle \overline{C} \rangle^{\mathrm{i}} \end{aligned} \tag{7.30}$$

where $\mu_{\mathrm{t}_\phi}$, $Sc_{\mathrm{t}_\phi}$ and the two coefficients $\mathbf{D}_{\mathrm{t}}$ and $\mathbf{D}_{\mathrm{disp,t}}$ have been defined before.

Final transport equations for $\langle k\rangle^{\mathrm{i}} = \langle\overline{\mathbf{u}'\cdot\mathbf{u}'}\rangle^{\mathrm{i}}/2$ and $\langle\varepsilon\rangle^{\mathrm{i}} = \mu\langle\overline{\nabla\mathbf{u}':(\nabla\mathbf{u}')^{\mathrm{T}}}\rangle^{\mathrm{i}}/\rho$, in their so-called high Reynolds number form, can now include the buoyancy generation terms due to temperature and concentration fluctuations as:

$$\rho\nabla\cdot(\bar{\mathbf{u}}_{\mathrm{D}}\langle k\rangle^{\mathrm{i}}) = \nabla\cdot\left[\left(\mu+\frac{\mu_{t_\phi}}{\sigma_k}\right)\nabla(\phi\langle k\rangle^{\mathrm{i}})\right] + P^{\mathrm{i}} + G^{\mathrm{i}} + G^{\mathrm{i}}_{\beta} + G^{\mathrm{i}}_{\beta_{\mathrm{C}}} - \rho\phi\langle\varepsilon\rangle^{\mathrm{i}} \tag{7.31}$$

$$\rho\nabla\cdot(\bar{\mathbf{u}}_{\mathrm{D}}\langle\varepsilon\rangle^{\mathrm{i}}) = \nabla\cdot\left[\left(\mu+\frac{\mu_{t_\phi}}{\sigma_\varepsilon}\right)\nabla\left(\phi\langle\varepsilon\rangle^{\mathrm{i}}\right)\right] + \frac{\langle\varepsilon\rangle^{\mathrm{i}}}{\langle k\rangle^{\mathrm{i}}}\left[c_1P^{\mathrm{i}} + c_2G^{\mathrm{i}} + c_1c_3(G^{\mathrm{i}}_{\beta} + G^{\mathrm{i}}_{\beta_{\mathrm{C}}}) - c_2\rho\phi\langle\varepsilon\rangle^{\mathrm{i}}\right] \tag{7.32}$$

where $c_1$, $c_2$, $c_3$ and $c_k$ are constants and the production terms have the following physical significance:

(1) $P^{\mathrm{i}} = -\rho\langle\overline{\mathbf{u}'\mathbf{u}'}\rangle^{\mathrm{i}} : \nabla\bar{\mathbf{u}}_{\mathrm{D}}$ is the production rate of $\langle k\rangle^{\mathrm{i}}$ due to gradients of $\bar{\mathbf{u}}_{\mathrm{D}}$.

(2) $G^{\mathrm{i}} = c_k\rho\dfrac{\phi\langle k\rangle^{\mathrm{i}}|\bar{\mathbf{u}}_{\mathrm{D}}|}{\sqrt{K}}$ is the generation rate of the intrinsic average of $\langle k\rangle^{\mathrm{i}}$ due to the action of the porous matrix.

(3) $G^{\mathrm{i}}_{\beta} = \beta^{\mathrm{k}}_{\phi}\phi\dfrac{\mu_{t_\phi}}{\sigma_{t_\phi}}\mathbf{g}\cdot\nabla\langle\bar{T}\rangle^{\mathrm{i}}$ is the generation of $\langle k\rangle^{\mathrm{i}}$ due to mean temperature variation within the fluid.

(4) $G^{\mathrm{i}}_{\beta_{\mathrm{C}}} = \beta^{\mathrm{k}}_{\mathrm{C}_\phi}\phi\dfrac{\mu_{t_\phi}}{Sc_{t_\phi}}\mathbf{g}\cdot\nabla\langle\bar{C}\rangle^{\mathrm{i}}$ is the generation of $\langle k\rangle^{\mathrm{i}}$ due to concentration gradients.

## 7.3. HYDRODYNAMIC STABILITY

For a system oriented in the upward direction with gravity acting downward, the hydrodynamic stability of a thermal system will depend on both the thermal and concentration drives acting on an REV, according to Equation (7.24). Depending on the direction of the property gradients, both such drives may induce instability leading eventually to turbulent flow. As such, unconditionally *unstable* situations are presented in Figure 7.2 where hotter fluid (Figure 7.2a) composed by a less-dense mixture is positioned at the bottom of the fluid layer (Figure 7.2b). For positive $\beta_\phi$ and $\beta_{\mathrm{C}_\phi}$ values, with negative gradients of $\langle\bar{T}\rangle^{\mathrm{i}}$ and $\langle\bar{C}\rangle^{\mathrm{i}}$, both drives expressed by Equations (7.25) and (7.30) will give $G^{\mathrm{i}}_{\beta} > 0$ and $G^{\mathrm{i}}_{\beta_{\mathrm{C}}} > 0$, respectively, causing positive sources term in the $\langle k\rangle^{\mathrm{i}}$ Equation (7.31). If a heavier component is positioned at the top of this heated-from-below layer (Figure 7.2c), hydrodynamic instability will also occur and a source term will appear in Equation (7.31). An initially laminar flow may then undergo transition and become turbulent.

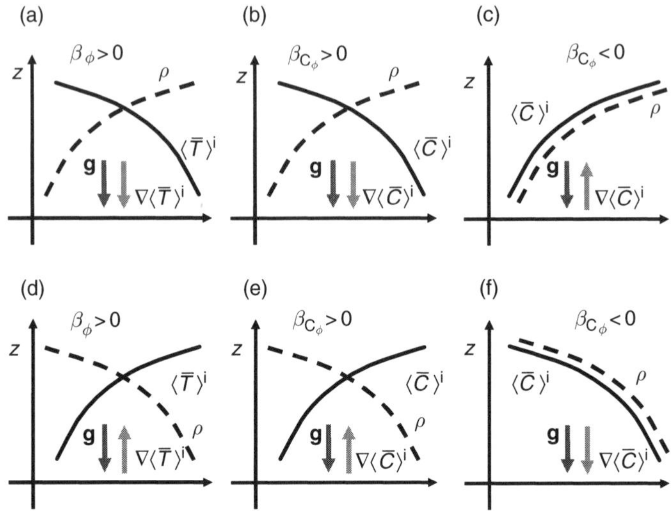


**Figure 7.2.** Stability analysis of a layer of fluid subjected to gradients of temperature and concentration. Unconditionally *unstable* cases: hotter fluid (a) with less dense mixture at the bottom (b), (c). Unconditionally *stable* cases: colder fluid (d) with denser mixture at the bottom (e), (f).

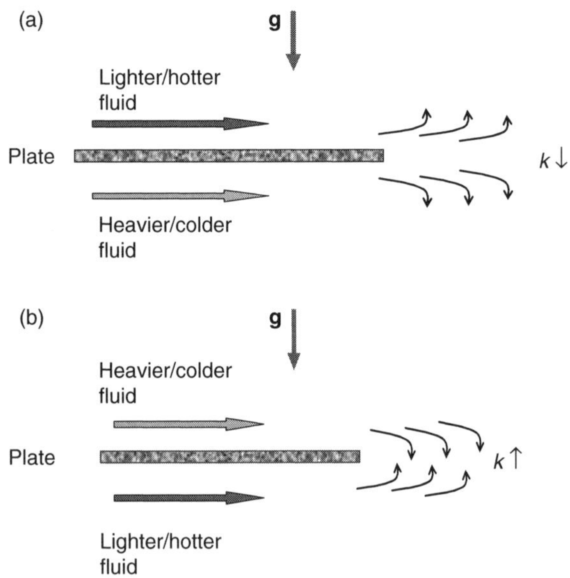


**Figure 7.3.** Flows separated by a finite plate: (a) unconditionally stable cases, (b) unconditionally unstable cases.

On the other hand, for a layer heated from above (Figure 7.2d) with lighter components flowing at the top (Figure 7.2e), both values of source terms $G^{\mathrm{i}}_{\beta}$ and $G^{\mathrm{i}}_{\beta_{\mathrm{C}}}$, in Equation (7.31), will be less than zero, leading to an unconditionally stable situation. Turbulence, if existing, might decay and the flow may relaminarize. Also in this category is the case of heated-from-above systems with heavier components flowing at the bottom (Figure 7.2f). Any other combination regarding a heavier or a lighter component flowing in a non-isothermal fluid may be conditionally unstable, depending upon the balance between source and sink terms that might appear as a result of temperature and concentration distributions within the flow.

Another example of flow instabilities induced by density differences is shown next. The mixing of two turbulent currents of distinct fluids flowing on each side of a finite plate is schematically presented in Figure 7.3. Unconditially stable cases are shown in Figure 7.3a as the overall level of $k$ will always decrease past the end of the plate. The opposite situation, namely, unconditionally unstable cases, is presented in Figure 7.3b.

# PART TWO

# Applications

Chapter 8

# Numerical Modeling and Algorithms

## 8.1. INTRODUCTION

Important advances in solving the coupled set for flow, heat and mass balance equations have been published during the last two decades with the aim of achieving the minimum computing time while preserving accuracy. Segregated and coupled schemes, block-implicit and pointwise algorithms have been used with this purpose. Many proposed solution sequences make use of modern computer architectures such as vector and parallel computers.

This increasing use of numerical calculations is certainly correlated with the associated computational cost of a determined solution. In the last few decades the computational speed has increased sharply reducing the associated computational cost. Figure 8.1 taken from Chapman (1979) indicates this tendency and shows that the relative computational cost for a certain problem has decreased by a factor of ten every eight years. This chapter overviews the most common engineering tools available for numerical calculation of flow fields based on the control-volume approach.

## 8.2. THE NEED FOR ITERATIVE METHODS

When the numerical solution of Equation (1.1) is sought, a system of simultaneous algebraic equations has to be solved. For multidimensional problems, arising from discretization of either *structured* or *non-structured* meshes, the matrix of coefficients will have the general form shown in Figure 8.2.

As indicated by the figure, convective/diffusive problems, governed by the differential equations recast into Equation (1.1), give rise to a matrix formed by non-zero elements close to the main diagonal and elements of null value elsewhere. Orthogonal grid layouts are generally ordered such that the coefficients are laid over *three*, *five* or *seven* diagonals depending upon the problem dimension. Non-structured grids, while permitting greater flexibility in local refining without increasing substantially the problem size, do not possess a matrix arrangement in a well-ordered form. In both cases, the larger the problem, the bigger the *matrix sparsity index*, defined as the ratio of the number of null terms to the total number of elements, where the total number of elements encompass null and non-null terms.

Methods available for solving systems of algebraic equations will be discussed later and depend on the nature of the equation. For linear equations, *direct methods* can be used whereas for non-linear equations one is limited to the use of *indirect* solvers.

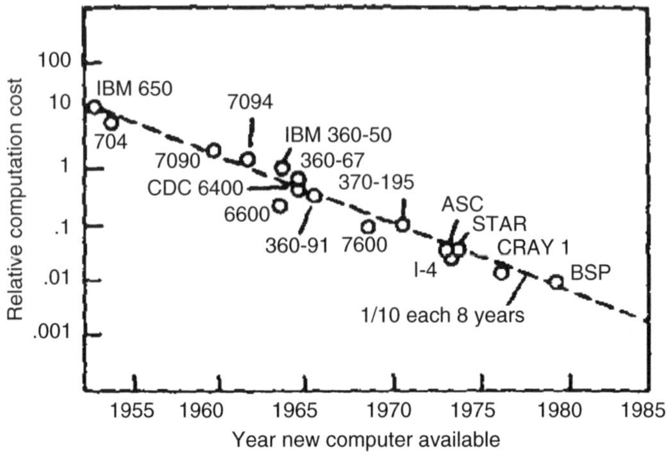


**Figure 8.1.** Trend of computational cost for a given flow and algorithm.

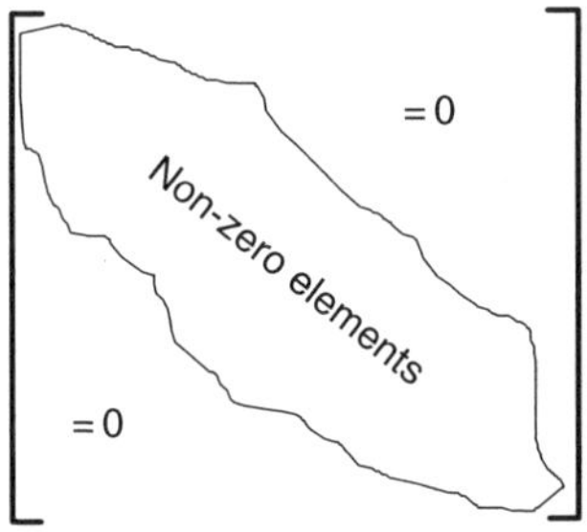


**Figure 8.2.** Sparse matrix of coefficients for numerical fluid dynamics problems.

A direct method is the one where the number of floating point operations to achieve the solution is known in advance, whereas in indirect schemes one cannot determine *a priori* the necessary computational effort to achieve convergence. Indirect solvers make use of recurrence formulas to update the variable at individual points, lines or planes of the computational domain. Convergence, if obtained, usually requires several sweeps over the entire grid. An overview on these methods can be found in texts on numerical analysis (Richtmyer and Norton, 1967).

In any case, the use of a direct method, which works on the entire matrix (including the zero-value elements), may lead to prohibitive computing time and cost, especially if one considers the inherent non-linearity of the problem. Inversion of the full matrix a number of times would be necessary before final convergence. The iterative methods, on the other hand, handle only the non-zero terms and can accommodate different physical phenomena with little reprogramming. Even when solving linear problems, iterative methods may sometimes be a better choice.

## 8.3. INCOMPRESSIBLE VS. COMPRESSIBLE SOLUTION STRATEGIES

Existing methods for solving the Navier-Stokes equations are roughly classified into density-based methods and pressure-based methods. Generally speaking, the former are historically associated with the solution of high speed gas flows whereas the latter comprises studies dealing with low speed or incompressible flows with heat transfer. Density-based methods have been used to analyze aerodynamic flows over bodies and in most cases solve the system with the matrix in Figure 8.2 with a simultaneous procedure.

In addition to obtaining the velocity field via the momentum equation ($\phi = u_i$ in Equation (1.1)), density-based methods make use of the continuity equation ($\phi = 1$) as the active equation to specify density and the equation of state to calculate pressure. So, in this formulation, all variables involved possess a respective equation (Beam and Warming, 1978; Pullian and Steger, 1980). Extension of the compressible formulation to all speed flows has been lately reported (Merkle and Choi, 1987; Withington *et al.*, 1991).

When density varies little with pressure, the use of an equation of state does not help much, since no pressure effects are felt on the value of the density $\rho$. The continuity equation would be of no use except that of constraining the calculated velocity field. The difficult can be overcome by using the numerical artifice of establishing an equation for pressure. This artifice combines the momentum and continuity equations. Several texts are available in the literature devoted to the analysis of pressure-based methods for solving fluid flow and heat/mass transfer problems. Among the most well-known, one can mention the text book of Patankar (1980), Peyret and Taylor (1983), Anderson *et al.* (1984), Minkowycz *et al.* (1988), Hirsch (1990), and the recent and excellent textbook of Maliska (1995) (the latter so far available in Portuguese limiting its use within the international community). Review articles on the subject of pressure-based methods have also been published (Patankar, 1988; Merkle *et al.*, 1992; Shyy, 1994) and efforts on extending the pressure-based methodology to all speed flows can also be found (Rhie, 1986; Shyy and Braaten, 1988; Shyy *et al.*, 1992). New titles on the control volume method include Ferziger and Periç (1999) and Versteeg and Malalasekera (1995).

## 8.4. GEOMETRY MODELING

### 8.4.1. Computational Grids

The numerical treatment of irregular geometries has particular relevance to the simulation of oil reservoirs and underground dispersion of contaminants. Finite-volume and finite-element techniques have been used for this purpose. The irregular shape of the reservoirs and their geological faults can be treated with the use of boundary-fitted coordinate systems and by means of non-structured grids. Examples of possible grids for use in fluid dynamics problems are shown in Figure 8.3.

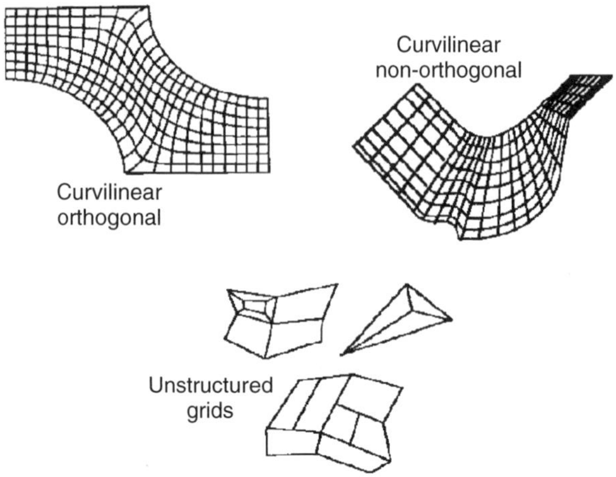


**Figure 8.3.** Examples of computational grids.

When a curvilinear coordinate system is used, all governing equations are usually transformed into the computational plane where the unknowns are actually solved. This transformation takes the general form:

$$\begin{aligned} \xi &= \xi(x, y, z, t) \\ \eta &= \eta(x, y, z, t) \\ \gamma &= \gamma(x, y, z, t) \\ \tau &= t \end{aligned} \tag{8.1}$$

More recently, the use of non-structured grids, such as the ones associated with the finite-element approach, has been extended to the finite-volume family of formulation strategies. Unstructured grids shall be referred to later.

### 8.4.2. Structured Grids

Computational grid layouts are of the *structured* type when the computational node is surrounded by lines (or planes in three dimensions) based on some coordinate system, or, say, along the lines (or planes) where one of the coordinates is of constant value. As a general rule, the use of *structured* grids for computing complex geometries of engineering equipment can always be accomplished in a numerical solution by means of any of the following systems: (a) *regular orthogonal* coordinates with stepwise approximation of boundary control-volumes that cannot be described by constant coordinate lines,

(b) *orthogonal curvilinear* coordinates also with approximations to accommodate non-orthogonal boundaries, and (c) *non-orthogonal curvilinear* coordinates fully complacent with the body contour.

The use of regular orthogonal coordinates (*cartesian*, *cylindrical* or *spherical*) implies the simplest form of governing equations but approximations at irregular boundaries generally lead to poor representation for the exchange rates of mass, momentum and energy between the domain of calculus and its surrounding environment (Gosman and Ideriah, 1976). The discretization process in methods using orthogonal curvilinear coordinate systems follows very closely its Cartesian counterpart leading to an algebraic equation also similar to that of a regular mesh. The computational grid, however, is difficult to generate and does not possess the property of a reasonable control on interior points location (Pope, 1978; Rapley, 1982; Rapley, 1985).

In a non-orthogonal curvilinear coordinate system, the grid curvature makes the formulation less *conservative* in the sense that results will be sensitive to how accurate terms depending on the grid curvature are evaluated (Vinokur, 1974). In spite of difficulties, the relative ease in implementing a non-orthogonal system in existing codes has favored its ever greater use, and examples of engineering flow analysis using body-fitted coordinates can be found in the literature for nuclear-fuel rod bundles (Sha and Thompson, 1979; Padila and de Lemos, 1992) non-circular ducts (de Lemos, 1992c), complex duct junctions (Leschziner and Dimitriadis, 1989), flow in cuspid ducts (de Lemos, 1992b,d, 1994b), internal combustion engines (Gosman *et al.*, 1984), separated flows (Nakayama, 1985), channels with longitudinal curvature (Peric, 1990) and in model dump combustors (Joshi and Vanka, 1991), among others. Large computer codes using body-fitted coordinate system have also been documented in Chen *et al.* (1980) and Chien *et al.* (1983). In addition, most advanced Computational Fluid Dynamics software nowadays makes use of boundary-fitted coordinates.

When Equation (1.1) is written based on the system (8.1) with an element illustrated in Figure 8.4, one has,

$$\frac{1}{J}\frac{\partial \rho}{\partial t}+\frac{\partial(\rho U)}{\partial \xi}+\frac{\partial(\rho V)}{\partial \eta}+\frac{\partial(\rho W)}{\partial \gamma}=0 \tag{8.2}$$

where $J$ is the Jacobian of the transformation given by (8.1) and $U$, $V$, $W$ are the contravariant velocity components defined as (for a fixed transformed grid),

$$\begin{aligned} U &= u(y_\eta z_\gamma - y_\gamma z_\eta)+v(z_\eta x_\gamma - z_\gamma x_\eta)+w(x_\eta y_\gamma - x_\gamma y_\eta)\\ V &= u(y_\gamma z_\xi - y_\xi z_\gamma)+v(z_\gamma x_\xi - z_\xi x_\gamma)+w(x_\gamma y_\xi - x_\xi y_\gamma)\\ W &= u(y_\xi z_\eta - y_\eta z_\xi)+v(z_\xi x_\eta - z_\eta x_\xi)+w(x_\xi y_\eta - x_\eta y_\xi) \end{aligned} \tag{8.3}$$

being the Jacobian given by,

$$J\left\{x_\xi y_\eta z_\gamma + x_\gamma y_\xi z_\eta - x_\eta y_\gamma z_\xi - x_\xi y_\gamma z_\eta - x_\gamma y_\eta z_\xi - x_\eta y_\xi z_\gamma\right\} \tag{8.4}$$

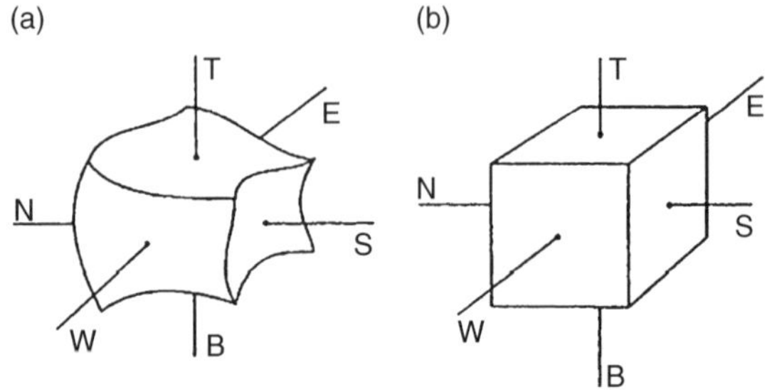


**Figure 8.4.** Element transformation: (a) generalized coordinates ($\xi$, $\eta$, $\gamma$), (b) Cartesian coordinates ($x$, $y$, $z$).

For a general variable $\varphi$, Equation (1.1) can be written in curvilinear coordinates as,

$$\begin{aligned}
&\frac{\partial}{\partial t}\left(\frac{\rho\varphi}{J}\right)+\frac{\partial}{\partial \xi}(\rho \mathrm{U}\varphi)+\frac{\partial}{\partial \eta}(\rho \mathrm{V}\varphi)+\frac{\partial}{\partial \gamma}(\rho \mathrm{W}\varphi)\\
&=\frac{\partial}{\partial \xi}\left\{\left(\mathrm{A}\frac{\partial\varphi}{\partial \xi}+\mathrm{D}\frac{\partial\varphi}{\partial \eta}+\mathrm{E}\frac{\partial\varphi}{\partial \gamma}\right)\frac{\Gamma_\varphi}{J}\right\}+\frac{\partial}{\partial \eta}\left\{\left(\mathrm{D}\frac{\partial\varphi}{\partial \xi}+\mathrm{B}\frac{\partial\varphi}{\partial \eta}+\mathrm{F}\frac{\partial\varphi}{\partial \gamma}\right)\frac{\Gamma_\varphi}{J}\right\}\\
&+\frac{\partial}{\partial \gamma}\left\{\left(\mathrm{E}\frac{\partial\varphi}{\partial \xi}+\mathrm{F}\frac{\partial\varphi}{\partial \eta}+\mathrm{C}\frac{\partial\varphi}{\partial \gamma}\right)\frac{\Gamma_\varphi}{J}\right\}+\frac{\mathrm{S}_\varphi}{J}
\end{aligned} \tag{8.5}$$

where the coefficients $A$-$F$ in Equation (8.5) depend on the metrics of the transformation and are given by,

$$\begin{aligned}
A &= \xi_x^2+\xi_y^2+\xi_z^2\\
B &= \eta_x^2+\eta_y^2+\eta_z^2\\
C &= \gamma_x^2+\gamma_y^2+\gamma_z^2\\
D &= \xi_x\eta_x+\xi_y\eta_y+\xi_z\eta_z\\
E &= \eta_x\gamma_x+\eta_y\gamma_y+\eta_z\gamma_z\\
F &= \gamma_x\xi_x+\gamma_y\xi_y+\gamma_z\xi_z
\end{aligned} \tag{8.6}$$

An application of (8.5) in porous media shall be discussed later.

### 8.4.3. Unstructured Grids

When using unstructured grids, the computational domain is divided into nodes, elements or volumes formed by lines (or planes) which do not have a constant value of some coordinate. An advantage of unstructured grids is the possibility of refining or *adapting* it in regions of high gradients without increasing the number of nodes elsewhere.

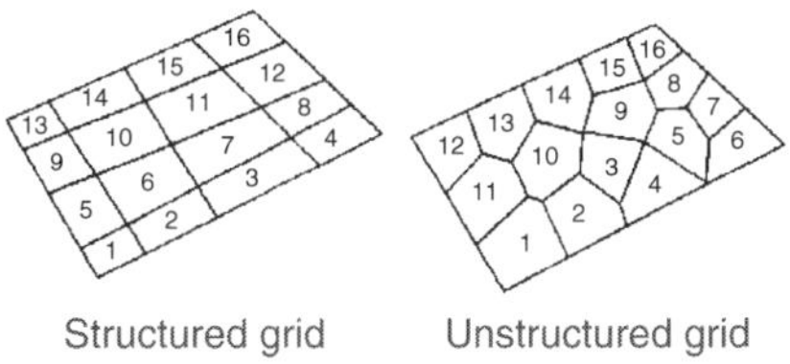


**Figure 8.5.** Grid numbering.

A comparison of typical node indexing used in both structured and non-structured grids is shown in Figure 8.5.

Due to the non-ordered numbering shown in the figure, the matrix of coefficients will not possess a fixed band. Among existing possibilities, non-structured grids can be based on Voronoi elements (Voronoi, 1908) and on the "control volume finite element method" (Baliga and Patankar, 1980; Baliga *et al.*, 1983). Examples of the Voronoi and the control-volume finite-element grids are illustrated in Figure 8.6.

### 8.4.4. Application to Reservoir Simulation

The subject of petroleum reservoir simulation has been considered by many research groups and has been the motive of detailed documentation (Aziz and Settari, 1980). The development of numerical methods for this particular application has inherent difficulties due to the complexity of the geometry analyzed. As a desired characteristic of an applicable numerical formulation one can mention the ability to handle irregular shapes including consideration of geological faults. Robustness, accuracy and stability are also essential requirements for suitable computational modeling.

Guided by the above ideas, flows in oil reservoirs have been treated with Cartesian coordinates (Yanosik and McCracken, 1979; Rubin and Blunt, 1991) in spite of the fact

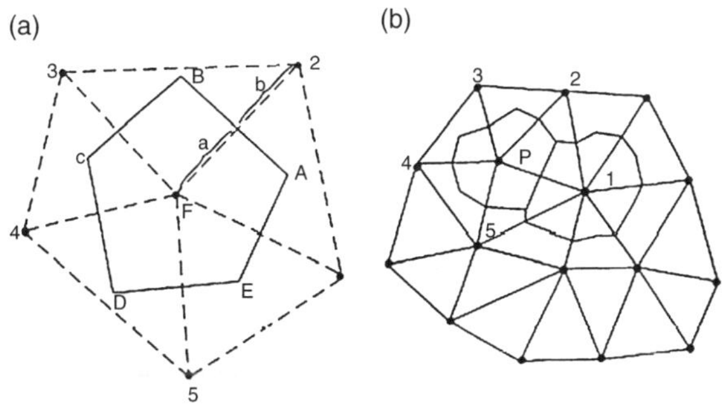


**Figure 8.6.** Unstructured grids: (a) Voronoi diagram, (b) control volume finite element.

that accurate treatment of boundaries and internal volumes cannot be done well with this sort of grid. Better results have been obtained with the use of a system of curvilinear coordinate (Ferguson and Wadsley, 1986; Sharpe and Anderson, 1991; Maliska *et al.*, 1994). Some of the basic equations for this case are shown below.

For three-dimensional numerical solution in curvilinear coordinates, Equation (1.3) needs to be solved in this new coordinate system. There are two possibilities in achieving this goal. Either one can integrate the governing equations written in the Cartesian coordinates in the irregular element of Figure 8.4a, or else one can transform the governing equation into the new system and proceed with the integration on the volume of Figure 8.4b.

Using the second alternative when applying the black-oil model shown in Chapter 1, one can get Equation (1.18), transformed into the new system, as

$$\frac{1}{J}\frac{\partial}{\partial t}\left(\frac{S_\kappa}{B_\kappa}\phi\right) = \frac{\partial}{\partial \xi}\left[D_{1\kappa}\frac{\partial p}{\partial \xi} + D_{2\kappa}\frac{\partial p}{\partial \eta} + D_{3\kappa}\frac{\partial p}{\partial \gamma}\right] + \frac{\partial}{\partial \eta}\left[D_{4\kappa}\frac{\partial p}{\partial \xi} + D_{5\kappa}\frac{\partial p}{\partial \eta} + D_{6\kappa}\frac{\partial p}{\partial \gamma}\right] + \frac{\partial}{\partial \gamma}\left[D_{7\kappa}\frac{\partial p}{\partial \xi} + D_{8\kappa}\frac{\partial p}{\partial \eta} + D_{9\kappa}\frac{\partial p}{\partial \gamma}\right] - \frac{q_\kappa}{J} \tag{8.7}$$

where the coefficients in Equation (8.7) can be found in Maliska *et al.*, (1994); and the phase identifier is now $\kappa$ to avoid confusion with the coordinate $\gamma$. One can see that for solving Equation (8.7) one needs to evaluate three direct derivatives and six cross-derivatives (see Maliska *et al.* (1994). According to the finite-volume procedure, Equation (8.7) needs now to be integrated in time and over the elemental control volume shown in Figure 8.4b. The integration gives:

$$\begin{aligned}\frac{1}{J}\frac{\Delta V}{\Delta t}\left[\left(\frac{S_\kappa}{B_\kappa}\phi\right) - \left(\frac{S_\kappa}{B_\kappa}\phi\right)^*\right] = & \left[D_{1\kappa}\frac{\partial p}{\partial \xi} + D_{2\kappa}\frac{\partial p}{\partial \eta} + D_{3\kappa}\frac{\partial p}{\partial \gamma}\right]_e \Delta\eta\Delta\gamma \\ & - \left[D_{1\kappa}\frac{\partial p}{\partial \xi} + D_{2\kappa}\frac{\partial p}{\partial \eta} + D_{3\kappa}\frac{\partial p}{\partial \gamma}\right]_w \Delta\eta\Delta\gamma \\ & + \left[D_{4\kappa}\frac{\partial p}{\partial \xi} + D_{5\kappa}\frac{\partial p}{\partial \eta} + D_{6\kappa}\frac{\partial p}{\partial \gamma}\right]_n \Delta\xi\Delta\gamma \\ & - \left[D_{4\kappa}\frac{\partial p}{\partial \xi} + D_{5\kappa}\frac{\partial p}{\partial \eta} + D_{6\kappa}\frac{\partial p}{\partial \gamma}\right]_s \Delta\xi\Delta\gamma \\ & + \left[D_{7\kappa}\frac{\partial p}{\partial \xi} + D_{8\kappa}\frac{\partial p}{\partial \eta} + D_{9\kappa}\frac{\partial p}{\partial \gamma}\right]_t \Delta\xi\Delta\eta \\ & - \left[D_{7\kappa}\frac{\partial p}{\partial \xi} + D_{8\kappa}\frac{\partial p}{\partial \eta} + D_{9\kappa}\frac{\partial p}{\partial \gamma}\right]_b \Delta\xi\Delta\eta - \frac{q_\kappa \Delta V}{J}\end{aligned} \tag{8.8}$$

The method called IMPES (IMplicit Pressure Explicit Saturation) consists in solving an equation for pressure after adding (8.8) written for both water and oil saturations under the constraint given by Equation (1.16). The pressure being calculated, the saturations can

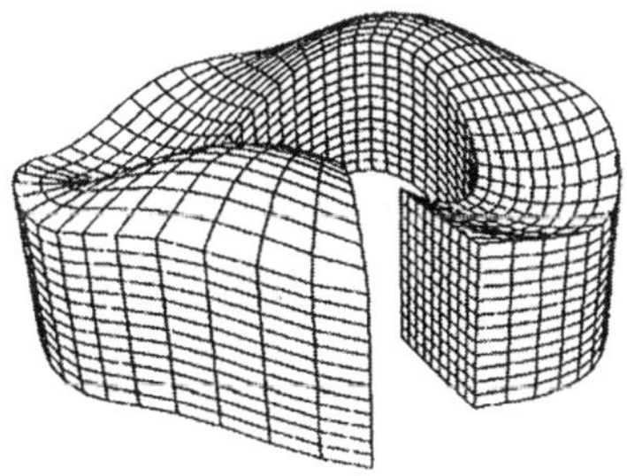

**Figure 8.7.** Three-dimensional structured grid for reservoir simulation.

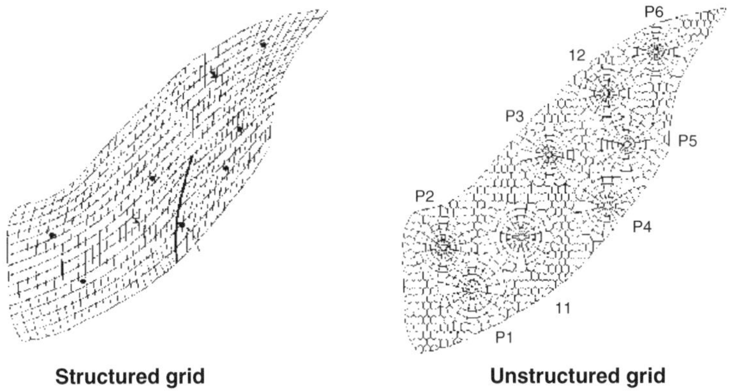


**Figure 8.8.** Numerical treatment of petroleum reservoirs (Maliska, 1995).

be found explicitly. A typical three-dimensional structured grid for reservoir simulation is shown in Figure 8.7.

Other methods making use of unstructured grids have been proposed. The same oil reservoir is analyzed in Figure 8.8 with structured and non-unstructured grids. Note the advantages of unstructured grids concentrating nodes around production wells.

Studies on preconditioning of the matrix of coefficients with particular application in petroleum reservoirs simulation have also been reported (Behie and Vinsome, 1982; Behie and Forsyth, 1984).

## 8.5. TREATMENT OF THE CONVECTION TERM

### 8.5.1. The Nature of the Numerical Solution

*Real-world* engineering problems can be tackled today with the numerical computation of available mathematical models completed with adequate information supplied in the form of constitutive and empirical equations.

A numerical solution of problems in fluid dynamics implies first that the partial differential equations governing the flow be transformed into a set algebraic equations, giving rise to the matrix shown in Figure 8.2. The procedure of transforming the character of the governing equation set is usually referred to as the *numerical formulation*. This first step includes the choice of a *method* (finite-difference, finite-volume, finite-element) and one or several *discretization schemes* for treating different terms in the original equations. Examples of grids used by different discretization methods are shown in Figure 8.9.

The algebraic equations come from approximating the original transport terms over non-overlapping finite-elements or finite-volumes. In this process, the discretization of the non-linear terms plays the most crucial role for stability and convergency of the solution. Once all participating equations have been converted to their appropriate discrete form, a numerical algorithm must be applied to the full set for finally getting the results.

One should note that the *formulation* in a numerical solution is associated with accuracy of the results whereas the *algorithm* is related to the required computing time. Further, the formulation should be consistent, in the sense that the *exact* solution should be approached as the number of grid points is increased. Finally, a *robust* algorithm, which leads to a non-divergent solution regardless of boundary conditions and input parameters, would further add to the solution the characteristic of stability. Convergence is then achieved when the numerical method is *stable*. Correctness of results requires that the numerical method employed be well formulated and uses a robust algorithm so that a *consistent* and *stable* solution is reached.

### 8.5.2. Interpolating Functions

When integrating Equation (1.1) around the control-volumes shown in Figure 8.4, information on the value of the general variable at the faces of the finite-volume is necessary. Simple interpolation schemes using a linear internodal variation (central differencing scheme – CDS) or the "upwinded" value for $\varphi$ is shown in Figure 8.10 (see Patankar, 1980). These schemes, although extremely easy to numerically implement, are known to cause oscillating and diffused solutions, respectively, as can be observed in the computation of a pulse shown in Figure 8.11. Oscillating solutions are non-physical and considered as a consequence of non-dissipative truncation errors. On the other hand, smoothed solutions are physically realistic but suffer from dissipative truncated approximations.

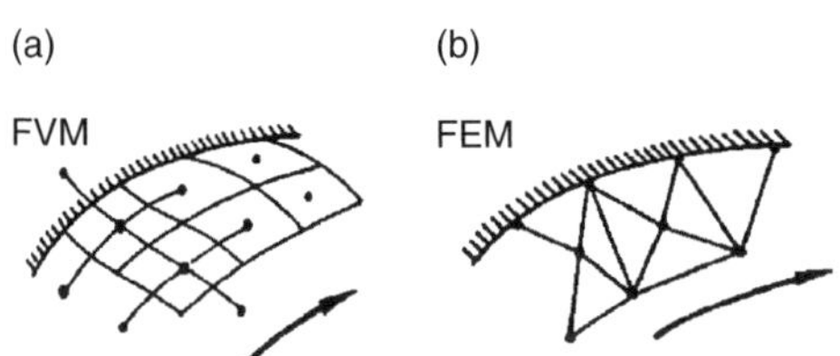


**Figure 8.9.** Grid layout for discretization methods: (a) finite-volume and (b) finite-element.

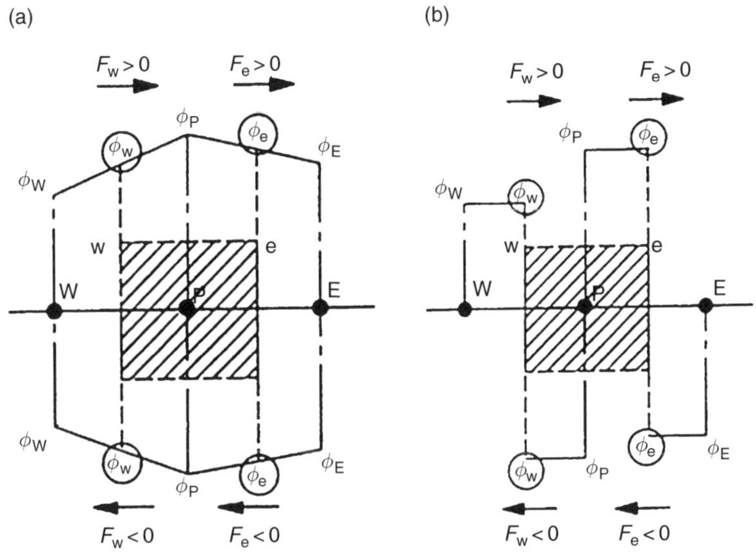


**Figure 8.10.** Simplified schemes: (a) central-differencing, (b) upwind.

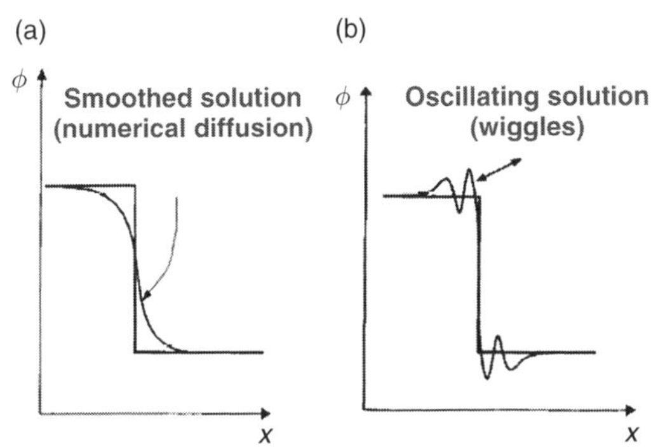


**Figure 8.11.** Effect of interpolation scheme: (a) UPWIND-numerical diffusion, (b) CDS-wiggles.

The analytical solution of a uni-dimensional convective–diffusive problem governed by

$$\frac{\partial}{\partial x}\left[\rho u \varphi - \Gamma_{\varphi}\frac{\partial \varphi}{\partial x}\right] = 0 \tag{8.9}$$

is the basis of the exponential scheme. This scheme considers both convection and diffusion as an exact function of the Peclet number, defined as the ratio of convection and diffusion strengths (see Patankar, 1980). One of the problems with the exponential scheme is the use of "hard-to-compute" exponential functions. Other schemes – hybrid scheme (Spalding, 1972), weighted upstream differencing scheme (Raithby and Torrance, 1974), power law scheme (Patankar, 1981), second-order upwind (Warming and Beam, 1976; Vanka, 1987) and QUICK (Leonard, 1979), still considering both convection and diffusion, try to avoid the use of exponential function. Examples of more advanced interpolating functions are shown in Figure 8.12.

The spurious oscillations appearing when computing sharp gradients in convection-dominated flows (see Figure 8.11) can be controlled by adding artificial viscosity to the governing equations (Roache, 1972; Hirsch, 1990) or by using intrinsic dissipative interpolating functions (Shyy, 1985). For density-based solvers, the concept of *total variation diminishing* (TVD) (Harten, 1984) has been widely used, permitting the computation of sharp gradients without solution smearing. Other studies on oscillation control have been reported (Harten, 1983; Chen *et al.*, 1991).

When the flow is inclined with respect to the grid lines, as shown in Figure 8.13, additional numerical diffusion is known to contaminate the solution. Schemes developed which take into consideration the slanting flow streamlines are the skew upstream differencing scheme and Skew Weighted Difference Scheme (Raithby, 1976). Further developments are known as the Modified Skew Upstream Scheme and Mass Weighting Upstream Scheme (Huget, 1985). More advanced discretization schemes for the convection term have been published, but a complete overview of their details is beyond the scope of this text. Here, only basic and more commonly used techniques were mentioned.

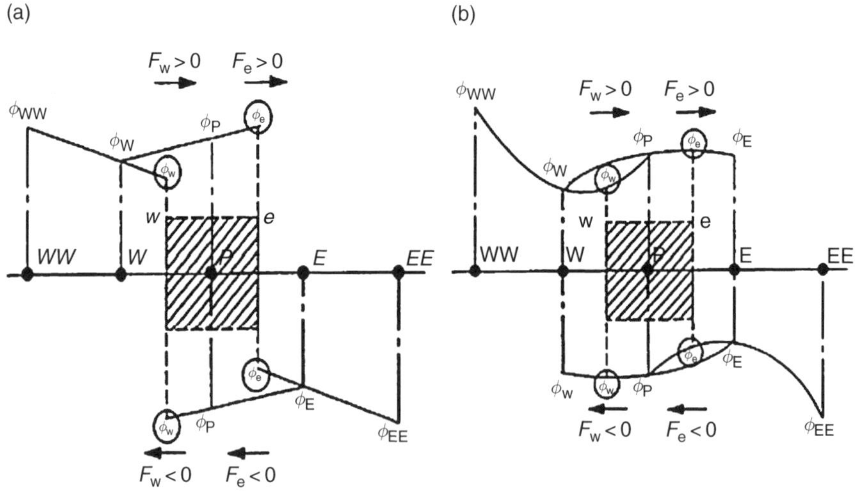


**Figure 8.12.** Advanced schemes: (a) second-order upwind, (b) QUICK scheme.

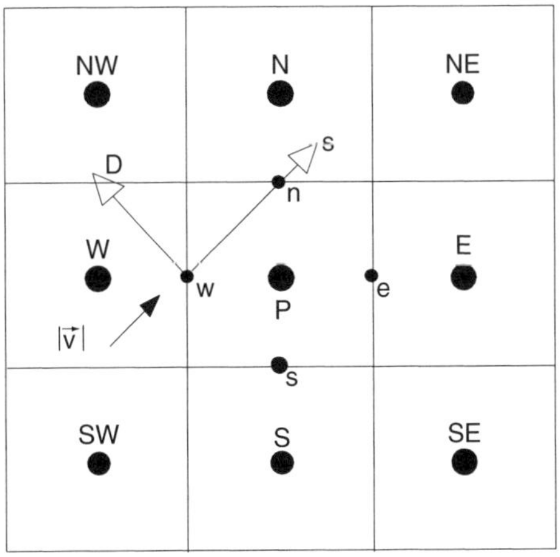


**Figure 8.13.** Inclination of grid lines with respect to velocity vector.

## 8.6. DISCRETIZED EQUATIONS FOR TRANSIENT THREE-DIMENSIONAL FLOWS

In the control-volume method, the focus of the present work, the equations governing the flow and heat/mass transfer are integrated, in time and space, over the volume shown in Figure 8.9. This operation takes the form,

$$\int_{v} \left\{ \frac{\partial}{\partial t}(\rho\varphi) + \frac{\partial}{\partial x_j}\left(\rho u_j \varphi - \Gamma_\varphi \frac{\partial}{\partial x_j}\right) - S_\varphi \right\} = 0 \tag{8.10}$$

The outcome of this manipulation is an algebraic equation connecting the nodal grid point in question to its neighbors in space and time. The coefficients take into consideration the local convection and diffusion strengths through the use of any of the interpolation functions given above. Depending on the discretization scheme used, the final equation set can be classified into *fully implicit*, *fully explicit* or *partially implicit* in time. Details of discretization procedures related to the control-volume approach can be found elsewhere (see for example Patankar and Spalding, 1972) and shall not be repeated here. The final discretized equation takes the following form for a general nodal point *P*:

$$a_P \varphi_P = \sum_{n=1}^{N} a_n \varphi_n + b \tag{8.11}$$

The number *N* in Equation (8.11) is the number of surrounding grid nodes and depends on the grid used and on the flow dimension. For a three-dimensional case, calculated with an orthogonal grid, $N$ is equal to 6. If non-orthogonal meshes are used, cross-derivative terms increase $N$ to a value of 9 for two-dimensional cases and 18 for three-dimensional situations.

Note that Equation (8.11) represents actually several systems of algebraic equations. For turbulent non-isothermal three-dimensional flows using the $k - \varepsilon$ model, the general variable is $\varphi = u_i$, $\Theta$, $k$, $\varepsilon$. These algebraic equations are coupled and highly non-linear, and the process of finding their numerical solution will necessarily make use of repetitive or iterative solvers. The different procedures most used in the literature for solving Equation (8.11) are discussed next.

## 8.7. SYSTEMS OF ALGEBRAIC EQUATIONS

### 8.7.1. Inter-linkage and Coupling among Variables

The coupled system of algebraic equations represented by Equation (8.11) is highly non-linear. Due to the elliptic nature of the original differential equations, a single variable depends on itself evaluated at another location, within the field, in addition to being connected to all other variables. These distinct dependencies are usually referred to as inter-linkage and coupling, respectively.

The traditional method to handle inter-linkage is to resolve the pertinent individual systems, for each variable, while holding all the others still. In this case, *block-wise* methods based on point (Jacobi, Gauss-Seidel, SOR), line (TDMA) or plane relaxation can be used (Patankar, 1980). Iterative methods which are based on incomplete factorization of the entire matrix of coefficients have also been applied to fluid flow problems (Stone, 1968; Schneider and Zedan, 1981). These methods are generally called *full-field* procedures. Generally speaking, block-wise methods offer advantage when used in conjunction with modern computer architectures (such as vector and parallel computers), whereas full-field solvers present better performance with scalar processors.

The handling of the coupling between variables will depend on the nature of the problem. Different velocity components, any scalar (temperature, mass concentration, etc.) or parameters for representing additional phenomena (turbulence, chemical reaction, etc.) will have their own algebraic equation and will be coupled with each other. For incompressible flows the pressure–velocity coupling constitutes the major difficulty whereas for compressible flows the pressure–velocity/density dependence will be of concern. For thermally driven flows, density will be affected by temperature (not pressure), the incompressible formulation still being applicable.

In the present context, when an error-smoothing operator is applied to individual variables in a sequential manner, for handling any sort of coupling among them, the entire procedure is here recalled as a segregated approach. If, by any artifice, more than one variable is updated in each sweep over the computational domain, a coupled method is being considered. With these ideas in mind, below is a summary of the most common algorithm sequences applied to the solution of incompressible flows. Since most applications of flows in porous media are within the incompressible limits, compressible algorithms shall not be discussed upon here.

### 8.7.2. Segregated Methods

One of the essential features of methods for incompressible flow is the establishment of an evolving equation for pressure. As seen above, for compressible flows the equation of state is the evolving equation for $p$ whereas for incompressible flows a combination of the continuity of mass and momentum equations gives rise to the so-called pressure equation. The calculated velocity field in the latter case is such that it has to satisfy continuity.

Early attempts to handle the pressure–velocity coupling are found in the works of Harlow and Welch (1965), Chorin (1967, 1971) and Patankar and Spalding (1972). With the publication of the textbook of Patankar in 1980, the SIMPLE method in Patankar and Spalding (1972) was revised and named SIMPLER. Maliska (1981) proposed the PRIME method using the same algebraic equation for correcting for pressure and velocities. The work by van Doormaal and Raithby (1984) improved the early Patankar's algorithm through a consistent form of the SIMPLE method. Galpin *et al.* (1985) have further increased the rate of convergency of all equations with their CELS method. Other methods handling incompressible flows have also been proposed, as for example in Issa (1986).

### 8.7.3. Coupled Methods

The dependence among the variables involved can also be treated implicitly at a point, line or plane. These procedures do not make use of a pressure or pressure-correction scheme since both momentum and continuity equations, in their discretized form, are solved locally by a direct method. The main motivation for these coupled methods is to reduce overall computing times by treating all dependences among variables in a direct manner.

Motivated by the foregoing, in de Lemos (1988a), a point-wise block-implicit numerical scheme based on the work of Vanka (1985–1987) for solving the continuity and momentum equations was applied to lid-driven cavity flows. Later, the technique was extended to fluid motion through a cylindrical tank (de Lemos, 1990), to buoyancy-driven streams (de Lemos, 1992a) and to the calculation of swirling flows (de Lemos, 1992b). A generalization of the method for one additional general scalar variable has also been documented (de Lemos, 1994a). Recently, the formulation in de Lemos (1992a,b) was extended to handle two scalar variables, as it occurs when computing axi-symmetric buoyant swirling flows (de Lemos, 1996). In this case, temperature and tangential velocity can be treated as scalar fields solved together with the cross-flow velocity components.

The point-wise aspect of the error smoothing operator used in those locally-implicit methods are promising regarding their applicability in vector and parallel computers.

## 8.8. TREATMENT OF THE u,w-T COUPLING

### 8.8.1. Introduction

The problem of heat transfer enhancement through gaps between walls in cavities finds many applications in science and engineering (Catton, 1978; Churchill, 1992). Enhancement or damping of heat transfer rates across layers of fluids have a wide range of employment spanning from simple insulating systems to sophisticated technological devices (Yang, 1987). When the system is closed and subjected to a temperature gradient, a recirculating flow field is established. If the cavity itself is further inclined with respect to the vertical direction, important changes occur in the heat transfer strength across the domain (Hollands *et al.*, 1976).

For two-dimensional cases, in addition to having gravity directly affecting both coordinates, the numerical solution of such flows imposes additional difficulties due to the intricate coupling between *temperature* and *cross-flow* fields. Linearization of governing equations followed by the use of iterative solvers is the common route found in the literature for solving such non-linear problems. Accordingly, the rate of convergence of any algorithm is essentially dictated by the degree in which physical coupling is mimicked by the method in question. Ultimately, this is an indication that numerical solutions of buoyant flows, in most cases, suffer from the disadvantage of longer computing times when compared to their non-buoyant counterpart.

Segregated methods, in which one individual flow variable is relaxed while holding the others still, are known to be rather sensitive when handling strong physical coupling. For that, the so-called *coupled* solvers, where all dependent variables are relaxed in the same domain location, have received much attention lately.

Benchmark solutions for the buoyancy-driven laminar flow field in a square cavity have been presented (de Vahl Davis, 1983) and a comparison of 37 different contributions for solving the same problem has been compiled also by the same author (de Vahl Davis and Jones, 1983). Multi-grid solution for this problem has also been published (Hortmann *et al.*, 1990). In the great majority of those works, a segregated method is generally employed with the repetitive solution of a pressure or pressure-correction equation followed by subsequent updates of the velocity and scalar fields. This strategy forms the basis of the SIMPLE family of algorithms (Patankar, 1980). Coupled line solvers for the temperature and velocity fields have shown improvements in computer time requirements for natural convection flows with large Rayleigh numbers (Galpin and

Raithby, 1986). The work in Galpin and Raithby (1986) is an indication of the advantage of coupled schemes for solving algebraic equations set with a high degree of inter-linkage among the variables. Recently, the block-implicit technique has also been applied to calculation of buoyant flows in a partially coupled manner (Tang and Joshi, 1999). In the present context, a fully implicit treatment is associated with the idea of simultaneously updating the flow and temperature fields at each step within the error-smoothing operator.

Following the aforementioned and based on Vanka's SGCS method (Vanka, 1986a,b), simulated lid-driven cavity fluid motion using a block-implicit numerical scheme was presented in de Lemos (1990). Later, the technique was extended to vertical (de Lemos, 1997a) and inclined cavities (de Lemos, 1997b). In those papers, a fully implicit treatment for temperature was made use of. The objective of this section is to present the numerical formulation and results of de Lemos (2000), which extended the block-implicit arrangement of de Lemos (1990–1997b) for including a fully implicit treatment of the energy equation. Heat cavity flows for vertical, horizontal and inclined geometry with respect to the horizontal direction are presented. Effects of tilt angle, Rayleigh number and aspect ratio are reported.

## 8.8.2. Analysis and Numerics

### *8.8.2.1. Geometry*

The geometry considered in this work is schematically shown in Figure 8.14. An enclosure of height $H$ and width $L$ is insulated at both top and bottom walls while constant-temperature conditions $T_1$ and $T_0$ prevail over its lateral faces, respectively. Depending upon the value of tilt angle $\alpha$, several cases can be identified as in Table 8.1. Cases in the table are referred to in Figure 8.14.

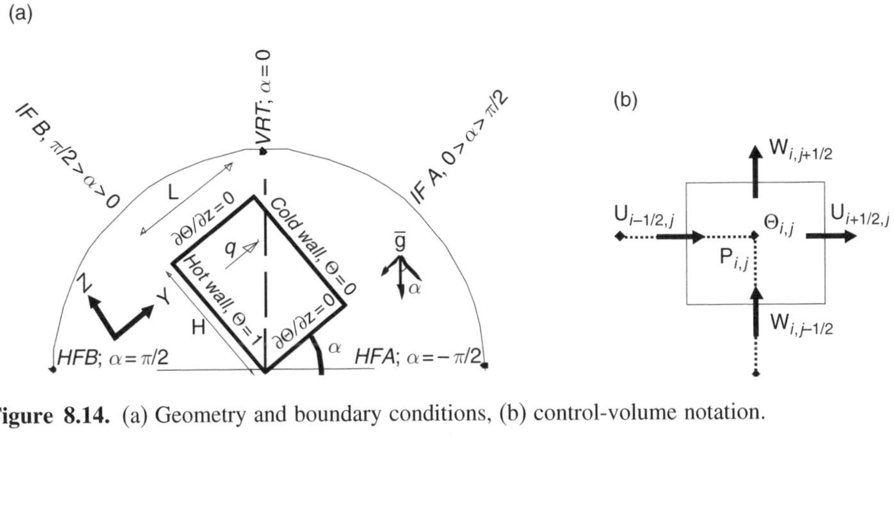


**Figure 8.14.** (a) Geometry and boundary conditions, (b) control-volume notation.

**Table 8.1.** Cases investigated.

| Case | Angle $\alpha$ | Description |
|---|---|---|
| HFB | $\pi/2$ | Horizontal, heated from below |
| IFB | $0 < \alpha < \pi/2$ | Inclined, heated from below |
| VRT | $\alpha = 0$ | Vertical, heated from left |
| IFA | $-\pi/2 < \alpha < 0$ | Inclined, heated from above |
| HFA | $-\pi/2$ | Horizontal, heated from above |

### *8.8.2.2. Compact Notation*

The conservation equations for mass, momentum and energy here analyzed can be written in a compact form if the existing analogies among the processes of accumulation, transport, convection and generation/destruction of those quantities is observed. This generic equation is commonly known in the literature as the *general transport equation* and can be written in its *conservative* two-dimensional laminar form as:

$$\frac{\partial}{\partial x_j}\left[\rho U_j \varphi - \Gamma_\varphi \frac{\partial \varphi}{\partial x_j}\right] = S_\varphi \tag{8.12}$$

In Equation (8.12), $\varphi$ can represent any quantity of vectorial or scalar nature (velocity or temperature), $\rho$ is the fluid density, $U_j$ are the velocity components $(U, W)$ in the $x_j$-directions $(y, z)$, respectively, $\Gamma_\varphi$ is the transport coefficient for diffusion and $S_\varphi$ is the source term. Table 8.2 identifies correspondent terms for the different equations represented by Equation (8.12). In Table 8.2 and Equation (8.12) gravity acts in both the $z$- and $y$-direction, $\mu$ is the fluid viscosity, $Pr$ the Prandtl number and $T$ the temperature.

### *8.8.2.3. Discretized Equations*

The set of equations for mass, momentum and energy summarized above is *differentiated* by means of the widely used control-volume approach of Patankar (1980). Equation (8.12) is integrated over the volume of Figure 8.14 yielding a set of algebraic equations. In the present work, for simplicity, the *upwind differencing scheme* is used to model convective fluxes across volume faces.

**Table 8.2.** Terms in general transport Equation (8.12).

| | $\varphi$ | $\Gamma_\varphi$ | $S_\varphi$ |
|---|---|---|---|
| Continuity | 1 | 0 | 0 |
| $z$ momentum | $W$ | $\mu$ | $-\frac{\partial P}{\partial z} - \cos\alpha \rho_0 g\beta(T - T_0)$ |
| $y$ momentum | $U$ | $\mu$ | $-\frac{\partial P}{\partial y} - \sin\alpha \rho_0 g\beta(T - T_0)$ |
| Energy | $T$ | $\mu/Pr$ | 0 |

Integrating then the continuity, momentum and energy equations around the point $(i, j)$ (see notation on Figure 8.14) one has:

$$F_{i+\frac{1}{2}}U_{i+\frac{1}{2},j}-F_{j-\frac{1}{2}}U_{i-\frac{1}{2},j}+F_{j+\frac{1}{2}}W_{i,j+\frac{1}{2}}-F_{j-\frac{1}{2}}W_{i,j-\frac{1}{2}}=0 \tag{8.13}$$

$$U_{i-\frac{1}{2},j}=\hat{U}_{i-\frac{1}{2},j}+\hat{d}_{i-\frac{1}{2}}\left[P_{i-1,j}-P_{i,j}\right]+\hat{g}_{i-\frac{1}{2}}\Theta_{i,j} \tag{8.14}$$

$$W_{i,j-\frac{1}{2}}=\hat{W}_{i,j-\frac{1}{2}}+\hat{d}_{j-\frac{1}{2}}\left[P_{i,j-1}-P_{i,j}\right]+\hat{g}_{j-\frac{1}{2}}\Theta_{i,j} \tag{8.15}$$

where

$$\hat{U}_{i-\frac{1}{2},j}=\left(\sum_{nb=1}^{4}a^{\mathrm{u}}_{nb}U_{nb}+f^{\mathrm{u}}_{i-\frac{1}{2},j}\right)\Big/a^{\mathrm{u}}_{i-\frac{1}{2},j} \tag{8.16}$$

$$\hat{W}_{i,j-\frac{1}{2}}=\left(\sum_{nb=1}^{4}u^{\mathrm{w}}_{nb}W_{nb}+f^{\mathrm{w}}_{i,j-\frac{1}{2}}\right)\Big/a^{\mathrm{w}}_{i,j-\frac{1}{2}} \tag{8.17}$$

$$\Theta_{i,j}=\hat{\Theta}_{i,j}=\left\{\sum_{nb=1}^{4}a^{\Theta}_{nb}\Theta_{nb}\right\}\Big/a^{\Theta}_{i,j} \tag{8.18}$$

and

$$\begin{aligned}\hat{d}_{i\frac{1}{2}}&=S^{\mathrm{u}}_{i\frac{1}{2}}/a^{\mathrm{u}}_{i\frac{1}{2},\mathrm{j}}\\ \hat{g}_{i\frac{1}{2}}&=\rho_0\sin\alpha g\beta(T_1T_0)/2a^{\mathrm{u}}_{i\frac{1}{2},\mathrm{j}}\end{aligned} \tag{8.19}$$

$$\begin{aligned}\hat{d}_{j-\frac{1}{2}}&=S^{\mathrm{w}}_{j-\frac{1}{2}}/a^{\mathrm{w}}_{i,j-\frac{1}{2}}\\ \hat{g}_{j-\frac{1}{2}}&=\rho_0\cos\alpha g\beta(T_1-T_0)/2a^{\mathrm{u}}_{i,j-\frac{1}{2}}\end{aligned} \tag{8.20}$$

where the geometric coefficients $F$s can be interpreted as (area of flow)/(volume of computational node). Further, when calculating free convection flows oriented as in Figure 8.14, the non-dimensional temperature is defined as:

$$\Theta=T-T_0/T_1-T_0 \tag{8.21}$$

and considers the maximum $T$ drop across the computational domain, $\Delta T=(T_1-T_0)$. The coefficients $a^{\varphi}_{nb}$ appearing in Equations (8.16)–(8.18), referent to *neighbor nodal points*, accounts for the contributions, at each face, of the mechanisms of convection and diffusion. For the general variable $\varphi$ and for the purely upwind formulation here considered its general form reads:

$$a^{\varphi}_{nb}=\frac{\Gamma_{\varphi}}{\rho}\frac{A_{nb}}{\delta r_{nb}}+\|A_{nb}U_{nb},0\| \tag{8.22}$$

where $A_{nb}$ is the area of transport, $U_{nb}$ the velocity prevailing over the control-volume face area $A_{nb}$ and $\delta r_{nb}$ the distance between adjacent control-volume centers. Also, in Equations (8.16) and (8.17), $f$ signifies all sources except the pressure gradient and buoyancy terms. It is important to notice that the source term in Equations (8.14) and (8.15) explicitly shows the contribution due to buoyancy. For the coupled treatment presented here, this arrangement is necessary as will be seen below.

### 8.8.2.4. Numerical Strategy

In order to smooth out errors due to initial guessed fields, corrections are defined as differences between *exact* and *approximate* variables. Residuals for momentum equations at each control-volume face, continuity of mass and energy equations are obtained by applying the just defined approximate values into Equations (8.14), (8.15) and (8.18). After some manipulation, a system connecting the residuals and corrections of Equations (8.7), (8.14), (8.15) and (8.18) can be written into matrix form as:

$$\begin{bmatrix} 1 & 0 & 0 & 0 & \hat{d}_{i-\frac{1}{2}} & \hat{g}_{i-\frac{1}{2}} \\ 0 & 1 & 0 & 0 & -\hat{d}_{i+\frac{1}{2}} & \hat{g}_{i+\frac{1}{2}} \\ 0 & 0 & 1 & 0 & \hat{d}_{j-\frac{1}{2}} & \hat{g}_{j-\frac{1}{2}} \\ 0 & 0 & 0 & 1 & -\hat{d}_{j-\frac{1}{2}} & \hat{g}_{j+\frac{1}{2}} \\ -F_{i-\frac{1}{2}} & F_{i+\frac{1}{2}} & -F_{j-\frac{1}{2}} & F_{j+\frac{1}{2}} & 0 & 0 \\ 0 & 0 & 0 & 0 & 0 & 1 \end{bmatrix} \begin{bmatrix} U'_{i-\frac{1}{2},j} \\ U'_{i+\frac{1}{2},j} \\ W'_{i,j-\frac{1}{2}} \\ W'_{i,j+\frac{1}{2}} \\ P'_{i,j} \\ \Theta'_{i,j} \end{bmatrix} = \begin{bmatrix} R_{i-\frac{1}{2},j} \\ R_{i+\frac{1}{2},j} \\ R_{i,j-\frac{1}{2}} \\ R_{i,j+\frac{1}{2}} \\ R_{i,j} \\ R^{\Theta}_{i,j} \end{bmatrix} \tag{8.23}$$

In Equation (8.23) the first four rows correspond to discretized forms of momentum equations applied to the four faces of the control-volume of Figure 8.14. The fifth row comes from the continuity equation and the last one has its origin in the discretization of Equation (8.18). Here also the subscripts identify locations in the grid, the superscript $'$ distinguishes corrections and the lhs residue vector corresponds to the one at previous iteration. In Equation (8.23) the influence of $\Theta$ on the flow field is directly accounted for by the $g$-terms. The reverse effect, or, say, the cross-flow influence on temperature, is here not treated implicitly. The solution of system (8.23) is easily obtained by finding first corrections for $\Theta$, then calculating the pressure $P$ and velocity components $U$ and $W$.

### 8.8.2.5. Partially Segregated Treatment

The algebraic equations for the velocity field were solved, in addition to the fully coupled scheme here described, by performing outer iterations for the non-dimensional temperature while keeping $U$, $W$ and $P$ from the previous iteration. A *line-by-line* smoothing operator, fully described elsewhere (Patankar, 1980), was used to relax $\Theta$ being the cross-flow field $(U, W)$ calculated by the locally coupled method seen above. This *partially segregated* or *semi-coupled* solution was set in such a way that the same number of sweeps per outer iteration, throughout the scalar ($\Theta$) and cross-flow fields ($U$, $W$), was obtained in both the coupled and semi-coupled methods. This procedure was found to be a reasonable way to fairly compare the two methods.

The reason for calling this second procedure a *partial* rather than a *full* segregated one lies in the fact that in full segregated methods all variables, including $U$, $W$ and $P$, are solved independently and in sequence along the entire algorithm. In the case presented here for comparison, only $\Theta$ is excluded from the implicit treatment implied by Equation (8.23).

#### *8.8.2.6. Boundary Conditions and Computational Details*

All velocity components were of null value at all boundaries. Interior velocities were also set to zero at start. For temperature, Figure 8.14 illustrates that both left and right walls were assigned the non-dimensional temperature of +1 and 0, respectively. At the upper and bottom plates the isolation condition $\partial\Theta/\partial z = 0$ was applied. A *single* grid of several sizes and equally distributed in the domain was used. The same relaxation parameters ($\gamma_\varphi = 0.40$ for $\varphi = U, W, P, T$) were used in all calculations. The Rayleigh number, appearing after non-dimensionalizing the buoyancy term, is given by $Ra = (\rho^2 L^3 g\beta\Delta T Pr)/\mu^2$ where, in all cases, the Prandtl number $Pr$ was set equal to 1.0.

Also, an essential feature of Vanka's algorithm (Vanka, 1986a–b), the multigrid technique, has not been used in the present study. Multigrid acceleration is known to be advantageous for mid- to large-size grids, but for the modest meshes here analyzed no substantial improvement on convergence rates was expected. Yet, another different domain sweep strategy was adopted. Distinct sweeping modes, namely horizontal and vertical, together with an alternating path in order to keep always physically connected cells, were followed when visiting each cell in the computational domain. For that, this sweeping strategy is here called *alternating symmetric coupled Gauss-Seidel*, or ASCGS for short. A comparison between SGCS and ASCGS is shown in Figure 8.15.

### 8.8.3. Results and Discussion

#### *8.8.3.1. Preliminary Results*

Although this work has not been mainly concerned with the accuracy of the solution (absolute value), but rather the algorithm to achieve it, a few results on the velocity fields are here presented for completeness. These preliminary results ensure program validation and were taken after reduction of all normalized residues for mass, momentum and energy equations to the pre-selected value of $1 \times 10^{-5}$. Residue history for the velocity calculations shown below are presented later in this paper and are referent to the HFB and VRT cases (see Table 8.1).

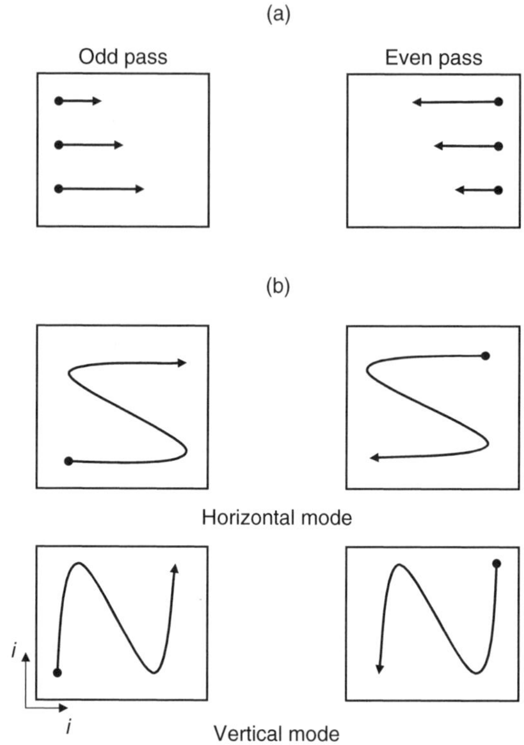


**Figure 8.15.** Different sweeping strategies: (a) SCGS, from node $(i, j)$ to $(i_{max}, j_{max})$ and back, (b) ASCGS, subsequent lines or columns keep always physically connected cells in both horizontal and vertical sweeping modes.

Grid-independence studies were conducted in order to determine a suitable grid size for the calculations. For square horizontal cavities heated from below, computations with different grids are presented in Figure 8.16a. The figure shows the results for the $U$-velocity at the chamber mid-plane for the HFB case ($y = L/2$, $\alpha = \pi/2$, $H/L = 1$, $Ra = 4 \times 10^4$, see Figure 8.14 and Table 8.1). One can see that for meshes larger than $45 \times 45$ the solution is nearly grid independent. Absolute values for velocities in buoyancy-driven flow in cavities have also been compared with benchmark solutions (de Vahl Davis, 1983) to check for code accuracy and correctness (de Lemos, 1990, 1992a, 1997a,b).

The Nusselt number $Nu$ and the film coefficient $h$ for the enclosure are calculated as an average of the values prevailing at the *east* and *west* faces. The equations used were:

$$Nu = \frac{hL}{k}; \quad h = \frac{h_e + h_w}{2} \tag{8.24}$$

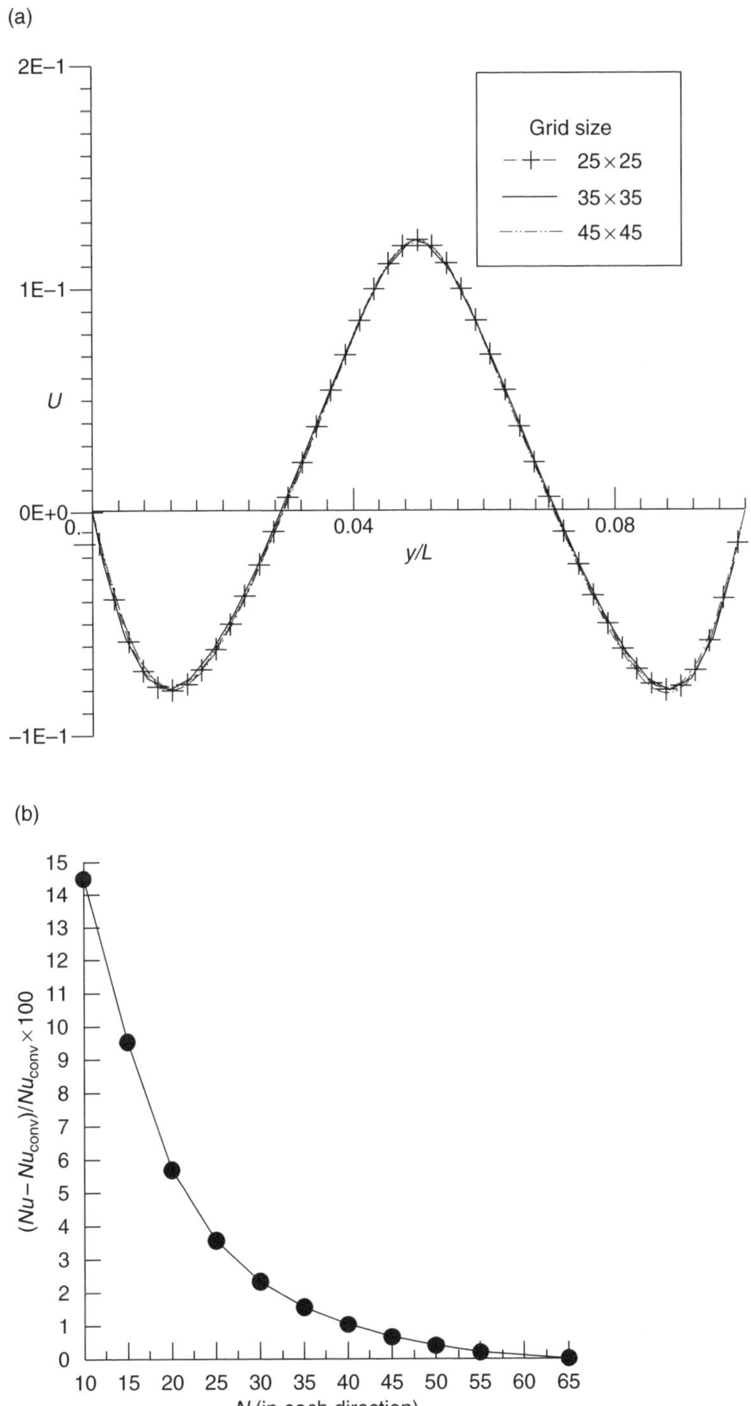


**Figure 8.16.** Grid independence studies: (a) vertical velocity component, (b) Nusselt number.

where $e$ and $w$ correspond to the non-isolated faces of the cavity shown in Figure 8.14 and $k_f$ is the fluid thermal conductivity. For the face $e$ at east,

$$h_e = \frac{\overline{q''_e}}{\Delta T}; \quad \overline{q''_e} = \frac{1}{A_e}\int_{A_e} q''_e \mathrm{d}A_e; \quad q''_e = -k\left.\frac{\partial T}{\partial z}\right|_e \tag{8.25}$$

Figure 8.16b presents results for $Nu$ as a function of the number of grid points. Results are presented in terms of $(Nu - Nu_{conv})/Nu_{conv} \times 100$ where the subscript "conv" refers to the asymptotic value as the grid increases. The figure indicates that for grids greater than $45 \times 45$, errors in $Nu$ are less that 1%. Hence, all results presented below considered this grid size. The absolute asymptotic value $Nu_{conv}$ calculated with equations (8.24) and (8.25) for $Ra = 4 \times 10^4$ and for the VRT case ($\alpha = 0$) was equal to 3.49. This value is only 0.85% higher than the one given by the correlation by Catton (1978). In spite of the close agreement with the correlation, one should point out that the use of an upwind differencing scheme to model convective fluxes may have contaminated the solution with some numerical diffusion. Also, an equally distributed grid has been used and no grid layout optimization, known to improve results accuracy, has been made use of.

Figure 8.17 presents again the $u$-component for the same case and position as in the previous Figure 8.16a, but now comparing the results calculated by the *partially segregated* and *coupled* approaches. Inspecting the figure one can conclude that, independently

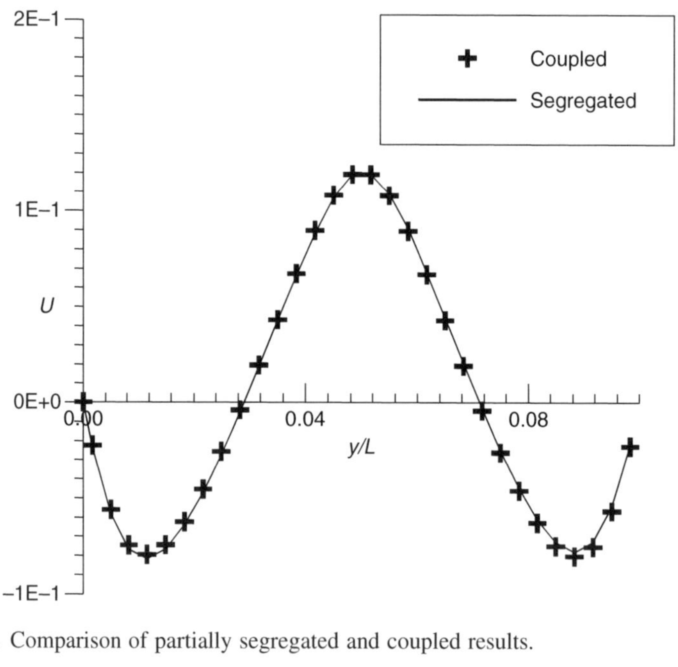


**Figure 8.17.** Comparison of partially segregated and coupled results.

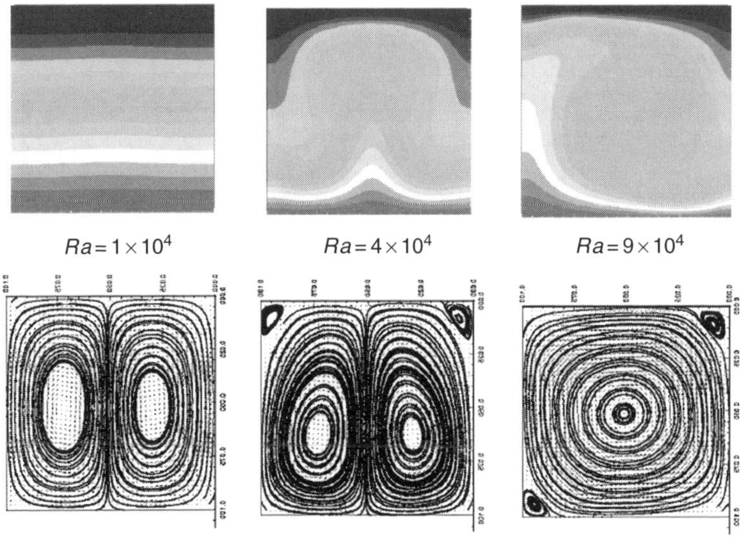


**Figure 8.18.** Effect of *Ra* on temperature and velocity fields – cavity heated from below (HFB).

of the error-smoothing technique used, final converged solutions are essentially equal as expected. The computational effort to achieve them, however, seems to be different, as it will be discussed later.

*The effect of* Ra. The influence of the Rayleigh number on the thermal and velocity fields is shown in Figure 8.18. For better visualization, streamlines are shown instead of vector plots. The simulation is concerned with the case of an enclosure heated from below (HFB case in Table 8.1) and with aspect ratio of $H/L = 1$. For large aspect ratios ($H/L \to \infty$), the so-called *Bénard Cells* are known to exist in the range $1708 \leq Ra \leq 50000$. For compact enclosures, such as the one in question, the critical *Ra* for the onset of motion increases due to the drag of the side walls (Churchill, 1992). Consequently, less heat is transferred across the same layer thickness and *Ra*.

Figure 8.18 shows that for small values of the Rayleigh number, only a slight departure from the purely conduction regime is detected, most likely due to the weak recirculation computed. As *Ra* increases to $4 \times 10^4$, the circulatory motion brings the bottom, hot temperature stream up to the top wall, substantially penetrating into the flow core. A further increase in *Ra* destroys the doubled-swirl pattern and the flow enters the transition regime. In the same figure, the corresponding streamlines further show that at $Ra = 4 \times 10^4$ a small recirculation bubble appears symmetrically attached at the bottom corners. An increase in the value of *Ra* ($9 \times 10^4$) seems to reach the instability regime and, as can be seen, a single cell pattern is obtained. Further increases in *Ra* would lead to turbulent regime, not possible to be computed with the mathematical model herein.

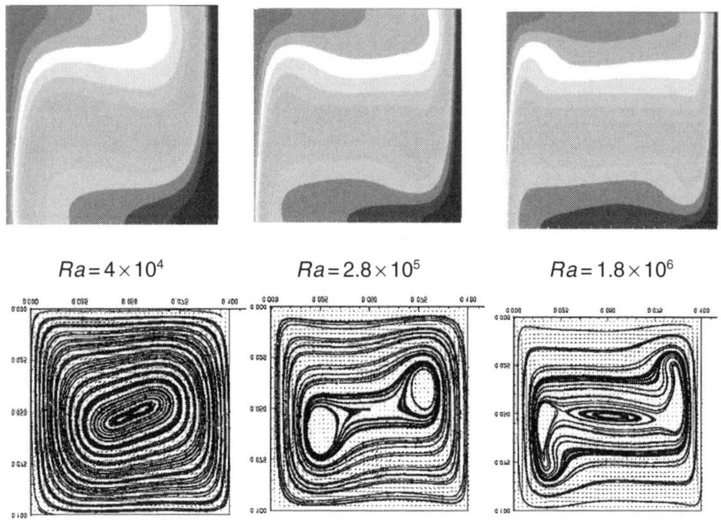


**Figure 8.19.** Temperature and velocity fields for cavity heated from left (VRT).

Figure 8.19 presents similar results for vertical enclosures, or, say, when the overall heat transfer rate crosses from left to right, being both horizontal surfaces insulated (VRT case, $\alpha = 0$). Here also, the figure shows a sequence of temperature and flow fields subjected to an increasing Rayleigh number. At low *Ra* (left), a small recirculating current distorts the temperature field bringing the hotter temperature at the upper left corner into the cavity mid plane. A further increase in *Ra* ($2.8 \times 10^5$) enhances convective transport across the layer increasing the strength of convection currents. At this stage, a larger recirculating bubble is formed over most of the flow core. Vertical stratification of temperature develops over the same region. At $Ra = 1.8 \times 10^6$ the boundary layer regime can be clearly detected. The temperature distribution in the core region reveals considerable stratification and the penetration of the wall layers near the horizontal plates are correctly simulated (Yang, 1987).

### *8.8.3.2. Effect of aspect ratio* H/L

The effect of the aspect ratio $H/L$ is shown in Figures 8.20 and 8.21 for a fixed Rayleigh number equal to $4 \times 10^4$ and $H/L$ up to 4. The three flow regimes known in the literature (Yang, 1987), namely the boundary-layer type, the transition and conduction regimes as $H/L$ increases, are not fully simulated due to the relatively small *Ra* used and the narrow range for $H/L$. The solution for $H/L = 4$ seems to be a representative of the boundary-layer regime for this *Ra*, having a somewhat stratified core and a unicell flow structure. As the $H/L$ ratio increases (not shown here), the flow stratification prevails over most of the domain and the conduction mechanism controls heat transfer (Yang,

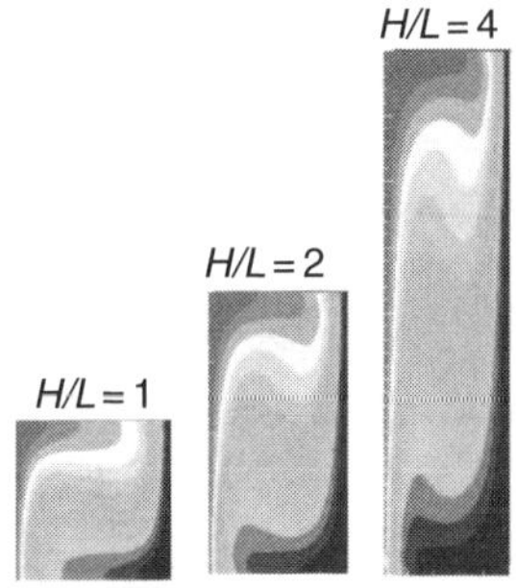


**Figure 8.20.** Effect of $H/L$ on temperature for vertical cavity heated from left.

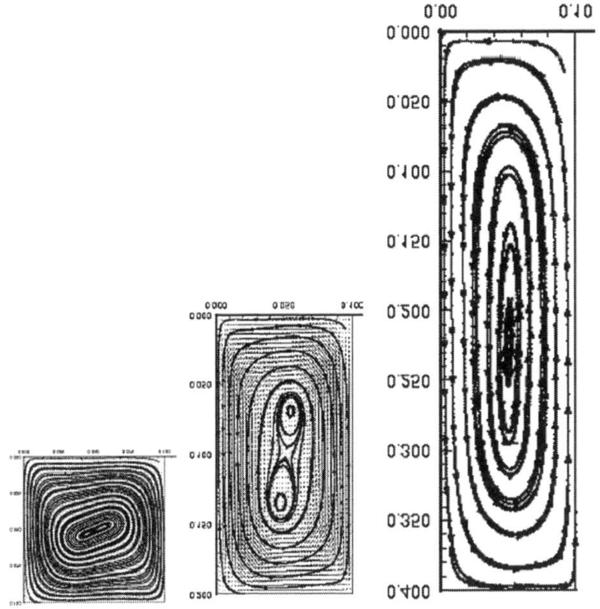

**Figure 8.21.** Streamlines for different aspect ratios for vertical cavities.

1987). It is also interesting to note that as $H/L$ increases the major flow current changes from horizontal to vertical and the core of the flow becomes mostly stratified. Also, a reduction on the average temperature gradient at the wall due to stratification, decreases the Nusselt number, ultimately leading to the mentioned conduction-dominated regime.

### *8.8.3.3. The effect of tilt angle* $\alpha$

Isotherms as a function of the tilt angle and for $Ra = 4 \times 10^4$ are shown in Figure 8.22. For the HFB case (*upper left corner*), one can clearly see the bulging of the isolines

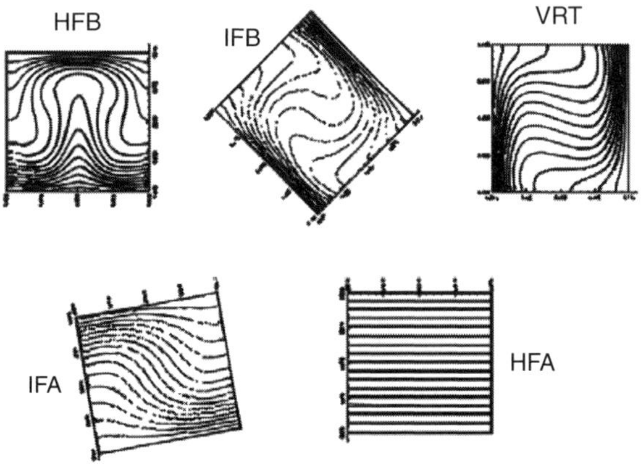


**Figure 8.22.** Isotherms for different cavity inclinations: HFB, $\alpha = 90°$; IFB, $\alpha = 45°$; VRT, $\alpha = 0°$; IFA, $\alpha = -85°$; HFA, $\alpha = -90°$.

deeply penetrating into the flow core. The circulatory motion provoking it is illustrated in Figure 8.23 and is akin to the *Bénard Cells* flow structure. Also interesting to note is the small recirculating bubble at the bottom sides of the cavity in Figure 8.23. Back to Figure 8.22, one can also see temperature fields for the tilt angle $\alpha$ spanning from an unconditionally unstable situation (HFB, $\alpha = 90°$) to stable no-flow distribution (HFA, $\alpha = -90°$). These results are in agreement with pertinent literature and indicate the correctness of the computer program developed.

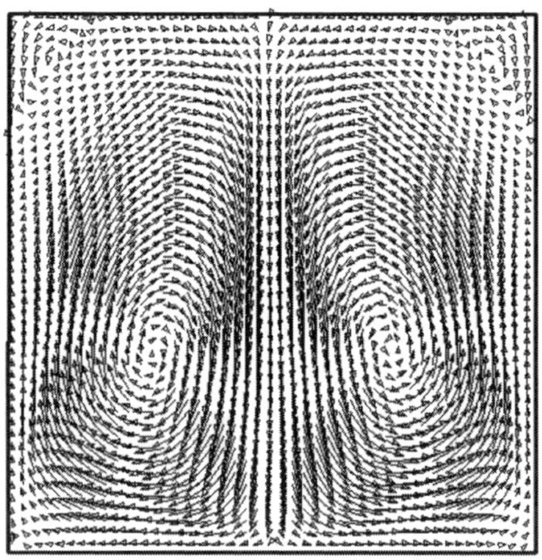

**Figure 8.23.** Vector plot for HFB case, $\alpha = 90°$.

#### 8.8.3.4. Residues

Normalized residue was defined as the norm of the cell mass and energy equation residues as:

$$R_{\text{abs}} = \left\{ \frac{\sum_{ij} R_{ij}^2}{N \cdot M} \right\}^{1/2} ; \quad R_T = \left\{ \frac{\sum_{ij} \left(R_{ij}^{\Theta}\right)^2}{N \cdot M} \right\}^{1/2} \tag{8.26}$$

where $N$ and $M$ are the number of cells in the $y$- and $z$-directions, respectively, and $R_{i,j}$ can be seen as the difference, for every cell, between the cell *outgoing* mass flux, $F_{\text{out}}$, and the *incoming* mass flux, $F_{\text{in}}$. Further, $R_T$ in Equation (8.26) is the residue for the energy Equation (8.18).

A *relative* mass residue can also be defined as,

$$R_{\text{rel}} = \left\{ \frac{\sum_{ij} \left( \frac{F_{\text{in}} - F_{\text{out}}}{F_{\text{in}} + F_{\text{out}}} \right)^2}{N \cdot M} \right\}^{1/2} \tag{8.27}$$

A discussion on the advantages in simultaneously monitoring $R_{\text{rel}}$ in addition to $R_{\text{abs}}$ is presented in de Lemos (1990, 1992a, 1997a,b) and it is based on the small range of the former (0, +1).

Mass residues for the HFB case calculated by Equations (8.26) and (8.27) are presented in Figure 8.24. The iteration counter refers to the total number of sweeps over the domain,

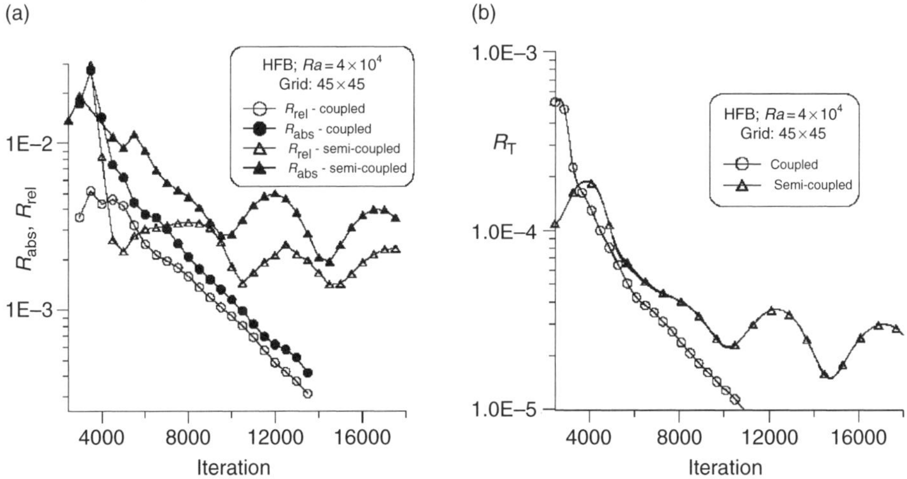


**Figure 8.24.** Residue history for HFB case: (a) relative and absolute mass residues, (b) residue for energy equation.

that is, the product of the *outer* counter times the number of *inner* sweeps. A quick word here on the numbers of iterations to convergence seems timely. Other schemes presented in the literature may indicate residue history as a function of *outer* iteration counters only. Some use the so-called *pseudo-transient* approach, and plot *time steps* instead. Each outer iteration, in turn, may consider a great number of internal sweeps, usually controlled by a specified residue reduction rate. Here, in this work, a fixed number of internal sweeps was considered. The relatively large number of necessary iterations seen in the figures below could be associated with the use of a single grid, the tightness of the relaxation parameters and the strong coupling among all variables involved.

Figure 8.24a indicates that, after an initial period of about 6000 overall iterations, a better convergence rate is obtained with the coupled scheme in either residue form. For Equation (8.18), the residue $R_T$ is presented in Figure 8.24b, also comparing the performance of both relaxation procedures. It is noteworthy that, although Figure 8.24 corresponds to the same case (HFB), they consider different independent variables requiring a different number of iterations to reduce the residue to the same pre-selected level. For temperature, 4000 iterations after the beginning of the relaxation suffice for detecting the advantages of the coupled scheme.

Figure 8.25 shows similar results for the vertical cavity case, where heat flows from left to right. The advantages of using a coupled scheme, rather than relaxing the temperature equation on separate (semi-coupling), are clear in the figures as residues always fall faster after a certain initial number of iteration. Both mass residues, in absolute and relative formats, show better performance for around 3000 iterations, in Figure 8.25a. The apparent advantage of the semi-coupled procedure over the coupled algorithm shown on the left of

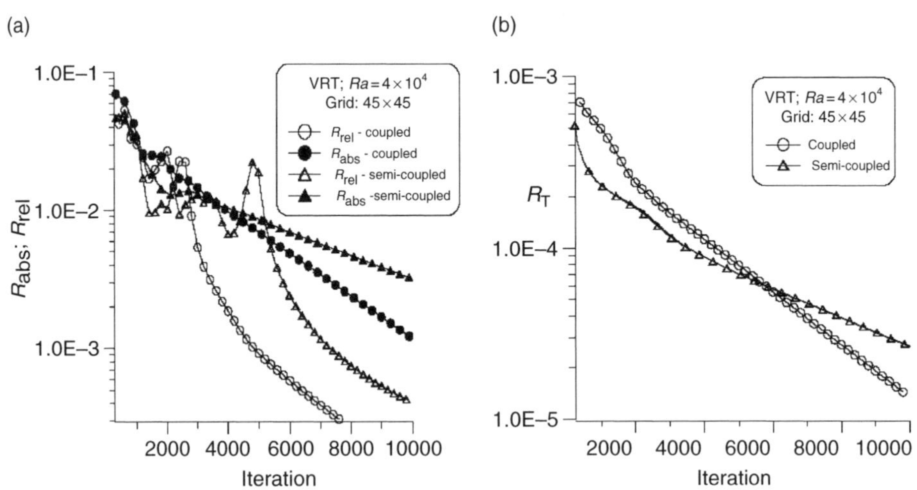


**Figure 8.25.** Residue history for VRT case: (a) mass residue, (b) energy equation residue.

Figure 8.25b, lasts only up to 6000 iterations. After that, changes in the temperature field are quickly transferred to the cross-flow which, in turn, mixes temperature at a faster rate and brings $R_T$ down to the pre-selected tolerance.

## 8.9. TREATMENT OF THE u,w-V COUPLING

### 8.9.1. Introduction

The use of numerical tools for solving real-world engineering problems has been a commonplace strategy in the past decade, mainly due to the accelerated advances in microprocessor technologies and substantial improvements in software development. For the particular case of modern swirl-induced gas turbine combustors, these two factors have led to the reduction of required time for design and analysis of such devices. Ultimately, many different configurations can be analyzed in a short period, reducing the early uncertainties in prototype conception and evaluation.

In spite of the ever-greater use of computational fluid dynamics and heat transfer, the numerical solution of swirling flows is still a challenging task due to the intricate coupling between tangential- and radial-momentum equations. Accordingly, the rate of convergence of any algorithm is essentially dictated by the degree in which physical coupling is mimicked by the method in question. In the end, this is an indication that numerical solutions of swirling flows, in most cases, suffer from the disadvantage of longer computing times when compared to their non-swirling counterparts. There seems to be, then, a greater need for the development of coupled algorithms for such problems.

Most solutions found in the literature for swirling flows solve each variable independently, in a segregated fashion, until final convergence is obtained (Hogg and Leschziner, 1989a,b; Nikjook and Mongia, 1991). Usually, velocity components and pressure field are decoupled from each other with the employment of repetitive solution of a pressure or pressure-correction equation. A subsequent update for the velocity field completes one cycle in this iterative process. This strategy forms the basis of the SIMPLE family of algorithms (Patankar, 1980, 1981).

As mentioned before, in de Lemos (1990, 1992b), a predictive technique based on the work of Vanka (1986a–b) for coupled solution of the momentum and continuity equations has been applied to several flow configurations. Therein the method solves, for each computational cell, all momentum and continuity equations in an implicit manner. An essential characteristic of Vanka's work, the multigrid artifice, has not been used in de Lemos (1990, 1992b) due to the relatively modest grids analyzed. Multigrid techniques are known to perform well with mid- to large-size grids, but are rather ineffective when applied to small size problems (Rabi and de Lemos, 2001). For this reason, no multigrid or any other large grid accelerating scheme was implemented in de Lemos (1990, 1992b).

The implicit handling of the pressure–velocity interaction, first proposed in Vanka (1986a,b), brings information from neighboring cells to the finite-volume immediately, increasing the overall convergence rates. Recently (de Lemos, 2000), the coupled method has been extended to buoyancy-driven cavity flows showing improvements in computer time requirements when an implicit scheme is also applied for handling the temperature–velocity coupling. In the present context, a fully implicit treatment is associated with the idea of simultaneously updating flow and scalar fields at each step within the error-smoothing operator. To the best of the author's knowledge, in any published work, no tangential velocity field, seen here as *scalar*, is numerically treated in a fully implicit manner in the same way as considered below.

Based on the foregoing, the objective of this section is to review the work of de Lemos (2003a), which further developed the early work on coupled schemes, extending it to a fully implicit algorithm for solving the axi-symmetric three-dimensional flow field occurring in model combustors. The tangential velocity $V$ is no longer decoupled from the cross-flow (components $U$ and $W$), so that the strong coupling due to the centripetal acceleration is directly accounted for in the method. In addition, the pointwise aspect of the error-smoothing operator proposed here makes it attractive for use in more advanced computer architectures such as vector and parallel computers.

### 8.9.2. Geometry and Flow Equations

The geometry here considered, for the solution of the internal flow field, is schematically shown in Figure 8.26. The combustor geometry is approximated by a circular duct of constant radius $R$. At inlet, the air–fuel mixture enters through a circular slot with clearance $r_1 - r_2$. At the opening downstream distant one diameter from the entrance, a circular baffle of radius $r_3$ is located. A swirling flow, impaired by vanes located at the

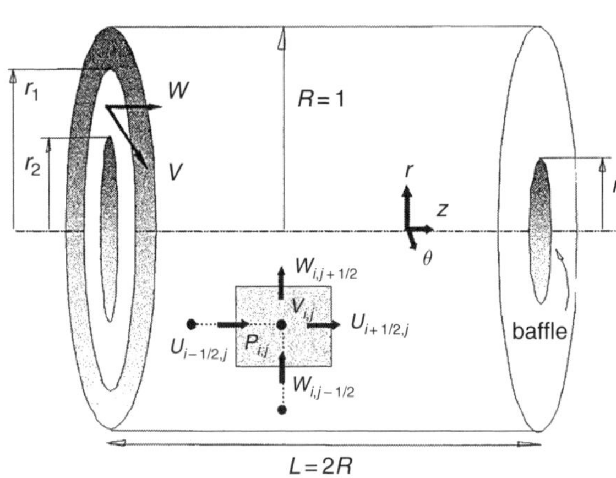


**Figure 8.26.** Model combustor geometry and control-volume notation.

chamber inlet, is simulated by assuming a constant tangential velocity $V_{in}$ at $z = 0$. At the inlet, a constant axial velocity $W_{in}$ is also considered.

The conservation equations for mass and momentum can then be written in a compact form if the existing analogies among the processes of accumulation, transport, convection and generation/destruction for the transported quantities are observed. This general equation can be written in its conservative two-dimensional laminar form as:

$$\frac{\partial}{\partial z}\left[\rho W\phi - \Gamma^{\phi}\frac{\partial \phi}{\partial z}\right] + \frac{1}{r}\frac{\partial}{\partial r}\left[r\left(\rho U\phi - \Gamma^{\phi}\frac{\partial \phi}{\partial r}\right)\right] - S^{\phi} \tag{8.28}$$

In Equation (8.28), the quantity represented by $\phi$ can either be a vector or a scalar, $\rho$ is the fluid density, $U$ and $W$ are the velocity components in the $r$- and $z$-directions, respectively, $\Gamma_{\phi}$ is the transport coefficient for diffusion and $S_{\phi}$ is the source term. Table 8.3 identifies corresponding terms for the different equations represented by Equation (8.28). In both Table 8.3 and Equation (8.28), axi-symmetry is considered.

### 8.9.3. Discretized Equations and Numerical Method

In this work, the set of equations for mass and momentum above is differentiated by means of the widely used control-volume approach of Patankar (1980). Herein, the computational domain is divided into finite non-overlapping regions containing a computational node in each region. A staggered grid arrangement is used in the present work due to its well-established advantages in calculating fluid flow problems (Patankar, 1981). Figure 8.26 also illustrates the control-volume used along with relevant notation.

The differential equations are then integrated over each volume yielding a set of algebraic equations for each dependent variable. The tangential velocity $V$ and pressure $P$ are centered around the computational node $(i, j)$. Cross-flow velocities $U$ and $W$ are assumed to prevail at cell faces as usually done in a staggered-grid arrangement. Internodal variation for the dependent variables can be of different kind corresponding to different finite-difference formulation. In the present work, for simplicity, the upwind differencing scheme is used to model convective fluxes across volume faces. However, the below formulation is presented in such way that no difficulties arise if another differencing scheme is employed.

**Table 8.3.** Terms in the general transport Equation (8.28).

| | $\phi$ | $\Gamma^{\phi}$ | $S^{\phi}$ |
|---|---|---|---|
| Continuity | 1 | 0 | 0 |
| Axial momentum | $W$ | $\mu$ | $-\frac{\partial P}{\partial z}$ |
| Radial momentum | $U$ | $\mu$ | $-\left(\frac{\partial P}{\partial r} + \frac{\mu U}{r^2} - \frac{\rho V^2}{r}\right)$ |
| Tangential momentum | $V$ | $\mu$ | $-\left(\frac{\mu V}{r^2} + \frac{\rho U V}{r}\right)$ |

The block-implicit arrangement given below for the flow and continuity equations, as mentioned, was first presented in Vanka (1986a,b). For the sake of completeness when extending it to swirling flows, the cross-flow equations for $U$ and $W$ are also included here. Integrating then the continuity equation around the point $(i, j)$ of Figure 8.26, following standard practices in numerical differentiation, one has (Patankar, 1981):

$$F_{i+\frac{1}{2}}U_{i+\frac{1}{2},j} - F_{i-\frac{1}{2}}U_{i-\frac{1}{2},j} + F_{j+\frac{1}{2}}W_{i,j+\frac{1}{2}} - F_{j-\frac{1}{2}}W_{i,j-\frac{1}{2}} = 0 \tag{8.29}$$

where the geometric coefficients $F$s can be interpreted as (area of flow)/(volume of computational node).

For the radial momentum equation, the final form for the $U_{i-\frac{1}{2},j}$ component contains coefficients representing influences by convection and diffusion mechanisms in addition to all sources and pressure gradient terms. The discretized equation reads:

$$a^{\mathrm{u}}_{1-\frac{1}{2}}U_{i-\frac{1}{2},j} = \sum_{nb=1}^{4} a^{\mathrm{u}}_{nb}U_{nb} + \frac{A_{i-\frac{1}{2}}}{\rho}\left[P_{i-1,j} - P_{i,j}\right] + S^{\mathrm{U}}_{\mathrm{C}} \tag{8.30}$$

where

$$a^{\mathrm{u}}_{1-\frac{1}{2}} = \sum_{nb=1}^{4} a^{\mathrm{u}}_{nb} - S^{\mathrm{U}}_{\mathrm{P}}; \quad S^{\mathrm{U}}_{\mathrm{P}} = -\left(\frac{\mu}{\rho}\right)\frac{1}{r_{i-\frac{1}{2}}}; \quad S^{\mathrm{U}}_{\mathrm{C}} = \frac{(V_{i-1,j} + V_{i,j})^2}{4r_{i-\frac{1}{2}}} \tag{8.31}$$

The source term in Table 8.3 was given the usual linearized form $S^{\mathrm{U}} = S^{\mathrm{U}}_{\mathrm{C}} + S^{\mathrm{U}}_{\mathrm{P}}U_{i-\frac{1}{2}}$. Also, the coefficients $a^{\mathrm{u}}_{nb}$ appearing in Equation (8.16) refer to *neighbor nodal points* and account for the contributions, at each face, of the mechanisms of convection and diffusion. For a variable $\phi$ and for the purely upwind formulation considered here its general form reads:

$$a^{\phi}_{nb} = \frac{\Gamma^{\phi}}{\rho}\frac{A_{nb}}{\delta r_{nb}} + \|A_{nb}U_{nb}, 0\| \tag{8.32}$$

where the operator $\|A, B\|$ means the greater of $A$ and $B$, $A_{nb}$ is the area of transport, $U_{nb}$ the velocity prevailing over the control-volume face area $A_{nb}$, and $\delta r_{nb}$ the distance between adjacent control-volume centers.

The last term on Equation (8.30) which is detailed in Equation 8.31, represents the discrete form of the *centripetal* acceleration shown in Table 8.3. For application in the numerical algorithm below, this source term, $S^{\mathrm{U}}_{\mathrm{C}}$, is split as,

$$S^{\mathrm{U}}_{\mathrm{C}} = \frac{(V_{i-1,j} + V_{i,j})^2}{4r_{i-\frac{1}{2}}} = \frac{(V^2_{i-1,j} - V^2_{i,j})}{4r_{i-\frac{1}{2}}} + \frac{(V_{i-1,j} + V_{i,j})}{2r_{i-\frac{1}{2}}}V_{i,j} \tag{8.33}$$

so that Equation (8.16) can be further manipulated to give:

$$U_{i-\frac{1}{2},j} = \hat{U}_{i-\frac{1}{2},j} + \hat{d}_{i-\frac{1}{2}}\left[P_{i-1,j} - P_{i,j}\right] + \hat{e}_{i-\frac{1}{2}} + \hat{f}_{i-\frac{1}{2}}V_{i,j} \tag{8.34}$$

where

$$
\begin{aligned}
\hat{U}_{i-\frac{1}{2},j} &= \frac{\sum_{nb=1}^{4} a^{\mathrm{u}}_{nb} U_{nb}}{a^{\mathrm{u}}_{i-\frac{1}{2}}} \\
\hat{d}_{i-\frac{1}{2}} &= \frac{A_{i-\frac{1}{2}}}{\rho a^{\mathrm{u}}_{i-\frac{1}{2}}} \\
\hat{e}_{i-\frac{1}{2}} &= \frac{(V^2_{i-1,j} - V^2_{i,j})}{4a^{\mathrm{u}}_{i-\frac{1}{2}} r_{i-\frac{1}{2}}} \\
\hat{f}_{i-\frac{1}{2}} &= \frac{(V_{i,j} + V_{i-1,j})}{2a^{\mathrm{u}}_{i-\frac{1}{2}} r_{i-\frac{1}{2}}}
\end{aligned}
\tag{8.35}
$$

For the coupled treatment presented here, the explicit contribution of $V$ in the source term of $U$ is necessary, as will be seen below.

A similar equation for the axial velocity component $W_{i,j-\frac{1}{2}}$ is given by:

$$
W_{i,j-\frac{1}{2}} = \hat{W}_{i,j-\frac{1}{2}} + \hat{d}_{j-\frac{1}{2}} \left(P_{i,j-1} - P_{i,j}\right) \tag{8.36}
$$

where

$$
\begin{aligned}
\hat{W}_{j-\frac{1}{2}} &= \frac{\sum_{nb=1}^{4} a^{\mathrm{w}}_{nb} W_{nb}}{a^{\mathrm{w}}_{j-\frac{1}{2}}} \\
\hat{d}_{j-\frac{1}{2}} &= \frac{A_{j-\frac{1}{2}}}{\rho a^{\mathrm{w}}_{j-\frac{1}{2}}}
\end{aligned}
\tag{8.37}
$$

Following a similar procedure for the velocity component $V$, a final finite-difference equation can be assembled in the following form:

$$
a^{\mathrm{V}}_{ij} V_{i,j} = \sum_{nb=1}^{4} a^{\mathrm{V}}_{nb} V_{nb} + S^{\mathrm{V}}_{\mathrm{C}} \tag{8.38}
$$

where

$$
\begin{aligned}
a^{\mathrm{V}}_{ij} &= \sum_{nb=1}^{4} a^{\mathrm{V}}_{nb} - S^{\mathrm{V}}_{\mathrm{P}} \\
S^{\mathrm{V}}_{\mathrm{P}} &= -\left\{ \frac{\mu}{\rho} \frac{1}{r_i^2} + \left\| \frac{U_{i-\frac{1}{2}} + U_{i+\frac{1}{2}}}{2r_i}, 0 \right\| \right\} \\
S^{\mathrm{U}}_{\mathrm{C}} &= \left\| - \frac{U_{i-\frac{1}{2}} + U_{i+\frac{1}{2}}}{2r_i}, 0 \right\| V_{i,j}
\end{aligned}
\tag{8.39}
$$

Here also the source term was given the linearized form $S^{\mathrm{V}} = S_{\mathrm{C}}^{\mathrm{V}} + S_{\mathrm{P}}^{\mathrm{V}} V_{ij}$, accounting further for the fact that the value of the *Corriollis* acceleration might be positive or negative. For positive $U$ values, this acceleration is considered in the $S_{\mathrm{P}}^{\mathrm{V}}$ coefficient whereas for negative radial velocity components an explicit treatment for the product $-UV$ is adopted. For simplicity, Equations (8.38) can be rearranged as:

$$V_{i,j} = \frac{\sum_{nb=1}^{4} a_{nb}^{\mathrm{V}} V_{nb} + S_{\mathrm{C}}^{\mathrm{V}}}{a_{ij}^{\mathrm{V}}}$$

All the rhs terms of the above equation are assembled for convenient use in the coupled scheme as,

$$\hat{V}_{i,j} = \frac{\sum_{nb=1}^{4} a_{nb}^{\mathrm{V}} V_{nb} + S_{\mathrm{C}}^{\mathrm{V}}}{a_{ij}^{\mathrm{V}}} \tag{8.40}$$

so that

$$V_{i,j} = \hat{V}_{i,j} \tag{8.41}$$

In Equation (8.41) all right-hand-side terms of Equation (8.40) are assembled for convenient use in the coupled scheme.

#### 8.9.3.1. *The Coupled Numerical Strategy*

In order to set up a scheme to annihilate the residues for the flow equations, *corrections* are then defined as differences between *exact* and *approximate* variables and can be written as:

$$\begin{aligned}
U'_{i-\frac{1}{2},j} &= U_{i-\frac{1}{2},j} - U^*_{i-\frac{1}{2},j} \\
U'_{i+\frac{1}{2},j} &= U_{i+\frac{1}{2},j} - U^*_{i+\frac{1}{2},j} \\
W'_{i,j-\frac{1}{2}} &= W_{i,j-\frac{1}{2}} - W^*_{i,j-\frac{1}{2}} \\
W'_{i,j+\frac{1}{2}} &= W_{i,j+\frac{1}{2}} - W^*_{i,j+\frac{1}{2}} \\
P'_{i,j} &= P_{i,j} - P^*_{i,j} \\
V'_{i,j} &= V_{i,j} - V^*_{i,j}
\end{aligned} \tag{8.42}$$

where the subscripts, as in Figure 8.26, identifies locations in the grid, the superscript $'$ (prime) distinguishes corrections and the symbol $*$ (asterisk) corresponds to the previous iteration.

Residuals for all four cross-flow momentum, tangential velocity and continuity of mass equations are readily obtained by plugging the *approximate* values, given by Equation (8.42), into Equations (8.34), (8.36) and (8.41). Then, defining residuals as the difference between the rhs and lhs of Equations (8.34), (8.36) and (8.41), one has:

$$
\begin{aligned}
R_{i-\frac{1}{2},j} &= \hat{U}_{i-\frac{1}{2},j}\hat{d}_{i-\frac{1}{2}}\left[P_{i-1,j} - P^*_{i,j}\right] + \hat{e}_{i-\frac{1}{2}} + \hat{f}_{i-\frac{1}{2}}V^*_{i,j} - U^*_{i-\frac{1}{2},j} \\
R_{i+\frac{1}{2},j} &= \hat{U}_{i+\frac{1}{2},j} + \hat{d}_{i+\frac{1}{2}}\left[P^*_{i,j} - P_{i,j+1}\right] + \hat{e}_{i+\frac{1}{2}} + \hat{f}_{i+\frac{1}{2}}V^*_{i,j} - U^*_{i+\frac{1}{2},j} \\
R_{i,j-\frac{1}{2}} &= \hat{W}_{i,j-\frac{1}{2}} + \hat{d}_{j-\frac{1}{2}}\left[P_{i,j-1} - P^*_{i,j}\right] - W^*_{i,j-\frac{1}{2}} \\
R_{i,j+\frac{1}{2}} &= \hat{W}_{i,j+\frac{1}{2}} + \hat{d}_{j+\frac{1}{2}}\left[P^*_{i,j} - P_{i,j+1}\right] - W^*_{i,j+\frac{1}{2}} \\
R_{i,j} &= -F_{i+\frac{1}{2}}U_{i+\frac{1}{2},j} + F_{i-\frac{1}{2}}U_{i-\frac{1}{2},j} - F_{j+\frac{1}{2}}W_{i,j+\frac{1}{2}} + F_{j-\frac{1}{2}}W_{i,j-\frac{1}{2}} \\
R^V_{i,j} &= -V^*_{i,j} + \hat{V}_{i,j}
\end{aligned}
\tag{8.43}
$$

Note that in Equation (8.43), velocities and pressure outside the $(i, j)$ volume are assumed exact, since the decomposition given by Equation (8.42) is not applied to them. Further, Equation (8.42) is not applied to the $\hat{e}$ and $\hat{f}$ terms of Equation (8.35).

To make this idea clear, one takes the west face of the control volume shown in Figure 8.26 as an example (see Equation (8.34)). For that particular face, one has:

$$
\begin{aligned}
(U^* + U')_{i-\frac{1}{2},j} = \hat{U}_{i-\frac{1}{2},j} &+ \hat{d}_{i-\frac{1}{2}}\left[P_{i-1,j} - (P^* + P')_{i,j}\right] \\
&+ \hat{e}_{i-\frac{1}{2}} + \hat{f}_{i-\frac{1}{2}}(V^* + V')_{i,j}
\end{aligned}
\tag{8.44}
$$

As mentioned, note that the decomposition detailed in Equation (8.42) was not applied to the "pseudo-velocity" $\hat{U}_{i-\frac{1}{2},j}$ nor to the coefficients $\hat{e}$ and $\hat{f}$ nor to the pressure located outside the control volume $(i, j)$. This is an essential feature of the method as will be seen below. For a given set of guessed or "starred" variables, Equation (8.34) yields a residue, $R_{i-\frac{1}{2},j}$, related to the incorrect velocity such that:

$$
U^*_{i-\frac{1}{2},j} = \hat{U}_{i-\frac{1}{2},j} + \hat{d}_{i-\frac{1}{2}}\left[P_{i-1,j} - P^*_{i,j}\right] + \hat{e}_{i-\frac{1}{2}} + \hat{f}_{i-\frac{1}{2}}V^*_{i,j} - R_{i-\frac{1}{2},j}
\tag{8.45}
$$

Subtracting now Equation (8.45) from Equation (8.44), an equation for the correction $U'$ is obtained in the form:

$$U'_{i-\frac{1}{2},j} + \hat{d}_{i-\frac{1}{2}} P'_{i,j} - \hat{f}_{i-\frac{1}{2}} V'_{i,j} = R_{i-\frac{1}{2},j} \tag{8.46}$$

An equivalent path to reach the same result is to start out with Equation (8.16). First expand the source term with a decompostion for $V_{i,j}$ as:

$$\begin{aligned} S_C^U &= \frac{(V_{i-1,j}+V_{i,j})^2}{4r_{i-\frac{1}{2}}} \\ &\approx \frac{(V_{i-1,j}+V^*_{i,j})^2 + 2(V_{i-1,j}+V^*_{i,j})V'_{i,j} + \overbrace{(V'_{i,j})^2}^{\approx 0}}{4r_{i-\frac{1}{2}}} \\ &= \frac{(V_{i-1,j}+V^*_{i,j})^2}{4r_{i-\frac{1}{2}}} + \frac{(V_{i-1,j}+V^*_{i,j})}{2r_{i-\frac{1}{2}}} V'_{i,j} \end{aligned} \tag{8.47}$$

where higher order terms such as $(V'_{i,j})^2$ have been neglected. Now, substituting Equation (8.47) into Equation (8.16), in addition to decomposing $U_{i-\frac{1}{2}}$ and $P_{i,j}$, one gets:

$$\begin{aligned} (U^* + U')_{i-\frac{1}{2},j} = \hat{U}_{i-\frac{1}{2},j} &+ \hat{d}_{i-\frac{1}{2}} \left[P_{i-1,j} - (P^* + P')_{i,j}\right] \\ &+ \frac{(V_{i-1,j}+V^*_{i,j})^2}{4a^{u}_{1-\frac{1}{2}} r_{1-\frac{1}{2}}} + \frac{(V_{i-1,j}+V^*_{i,j})}{2a^{u}_{1-\frac{1}{2}} r_{1-\frac{1}{2}}} V'_{i,j} \end{aligned} \tag{8.48}$$

Again for a given set of guessed or "starred" variables, Equation (8.16) yields a residue related to this approximate value such that:

$$U^*_{i-\frac{1}{2},j} = \hat{U}_{i-\frac{1}{2},j} + \hat{d}_{i-\frac{1}{2}} \left[P_{i-1,j} - (P^* + P')_{i,j}\right] + \frac{(V_{i-1,j}+V^*_{i,j})^2}{4a^{u}_{i-\frac{1}{2}} r_{i-\frac{1}{2}}} - R_{i-\frac{1}{2},j} \tag{8.49}$$

Subtracting Equation (8.49) from Equation (8.48), Equation (8.46) is obtained.

Now, substituting Equations (8.42) into corresponding Equations (8.43), a system connecting the residuals and corrections can be written into matrix form as:

$$\begin{bmatrix} 1 & 0 & 0 & 0 & \hat{d}_{i-\frac{1}{2}} & -\hat{f}_{i-\frac{1}{2}} \\ 0 & 1 & 0 & 0 & -\hat{d}_{i+\frac{1}{2}} & -\hat{f}_{i+\frac{1}{2}} \\ 0 & 0 & 1 & 0 & \hat{d}_{j-\frac{1}{2}} & 0 \\ 0 & 0 & 0 & 1 & -\hat{d}_{j+\frac{1}{2}} & 0 \\ -F_{i+\frac{1}{2}} & F_{i-\frac{1}{2}} & -F_{j+\frac{1}{2}} & F_{j+\frac{1}{2}} & 0 & 0 \\ 0 & 0 & 0 & 0 & 0 & 1 \end{bmatrix} \begin{bmatrix} U'_{i-\frac{1}{2},j} \\ U'_{i+\frac{1}{2},j} \\ W'_{i,j-\frac{1}{2}} \\ W'_{i,j+\frac{1}{2}} \\ P'_{i,j} \\ V'_{i,j} \end{bmatrix} = \begin{bmatrix} R_{i-\frac{1}{2},j} \\ R_{i+\frac{1}{2},j} \\ R_{i,j-\frac{1}{2}} \\ R_{i,j+\frac{1}{2}} \\ R_{i,j} \\ R^V_{i,j} \end{bmatrix} \tag{8.50}$$

In Equation (8.50) the influence of $V$ on the cross-flow field is directly accounted for by the $\hat{f}$-terms in the $U$-equation. For the axial direction, the $\hat{f}$-terms are of null value. The reverse effect, or, say, the cross-flow influence on the tangential velocity field, is here not treated implicitly. This fact accounts for the "zeros" in the last row of the matrix in Equation (8.50). The solution of the system (8.50) is then easily obtained by first finding corrections for $V$, then calculating the pressure $P$ and later calculating the velocity components $U$ and $W$.

Boundary conditions used were given velocity at the flow inlet and non-slip condition at chamber walls. For cells facing the outlet plane, overall mass-conservation balance in each computational cell was used to calculate the control-volume outgoing velocity. Initial null values were aligned for all velocities. Numerical implementation of boundary conditions was achieved, as usual, by maintaining the constant initial values at the boundaries, where applicable, or by updating them in each iterative sweep, as in the cases of outlet surfaces. Also, all computations below used an $18 \times 36$ grid, distributed equally in the domain of calculation.

### 8.9.4. Some Numerical Results

The algebraic equations for the three-dimensional velocity field were solved, in addition to the *fully coupled* scheme here described, by performing outer iterations for the components $V$ while keeping $U$, $W$ and $P$ from previous iteration. A *line-by-line* smoothing operator, fully described elsewhere (e.g. Patankar, 1980), was used to relax $V$, the cross-flow field ($U$, $W$) being calculated by the locally coupled method of de Lemos (1992b). This *segregated* solution was set in such a way that the same number of sweeps throughout the tangential and cross-flow fields were finally obtained. Since in the coupled scheme every sweep for $U$, $W$, and $P$ also implies smoothing out $V$ errors, this procedure was found to be a reasonable way to fairly compare the two methods. Further, the same relaxation parameter $\alpha$, defined by the expression $\phi = \phi^{\text{old}} + \alpha(\phi^{\text{new}} - \phi^{\text{old}})$ ($\alpha = 0.45$ for $P$, $V$ and $\alpha = 0.55$ for $U$, $W$), was used in both schemes for the two inlet swirling strengths investigated, namely $\omega^2 = 0.1$ and $\omega^2 = 1.0$, where $\omega^2 = (V_{in}/W_{in})$ with $V_{in}$ and $W_{in}$ are the tangential and axial velocities at the inlet, respectively. The values adopted for $W_{in}$ and Reynolds number $Re(= \rho W_{in} R/\mu)$ were 1 m/s and $10^3$, respectively.

Normalized residues for the tangential velocity field, $R_V$, and for the mass continuity equation, in its absolute ($R_{ij}^{abs}$) and relative form ($R_{ij}^{rel}$), can be defined as:

$$R_V = \left\{ \frac{\sum_{ij} \left(R_{i,j}^{V}\right)^2}{(N \cdot M)} \right\}^{\frac{1}{2}}$$

$$R_{ij}^{abs} = \left\{ \frac{\sum_{ij} (F_{out} - F_{in})_{ij}^2}{(N \cdot M)} \right\}^{\frac{1}{2}} \tag{8.51}$$

$$R_{ij}^{rel} = \left\{ \frac{\sum_{ij} \left(\frac{F_{out} - F_{in}}{F_{out} + F_{in}}\right)_{ij}^2}{(N \cdot M)} \right\}^{\frac{1}{2}}$$

where $N$ and $M$ are the number of cells in the $r$- and $z$-directions, respectively. Also, $F_{out}$ and $F_{in}$ are the *outgoing* and *incoming* mass fluxes of the cell, respectively.

Residue histories for the tangential velocity field are presented in Figure 8.27a. *Implicit* treatment of the tangential cross-flow coupling implies a greater error reduction during the first few iterations, but subsequently a more or less equivalent rate of decrease has

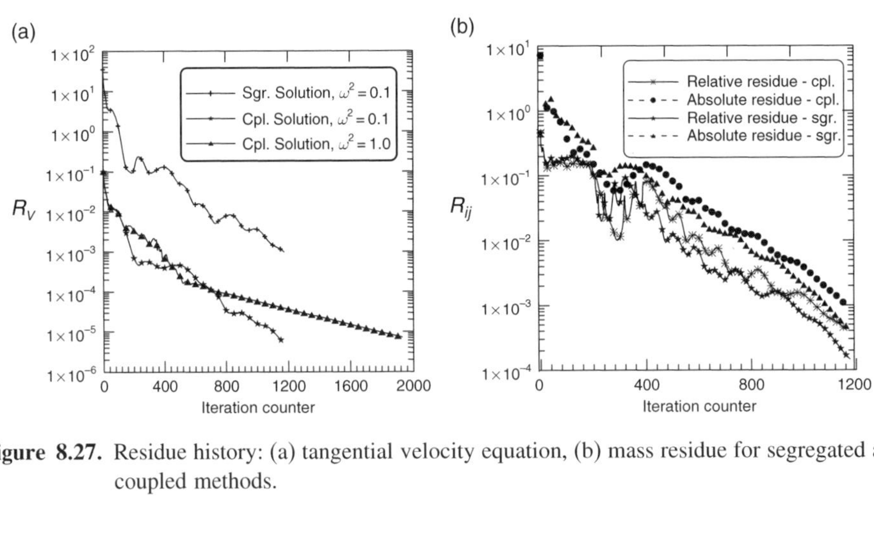


**Figure 8.27.** Residue history: (a) tangential velocity equation, (b) mass residue for segregated and coupled methods.

been obtained by the two methods. A primary consequence of this is a relatively longer computer run time when the decoupled solution is used with the same convergency criterion applied to the $V$-equation. The effect of different swirling strengths is also shown in the figure. Difficulties in obtaining numerical solutions at a higher swirl are indicated by the reduction in the rate of decrease of $R_V$ for iteration numbers greater than about 700, in Figure 8.27a.

Accompanying results for the mass residuals in both absolute and relative forms are presented in Figure 8.27b. Residuals for both the *coupled* and *segregated* solutions present a consistent reduction throughout the iterations. Recalling now the results for $R_V$ (Figure 8.27a), the figures seem to indicate that the flow field can quickly adjust itself to changes in the tangential velocity profile, and that, in the segregated case, those changes are too slowly transferred to the tangential component $V$. The coupled solution, however, quickly transmits back to the $V$-equation changes in the cross-flow pattern, more realistically simulating the strong interaction among the variables involved.

## 8.10. TREATMENT OF THE u, w-V-T COUPLING

### 8.10.1. Introduction

The growing trend in environment regulations points toward ever-tighter pollutant release limits of modern combustion systems. The new technologies of the present day for efficient energy production are based on the so-called "lean and low-NOx combustion". Accordingly, most flow fields in such systems are characterized by an ascending stream with an induced swirling motion. Swirling induces flame stabilization, allowing reduction in peak temperature, ultimately reducing pollutant-formation rates. Recently, the design and analysis of such systems, driven by increased microprocessor performance and associated low computational costs, have made systematic use of numerical tools. Ultimately, the growing number of available software packages for fluid system design have helped engineers in reducing the necessary time before a new concept is finally made available in the market.

In spite of the increasing use of CFD (Computational Fluid Dyanamics) tools, the numerical solution of such flows imposes additional difficulties due to the intricate coupling between *temperature*, *tangential velocity* and *cross-flow* fields. Buoyancy term, centripetal and Coriollis accelerations make the system of governing equations of a high degree of coupling. For instance, for a system oriented vertically, with the $W$-component of velocity in the vertical direction, the temperature $T$ appears in the $W$-equation. In addition, the radial $U$-component of velocity is governed by an equation that includes the $V$-component in the centripetal acceleration term. In addition, the tangential velocity $V$ is affected by $U$ through the Corriolis acceleration. Therefore, $U$, $W$, $V$ and

$T$ are further connected to each other when buoyancy and rotation are simultaneously present in the flow, increasing the aforementioned degree of coupling among all variables involved.

Linearization of governing equations followed by the use of iterative solvers is the common route found in the literature for solving such non-linear problems. Accordingly, the rate of convergence of any algorithm is essentially dictated by the degree in which physical coupling is mimicked by the method in question. Ultimately, this is an indication that numerical solutions of swirling flows, in most cases, suffer from the disadvantage of longer computing times when compared to their non-swirling counterpart.

*Segregated* methods, in which one individual flow variable is relaxed while holding the others still, are known to be rather sensitive when handling strong physical coupling. In such cases, the so-called *coupled* solvers, where all dependent variables are relaxed in the same domain location, have received much attention lately.

For buoyancy-driven laminar flows, benchmark solutions for the field in a square cavity have been presented (de Vahl Davis, 1983). Multi-grid solution for this problem has also been published (Hortmann *et al.*, 1990). In the great majority of these works, a segregated method is generally employed with the repetitive solution of a pressure or pressure-correction equation, followed by subsequent updates of the velocity and scalar fields. This strategy forms the basis of the SIMPLE family of algorithms (Patankar, 1980). Coupled line solvers for the temperature and velocity fields have shown improvements in computer time requirements for natural convection flows with large Rayleigh numbers (Galpin and Raithby, 1986). The work in Galpin and Raithby (1986) is an indication of the advantage of coupled schemes for solving algebraic equations set with a high degree of interlinkage among the variables. Recently, the block implicit technique has also been applied in the calculation of buoyant flows in a partially coupled manner (Tang and Joshi, 1999).

For swirling flows, most solutions found in the literature are based on segregated relaxation procedures (Hogg and Leschiziner, 1989a,b; Nikjook and Mongia, 1991). In the present context, a fully implicit treatment is associated with the idea of simultaneously updating flow and scalar fields at each step within the error-smoothing operator. To the best of the author's knowledge, in all the published work, neither temperature nor tangential velocity field, seen here as *scalars*, is treated in a fully implicit manner.

The objective of this section is to review the simulations in de Lemos (2003b), who further extended the technique presented in de Lemos (1992b, 2003a) for the azimuthal velocity and in de Lemos (1992b–1997b, 2000) for temperature, combining then the solution of both scalars into a single, fully implicit numerical treatment. To this end, computations are presented for a model furnace comprising incompressible laminar flow simultaneously heated and subjected to an incoming flow with swirl. Effects of Reynolds number, Rayleigh number and swirling strength on temperature patterns and convergence rates are reported.

## 8.10.2. Governing Equations and Numerical Method

### 8.10.2.1. Geometry

The geometry here considered is schematically shown in Figure 8.28. A typical furnace combustion zone is approximated by a model consisting of a circular chamber of constant radius $R$. At inlet, the mixture of air and fuel enters through a circular slot of clearance $r_1 - r_2$. At one diameter downstream the entrance, combustion gases are able to exit through an annulus of thickness $r_3 - r_4$. The temperature level is prescribed over the entire lateral wall and on the bottom and top lids, except at the exit area where a null temperature gradient is assumed to be established by the outward motion of the fluid.

Although it is recognized that the geometry of Figure 8.28 might be an oversimplification of modern industrial furnaces, essential elements, namely *swirl*, *buoyancy* and *recirculating zones* provide a good test case for the numerical method discussed here.

### 8.10.2.2. Compact Notation

The conservation equations for mass, momentum and energy here analyzed can be written in a compact form if the existing analogies among the processes of accumulation, transport, convection and generation/destruction of those quantities are observed. This generic equation is commonly known in the literature as the *general transport equation* and can be written in its *conservative* two-dimensional laminar for *axi-symmetric* cases as:

$$\frac{\partial}{\partial z}\left[\rho W\phi - \Gamma^{\phi}\frac{\partial \phi}{\partial z}\right] + \frac{1}{r}\frac{\partial}{\partial r}\left[r\left(\rho U\phi - \Gamma^{\phi}\frac{\partial \phi}{\partial r}\right)\right] = S^{\phi} \tag{8.52}$$

In Equation (8.52) $\varphi$ can represent any quantity of vectorial or scalar nature (velocity or temperature), $\rho$ is the fluid density, $U$ and $W$ are the velocity components in the $r$- and

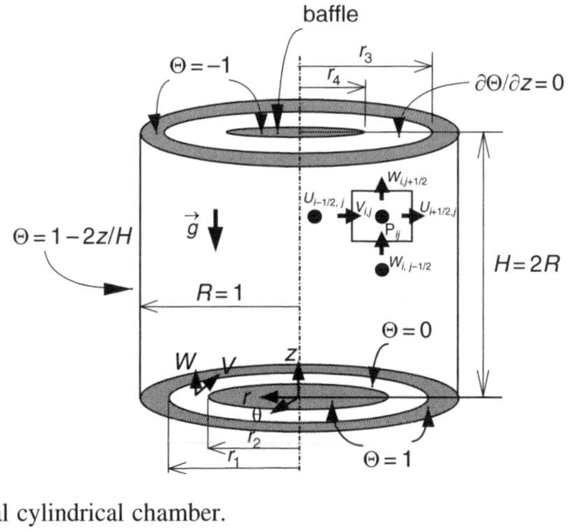


**Figure 8.28.** Vertical cylindrical chamber.

**Table 8.4.** Terms in the general transport Equation (8.52).

| | $\varphi$ | $\Gamma^{\phi}$ | $S^{\phi}$ |
|---|---|---|---|
| Continuity | 1 | 0 | 0 |
| Axial momentum | $W$ | $\mu$ | $-\frac{\partial P}{\partial z}+\rho g_z \beta(T-T_0)$ |
| Radial momentum | $U$ | $\mu$ | $-\left(\frac{\partial P}{\partial r}+\frac{\mu U}{r^2}-\frac{\rho V^2}{r}\right)$ |
| Azimuthal momentum | $V$ | $\mu$ | $-\left(\frac{\mu V}{r^2}+\frac{\rho U V}{r}\right)$ |
| Energy | $T$ | $\mu/\mathrm{Pr}$ | 0 |

$z$-directions, respectively, $\Gamma_{\varphi}$ is the transport coefficient for diffusion and $S_{\varphi}$ is the source term. Table 8.4 identifies corresponding terms for the different equations represented by (8.52). In both Table 8.4 and Equation (8.52), gravity acts in the $z$-direction, $\mu$ is the fluid viscosity, $Pr$ the Prandtl number, $T$ the temperature and $V$ the tangential velocity component.

#### *8.10.2.3. Boundary Conditions*

Boundary conditions used for all velocity components were given value at the flow inlet and non-slip condition at chamber walls. For cells facing the outlet plane, overall mass-conservation balance at each computational cell was used to calculate the control-volume outgoing axial velocity (at the top lid). Initial null values were set for all velocities.

For temperature, a linear profile along the vertical direction was assumed to prevail over the lateral wall (see Figure 8.28). Except in the opened areas – the bottom and the top – the non-dimensional temperature took the values $+1$ and $-1$, respectively. Through the inlet and outlet areas, the applied boundary conditions for the temperature were $\Theta = 0$ and $\partial\Theta/\partial z = 0$, respectively.

Numerical implementation of boundary conditions was achieved by maintaining the constant initial values at the boundaries, where applicable, or by updating them in each iteration, as in the cases of outlet surfaces or symmetry line.

All computations below used an $18 \times 36$ single grid distributed equally in the calculation domain. An essential characteristic of Vanka's work, the *multigrid* artifice, has *not* been used in the present work due to the relatively modest grid analyzed here. Multigrid techniques are known to perform well with mid- to large-size grids (Rabi and de Lemos, 2001), but are rather ineffective when applied to small size problems. For this reason, no multigrid or any other large grid accelerating scheme was implemented.

#### *8.10.2.4. Computational Grid and Finite-Difference Formulation*

In this work, the set of equations for mass, momentum and energy given above is *differentiated* by means of the widely used control-volume approach of Patankar (1980). The differential equations are integrated over each volume yielding a set of algebraic equations. Internodal variation for the dependent variables can be of different kind corresponding to different *finite-difference* formulations. In the present work, for simplicity, the *upwind differencing scheme* is used to model convective fluxes across volume faces. However, the formulation below is presented in such a way that no difficulties arise if another differencing scheme is employed.

#### 8.10.2.5. Discretized Equations

The block-implicit arrangement below for the flow and continuity equations, as mentioned, was first presented by Vanka (1986a,b). For the sake of completeness when extending it to buoyant and swirling problems, the flow equations are also included here. Integrating then the continuity equation around the point $(i, j)$ (see notation in Figure 8.28.) following standard practices in numerical differentiation, one has (Patankar, 1980):

$$F_i^1 U_{i+1/2,j} - F_i^2 U_{i-1/2,j} + F_j^1 W_{i,j+1/2} - F_j^2 W_{i,j-1/2} = 0 \tag{8.53}$$

where the geometric coefficients $F$s make computations convenient and efficient and can be interpreted as (area of flow)/(volume of computational node).

For the radial momentum equation, the final form for the $U_{i-\frac{1}{2},j}$ component contains coefficients representing influences by convection and diffusion mechanisms in addition to all sources and pressure-gradient terms. For application in the numerical algorithm, the equation can be written in such a way that (de Lemos, 1992b):

$$U_{i-\frac{1}{2},j} = \hat{U}_{i-\frac{1}{2},j} + \hat{d}_{i-\frac{1}{2}} \left[P_{i-1,j} - P_{i,j}\right] + \left(\frac{V_{i,j} + V_{i-1,j}}{2}\right)^2 \frac{1}{a^{\mathrm{u}}_{1-\frac{1}{2}} r_{1-\frac{1}{2}}} \tag{8.54}$$

where

$$\hat{d}_{i-\frac{1}{2}} = \frac{A_{i-\frac{1}{2}}}{\rho a^{\mathrm{u}}_{1-\frac{1}{2}}} \tag{8.55}$$

$$\hat{U}_{i-\frac{1}{2},j} = \frac{\sum_{nb=1}^{4} a^{\mathrm{u}}_{nb} U_{nb} + f^{\mathrm{u}}_{i-\frac{1}{2}}}{a^{\mathrm{u}}_{i-\frac{1}{2}}} \tag{8.56}$$

and the last term on (8.54) represents the discrete form of the *centripetal* acceleration shown in Table 8.4. Any other discrete term is accounted for in the "$f$" parameter. Equation (8.54) can further be manipulated to give,

$$U_{i-\frac{1}{2},j} = \hat{U}_{i-\frac{1}{2},j} + \hat{d}_{i-\frac{1}{2}} \left[P_{i-1,j} - P_{i,j}\right] + \hat{e}_{i-\frac{1}{2}} V_{i,j} \tag{8.57}$$

where

$$\hat{e}_{i-\frac{1}{2}} = \frac{(V_{i,j} + V_{i-1,j})}{2 r_{1-\frac{1}{2}} a^{\mathrm{u}}_{1-\frac{1}{2}}} \tag{8.58}$$

and the pseudo-velocity $\hat{U}_{i-\frac{1}{2},j}$ has been modified for accommodating all remaining terms. Further, in Equation (8.57) the last term on the rhs represents the influence of $V$ on the radial velocity $U$ and entails a linearization of the *centripetal* acceleration (see details in de Lemos, 1992b–2003a). All source terms, except the pressure gradient and the contribution due to the tangential velocity, are compacted in the first term on the rhs. For the coupled treatment presented here, the explicit contribution of $V$ in the source term of $U$ is necessary, as it will be seen below.

A similar equation for the axial velocity component $W_{i,j-\frac{1}{2}}$ is given by:

$$W_{i,j-\frac{1}{2}} = \hat{W}_{i,j-\frac{1}{2}} + \hat{d}_{j-\frac{1}{2}}\left[P_{i,j-1} - P_{i,j}\right] + \frac{g_z\beta(T_1 - T_0)}{a^{\mathrm{w}}_{j-\frac{1}{2}}}\left(\frac{\Theta_{i,j} + \Theta_{i,j-1}}{2}\right) \tag{8.59}$$

where

$$\hat{d}_{i-\frac{1}{2}} = \frac{A_{j-\frac{1}{2}}}{\rho a^{\mathrm{w}}_{j-\frac{1}{2}}} \tag{8.60}$$

$$\hat{W}_{j-\frac{1}{2}} = \frac{\sum_{nb=1}^{4} a^{\mathrm{w}}_{nb} U_{nb} + f^{\mathrm{w}}_{j-\frac{1}{2}}}{a^{\mathrm{w}}_{j-\frac{1}{2}}} \tag{8.61}$$

In Equation (8.59) $\beta$ is the thermal expansion coefficient and $g_z$ is the $z$-component of the gravity vector. For natural convection flows oriented as in Figure 8.28, the non-dimensional temperature $\Theta$ appearing in Equation (8.59) is defined as $\Theta = (T - T_0)/(T_1 - T_0)$ and is based on the maximum temperature drop across the computational domain $\Delta T = (T_1 - T_0)$. Further rearranging Equation (8.59), one has:

$$W_{i,j-\frac{1}{2}} = \hat{W}_{i,j-\frac{1}{2}} + \hat{d}_{j-\frac{1}{2}}\left[P_{i,j-1} - P_{i,j}\right] + \hat{g}_{j-\frac{1}{2}}\Theta_{i,j} \tag{8.62}$$

where

$$\hat{g}_{j-\frac{1}{2}} = \frac{g_z\beta(T_1 - T_0)}{2a^{\mathrm{u}}_{i,j-1/2}} \tag{8.63}$$

Here also, the pseudo-velocity $\hat{W}_{i,j-\frac{1}{2}}$ has been modified for including all additional terms (details in de Lemos, 2000). Similar to what has been mentioned above, it is important to notice that the source term in Equation (8.62) explicitly shows the contribution of $T$ (or $\Theta$) on $W$. For the coupled treatment presented here, this explicit arrangement also is shown later.

Following a similar procedure for the $\Theta$ and $V$ equations, final finite-difference equations can be assembled in the following form:

$$a^{\Theta}_{ij}\Theta_{i,j} = b^{\Theta}_{ij}\Theta_{i+1,j} + c^{\Theta}_{ij}\Theta_{i-1,j} + d^{\Theta}_{ij}\Theta_{i,j+1} + e^{\Theta}_{ij}\Theta_{i,j-1} \tag{8.64}$$

$$a^{\mathrm{V}}_{ij}V_{i,j} = b^{\mathrm{V}}_{ij}V_{i+1,j} + c^{\mathrm{V}}_{ij}V_{i-1,j} + d^{\mathrm{V}}_{ij}V_{i,j+1} + e^{\mathrm{V}}_{ij}V_{i,j-1} + f^{\mathrm{V}}_{ij} + g^{\mathrm{V}}_{ij}U_{i,j} \tag{8.65}$$

It is interesting to observe that the last term in (8.65) comes from discretization of the Coriollis acceleration in the $V$-equation and represents the feedback effect of the cross-flow on the tangential velocity (see Table 8.4). Consideration of the Corriollis acceleration in the block-implicit treatment shown below would hamper a proper matrix arrangement suitable for fast inversion. This point will be clear after presenting Section 8.1.2.6. In this work, however, this term is not treated implicitly and, when solving for $V$, it is compacted in the explicitly treated source term. The *centripetal* acceleration, however (see Table 8.4 and Equation (8.57)), is here implicitly handled. The explicit treatment is also employed when discretizing the *convection* terms in the $T$ and $V$ equations since no particular terms with $U$s or $W$s are shown in Equations (8.64) and (8.65).

For simplicity, Equations (8.64) and (8.65) can be rearranged such that

$$\Theta_{i,j} = \hat{\Theta}_{i,j}; \quad V_{i,j} = \hat{V}_{i,j} \tag{8.66}$$

where

$$\begin{aligned}\hat{\Theta}_{i,j} &= \frac{b^{\Theta}_{i,j}\Theta_{i+1,j} + c^{\Theta}_{i,j}\Theta_{i-1,j} + d^{\Theta}_{i,j}\Theta_{i,j+1} + e^{\Theta}_{i,j}\Theta_{i,j-1}}{a^{\Theta}_{i,j}} \\ \hat{V}_{i,j} &= \frac{b^{V}_{i,j}V_{i+1,j} + c^{V}_{i,j}V_{i-1,j} + d^{V}_{i,j}V_{i,j+1} + e^{V}_{i,j}V_{i,j-1} + f^{V}_{i,j} + g^{V}_{i,j}U_{i,j}}{a^{V}_{i,j}}\end{aligned} \tag{8.67}$$

#### 8.10.2.6. Numerical Strategy

In order to smooth out errors due to initial guessed fields, *corrections* are defined as differences between *exact* and *not-yet-converged* variables. Residuals for momentum transport at each control volume face, continuity of mass and $\varphi$ equations are obtained by applying the just defined *approximate* values into Equations (8.57), (8.62), (8.64) and (8.65).

Taking the west face of the control volume shown in Figure 8.28 as an example (see Equation (8.57)), and assuming that for a general variable $\varphi$ a decomposition $\varphi = \varphi^* + \varphi'$ applies, where $\varphi^*$ is a guessed value and $\varphi'$ a correction, one has:

$$(U^* + U')_{i-\frac{1}{2},j} = \hat{U}_{i-\frac{1}{2},j} + \hat{d}_{i-\frac{1}{2}}\left[P_{i-1,j} - (P^* + P')_{i,j}\right] + \hat{e}_{i-\frac{1}{2}}(V^* + V')_{i,j} \tag{8.68}$$

For a given set of guessed or "starred" variables, Equation (8.57) yield a residue, $R_{i-\frac{1}{2},j}$, related to the incorrect velocity such that:

$$U^*_{i-\frac{1}{2},j} = \hat{U}_{i-\frac{1}{2},j} + \hat{d}_{i-\frac{1}{2}}\left[P_{i-1,j} - P^*_{i,j}\right] + \hat{e}_{i-\frac{1}{2}}V^*_{i,j} - R_{i-\frac{1}{2},j} \tag{8.69}$$

Subtracting now (8.69) from (8.68), an equation for the correction $U'$ is obtained in the form:

$$U'_{i-\frac{1}{2},j} + \hat{d}_{i-\frac{1}{2}}P'_{i,j} - \hat{e}_{i-\frac{1}{2}}V'_{i,j} = R_{i-\frac{1}{2},j} \tag{8.70}$$

Writing similar equations for all four faces, a system connecting the residuals and corrections can be written into matrix form as:

$$\begin{bmatrix} 1 & 0 & 0 & 0 & \hat{d}_{i-\frac{1}{2}} & 0 & -\hat{e}_{i-\frac{1}{2}} \\ 0 & 1 & 0 & 0 & -\hat{d}_{i+\frac{1}{2}} & 0 & -\hat{e}_{i+\frac{1}{2}} \\ 0 & 0 & 1 & 0 & \hat{d}_{j-\frac{1}{2}} & -\hat{g}_{j-\frac{1}{2}} & 0 \\ 0 & 0 & 0 & 1 & -\hat{d}_{j+\frac{1}{2}} & -\hat{g}_{j+\frac{1}{2}} & 0 \\ -F_i^1 & F_i^2 & -F_j^1 & F_j^2 & 0 & 0 & 0 \\ 0 & 0 & 0 & 0 & 0 & 1 & 0 \\ 0 & 0 & 0 & 0 & 0 & 0 & 1 \end{bmatrix} \begin{bmatrix} U'_{i-\frac{1}{2},j} \\ U'_{i+\frac{1}{2},j} \\ W'_{i,j-\frac{1}{2}} \\ W'_{i,j+\frac{1}{2}} \\ P'_{i,j} \\ \Theta'_{i,j} \\ V'_{i,j} \end{bmatrix} = \begin{bmatrix} R_{i-\frac{1}{2},j} \\ R_{i+\frac{1}{2},j} \\ R_{i,j-\frac{1}{2}} \\ R_{i,j+\frac{1}{2}} \\ R_{i,j} \\ R^{\Theta}_{i,j} \\ R^{V}_{i,j} \end{bmatrix} \quad (8.71)$$

where the subscripts identify locations in the grid, the superscript $'$ (prime) distinguishes corrections and the lhs represents the residue vector calculated at previous iteration.

In Equation (8.71) the influence of $\Theta$ on the flow field is directly accounted for by the $e$-terms. Similarly, the influence of $V$ on $U$ is implicitly considered by the $g$-terms. For the radial and axial directions, the $g$- and $e$-terms, respectively, are of null value. As mentioned before, the reverse effect, or, say, the cross-flow influence on the $\Theta$ and $V$ fields is here not treated implicitly. The solution of the system (8.71) is then easily obtained by first finding corrections for $\Theta$ and $V$, later calculating the pressure $P$ and velocity components $U$ and $W$. Essentially, the method consists of finding the corrective values for $U$, $V$ and $P$, such that the balance equations are correctly satisfied.

### 8.10.2.7. Computational Parameters

The same relaxation parameters ($\alpha = 0.55$ for $U$, $W$, $P$, $V$ and $\Theta$) were used in all calculations. The swirling strength, $S$, Reynolds number, $Re$, and Rayleigh number, $Ra$, are defined as:

$$S = \left.\frac{V}{W}\right|_{\text{in}}; \quad Re = \frac{\rho W_{\text{in}} 2R}{\mu}; \quad Ra = \frac{Pr\rho^2 \beta g_z \Delta T R^3}{\mu \mu^2} \quad (8.72)$$

These three parameters were varied as follows: $1 < S < 10^3, 2 < Re < 10^3$ and $10^2 < Ra < 10^5$. The incoming axial velocity at inlet, $W_{\text{in}}$, was such that the Reynolds number, in most of the cases run, took the value in the range 2–200. This relatively small input value for $W_{\text{in}}$ indicates that although the flow comes inside the chamber with appreciable rotation ($S$ up to $10^3$), it carries almost no momentum in the axial direction. This incoming velocity level was found to be consistent with the weak currents driven in a thermally driven flow. With that, cases with balanced *natural* and *forced* convection mechanisms could be analyzed.

### *8.10.2.8. Partially Segregated Scheme*

The algebraic equations for the velocity field were solved, in addition to the fully coupled scheme here described, by performing outer iterations for the components $\Theta$ and $V$ while keeping $U$, $W$, $P$ from the previous iteration. A *line-by-line* smoothing operator, fully described elsewhere (e.g. Patankar, 1980), was used to relax $\Theta$ and $V$, the secondary flow field ($U$, $W$) being calculated by the locally coupled method seen above. This *partially segregated* solution was set in such a way that the same number of sweeps throughout the scalar ($\Theta$, $V$) and cross-flow fields ($U$, $W$, $P$) were obtained. Since in the coupled scheme every sweep for $U$, $W$, $P$ also implies smoothing out $\Theta - V$ errors, this procedure was found to be a reasonable way to fairly compare the two methods. In all partially segregated computations, four sweeps per scalar per outer iteration was performed.

The reason for calling this second procedure a *partially* rather than a *fully* segregated scheme lies in the fact that in full segregated methods all variables, including $U$, $W$ and $P$, are solved independently and in sequence along the entire algorithm. In the case here presented for comparison, only $\Theta$ and $V$ are excluded from the implicit treatment implied by Equation (8.71).

## 8.10.3. Some Numerical Results

The results shown below were obtained with a previously developed numerical tool (de Lemos, 2000, 2003a,b). The justification for using this code instead of running one of the many existing programs lies in the fact that adapting the full block-implicit arrangement shown above, involving all vector and scalar variables, would demand too much reprogramming since no pressure correction or pressure equation is used here. Yet, to the authors' knowledge, no available source code entails exactly the same numerical formulation shown here before.

Previous computations with the numerical model herein have assured the correctness and validation of the computer code developed (de Lemos, 1990, 2003a). Those previous results were taken after reduction of all normalized residues for mass, momentum and energy equations to the pre-selected value of $1 \times 10^{-6}$. Accuracy of the present computer program has been checked by comparing calculations for thermally driven flows in a square cavity with benchmark solutions reported by de Vahl Davis (1983). Results in de Lemos (2000) reproduced de Vahl Davis' calculations with less than 0.01% discrepancy for grids greater than $25 \times 25$ and $Ra = 10^4$.

Similar comparisons for the exact geometry and flow conditions here analyzed are more difficult to be worked on due to the scarcity of experimental data and corresponding calculations in the literature. Nevertheless, grid-independence studies for the solutions herein have also been carried out (not shown here due to lack of space). For grids greater than $18 \times 36$, discrepancies in the calculations fall within 0.1%. Since the main objective of this work is to further test the proposed formulation rather than obtaining a detailed calculation of the flow, the chosen grid size was found to give accurate enough results. In summary, the results shown below should be regarded as an aid in comparing *segregated* and *coupled* solution sequences since comparisons between the performances of these two algorithms is the main focus herein.

Figure 8.29 shows results for the velocities at the chamber mid-planes. The figures present velocities $U$ along the vertical $z$-direction (Figure 8.29a, $R = 0.5$) and $W$ along $r$-direction (Figure 8.29b, $z = h/2 = R$), respectively. Inspecting those results, one can conclude that, independently of the smoothing technique used, final converged solutions are essentially equal. The computational effort to achieve them, however, seems to be different, as it is discussed below.

Figure 8.30 shows results for the temperature field when subjected to an increase in the incoming mass flow rate (increase in $Re$, see Equation 8.72). The figure indicates that the core of the flow becomes homogenized as more fluid comes into the chamber due to higher recirculating motion in the $r$–$z$ plane. Outlet temperatures are correctly increased when the higher axial mass flow rate sweeps hot fluid from bottom layers through the exit (see Figure 8.28 for geometry details). Increase of the size of central recirculating bubble is also clearly detected by the downward wash of isolines at the centerline.

Figure 8.31 shows calculations for the temperature field done with different values for $Ra$, spanning from $10^2$ to $10^4$. Distortion of the temperature profiles also indicates strength of convective ascending currents close to the wall with corresponding downward motion at the central region. Interesting to note is the increase in temperature gradients close to the center and at the bottom lid, due to the just mentioned downward stream. When analyzing real equipment, not subjected to the imposed boundary conditions here used, steep gradients of temperature close to the walls might be an indication of possible

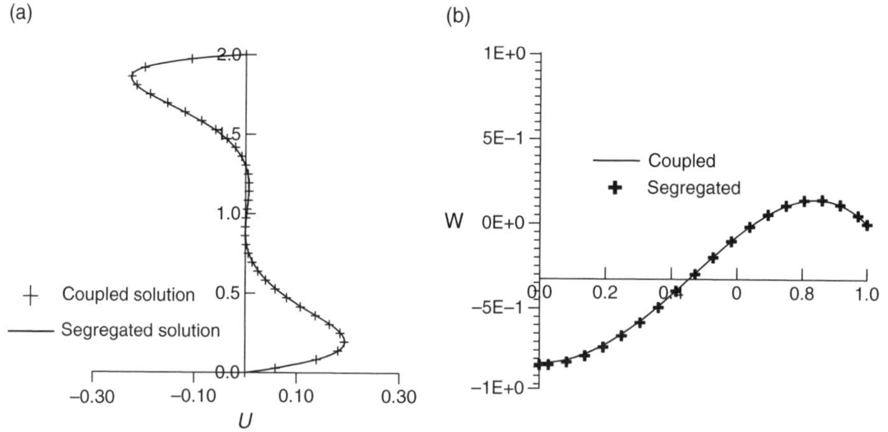


**Figure 8.29.** Radial velocity $U$ along the $z$-direction at $r = 0.5$, axial velocity along radius at $z = L/2 = R$.

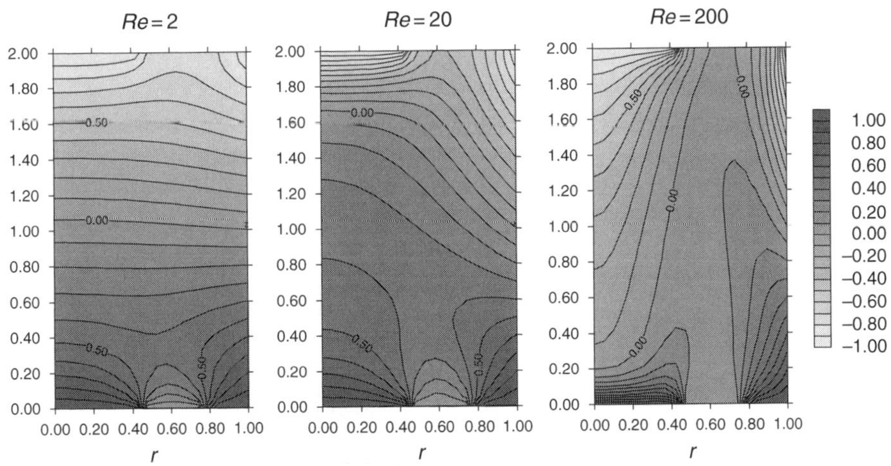


**Figure 8.30.** Effect of $Ra$ on temperature field, $S = 1$, $Re = 10^2$.

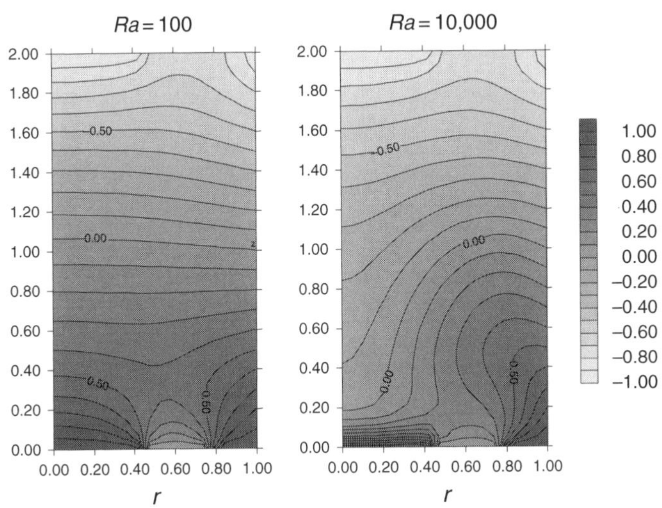


**Figure 8.31.** Effect of $Ra$ on temperature field, $Re = 2, S = 1$.

temperature raise at some particular locations. Design engineers may then use this sort of information to overcome potential material damage when performing preliminary thermal design.

Figure 8.32 presents the temperature pattern for different values of $S$. It is interesting to note that $T$ changes only slighthy, even though $S$ changes by such a large factor of $10^3$. Considering the assumed axi-symmetry of the flow, a strong rotation will carry

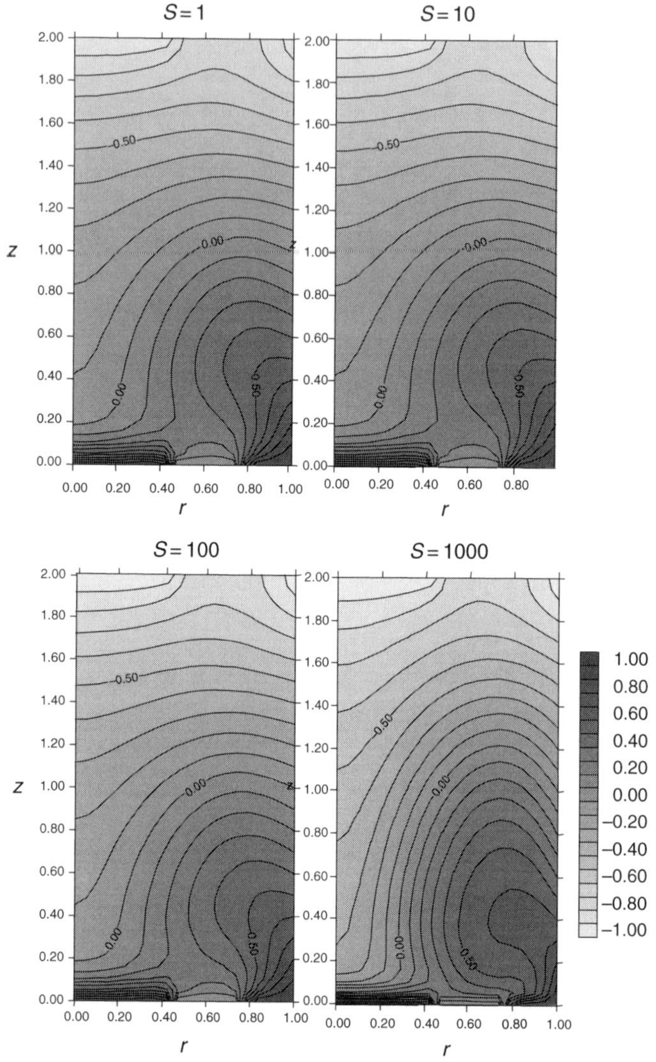


**Figure 8.32.** Effect of swirling strength on temperature, $Ra = 10^4$, $Re = 2$.

fluid tangentially, essentially through isothermal zones. On the other hand, an increase in *Re* or *Ra*, shown in Figures 9.8 and 8.31, substantially distort's the temperature by increasing the ascending cross-flow currents. One should note that in a real, fully three-dimensional flows in industrial equipment, ascending currents are quite strong, playing certainly a definite role in establishing the temperature pattern inside such domains. For the simplified flow and geometry here analyzed, however, no such effect was expected.

Normalized residues were defined as the norm of the cell residue for mass, energy and tangential velocity equations as:

$$R_{abs} = \left( \frac{\sum_{ij} (R_{ij})^2}{(N \cdot M)} \right)^{\frac{1}{2}} \tag{8.73}$$

where $\phi$ in (8.73) refers to the general transport variable as defined in Table 8.4, $N$ and $M$ are the number of cells in the $r$- and $z$-directions, respectively. For continuity equation, $R_{ij}$ can be seen as the difference, for every cell, between the cell *outgoing* mass flux, $F_{out}$, and the *incoming* mass flux, $F_{in}$. A *relative* mass residue can then be defined as:

$$R_{rel} = \left( \frac{\sum_{I\,J} \left( \frac{F_{out}-F_{in}}{F_{out}+F_{in}} \right)^2}{(N \cdot M)} \right)^{\frac{1}{2}} \tag{8.74}$$

A discussion on the advantages in simultaneously monitoring $R_{rel}$ in addition to $R_{abs}$ is presented in de Lemos (1990, 1992a,b, 2000, 2003a,b), and it is based on the small range of the former – $(0, +1)$. Equation (8.74) can be seen as normalization for the residue values for cases ranging from a totally "incorrect" field ($F_{out} = -F_{in}$, $R_{rel} = 1$) to a perfectly "converged" solution ($F_{out} = F_{in}$, $R_{rel} = 0$).

The three parameters here investigated, namely *Re*, *Ra* and *S*, were analyzed in terms of their influence on the overall convergence rates. Results are shown in Figures 8.33, 8.34 and 8.35, respectively. The iteration counter refers to the total number of sweeps over the domain, that is, the product of the outer counter times the number of inner sweeps. A quick word here on the numbers of iterations to convergence seems timely. Other schemes presented in the literature may indicate residue history as a function of outer iteration counters only. Some use the so-called pseudo-transient approach, and plot time steps instead. Each outer iteration, in turn, may consider a great number of internal sweeps, usually controlled by a specified residue reduction rate. Here, in this work, a fixed number of internal sweeps was considered. The relatively large number of necessary iterations seen in the figures below could be associated with the use of a single grid, the tightness of the relaxation parameters and the strong coupling among all variables involved. Ultimately, all of these factors together tend to delay convergence.

It is interesting to note that a faster residue reaction for higher Reynolds numbers (faster convergence rates, see Figure 8.33), possibly reflects the fact that, as *Re* increases, the flow becomes more *forced-convection* dominated decreasing the $U$–$W$–$T$ coupling in relation to the $U$–$W$–$P$ connection. This, in turn, facilitates the solution of the energy equation once the velocity field is calculated. This idea is supported when Figure 8.34 is inspected, showing worse convergence rates for a higher *Ra*. There, the higher degree of

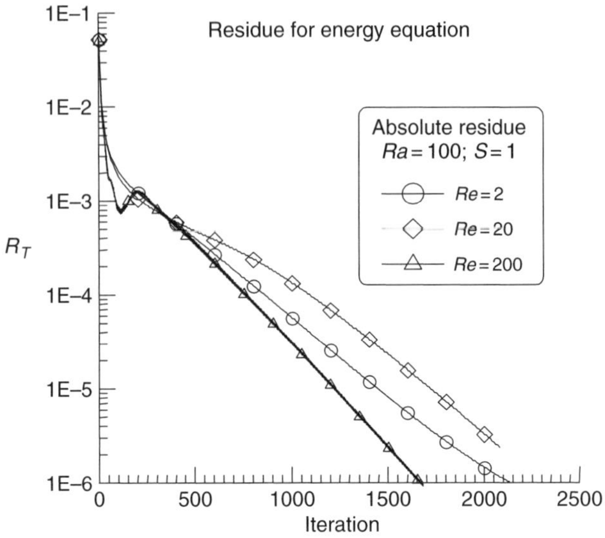


**Figure 8.33.** Influence of *Re* on convergence rate of *T*-equation.

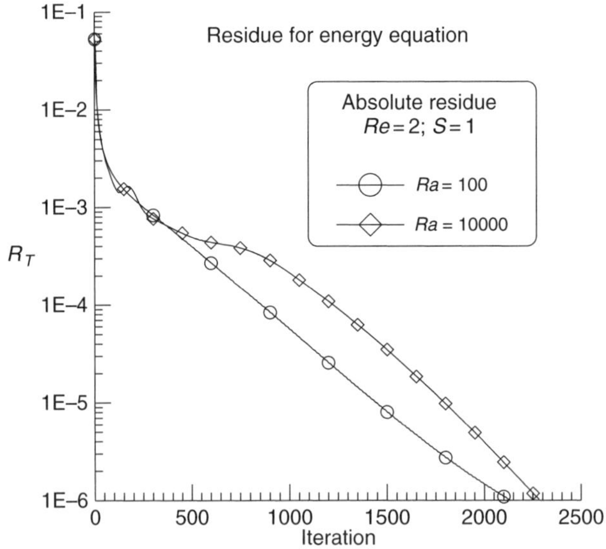


**Figure 8.34.** Effect of *Ra* on $R_T$.

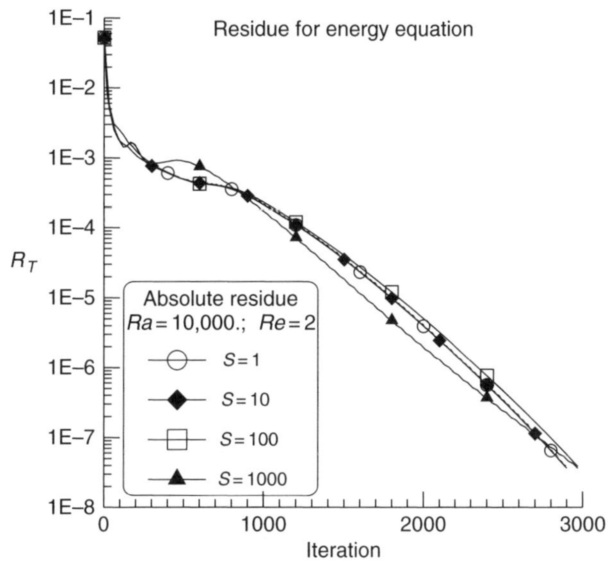


**Figure 8.35.** Effect of swirling strength $S$ on $R_T$.

coupling between temperature and cross-flow fields makes computation more demanding, reflecting the increase in physical coupling among the flow variables and temperature.

On the other hand, when $S$ is varied, Figure 8.35 shows a weak dependency of $R_T$ on the swirling strength. According to Figure 8.32, no substantial change on temperature patterns was detected when inlet rotation increases (except for $S = 10^3$). For such small inlet mass flow rates ($Re = 2$), viscous shear driven by the incoming swirling motion enhances the cross-flow field which, in turn, distorts isothermal lines. Such an indirect or second-order relationship between $V$ and $T$ is apparently reflected on the residue histories shown in Figure 8.35.

Mass residues calculated by Equations (8.73) and (8.74) are presented in Figure 8.36 indicating that, after an initial period when low frequency errors are damped, a better convergence rate is obtained with the coupled method in either form of residue.

Residue histories for the temperature and tangential velocity fields are presented in Figures 8.37 and 8.38, respectively. For the coupled strategy, a quicker reduction for the residue is obtained, since in each sweep, for every cell, information on all variables propagates at the same "time rate". Also, the figures show that the segregated method, for the case analyzed here, use as many as 20–30% more iterations to bring $R_V$ down to the same level as the coupled scheme and nearly two times as many outer sweeps to reduce to the same value of $R_T$. Although the computational effort per field sweep in both schemes are not necessarily equal, due to the fact that the number of floating point operations per iteration in the two algorithms are not the same, a primary consequence of

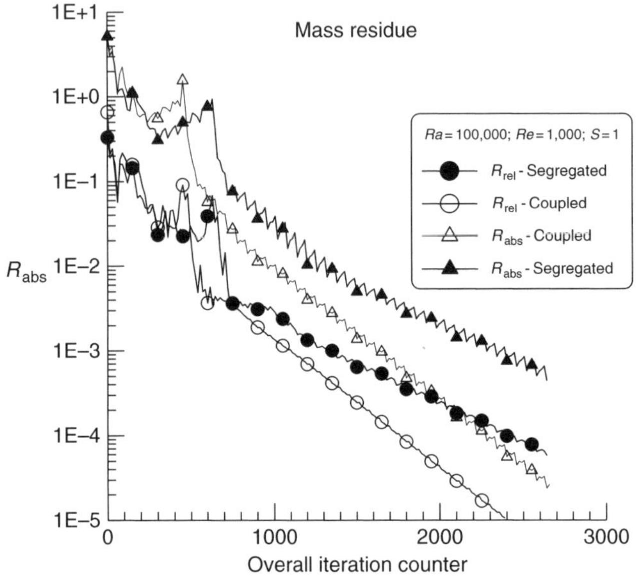


**Figure 8.36.** Mass residue history for segregated and coupled approaches.

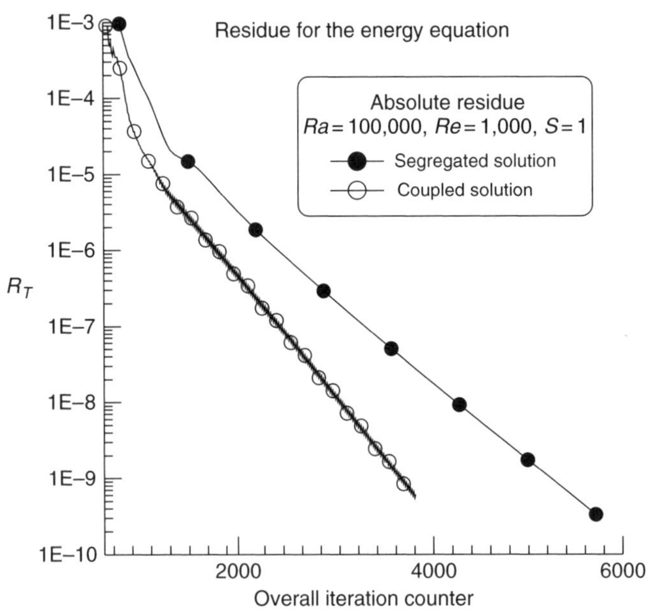


**Figure 8.37.** Influence of solution scheme on residue history of energy equation.

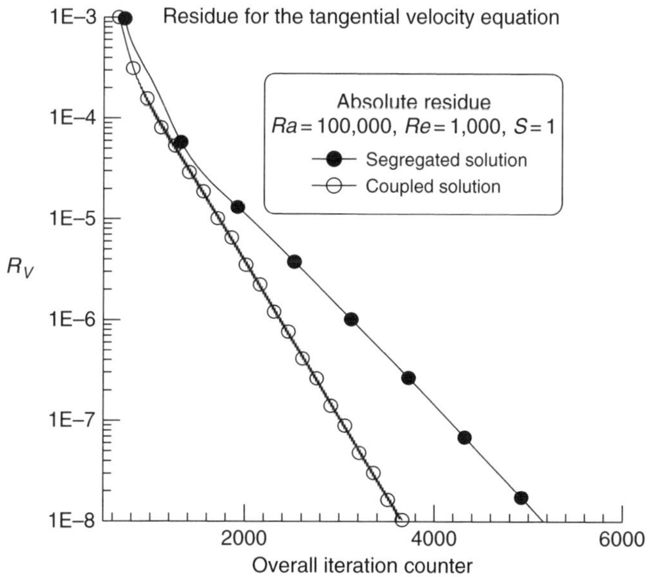


**Figure 8.38.** Influence of solution scheme on convergence of $V$-equation.

the foregoing is that a relatively longer computing time is expected when the decoupled solution is used with the same convergence criterion applied to the scalar equations.

Finally, Figures 8.37 and 8.38 seem to indicate also that the cross-flow field can quickly adjust itself to changes in the scalar profiles, and that, in the segregated case, those changes are too slowly transferred back to $\Theta - V$. The coupled solution, however, quickly transmits back to $\Theta$- and $V$-equations changes in the cross-flow pattern, more realistically simulating the strong interaction among the variables involved.

Chapter 9

# Applications in Hybrid Media

When a macroscopic interface exists between a porous media and a clear fluid, the configuration so formed is called a "hybrid medium". A number of engineering and environmental systems can be modeled by this particular configuration, including flow in channels with porous baffles and obstacles and atmospheric boundary layer over thick and dense rain forests. This chapter shall present and discuss some numerical results of macroscopic modeling considering the following topics.

## 9.1. FORCED FLOWS IN COMPOSITE CHANNELS

This section covers flow behavior along and across a composite channel. By this we mean a two-dimensional channel having a layer of porous material inside. As such, the flows here investigated are schematically shown in Figure 9.1. The channels are partially filled with a layer of porous material. A constant-property fluid flows longitudinally from left to right permeating through both the clear region and the porous structure. The case in Figure 9.1a uses symmetry boundary condition at the channel center ($y = 0$) whereas in Figure 9.1b a solid wall is assumed at the bottom surface. Also, in Figure 9.1, $H$ is the distance in between the channel walls and $s$ the clearance for the non-obstructed flow passage.

### 9.1.1. Numerical Implementation of Jump Condition for Laminar Flow

#### *9.1.1.1. Macroscopic Model*

*Governing Equations* A macroscopic form of the governing equations is obtained by taking the volumetric average of the entire equation set. In this development, the porous medium is considered rigid, undeformable and saturated by an incompressible fluid.

The microscopic continuity equation for the fluid phase is given by

$$\nabla \cdot \mathbf{u} = 0 \tag{9.1}$$

Applying the volume average operator to Equation (9.1), one has (see Pedras and de Lemos, 2001a).

$$\nabla \cdot \mathbf{u}_\mathrm{D} = 0 \tag{9.2}$$

where the local velocity vector $\mathbf{u}$ is of null value at the local interfacial area $A_\mathrm{i}^\mathrm{m}$ and the Dupuit-Forchheimer relationship, $\mathbf{u}_\mathrm{D} = \phi\langle\mathbf{u}\rangle^\mathrm{i}$, has been used. Note that for simplicity of notation in this section, the macroscopic interface area between the finite porous substrate

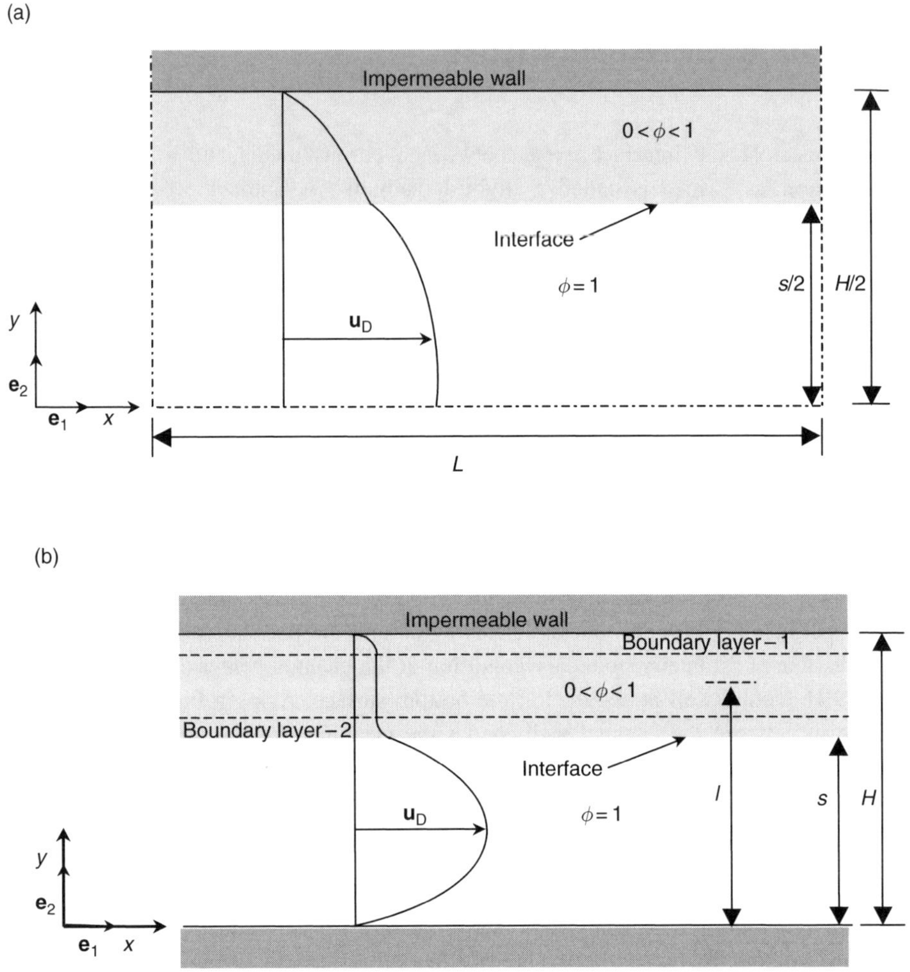


**Figure 9.1.** Model for channel flow with porous material: (a) without the Forchheimer term (b) with the Forchheimer term.

and the surrounding fluid is named $A_i$, whereas the interfacial area between the solid and the fluid phase within the porous matrix is renamed $A_i^m$ (instead of $A_i$ in Chapter 2). Equation (9.2) represents the macroscopic continuity equation for an incompressible fluid in a rigid porous medium.

The microscopic Navier-Stokes equation for an incompressible fluid with constant properties can be written as,

$$\rho\left[\frac{\partial \mathbf{u}}{\partial t}+\nabla\cdot(\mathbf{uu})\right]=-\nabla p+\mu\nabla^2\mathbf{u} \tag{9.3}$$

Hsu and Cheng (1990) have applied the volume averaging procedure to Equation (9.3) obtaining,

$$\rho\left[\frac{\partial}{\partial t}\left(\phi\langle\mathbf{u}\rangle^{\mathrm{i}}\right)+\nabla\cdot\left(\phi\langle\mathbf{uu}\rangle^{\mathrm{i}}\right)\right]=-\nabla\left(\phi\langle p\rangle^{\mathrm{i}}\right)+\mu\nabla^{2}\left(\phi\langle\mathbf{u}\rangle^{\mathrm{i}}\right)+\mathbf{R} \tag{9.4}$$

where

$$\mathbf{R}=\frac{\mu}{\Delta V}\int_{A_{\mathrm{i}}^{m}}\mathbf{n}\cdot(\nabla\mathbf{u})\,dS-\frac{1}{\Delta V}\int_{A_{\mathrm{i}}^{m}}\mathbf{n}p\,dS \tag{9.5}$$

As before, the term **R** represents the total drag per unit volume acting on the fluid by the action of the porous structure. A common model for it is the Darcy–Forchheimer extended model and is given by

$$\mathbf{R}=-\left[\frac{\mu\phi}{K}\mathbf{u}_{\mathrm{D}}+\frac{c_{\mathrm{F}}\phi\rho|\mathbf{u}_{\mathrm{D}}|\mathbf{u}_{\mathrm{D}}}{\sqrt{K}}\right] \tag{9.6}$$

where the constant $c_{\mathrm{F}}$ is known in the literature as the non-linear Forchheimer coefficient.

Then, making use again of the expression $\mathbf{u}_{\mathrm{D}}=\phi\langle\mathbf{u}\rangle^{\mathrm{i}}$, Equation (9.6) can be rewritten as,

$$\rho\left[\frac{\partial\mathbf{u}_{\mathrm{D}}}{\partial t}+\nabla\cdot\left(\frac{\mathbf{u}_{\mathrm{D}}\mathbf{u}_{\mathrm{D}}}{\phi}\right)\right]=-\nabla\left(\phi\langle p\rangle^{\mathrm{i}}\right)+\mu\nabla^{2}\mathbf{u}_{\mathrm{D}}-\left[\frac{\mu\phi}{K}\mathbf{u}_{\mathrm{D}}+\frac{c_{\mathrm{F}}\phi\rho|\mathbf{u}_{\mathrm{D}}|\mathbf{u}_{\mathrm{D}}}{\sqrt{K}}\right] \tag{9.7}$$

#### *9.1.1.2. Interface Condition between the Clear Fluid and the Porous Medium*

The equation proposed in Ochoa-Tapia and Whitaker (1995a,b) for describing the stress jump at the interface between the clear flow region and the porous structures is given by,

$$\mu_{\mathrm{eff}}\left.\frac{\partial u_{\mathrm{D}_{\xi}}}{\partial\eta}\right|_{\text{Porous medium}}-\mu\left.\frac{\partial u_{\mathrm{D}_{\xi}}}{\partial\eta}\right|_{\text{Clear fluid}}=\beta\frac{\mu}{\sqrt{K}}\left.u_{\mathrm{D}_{\xi}}\right|_{\text{Interface}} \tag{9.8}$$

where $u_{\mathrm{D}_{\xi}}$ is the Darcy velocity component parallel to the interface aligned with the direction $\xi$ and normal to the direction $\eta$, $\mu_{\mathrm{eff}}=\mu/\phi$ is the effective viscosity for the porous region according to Ochoa-Tapia and Whitaker (1995a,b) $\mu$ is the fluid dynamic viscosity, $K$ is the permeability and $\beta$ an adjustable coefficient which accounts for the stress jump at the interface. Equation (9.8) will be later adapted to the geometry and coordinate system here employed.

#### *9.1.1.3. Numerical Model*

The numerical method used for discretizing the system of equations is the control volume method of Patankar (1980). In the implementation herein, a system of generalized coordinates was used although all simulation to be shown employed only Cartesian coordinates. Nevertheless, the use of a general system $\eta$-$\xi$ for discretizing the equations was found to be adequate for future simulations.

Since the entire derivation herein is set up for solving two-dimensional flows, both cases employ the spatially periodic boundary condition along the $x$ coordinate. This is done in order to simulate fully developed flow for which analytical solutions are available for comparison. The spatially periodic condition is implemented by running the solution repetitively, until outlet profiles in $x = L$ match those at the inlet ($x = 0$). Figure 9.2a shows a general control volume in a two-dimensional configuration. The faces of the volume are formed by lines of constant coordinates $\eta$-$\xi$.

For steady state, a general form of the discrete equations for a general variable $\varphi$ becomes,

$$I_e + I_w + I_n + I_s = S_\varphi \tag{9.9}$$

where $I_e$, $I_w$, $I_n$ e $I_s$ are the fluxes of $\varphi$ at faces east, west, north and south of the control volume of Figure 9.2a, respectively, and $S_\varphi$ is a source term. Details on the numerical

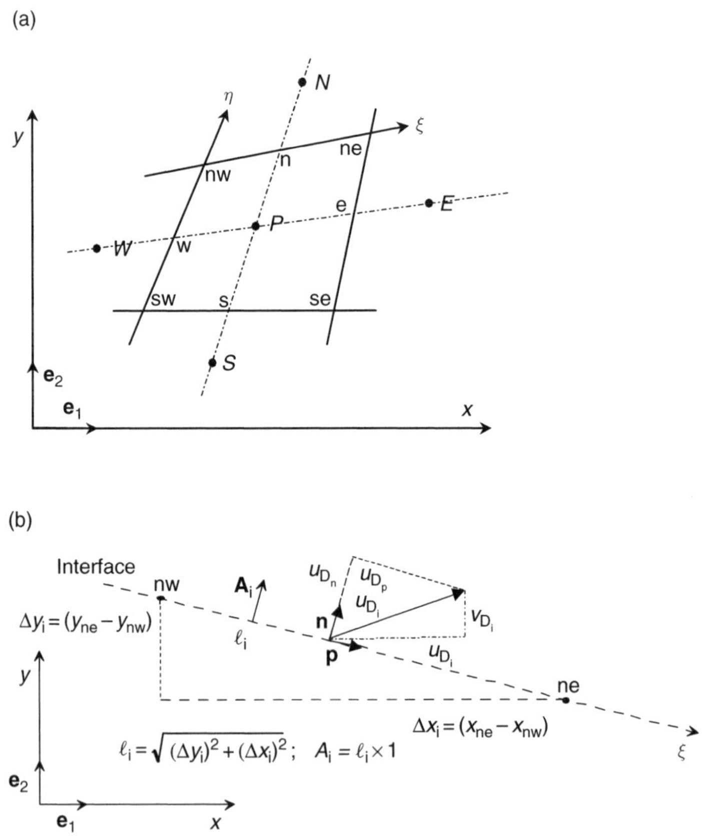


**Figure 9.2.** Notation for: (a) control volume discretization, (b) interface treatment.

methodology employed in obtaining (9.9) can be found in Pedras and de Lemos (2001b). Here, all computations were carried out until the residue of the algebraic equations were brought down to $10^{-7}$, where the residue was defined as the difference between the right and left sides of the discretized equations.

*Implementation of the Interface Condition* For hybrid domains, in addition to Equation (9.8), continuity of velocity and pressure fields prevailing at the interface are given by,

$$\mathbf{u}_D\big|_{0<\phi<1} = \mathbf{u}_D\big|_{\phi=1} \tag{9.10}$$

$$\langle p\rangle^i\big|_{0<\phi<1} = \langle p\rangle^i\big|_{\phi=1} \tag{9.11}$$

Conditions (9.8), (9.10) and (9.11) were proposed in Ochoa-Tapia and Whitaker (1995a) using the concept of stress jump at the interface.

Figure 9.2b shows details of the interface dividing two control volumes, one being located in the porous region and the other lying in the clear fluid. The computational grid based on generalized coordinate system $\eta$-$\xi$ is such that the interface coincides with a line of constant $\eta$, extending along the $\xi$ coordinate. In this arrangement, the interface between the two neighbor volumes, each one located at each side of the interface, belongs to both faces of the two volumes. Thus, according to the Figure 9.2b, $\mathbf{u}_{D_i}$ is the Darcy velocity at the interface and $\mathbf{u}_{D_p}$ its component parallel to the interface itself. It is interesting to point out that although the results here presented are based on a orthogonal Cartesian coordinate system, the derivation to follow is extended to a general system of coordinates $\eta$-$\xi$, the only restriction being the alignment of the interface with a line of constant coordinate. The motivation behind this generalization is to prepare the numerical tool for future use in a complex geometry.

The terms on the left of (9.8) were discretized according to the nomenclature shown in Figure 9.2a. Details of such derivation can be found in Silva and de Lemos (2002, 2003a) and hence not repeated here. This work focuses on the handling of the term on the right of Equation (9.8) for which a detailed study is presented below.

Back to Figure 9.2b, one can identify all variables located at the interface. In the figure, the Darcy velocity at the interface is given by, $\mathbf{u}_{D_i}$. It can be written in either the $x$-$y$ or $\eta$-$\xi$ coordinate systems as,

$$\mathbf{u}_{D_i} = u_{D_i}\mathbf{e}_1 + v_{D_i}\mathbf{e}_2 = u_{D_n}\mathbf{n} + u_{D_p}\mathbf{p} \tag{9.12}$$

where $u_{D_i}$ and $v_{D_i}$ are the components of $\mathbf{u}_{D_i}$ in the $x$ and $y$ directions, respectively. Likewise, $u_{D_n}$ and $u_{D_p}$ are the $\mathbf{u}_{D_i}$ components along the $\eta$ and $\xi$, respectively.

The macroscopic interfacial area vector, normal to the surface, can be expressed as,

$$\mathbf{A}_i = \mathbf{n}A_i = -(y_{ne} - y_{nw})\mathbf{e}_1 + (x_{ne} - x_{nw})\mathbf{e}_2 = -\Delta y_i\mathbf{e}_1 + \Delta x_i\mathbf{e}_2 \tag{9.13}$$

The unit vector normal to the interface, $\mathbf{n}$, is given by,

$$\mathbf{n} = \frac{\mathbf{A}_i}{|\mathbf{A}_i|} \tag{9.14}$$

and therefore its orthogonal unit vector, parallel to the interface, is,

$$\mathbf{p} = \left[ \frac{(x_{ne} - x_{nw})\,\mathbf{e}_1 + (y_{ne} - y_{nw})\,\mathbf{e}_2}{\sqrt{(x_{ne} - x_{nw})^2 + (y_{ne} - y_{nw})^2}} \right] \tag{9.15}$$

Since the considered geometry has two dimensions, one has $|\mathbf{A}_i| = A_i = \ell_i \times 1$, giving further,

$$\mathbf{p} = \frac{\Delta x_i \mathbf{e}_1 + \Delta y_i \mathbf{e}_2}{\ell_i} \tag{9.16}$$

Therefore, the velocity component parallel to the interface, $u_{D_p}$, can be calculated as the scalar product of Equations (9.12) and (9.15) in the form,

$$u_{D_p} = \mathbf{u}_{D_i} \cdot \mathbf{p} \tag{9.17}$$

$$u_{D_p} = \left[ \frac{u_{D_i}(x_{ne} - x_{nw}) + v_{D_i}(y_{ne} - y_{nw})}{\sqrt{(x_{ne} - x_{nw})^2 + (y_{ne} - y_{nw})^2}} \right] = \frac{u_{D_i}\Delta x_i + v_{D_i}\Delta y_i}{\ell_i} \tag{9.18}$$

A Darcy velocity vector parallel to the interface, $\mathbf{u}_{D_p}$, is then given by,

$$\mathbf{u}_{D_p} = u_{D_p}\mathbf{p} = \frac{u_{D_i}\Delta x_i + v_{D_i}\Delta y_i}{\ell_i} \left[ \frac{\Delta x_i \mathbf{e}_1 + \Delta y_i \mathbf{e}_2}{\ell_i} \right] \tag{9.19}$$

Integrating the lhs of Equation (9.8) over the macroscopic interfacial area $A_i$, and considering further constant velocity $\mathbf{u}_{D_p}$ and constant properties prevailing over the integration area, one has,

$$\mathbf{I}_i^{\beta_{x,y}} = \int_{A_i} \mu \frac{\beta}{\sqrt{K}} \mathbf{u}_{D_p} dA_i \approx \mu_i \frac{\beta}{\sqrt{K}} \mathbf{u}_{D_p} A_i = \mu_i \frac{\beta}{\sqrt{K}} \mathbf{u}_{D_p} \ell_i \tag{9.20}$$

Making use of (9.19), one has further,

$$\mathbf{I}_i^{\beta_{x,y}} = \mu_i \frac{\beta}{\sqrt{K}} \frac{(u_{D_i}\Delta x_i + v_{D_i}\Delta y_i)}{\ell_i} (\Delta x_i \mathbf{e}_1 + \Delta y_i \mathbf{e}_2) \tag{9.21}$$

For numerical solution in a general two dimensional geometry, the momentum equation components in the $x$ and $y$ direction are obtained by decomposing (9.21) such that,

$$I_{\mathrm{i}}^{\beta_x} = \mu_{\mathrm{i}} \frac{\beta}{\sqrt{K}} \frac{\left(u_{\mathrm{D_i}} \Delta x_{\mathrm{i}} + v_{\mathrm{D_i}} \Delta y_{\mathrm{i}}\right)}{\ell_{\mathrm{i}}} \Delta x_{\mathrm{i}} \tag{9.22}$$

and

$$I_{\mathrm{i}}^{\beta_y} = \mu_{\mathrm{i}} \frac{\beta}{\sqrt{K}} \frac{\left(u_{\mathrm{D_i}} \Delta x_{\mathrm{i}} + v_{\mathrm{D_i}} \Delta y_{\mathrm{i}}\right)}{\ell_{\mathrm{i}}} \Delta y_{\mathrm{i}} \tag{9.23}$$

Terms on the rhs of Equations (9.22) and (9.23) are added to the discretized momentum equation components in the $x$ and $y$ directions, respectively, when the nodal point in question has a face coincident with the interface. For ease of implementation, these additional terms are treated in an explicit form and are added to the rhs of Equation (9.9).

#### *9.1.1.4. Some Numerical Results*

The two cases pictured in Figure 9.1 are associated with solutions of different forms of Equation (9.7). The case shown in Figure 9.1a is solved without the last term on the right of Equation (9.7). An analytical solution for this case was first proposed in Kuznetsov (1996). On the other hand, the case shown in Figure 9.1b considers such non-linear term, also referred to in the literature as Forchheimer term. Analytical distributions for the velocity field were presented in Kuznetsov (1996–1999). In both cases, numerical predictions use analytical profiles for validation of the numerical implementation herein.

Grid independence studies are shown in Figure 9.3. The figure shows several non-dimensional velocity profiles for $\phi = 0.6$, $K = 4 \times 10^{-7}\,\mathrm{m}^2$ and Darcy number $Da = 4 \times 10^{-3}$ where $Da = K/H^2$. Case (a) mentioned above considers solution of Equation (9.7) with $\phi = 1$. The value for the coefficient $\beta$ is set equal to zero for all solutions presented in the figure. The curves indicate that for more than 40 nodal points in the cross-stream direction, the solution is essentially grid independent. One should point out that the numerical methodology here considered was focused on two dimensional flows, so that simulating a fully developed situation shown in the figures required the use of nodal points along the axial direction and the employment of the spatially periodic condition mentioned earlier. For all runs here studied, a total of 50 nodes in the axial direction was found to suffice.

Figure 9.4 shows the effect of the permeability $K$ reproduced by both the numerical solution and the one-dimensional theoretical treatment. One can see that greater the permeability, more flow crosses the porous substratum located in the region $0.5 < (y/H) < 1$. The agreement between numerical and analytical solution can be noted in the figure.

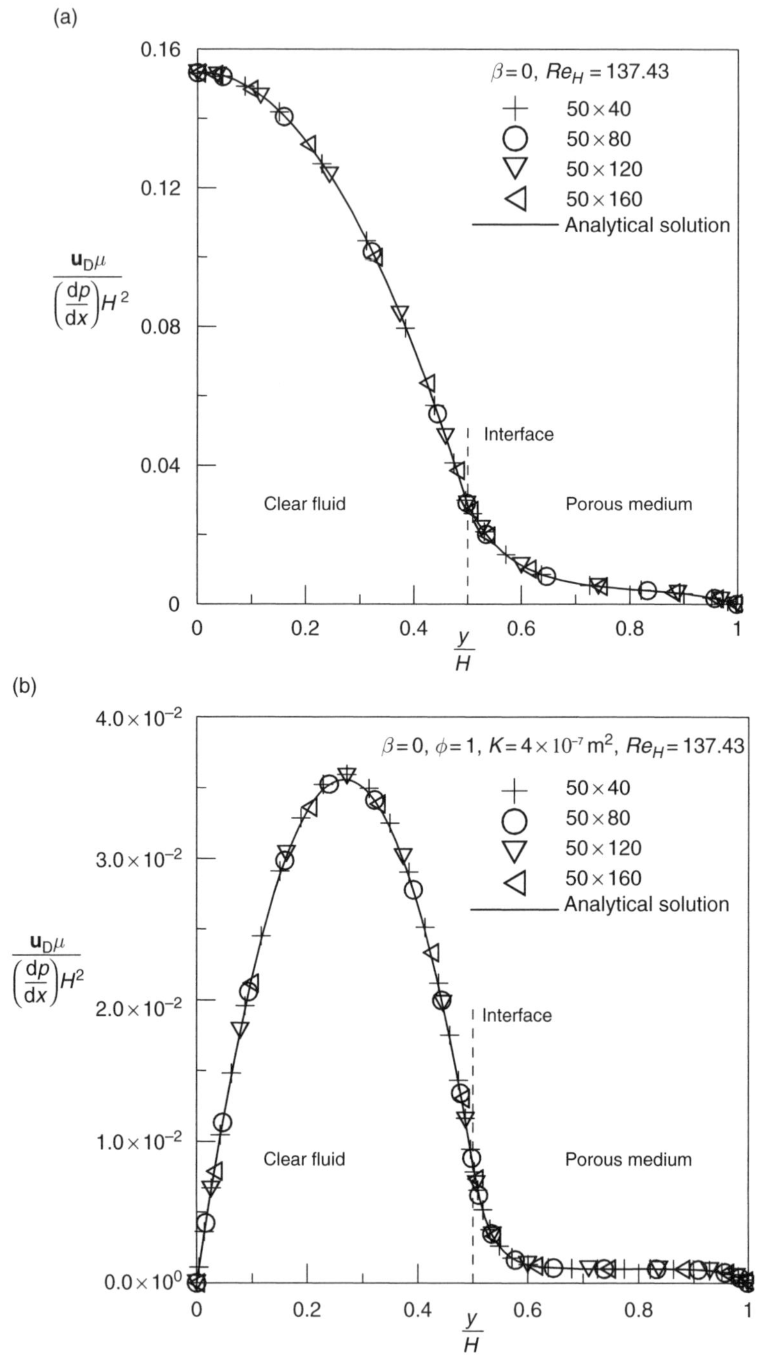


**Figure 9.3.** Effect of grid size on velocity field: (a) without the Forchheimer term, (b) with the Forchheimer term.

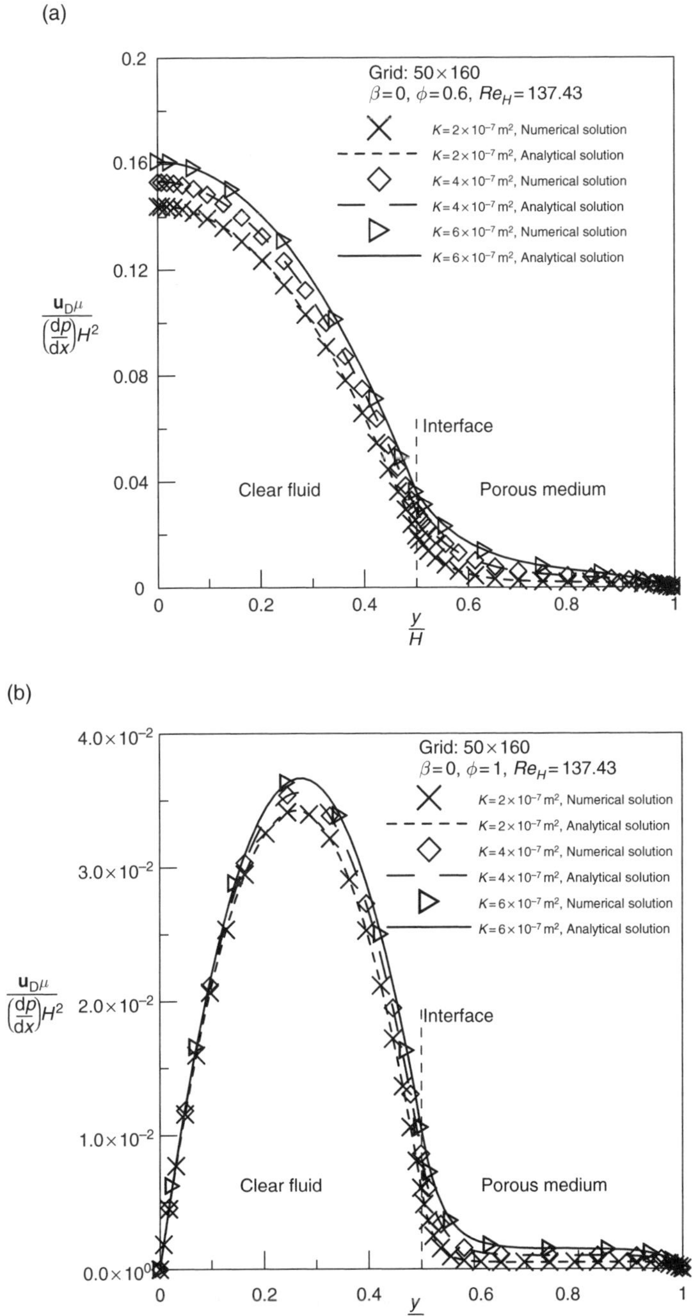


**Figure 9.4.** Comparison between analytical and numerical solution for different values of permeability, $K$: (a) without the Forchheimer term, (b) with the Forchheimer term.

Figure 9.5 investigates the effect of the value of $\phi$ on the behavior of the velocity field. Also here, as expected, the greater the porosity the higher the mass flow rate within the permeable layer. One should point out, however, that all solutions presented in Figures 9.4 and 9.5 are obtained for a fixed Reynolds number, so that overall mass flow rate through the channel was held constant. The use of the pressure gradient when setting up the non-dimensional velocity profiles in the figures may misleadingly indicate an increase in the flow rate in the clear passage, at the channel center, for the cases of increasing $K$ or $\phi$. Also to note is that results in Figures 9.4 and 9.5 are for $\beta = 0$, indicating that even without considering the numerical implementation of the jump condition, which is the main object of this investigation, the computer code developed seems to correctly reproduce the exact solution. With these preliminary tests done, further results including the jump at the interface can be better assessed.

Extending the foregoing results, Figure 9.6 finally compares both analytical and the numerical solution for $\beta$ varying from $-0.5$ to $0.5$ for a fixed porosity $\phi = 0.6$ and constant Darcy number $Da = 4 \times 10^{-3}$. Here also results seem to indicate the correctness of the numerical implementations for the range of parameters investigated. Ultimately, results in Figure 9.6 show the appropriateness on the numerical methodology here employed for considering the stress jump at the interface between a porous medium and a clear region.

### 9.1.2. Jump Condition for Mean Turbulent Flow

Transport equations for $\langle k\rangle^{\mathrm{i}} = \langle\overline{\mathbf{u}'\cdot\mathbf{u}'}\rangle^{\mathrm{i}}/2$ and $\langle\varepsilon\rangle^{\mathrm{i}} = \mu\langle\overline{\nabla\mathbf{u}' : (\nabla\mathbf{u}')^{\mathrm{T}}}\rangle^{\mathrm{i}}/\rho$ in their so-called "high Reynolds number" form shown in Chapter 4 are:

$$\rho\left[\frac{\partial}{\partial t}\left(\phi\langle k\rangle^{\mathrm{i}}\right) + \nabla\cdot\left(\bar{\mathbf{u}}_{\mathrm{D}}\langle k\rangle^{\mathrm{i}}\right)\right] = \nabla\cdot\left[\left(\mu + \frac{\mu_{t_\phi}}{\sigma_k}\right)\nabla\left(\phi\langle k\rangle^{\mathrm{i}}\right)\right] + P^{\mathrm{i}} + G^{\mathrm{i}} - \rho\phi\langle\varepsilon\rangle^{\mathrm{i}} \tag{9.24}$$

where $P^{\mathrm{i}} = -\rho\langle\overline{\mathbf{u}'\mathbf{u}'}\rangle^{\mathrm{i}} : \nabla\bar{\mathbf{u}}_{\mathrm{D}}$, $G^{\mathrm{i}} = c_k\rho\dfrac{\phi\langle k\rangle^{\mathrm{i}}|\bar{\mathbf{u}}_{\mathrm{D}}|}{\sqrt{K}}$ and

$$\rho\left[\frac{\partial}{\partial t}\left(\phi\langle\varepsilon\rangle^{\mathrm{i}}\right) + \nabla\cdot\left(\bar{\mathbf{u}}_{\mathrm{D}}\langle\varepsilon\rangle^{\mathrm{i}}\right)\right] = \nabla\cdot\left[\left(\mu + \frac{\mu_{t_\phi}}{\sigma_\varepsilon}\right)\nabla\left(\phi\langle\varepsilon\rangle^{\mathrm{i}}\right)\right] + c_1 P^{\mathrm{i}}\frac{\langle\varepsilon\rangle^{\mathrm{i}}}{\langle k\rangle^{\mathrm{i}}} + c_2\frac{\langle\varepsilon\rangle^{\mathrm{i}}}{\langle k\rangle^{\mathrm{i}}}\left(G^{\mathrm{i}} - \rho\phi\langle\varepsilon\rangle^{\mathrm{i}}\right) \tag{9.25}$$

where $c_1$, $c_2$ and $c_k$ are constants, $P^{\mathrm{i}}$ is the production rate of $\langle k\rangle^{\mathrm{i}}$ due to gradients of $\mathbf{u}_{\mathrm{D}}$, and $G^{\mathrm{i}}$ is the generation rate of the intrinsic average of $k$ due to the action of the porous matrix.

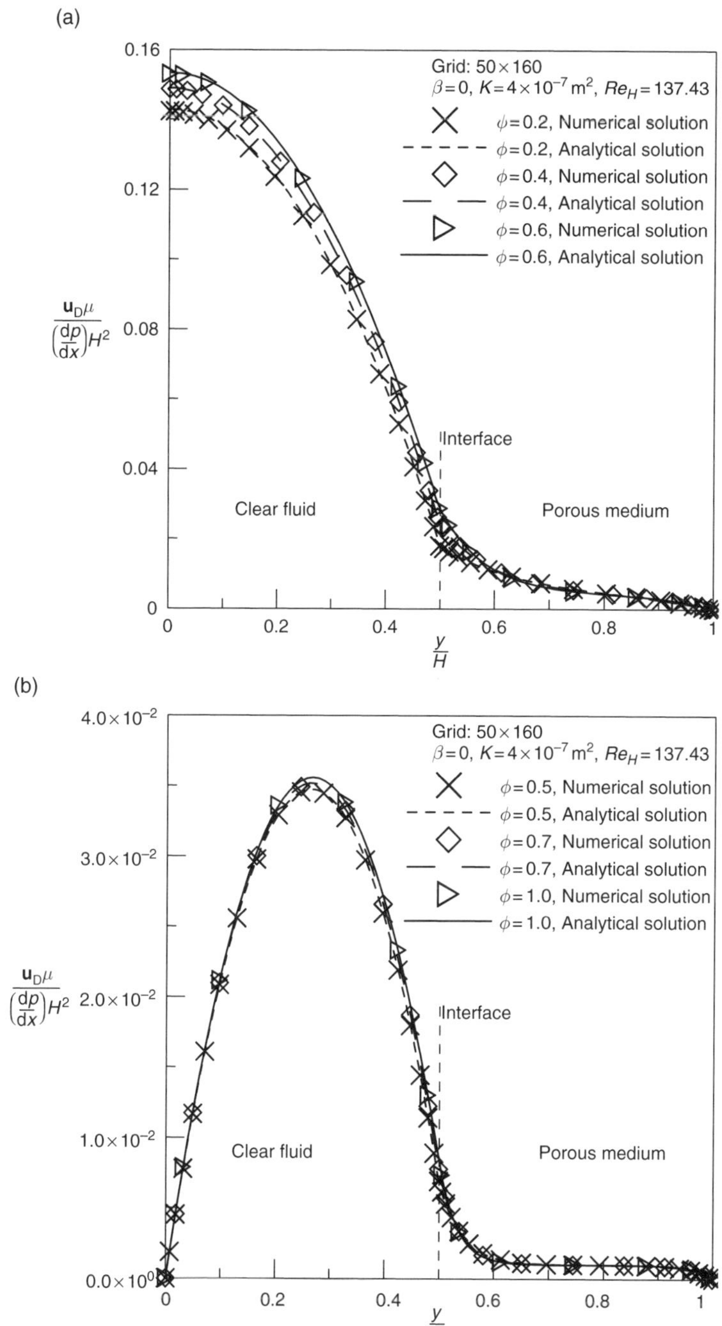


**Figure 9.5.** Comparison between analytical and numerical solution for different porosities, $\phi$: (a) without the Forchheimer term, (b) with the Forchheimer term.

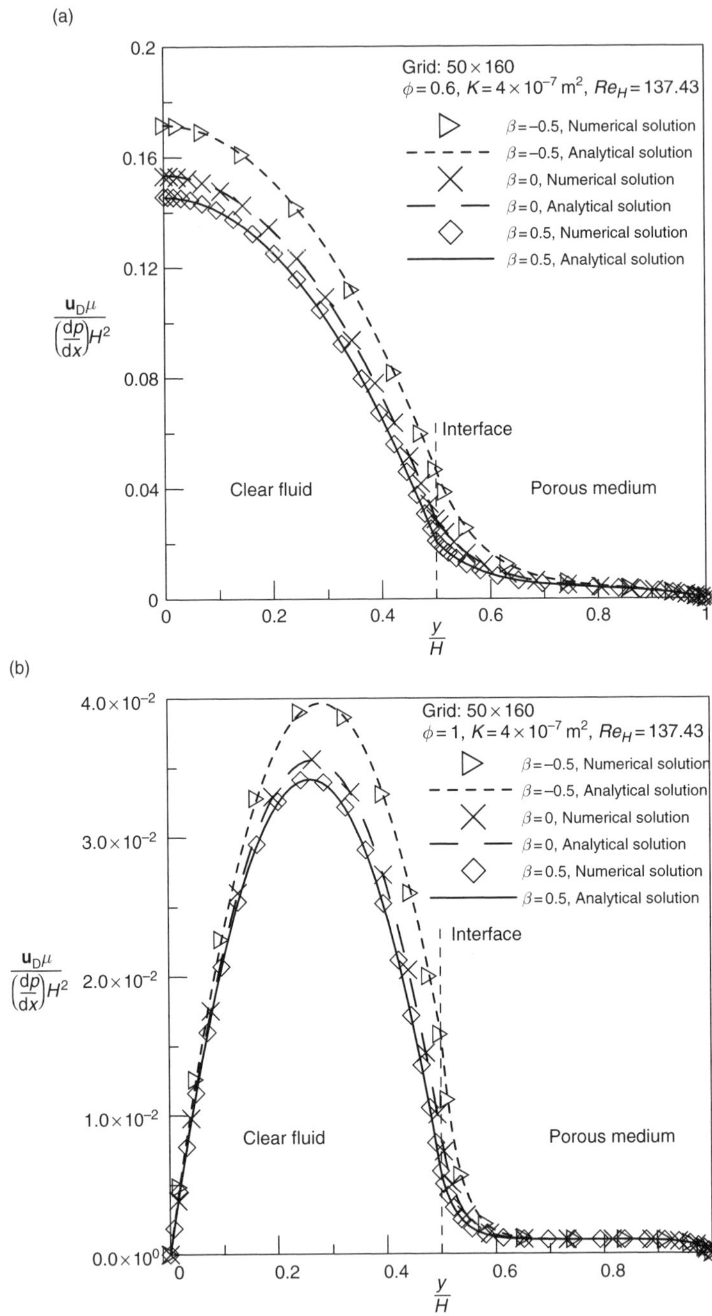


**Figure 9.6.** Comparison between analytical and numerical solution for different values of $\beta$: (a) without the Forchheimer term (b) with the Forchheimer term.

#### 9.1.2.1. *Interface and Boundary Conditions*

Equation (9.8) proposed in Ochoa-Tapia and Whitaker (1995a,b) describes here also the stress jump at the interface between the clear flow region and the porous structure. For clarity, it is repeated here and the velocity parallel to the interface is renamed as $u_{D_p}$:

$$\mu_{\text{eff}} \left. \frac{\partial u_{D_p}}{\partial \eta} \right|_{\text{Porous medium}} - \mu \left. \frac{\partial u_{D_p}}{\partial \eta} \right|_{\text{Clear fluid}} = \beta \frac{\mu}{\sqrt{K}} \left. u_{D_p} \right|_{\text{Interface}} \tag{9.26}$$

where again the notation in Figures 9.1 and 9.2 applies.

It is important to emphasize that the macroscopic model for the interface here employed makes no assumption about the topology of the surface, nor is this interface the one existing in transpired solid walls, for example. Or say, although the microscopic interfacial area surrounding the irregular geometry of solid particles facing the clear medium may be characterized by statistical values, such as average thickness or roughness, in the present macroscopic view, no such thickness or roughness is associated with the interface. In fact, in Kaviany (1995, p. 71), the order of magnitude of the roughness of the interface is of the order of $d$ (pore/particle diameter), which is much higher than $\sqrt{K}$, another length associated with permeable media. Had the interface roughness been considered, that would be of the order of $d$, the mean particle/pore diameter. Here, irregular or rough boundaries between the porous medium and the clear fluid are treated under the macroscopic view and, as such, no statistical value of interface thickness is attributed to the modeled surface separating the two media. Likewise, transpired walls made of a porous substrate with extremely small pore sizes are not treated here. Also, the macroscopic velocity at the interface and on its surroundings is assumed to be of sufficient value so that a viscous sub-layer similar to the one existing over impermeable surfaces are not present in the context herein.

In addition to Equation (9.26), continuity of velocity, pressure, statistical variables and their fluxes across the interface are given by:

$$\left. \bar{\mathbf{u}}_D \right|_{0<\phi<1} = \left. \bar{\mathbf{u}}_D \right|_{\phi=1} \tag{9.27}$$

$$\left. \langle \bar{p} \rangle^i \right|_{0<\phi<1} = \left. \langle \bar{p} \rangle^i \right|_{\phi=1} \tag{9.28}$$

$$\left. \langle k \rangle^v \right|_{0<\phi<1} = \left. \langle k \rangle^v \right|_{\phi=1} \tag{9.29}$$

$$\left(\mu + \frac{\mu_{t_\phi}}{\sigma_k}\right) \left. \frac{\partial \langle k \rangle^v}{\partial y} \right|_{0<\phi<1} = \left(\mu + \frac{\mu_t}{\sigma_k}\right) \left. \frac{\partial \langle k \rangle^v}{\partial y} \right|_{\phi=1} \tag{9.30}$$

$$\left. \langle \varepsilon \rangle^v \right|_{0<\phi<1} = \left. \langle \varepsilon \rangle^v \right|_{\phi=1} \tag{9.31}$$

$$\left(\mu + \frac{\mu_{t_\phi}}{\sigma_\varepsilon}\right) \left. \frac{\partial \langle \varepsilon \rangle^v}{\partial y} \right|_{0<\phi<1} = \left(\mu + \frac{\mu_t}{\sigma_\varepsilon}\right) \left. \frac{\partial \langle \varepsilon \rangle^v}{\partial y} \right|_{\phi=1} \tag{9.32}$$

Further, the extension of Equation (9.26) to the case of turbulent flow can be given as:

$$\left(\mu_{\mathrm{eff}}+\mu_{\mathrm{t}_\phi}\right)\frac{\partial \bar{u}_{\mathrm{D_p}}}{\partial y}\bigg|_{0<\phi<1}-(\mu+\mu_{\mathrm{t}})\,\frac{\partial \bar{u}_{\mathrm{D_p}}}{\partial y}\bigg|_{\phi=1}=(\mu+\mu_{\mathrm{t}})\,\frac{\beta}{\sqrt{K}}\,\bar{u}_{\mathrm{D_p}}\bigg|_{\mathrm{Interface}} \tag{9.33}$$

Equations (9.27) and (9.28) were also proposed in Ochoa-Tapia and Whitaker (1995a,b) whereas relationships (9.29) through (9.33) were used by Lee and Howell (1987).

Before proceeding, a word about the use of expression (9.26) seems timely. In Ochoa-Tapia and Whitaker (1995a,b), such an equation was proposed in order to accommodate a possible discontinuity in the diffusion flux of momentum across the interface. This model has already been applied to laminar flows involving analytical (Kuznetsov, 1996–1999) and purely numerical solutions (Silva and de Lemos, 2003a). As already mentioned, an extension to turbulent has also been published (Lee and Howell, 1987) which adopted the continuity of diffusion fluxes for $u_{\mathrm{D}}$ (without any stress jump), for $\langle k\rangle^{\mathrm{v}}$ (Equation (9.30)) as well as for $\langle \varepsilon\rangle^{\mathrm{v}}$ (Equation (9.32)). When the rhs of Equation (9.26) is equal to zero, or say, when no stress jump is accounted for by setting $\beta=0$, for example, one has a matching of diffusion fluxes across the interface, much like Equations (9.30) and (9.32) for $\langle k\rangle^{\mathrm{v}}$ and $\langle \varepsilon\rangle^{\mathrm{v}}$, respectively. Or say, due to the fact that a second derivate exists in Equations (9.7), (9.24) and (9.25), it is possible, but not mandatory, that diffusion fluxes match at the interface (see Kaviany, 1995, p. 92). Had a similar "jump" been considered in order to account for some extra effect in the $\langle k\rangle^{\mathrm{v}}$-equation, for example, we would have had an interface condition of the form:

$$\left(\mu+\frac{\mu_{\mathrm{t}_\phi}}{\sigma_k}\right)\frac{\partial\langle k\rangle^{\mathrm{v}}}{\partial y}\bigg|_{0<\phi<1}-\left(\mu+\frac{\mu_{\mathrm{t}}}{\sigma_k}\right)\frac{\partial\langle k\rangle^{\mathrm{v}}}{\partial y}\bigg|_{\phi=1}=[\text{Jump Term } \mathit{for}\ \langle k\rangle^{\mathrm{v}}] \tag{9.34}$$

instead of Equation (9.30). That was the proposition of de Lemos and Silva (2006) to be shown in section "Jump Condition for Turbulence Kinetic Energy" on p. 203. Here, however, no such discontinuity on diffusion fluxes for the statistical quantities is assumed.

Also, standard wall functions have been employed for calculating the flow in the proximity of channel walls. Detailed information on such numerical treatment can be found in Silva and de Lemos (2003a,b). Justification for using such simpler treatment is twofold:

(1) Final velocity values close to the interface will be a function not only of the inertia and viscous effects in the full Navier-Stokes equation, but also due to the Darcy and Forchheimer resistance terms. Therefore, eventual errors coming from inaccurate use of a more appropriate boundary conditions will have little influence on the final value for the velocity close to the wall since drag forces, caused by the porous structure, will play an important role in determining the final value for the wall velocity.

(2) Logarithm wall laws are simple to be incorporated when simulating flow over rigid surfaces and for that they have been modified to include surface roughness and to simulate flows over irregular surfaces at the bottom of rivers (Lane and Hardy, 2002).

In addition, is it interesting to emphasize that the class of flows under consideration is akin to having a sequence of closely spaced grids in a flow with a flat macroscopic Darcy velocity profile. Mechanical energy is transformed into turbulence kinetic energy as the flow crosses and is perturbed by the porous matrix. This interpretation of the model here used (Pedras and de Lemos, 2001a) has been detailed in de Lemos and Pedras (2001).

Further, one should point out that the condition given in Equation (9.33) was first proposed by Lee and Howell (1987) and is valid along the macroscopic surface area dividing the clear and the porous regions. Application of the volume average operators to an REV (Whitaker, 1969; Gray and Lee, 1977) gives rise to terms such as the Darcy and Forchheimer flow resistances which, according to the literature (Ochoa-Tapia and Whitaker, 1995a,b; Kuznetsov, 1996–1999), are not presented when analyzing macroscopic interfacial areas as here considered.

The flow in Figure 9.1 was computed with the set of Equations (9.7), (9.24) and (9.25) with additional constitutive Equation (7.2) and the Kolmogorov–Prandtl Equation (7.4). Grid independence studies were conducted, and for more than 40 nodal points in the cross-stream direction, the solution was essentially grid independent. The wall function approach was used for treating the flow close to the wall. One should emphasize that the numerical methodology here considered was focused on two dimensional flows, so that simulating the fully developed situation shown in the figure required the use of nodal points along the axial direction as well as the employment of the spatially periodic condition mentioned earlier. For all runs here studied, a total of 50 nodes in the axial direction was found to suffice. Also interesting to emphasize is that the sign of coefficient $\beta$ in Equation (9.26) will depend on the orientation of the $y$-axis in relation to the porous layer location. Here, the same orientation was used as the one given by Kuznetsov (1996–1999), which considers the porous layer at the top of the channel with its normal pointing toward the negative $y$-direction. As such, coherent computations for laminar flow (Silva and de Lemos, 2003a) were obtained. Further, grid independence studies were carried out by Silva and de Lemos, 2003a) indicating the proper size of the number of nodal points used around the interface. There, the authors correctly reproduced, with their numerical tool, the boundary layers around the interface proposed by the analytical study of Kuznetsov (1996–1999). Also for the case of turbulent flow, as seen in Figure 9.7, the number of grid points used seems to be appropriate.

The effect of $Re_H$ is shown in Figure 9.8. The mean velocity in Figure 9.8a indicates the increasing mass flow rate, within either the porous material or the clear passage, as the Reynolds number increases. In Figure 9.8b the collapse of curves for the turbulent kinetic energy divided by the mean mechanical energy shows that, for the range of $Re_H$ analyzed, the percentage of energy transformed into turbulence remains the same.

Figure 9.9 shows the effect of the permeability $K$ on both the mean and statistical fields. One can see that the greater the permeability, more flow crosses the porous substratum located in the region $0.5 < y/H < 1$ (Figure 9.9a). The curves representing the statistical

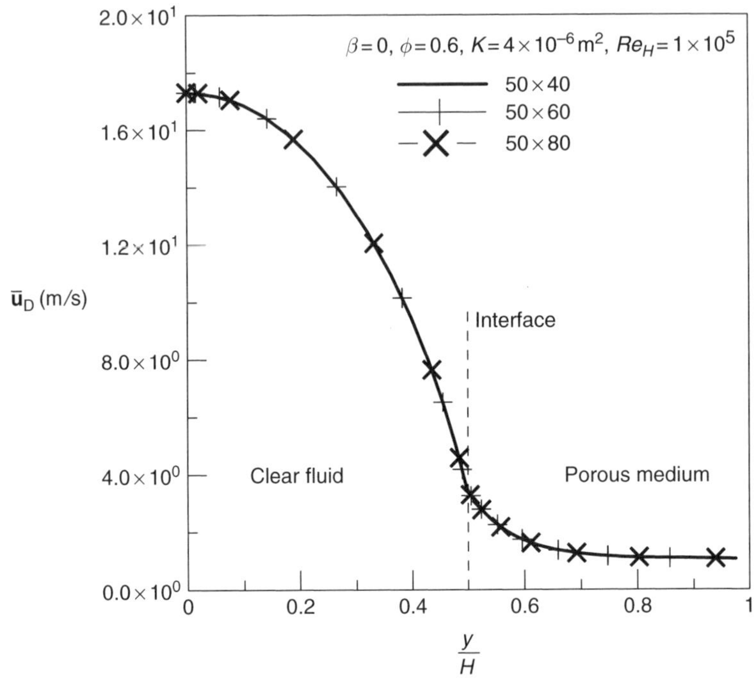


**Figure 9.7.** Effect of mesh size on numerical solution.

field in Figure 9.9b show that, except close to the interface, the levels of $k$ increase with increasing $K$. Within the clear fluid the production of turbulent kinetic energy is known to be dictated by gradients of the mean velocity ($P^{\mathrm{i}}$ on the rhs of (9.24)) whereas inside the permeable structure, the model of Pedras and de Lemos (2001a) proposes a factor proportional to $\mathbf{u}_{\mathrm{D}}$ as a generating mechanism for $k$ ($G^{\mathrm{i}}$ in Equation (9.24)).

Figure 9.10a investigates the effect of the value of $\phi$ on the behavior of the mean velocity field. One can note that close to the interface, the greater the porosity, the higher the mass flow rate within the permeable layer. At the center of the channel, the velocity decreases in order to keep the imposed mass flow rate the same. It is interesting to observe that since the overall mass flow rate is forced to be constant, instead of the overall pressure loss along the channel, an increase in mass flow rate along the porous bed in the interface region is compensated by a slight reduction on local velocities close to the wall. Figure 9.10b shows corresponding curves for the behavior of the turbulent kinetic energy. Away from the interface, values of $k$ present a slight increase as $\phi$ is incremented. Greater values for the turbulence level within the porous layer ($y/H > 0.6$) are coherent with the model of Equation (9.24) for the extra generation rate due to the porous matrix. As said, this extra $G^{\mathrm{i}}$ term (3rd on the right of (9.24)) was modeled as

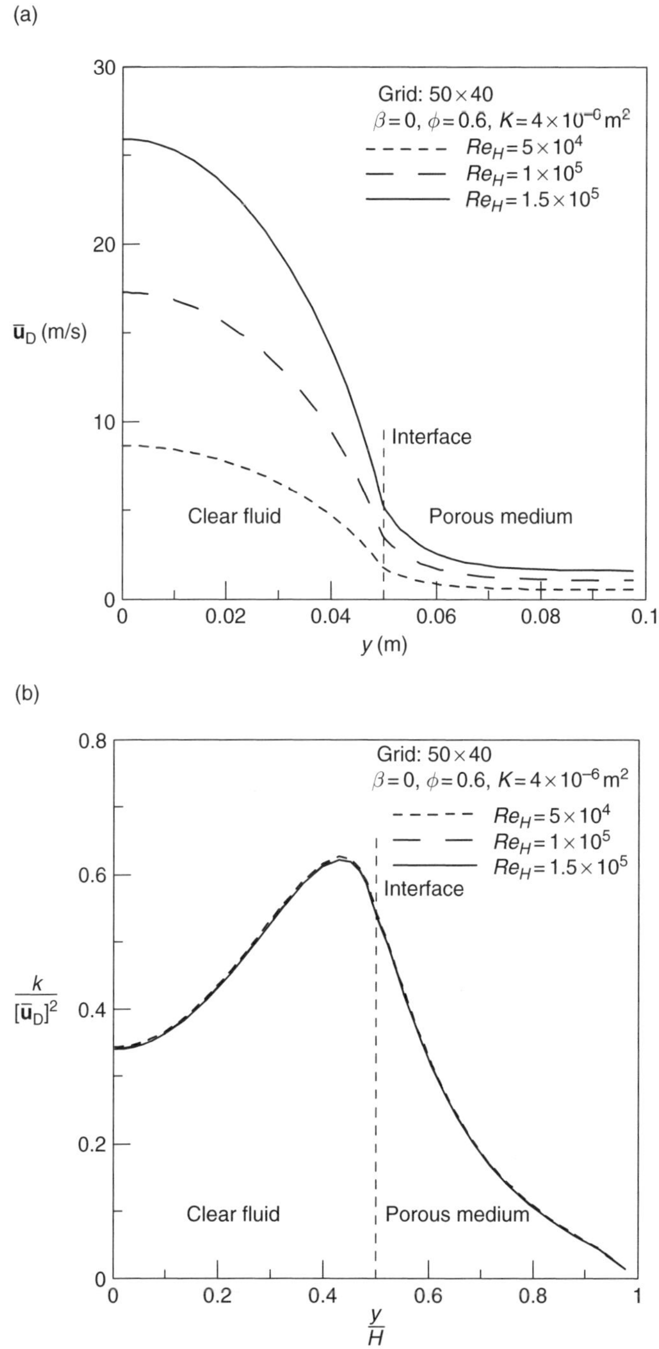


**Figure 9.8.** Effect of Reynolds number, $Re_H$, on macroscopic field: (a) mean velocity, (b) turbulent kinetic energy.

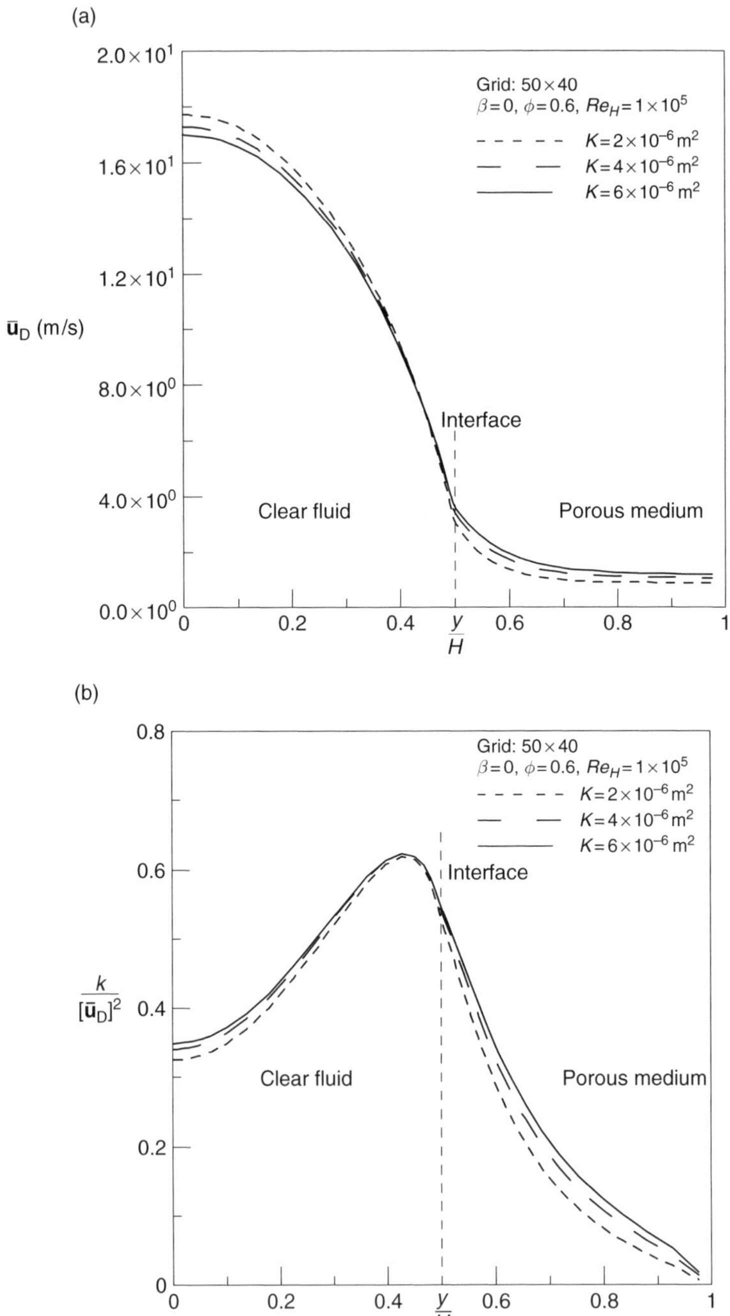


**Figure 9.9.** Effect of permeability $K$ on macroscopic field: (a) mean velocity, (b) turbulent kinetic energy.

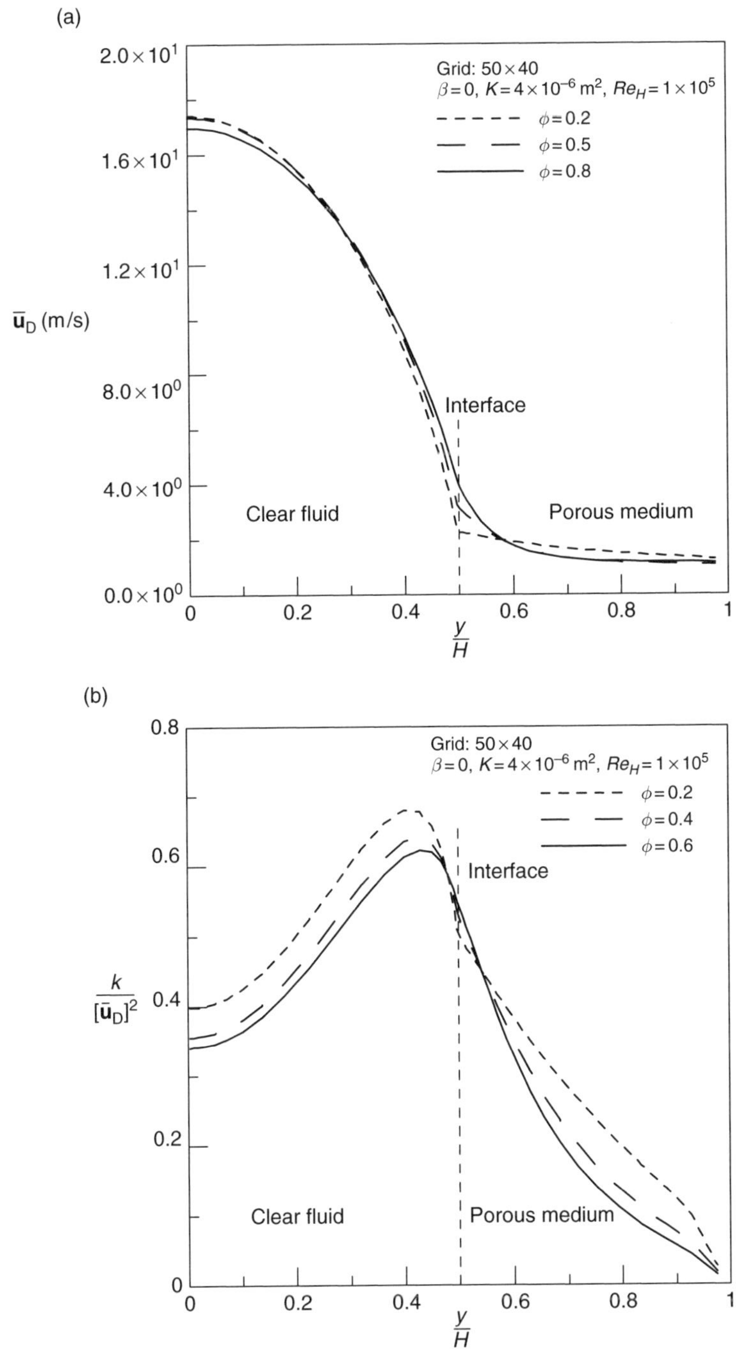


**Figure 9.10.** Effect of porosity on macroscopic field: (a) mean velocity, (b) turbulent kinetic energy.

proportional to $\mathbf{u}_D$ and, inside the porous layer ($y/H > 0.6$), the mean Darcy velocity increases as $\phi$ is reduced.

Figure 9.11 finally presents numerical solutions for $\beta$ varying from $-0.5$ to 0.5 for a fixed porosity $\phi = 0.6$, permeability $K = 4 \times 10^{-4}\,\mathrm{m}^2$ and $Re_H = 1 \times 10^5$. Profiles for $\mathbf{u}_D$ change substantially as the factor $\beta$ is varied, from a smooth variation across the interface for a negative $\beta$, to an abrupt change in the velocity profiles when $\beta > 0$. For positive values, the Darcy velocity $\mathbf{u}_D$ is higher inside the permeable structure. In turn, turbulent kinetic energy is generated at a faster rate by the model proposed in Equation (9.24). Although mean velocity profiles are flat in this same porous region, reducing the production rate $P^i$, the generating mechanism $G^i$ increases the overall value of $k$. In the clear fluid, steeper gradients in the fluid layer also contributes for increasing the value of the turbulent kinetic energy. Then, either by $P^i$ in the clear fluid or by $G^i$ in the porous layer, turbulent kinetic energy is generated at a faster rate for positive values of $\beta$. Ultimately, results in Figure 9.11 indicate that for flows where models with $\beta > 0$ are suitable, a greater portion of the mean mechanical energy of the flow is converted into turbulence. If that is the case of environmental flows over dense and thick rain forests, for example, results herein might be useful to environmentalists and engineers analyzing important natural and engineering flows.

### 9.1.3. Jump Condition for Turbulence Kinetic Energy

#### *9.1.3.1. Introduction*

Investigation of flow over layers of permeable media has many applications in several environmental and engineering analyses. Turbulent atmospheric boundary layer over forests under fire (Zhou and Pereira, 2000), flow over vegetation and crop fields (Hoffmann, 2004), currents at bottom of rivers (Lane and Hardy, 2002), as well as grain storage and drying, can be characterized by some sort of porous layer over which a fluid permeates. Also, practical analysis of engineering flows can further benefit from more realistic mathematical and numerical modeling, as in the case of shell-and-tube heat exchangers (Prithiviraj and Andrews, 1998) and nuclear reactor core (Sha, 1981), for example, where the rod bundles can be seen, in a macroscopic view, as a permeable medium.

When the domain of analysis presents a macroscopic interfacial area between a porous substrate and a clear flow region, the literature proposes the existence of a discontinuity in the momentum diffusion flux between the two media (Ochoa-Tapia and Whitaker, 1995a,b). Analytical solutions involving such models have been published (Kuznetsov, 1996, 1997, 1999). Also, in such works volume average properties for a homogenous treatment of flow in porous media are obtained by means of the Volume-Average Theorem (VAT) (Gray and Lee, 1977; Whitaker, 1999).

Recently, the interface jump condition has been investigated for laminar flows, either considering non-linear effects in momentum equation or neglecting the Forchheimer term

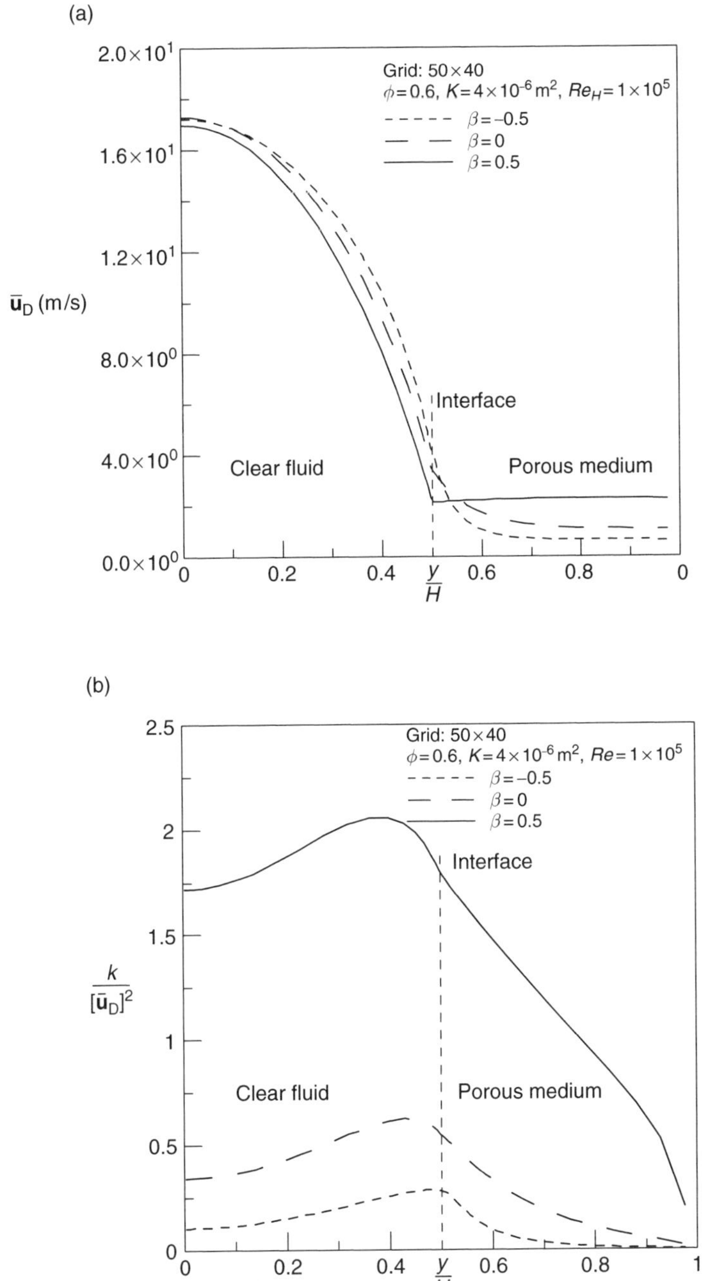


**Figure 9.11.** Effect of parameter $\beta$ on hydrodynamic field: (a) mean velocity, (b) turbulent field.

in the macroscopic model (Silva and de Lemos, 2003a). Therein, the authors simulated laminar flow over such interfaces and validated their results against analytical solutions by Kuznetsov (1996–1999). Such work is based on the numerical methodology proposed for hybrid media and applied by de Lemos and Pedras (2000b) and Assato *et al.* (2005). The same numerical technique has been applied for computing turbulent flow (Silva and de Lemos, 2003b) in a channel partially filled with a flat layer of porous material. Flows over wavy interfaces were also computed for both laminar (Silva and de Lemos, 2003c) and turbulent flows (de Lemos and Silva, 2003). There, the authors made use of the shear stress jump condition at the interface. Those works were based on a numerical methodology specifically proposed for hybrid media (de Lemos and Pedras, 2000b; Assato *et al.*, 2005).

A distinct line of investigation on turbulent flow over permeable media is based on the assumption that within the porous layer the flow remains laminar (Kuznetsov *et al.*, 2002; Kuznetsov and Xiong, 2003; Kuznetsov, 2004; Kuznetsov and Becker, 2004), which, in turn, precludes application of such methodology to flows through highly permeable media as atmospheric boundary layer over forests or crop fields.

Further, fine flow computations and experiments of flow over and inside a bed of rods in a two-dimensional channel have been presented (Prinos *et al.*, 2003). Three-dimensional computational studies simulating flow over a layer formed by cubic blocks (Breugem *et al.*, 2004; Breugem and Boersma, 2005) also emphasize that depending on the permeable structure shape, turbulence may exist inside the porous bed and, as such, a turbulence model must be employed.

As seen, all models above considered either a flat or a rough (wavy) macroscopic interface limiting the porous substrate. The stress jump condition for the momentum equations was applied, but in most publications so far, no such flux discontinuity for the $\langle k \rangle^{\mathrm{v}}$ equation has been considered. Motivated by that, de Lemos (2004–2005) proposed a model that assumes diffusion fluxes of turbulent kinetic energy on both sides of the interface to be unequal, which differs from all studies presented up to now. The purpose of this contribution is to explore and further document such proposal, investigating how its behavior as medium properties, such as permeability and porosity, is varied.

#### *9.1.3.2. Interface and "Jump" Conditions*

Equation (9.33) is here repeated for convenience,

$$(\mu_{\mathrm{eff}} + \mu_{t_\phi}) \frac{\partial \bar{u}_{\mathrm{D_p}}}{\partial y}\bigg|_{\text{Porous medium}} - (\mu + \mu_{\mathrm{t}}) \frac{\partial \bar{u}_{\mathrm{D_p}}}{\partial y}\bigg|_{\text{Clear fluid}} = (\mu + \mu_{\mathrm{t}}) \frac{\beta}{\sqrt{K}} \bar{u}_{\mathrm{D_p}}\bigg|_{\text{Interface}} \tag{9.35}$$

where all notation in Figures 9.1 and 9.2 applies.

It is interesting to note that Equation (9.35) comes from the sole extension, to turbulent flows, of the proposal in Ochoa-Tapia and Whitaker (1995a,b), which for laminar flow reads:

$$\mu_{\text{eff}} \frac{\partial u_{D_p}}{\partial y}\bigg|_{\text{Porous medium}} - \mu \frac{\partial u_{D_p}}{\partial y}\bigg|_{\text{Clear fluid}} = \mu \frac{\beta}{\sqrt{K}} u_{D_p}\bigg|_{\text{Interface}} \tag{9.36}$$

Local instantaneous velocities in (9.36) were replaced by time averaged values and a "total" diffusivity is used as a substitute for the molecular diffusivity. Physically, Equation (9.35) is guided by the same arguments that standard well-known "eddy diffusivity" models rely on, which is the use of a diffusion-like expression having, instead, gradients of time averaged values and a total (laminar plus turbulent) viscosity. Although it is recognized that Equation (9.35) needs further validation against experimental values, its application herein is assumed in lieu of better information.

Continuity of velocity, pressure, statistical variables and their fluxes across the interface are given by (see Silva and de Lemos, 2003b, for details),

$$\bar{\mathbf{u}}_D\big|_{\text{Porous medium}} = \bar{\mathbf{u}}_D\big|_{\text{Clear fluid}} \tag{9.37}$$

$$\langle \bar{p} \rangle^i\big|_{\text{Porous medium}} = \langle \bar{p} \rangle^i\big|_{\text{Clear fluid}} \tag{9.38}$$

$$\langle k \rangle^v\big|_{\text{Porous medium}} = \langle k \rangle^v\big|_{\text{Clear fluid}} \tag{9.39}$$

$$\left(\mu + \frac{\mu_{t_\phi}}{\sigma_k}\right) \frac{\partial \langle k \rangle^v}{\partial y}\bigg|_{\text{Porous medium}} = \left(\mu + \frac{\mu_t}{\sigma_k}\right) \frac{\partial \langle k \rangle^v}{\partial y}\bigg|_{\text{Clear fluid}} \tag{9.40}$$

$$\langle \varepsilon \rangle^v\big|_{\text{Porous medium}} = \langle \varepsilon \rangle^v\big|_{\text{Clear fluid}} \tag{9.41}$$

$$\left(\mu + \frac{\mu_{t_\phi}}{\sigma_\varepsilon}\right) \frac{\partial \langle \varepsilon \rangle^v}{\partial y}\bigg|_{\text{Porous medium}} = \left(\mu + \frac{\mu_t}{\sigma_\varepsilon}\right) \frac{\partial \langle \varepsilon \rangle^v}{\partial y}\bigg|_{\text{Clear fluid}} \tag{9.42}$$

Equations (9.37) and (9.38) were also proposed by Ochoa-Tapia and Whitaker (1995a) whereas relationships (9.39) through (9.42) were used by Lee and Howell (1987).

In Silva and de Lemos (2003b), no "jump" condition was considered when treating the diffusion flux of $\langle k \rangle^v$ across the interface, as can be seen by Equation (9.40). In de Lemos (2005a), such discontinuity in the diffusion transport of $\langle k \rangle^v$ between the two media was first considered. Such "jump" might be a model for accounting for interface roughness or be a way to comply with irregular interfaces. In addition, it can also be seen as an accommodation of the fact that close to the interface, the permeability $K$ attains higher values than those used within the porous substrate. For that, the interface condition of de Lemos (2005a) is here applied:

$$\left(\mu_{\text{eff}} + \frac{\mu_{t_\phi}}{\sigma_k}\right) \frac{\partial \langle k \rangle^v}{\partial y}\bigg|_{\text{Porous medium}} - \left(\mu + \frac{\mu_t}{\sigma_k}\right) \frac{\partial \langle k \rangle^v}{\partial y}\bigg|_{\text{Clear fluid}} = (\mu + \mu_t) \frac{\beta}{\sqrt{K}} \langle k \rangle^v\bigg|_{\text{Interface}} \tag{9.43}$$

instead of Equation (9.40). Condition (9.43) is imposed along the interface shown in Figure 9.2b.

Equation (9.43) results from the following reasoning. If interface condition (9.35) is written in its instantaneous form, it gives:

$$(\mu_{\text{eff}}+\mu_{t_\phi})\frac{\partial u_{D_p}}{\partial y}\bigg|_{\text{Porous medium}} - (\mu+\mu_t)\frac{\partial u_{D_p}}{\partial y}\bigg|_{\text{Clear fluid}} = (\mu+\mu_t)\frac{\beta}{\sqrt{K}}\,u_{D_p}\bigg|_{\text{Interface}} \tag{9.44}$$

Assuming that the component of the Darcy velocity along the interface varies with time, a standard time decomposition for it can be written as $u_{D_p}=\overline{u}_{D_p}+u'_{D_p}$. Next, considering that only velocities fluctuate in time, subtracting (9.35) from (9.44) results in the following relationship for the fluctuating interface velocity $u'_{D_p}$:

$$(\mu_{\text{eff}}+\mu_{t_\phi})\frac{\partial u'_{D_p}}{\partial y}\bigg|_{\text{Porous medium}} - (\mu+\mu_t)\frac{\partial u'_{D_p}}{\partial y}\bigg|_{\text{Clear fluid}} = (\mu+\mu_t)\frac{\beta}{\sqrt{K}}\,u'_{D_p}\bigg|_{\text{Interface}} \tag{9.45}$$

Taking now the scalar product of $u'_{D_p}$ and Equation (9.45), one gets,

$$(\mu_{\text{eff}}+\mu_{t_\phi})\frac{\partial\left((u'_{D_p}\cdot u'_{D_p})/2\right)}{\partial y}\Bigg|_{\text{Porous medium}} - (\mu+\mu_t)\frac{\partial\left((u'_{D_p}\cdot u'_{D_p})/2\right)}{\partial y}\Bigg|_{\text{Clear fluid}}$$
$$= (\mu+\mu_t)\frac{\beta}{\sqrt{K}}\left((u'_{D_p}\cdot u'_{D_p})/2\right)\bigg|_{\text{Interface}} \tag{9.46}$$

If one now apply the time averaging operation to Equation (9.46) and for flows mostly parallel to the interface approximate the turbulence kinetic energy as $\langle k\rangle^{v}\approx\overline{u'_{D_p}\cdot u'_{D_p}}/2$, condition (9.43) is recovered after introducing the constant $\sigma_k$.

One should further mention that a different coefficient $\beta$, on the rhs of (9.43), might be necessary to accommodate real engineering flows over porous substrates. Proposition (9.43), as such, should be regarded as a first step towards realistic modeling subjected to improvements as experimental data on macroscopic interfaces become available.

In order to check for the correctness of computer code developed, two results were compared. Calculations applying Equation (9.40) used in Silva and de Lemos (2003b) were compared with simulation with (9.43) setting $\beta=0$. Figure 9.12 correctly shows superposition of results for the mean and turbulent fields, indicating proper computer implementation of the diffusion-jump model for the $k$-equation.

Figure 9.13a presents numerical solutions for $\beta$ varying from $-0.5$ to 0.5 for a fixed porosity $\phi=0.6$, permeability $K=4\times10^{-4}\,\text{m}^2$ and $Re_H=1\times10^5$. Results are compared with those by Silva and de Lemos (2003b), who used interface conditions (9.35) and (9.40) for the mean and turbulent fields, respectively. When condition (9.43) is used

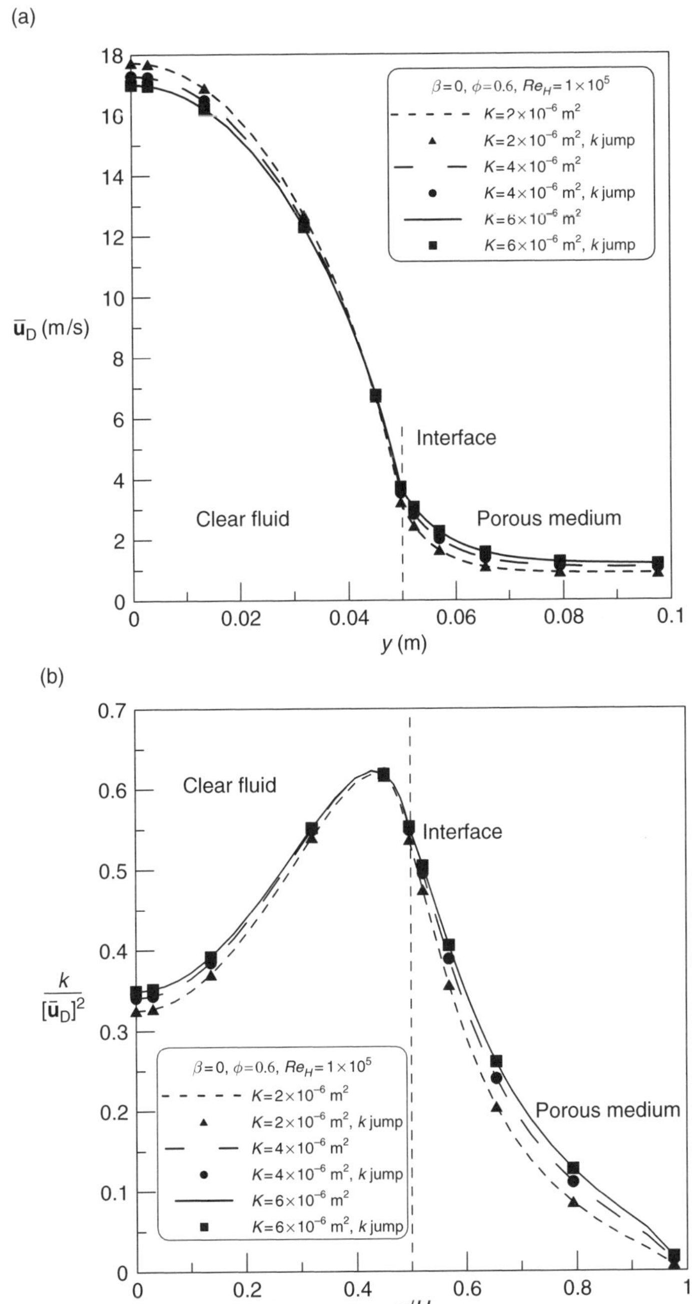


**Figure 9.12.** Calculations using Equation (9.40) (*lines*) compared with simulations with Equation (9.43) (*symbols*) and $\beta = 0$: (a) mean field, (b) turbulent field.

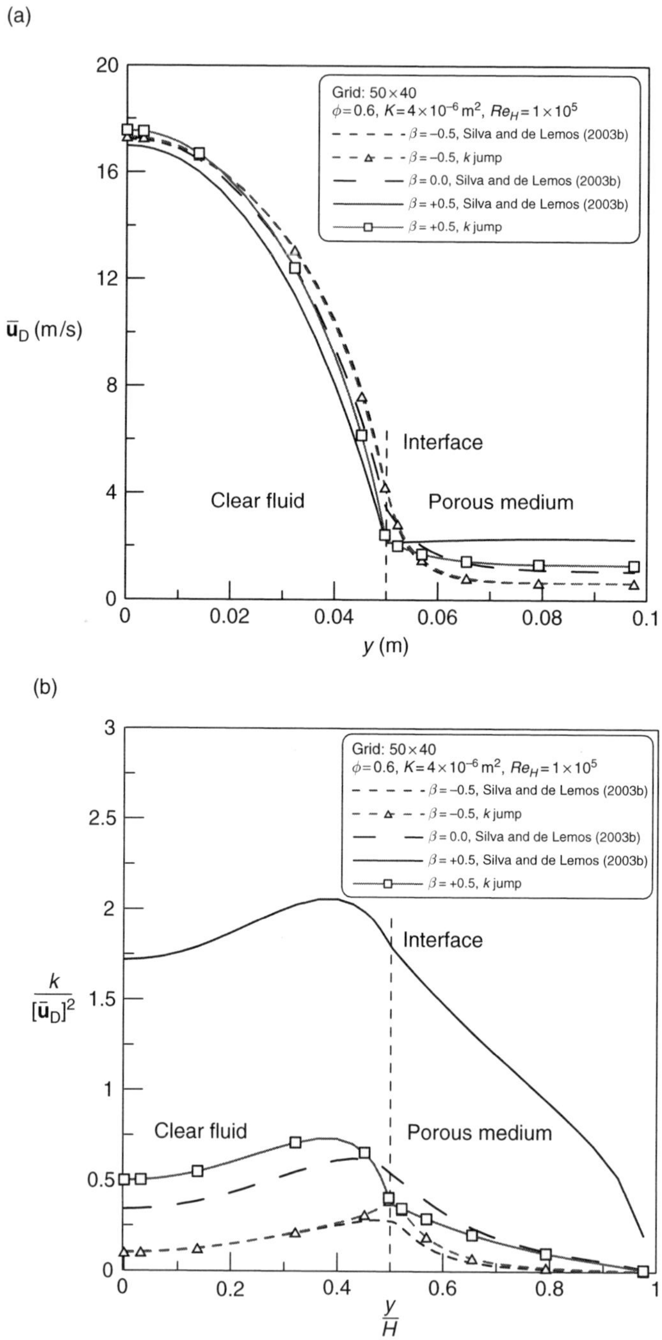


**Figure 9.13.** Effect of jump conditions on mean and turbulent fields: (a) mean velocity **u**, (b) non-dimensional turbulent kinetic energy.

in place of (9.40), profiles for $\mathbf{u}_D$ change substantially as the factor $\beta$ is varied from a smooth variation across the interface for a negative $\beta$ (small dashed line without symbols, Figure 9.13a) to an abrupt change in the velocity profiles when $\beta > 0$ (solid line without symbols, Figure 9.13a). For positive $\beta$ values, the Darcy velocity $\mathbf{u}_D$ is slightly higher inside the permeable structure than at the interface, indicating that flow resistance at this position would be higher than everywhere across the porous layer (de Lemos, 2005a). This unrealistic result is not obtained when condition (9.43) is applied for $\langle k \rangle^v$ (solid line with symbols, Figure 9.13a). In this case, velocities close to the wall region are higher than for $\beta = 0$ ($y > 0.064$ m, large dashed line), but around the interface no such minimum in the value of $\mathbf{u}_D$ is observed. For $\beta < 0$ (small dashed lines) no substantial difference in the calculated profiles, with and without a jump condition for $\langle k \rangle^v$, is observed.

Distributions for turbulent kinetic energy as a function of the interface boundary condition are shown in Figure 9.13b. The clear separation for the two distributions for positive and negative values of $\beta$ calculated by Silva and de Lemos (2003b) (solid and small dashed lines, without symbols) is not seen when interface condition (9.43) is applied (solid and small dashed lines, with symbols). The region of maximum turbulent kinetic energy is within the clear flow ($y < 0.05$ m) for $\beta > 0$ (solid line with symbols), whereas the use of a negative value for the jump parameter causes a peak for $\langle k \rangle^v$ at the interface (dashed curve with symbols). If one compares with experimental values by Prinos *et al.* (2003) (not shown here), one can conclude that models with negative $\beta$ values are closer to representing reality for turbulent flow around a porous medium.

The effect of $Re_H$ is shown in Figure 9.14. Plots on the left (a, c) are for mean velocity whereas curves on the right of the figure (b, d) details the behavior of the turbulent filed. Also, drawings on the top (a, b) were calculated for $\beta < 0$ whereas that on the bottom (c, d) used positive values of $\beta$. The mean velocity profiles in Figure 9.14a, c confirms the increasing mass flow rate within either the porous material or the clear passage as $Re_H$ increases. In Figure 9.8b,d the collapse of curves for the turbulent kinetic energy divided by the mean mechanical energy shows that, for the range of $Re_H$ here analyzed, the percentage of energy transformed into turbulence remains the same, regardless of the diffusion-jump model used. The most striking feature in Figure 9.14 is the different response, in the turbulence field (b, d), when using values for $\beta$ of different sign. Negative values for $\beta$ (Figure 9.14b) indicate that the peaks in the curves lie lower than when no jump condition is used, and that this peak is at the interface. On the other hand, for a positive $\beta$, the levels of $k$ are higher than if no jump condition is applied. In addition, the peaks are moved towards the center of the channel. The behavior of the curves is associated with corresponding mean velocity profiles. Within the clear fluid, the production of turbulent kinetic energy is known to be dictated by gradients of the mean velocity ($P^i$ on the right of (9.24)) whereas inside the permeable structure, the model of Pedras and de Lemos (2001a) proposes a factor proportional to $\mathbf{u}_D$ as a generating mechanism for $k$ ($G^i$ in Equation (9.24)).

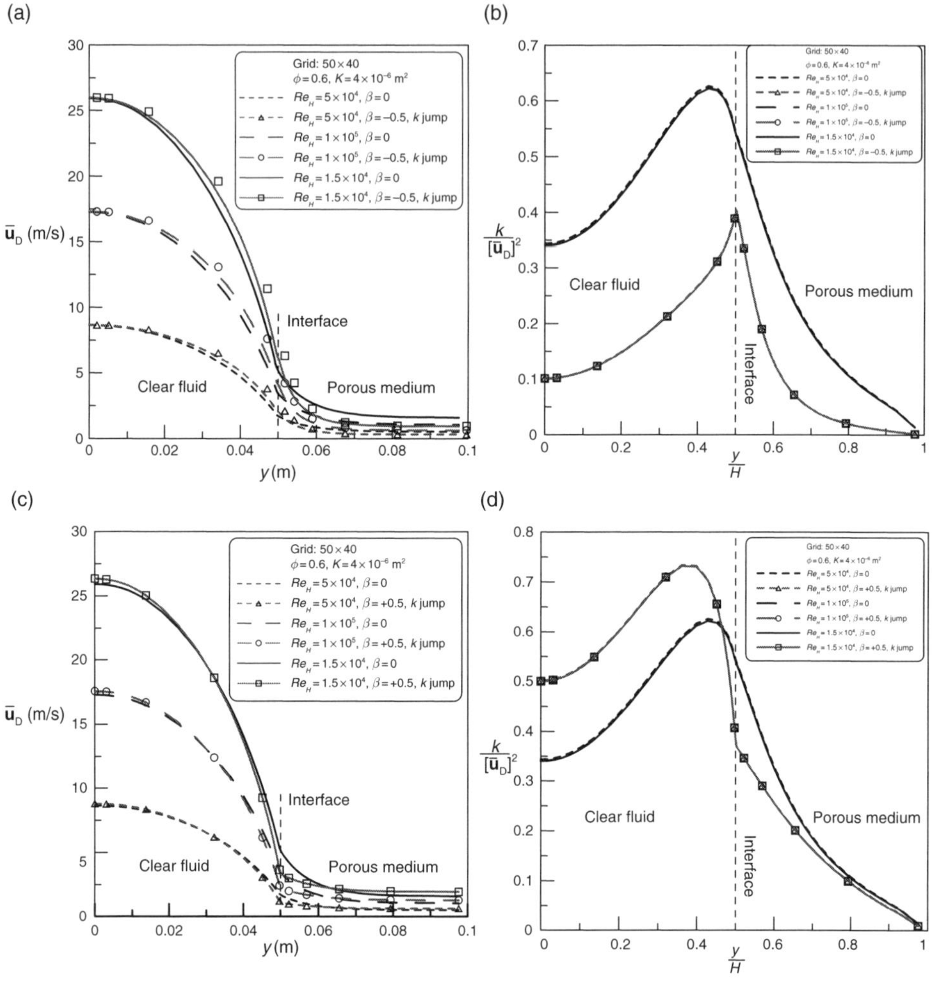


**Figure 9.14.** Effect of Reynolds number, $Re_H$, on macroscopic field. $\beta < 0$: (a) mean velocity, (b) turbulent kinetic energy; $\beta > 0$: (c) mean velocity, (d) turbulent kinetic energy.

Figure 9.15 shows the effect of the permeability $K$ on both the mean and statistical fields. Plots a, b, c, d follow the same convention described in Figure 9.14. The figure indicates that the greater the permeability, more flow crosses the porous substratum located in the region $0.5 < (y/H) < 1$ (Figure 9.15a–c). The curves representing the statistical field in Figure 9.15b–d show that the levels of $k$ increase with increasing $K$. As more fluid flows in a less resistant medium, more mean mechanical energy is transformed into turbulence increasing the overall level of $k$.

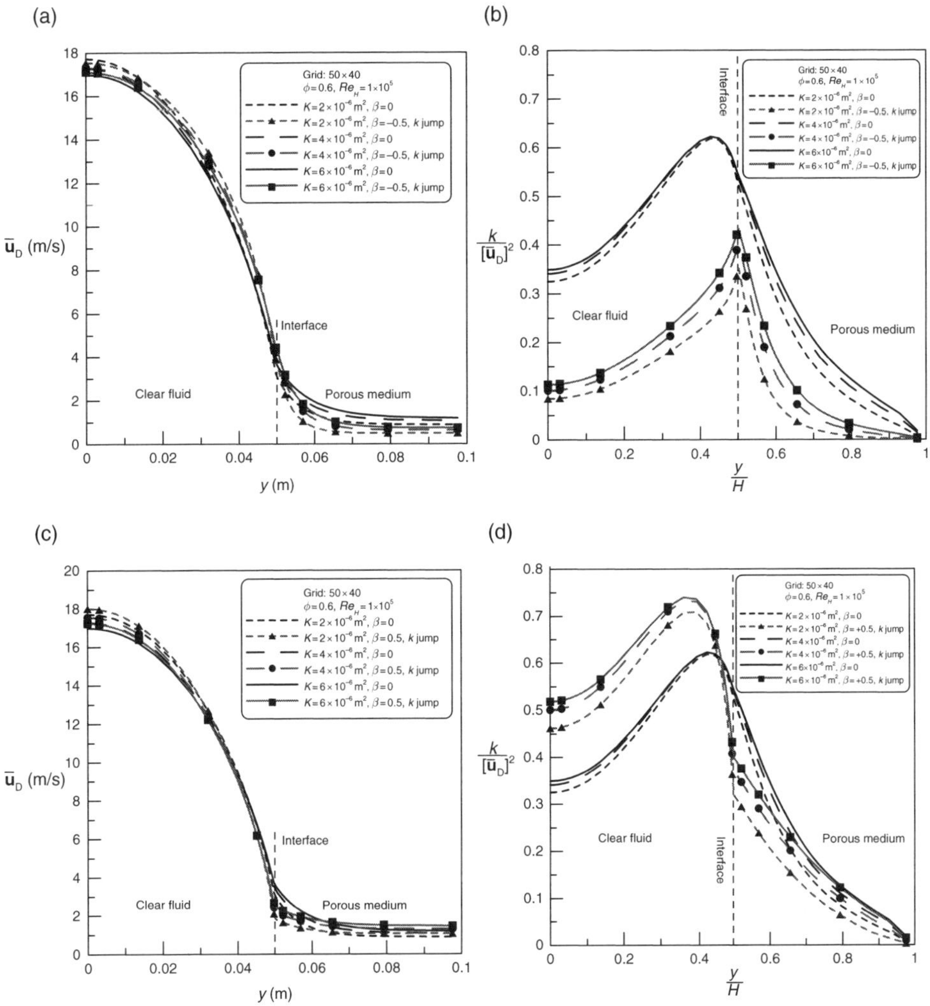


**Figure 9.15.** Effect of permeability, $K$, on macroscopic field. $\beta < 0$: (a) mean velocity, (b) turbulent kinetic energy; $\beta > 0$: (c) mean velocity, (d) turbulent kinetic energy.

Finally, Figure 9.16 investigates the effect of the value of $\phi$ on the behavior of the mean and turbulent fields, following, again, the same convention established when presenting Figure 9.14 (plots a, b, c, d). For the mean field (a, c), one can note that close to $y/H = 0.5$ the greater the porosity, the higher the velocity at the interface and the greater the mass flow rate closer to this region. At the center of the channel, the velocity decreases in order to keep the imposed mass flow rate the same. It is interesting to observe that since the overall mass flow rate is forced to be constant, instead of the overall pressure loss

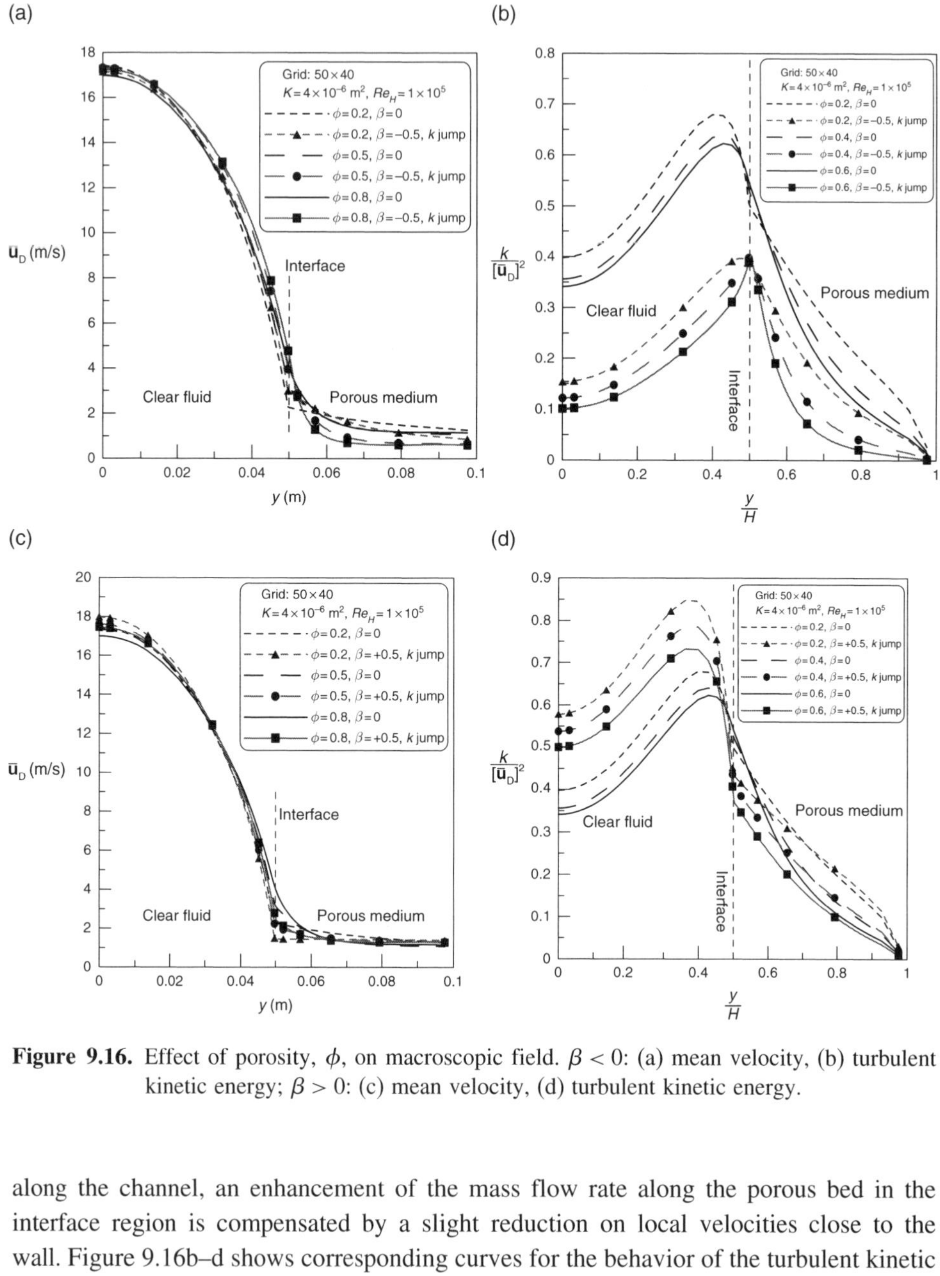


**Figure 9.16.** Effect of porosity, $\phi$, on macroscopic field. $\beta < 0$: (a) mean velocity, (b) turbulent kinetic energy; $\beta > 0$: (c) mean velocity, (d) turbulent kinetic energy.

along the channel, an enhancement of the mass flow rate along the porous bed in the interface region is compensated by a slight reduction on local velocities close to the wall. Figure 9.16b–d shows corresponding curves for the behavior of the turbulent kinetic energy. Values of $k$ present a slight reduction as $\phi$ is incremented. Lower values for the turbulence level within the porous layer are coherent with the model of Equation (9.24) for the extra generation rate due to the porous matrix. As said, this extra $G^{\mathrm{i}}$ term (3rd on the right of (9.24)) was modeled as proportional to $\mathbf{u}_D$ and, inside the porous layer, the mean Darcy velocity is reduced as $\phi$ increases.

Ultimately, results in Figures 9.8, 9.4 and 9.10 indicate that for flows where models with $\beta < 0$ are suitable, a smaller portion of the mean mechanical energy of the flow is converted into turbulence. Results herein might be useful to environmentalists and engineers analyzing important natural and engineering flows. Although in the porous substrate mean velocity profiles are flatter, reducing the production rate $P^i$, the generating mechanism $G^i$ is proportional to $\mathbf{u}_D$, increases the overall value of $k$. In the clear fluid, steeper gradients in the fluid layer also contribute for increasing the value of the turbulent kinetic energy. Then, either by $P^i$ in the clear fluid or by $G^i$ in the porous layer, turbulent kinetic energy is generated at a faster rate for positive values of $\beta$.

## 9.2. CHANNELS WITH POROUS AND SOLID BAFFLES

### 9.2.1. General Remarks

Flow and heat transfer in channels with solid or impermeable baffles have attracted attention of researchers due to the technological advantages in increasing energy transfer in engineering equipment (Berner *et al.*, 1984; Webb and Ramadhyani, 1985; Kelkar and Patankar, 1987; Huang and Vafai, 1994; Lopez *et al.*, 1996; Hwang, 1997; Ko and Anand, 2003; Yang and Hwang, 2003; Da Silva Miranda and Anand, 2004; Haji-Sheikh *et al.*, 2004). If the baffles are made of a permeable or perforated material, flow head losses might be substantially reduced. Accordingly, due to the growing applicability of porous media in many fields of engineering and science – such as petroleum and gas engineering, heat exchangers, combustion in porous matrices, cooling of electronic devices, to mention only a few – a good understanding of transport processes in such media is of advantage to the design and analysis of engineering equipment.

Berner *et al.* (1984) presented the main features of the flow in a channel with solid baffles (Figure 9.17a). This was obtained using flow visualization techniques, manometry and laser-Doppler anemometry. The experiment was applied to a two-dimensional water flow around baffles, with $L/H = 1$, $h/H$ from 0.5 to 0.9, and Reynolds number $Re$ ranging from 600 to 10,500. They suggested that laminar flow occurred for $Re \leq 600$. Following the work in Berner *et al.* (1984), Kelkar and Patankar (1987) presented numerical simulations for laminar flow in a channel with solid baffles for $Re$ raging from 100 to 500. They concluded that for high Prandtl number fluids, heat transfer within the channel is significantly enhanced, a result that is in agreement with similar predictions by Webb and Ramadhyani (1985). Lopez *et al.* (1996) investigated three-dimensional effects for this same geometry and obtained similar results as those taken at the center of the channel. Further, Huang and Vafai (1994) studied heat transfer enhancement in channels containing porous cavities and block obstacles.

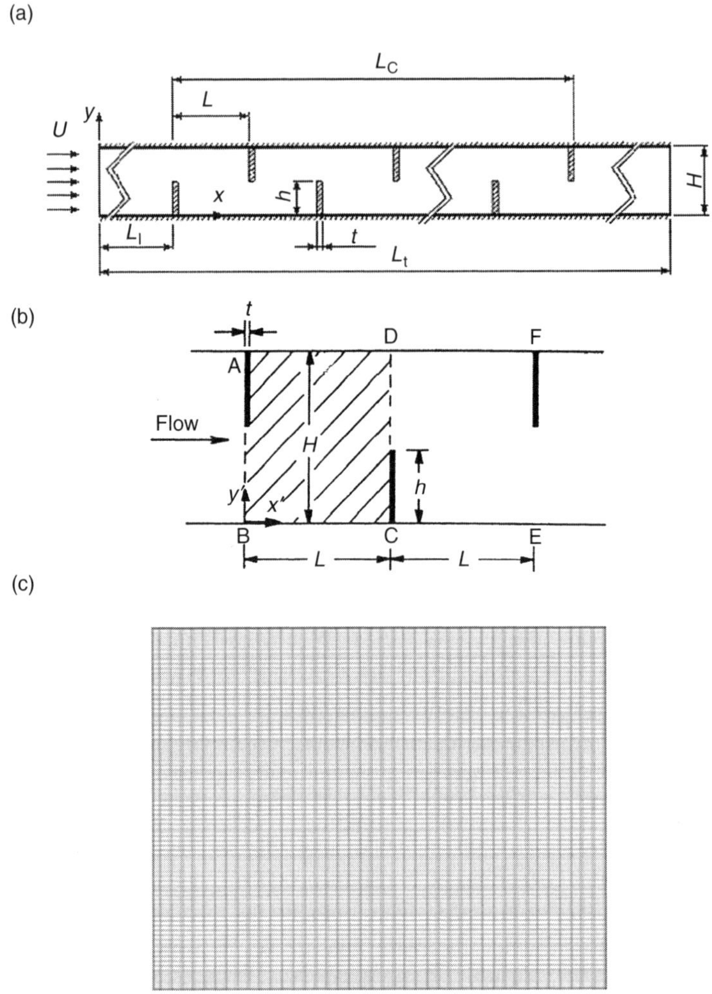


**Figure 9.17.** Problem under consideration: (a) geometry, (b) channel cell, (c) section of computational grid of length $2L$.

The first experimental work investigating the use of permeable baffles, instead of using solid plates, was presented by Hwang (1997). In that work, the author found out that heat transfer between the channel walls and the fins was enhanced, showing then that for turbulent flow the use of porous baffles benefited heat transfer, i.e., the use of porous baffles in substitution to solid (impermeable) material represented a net gain. Motivated by this, Yang and Hwang (2003) carried out a numerical investigation of the problem, reaching similar conclusions. For corroborating Hwang's (1997) conclusion,

Ko and Anand (2003) carried out an experimental program for turbulent flow with porous baffles made of different materials. However, results in Ko and Anand (2003) were not very promising. In their analysis, the porous baffles presented in most cases a flow behavior as good as the one with solid baffles. Da Silva Miranda and Anand (2004) presented a numerical analysis of laminar air flow in a channel with porous baffles having $h/H = 0.33$, $t/H$ from 0.09 to 0.25, permeability $K$ from $10^{-7}$ to $7.6 \times 10^{-8}\,\mathrm{m}^2$, porosity $\phi = 0.92$ and $\lambda = k_s t/(k_f L)$ ranging from 0.09 to 330. More recently, Haji-Sheikh *et al.* (2004) published a detailed numerical study of heat transfer to a fluid in a saturated porous medium using the Green's function method.

In summary, Ko and Anand (2003) and Da Silva Miranda and Anand (2004) noticed that in none of the cases analyzed (with the substitution of the solid baffles for porous baffles) an increase in the overall heat transfer along the channel was obtained. Motivated by such engineering application, Rocamora and de Lemos (2000b,c) presented numerical solutions for laminar flow in hybrid (clear/porous) media. One of the greatest difficulties in numerically solving such hybrid systems arises in the treatment given at the interface between the two media. Most models consider constant properties for the porous substrate but, close to the interface, the permeability of the medium is known to increase. This effect is commonly modeled by a difference in the macroscopic shear stress at both sides of the macroscopic interface. Therefore, due to the importance of this particular theme, a literature review on this subject is also briefly included here. In fact, Rocamora and de Lemos (2000b,c) did not consider a stress jump at the interface between the porous media and the clear fluid. In the literature, for laminar flow, Ochoa-Tapia and Whitaker (1995a,b) proposed an adjustable coefficient for modeling the stress jump at the interface. Kuznetsov (1996–1999) presented analytical solutions for laminar velocity profiles in a channel partially filled with porous material, taking into consideration such a jump condition. Turbulent flow in a parallel-plate composite channel has also been investigated (Kuznetsov *et al.* 2002). Recently, Silva and de Lemos (2003a) presented numerical solutions for this same geometry, also considering the jump at the interface.

Tofaneli and de Lemos (2002a) investigated isothermal laminar flow in a channel containing porous fins. The channel, schematically shown in Figure 9.17a, had baffles made of porous and solid material attached to both walls. In that work, the effects of inlet Reynolds number and jump coefficients were considered and the numerical methodology proposed in Silva and de Lemos (2003a) was used. Later, Tofaneli and de Lemos (2002b) investigated the influence of porosity and permeability on the flow pattern in the same geometry (Figure 9.17a). Subsequently, de Lemos and Tofaneli (2003a,b) further extended the earlier results in Tofaneli and de Lemos (2002a,b) comparing then the pressure drop for turbulent flow along the channel of Figure 9.17.

Following up with the work in baffled channels, Santos and de Lemos (2004a) and de Lemos and Santos (2004) carried out a numerical analysis of the flow with heat transfer in a parallel plate channel containing 16 porous baffles made with material having permeability $K = 10^{-9}\,\mathrm{m}^2$ and porosity $\phi = 0.4$. Their results indicated that, in spite of using a permeable medium, total drag and heat transfer along the channel were close to those obtained with solid plates, possibly due to the extremely low permeability used. Santos and de Lemos (2004b) further investigated the influence of permeability and porosity on flow and heat transfer in a porous baffled channel with $h/H = 0.5$, $Pr = 0.7$ and $\lambda = 0.4$. They concluded that there was not an effective increase on heat transfer when porous baffles were used.

The purpose of this section is to review findings in Santos and de Lemos (2006), who documented a numerical analysis on laminar flow and heat transfer along a channel with porous and solid baffles. The influence of the permeability and porosity of the material is analyzed, in addition to the effect of the fluid properties ($0.7 < Pr < 7.0$), baffle material ($0.04 < \lambda < 40$) and geometry ($0.25 < h/H < 0.75$). Here, only one unique set of transport equations is used for both the clear domain and the porous medium.

### 9.2.2. Friction Factor

The friction factor in a section of the channel, shown in Figure 9.17b, can be calculated as:

$$f = \frac{\Delta P}{\rho}\frac{D_h}{L}\frac{2}{\overline{U}^2} \tag{9.47}$$

where $\Delta P$ is the pressure loss for the cell, $\overline{U}$ is the mean velocity and $D_h = 2H$ is the hydraulic diameter. Also, for flow in between infinite parallel plates without baffles, the friction factor is given by:

$$(fRe)_0 = 96$$

where the subscript "0" identifies values for unobstructed channels and the Reynolds number is defined as $Re = \rho\overline{U}D_h/\mu$.

### 9.2.3. Nusselt Number

Heat transfer between the channel walls and the fluid, in the cell shown in Figure 9.17b, is numerically calculated using the Nusselt number defined by:

$$\overline{Nu} = \frac{\overline{h}D_h}{k_\mathrm{f}} \tag{9.48}$$

where

$$\overline{h} = \frac{Q_\mathrm{cell}}{(2L \times 1)\Delta T_\mathrm{ml}} \tag{9.49}$$

$$Q_\mathrm{cell} = \rho\overline{U}c_\mathrm{p}(H \times 1)\,(T_\mathrm{b}(0) - T_\mathrm{b}(L)) \tag{9.50}$$

$$\Delta T_{\mathrm{ml}} = \frac{(T_{\mathrm{w}} - T_{\mathrm{b}}(L)) - (T_{\mathrm{w}} - T_{\mathrm{b}}(0))}{\ln[(T_{\mathrm{w}} - T_{\mathrm{b}}(L)) / (T_{\mathrm{w}} - T_{\mathrm{b}}(0))]} \tag{9.51}$$

$$T_{\mathrm{b}}(x') = \frac{\int_0^{\mathrm{H}} uT \,\mathrm{d}y'}{\int_0^{\mathrm{H}} u \,\mathrm{d}y'} \tag{9.52}$$

This formulation takes into consideration the heat transferred to the fluid, $Q_{\mathrm{cell}}$, through the surfaces at a constant temperature, $T_{\mathrm{w}}$, and the energy absorbed by the fluid via variation of bulk temperature along the cell, $T_{\mathrm{b}}$. In this study, the channel wall temperature, $T_{\mathrm{w}}$, is kept constant so that the definitions above apply.

Before proceeding, a word about the nomenclature, simulating conditions and grid size here employed seems timely.

In the context herein, the expression "solid baffle" means a structure through which no fluid permeates. On the other hand, "porous baffles" are here characterized by a solid matrix, not fully compacted so that fluid can freely flow within the existing void space. In addition, all results in the figures are divided by corresponding values obtained in channels without baffles, which are here identified by subscript "0".

Further, only steady state solutions are sought herein and no consideration for unsteady or transient behavior is here made. Accordingly, for the *Re* range here considered, most numerical results found in the literature discard the appearance of unsteady vortex motion.

Different grid sizes were employed in order to check their influence on the numerical values of $f$ and $Nu$ calculated in a fully developed channel cell, which is schematically shown in Figure 9.17b. Table 9.1 shows the percent errors $\delta$ and $\varepsilon$, defined as:

$$\delta = \left( \frac{f|_{\mathrm{ref}} - f}{f|_{\mathrm{ref}}} \right) \times 100 \tag{9.53}$$

$$\varepsilon = \left( \frac{\overline{Nu}\big|_{\mathrm{ref}} - \overline{Nu}}{\overline{Nu}\big|_{\mathrm{ref}}} \right) \times 100 \tag{9.54}$$

for grids of several sizes, taken as reference calculations run on a $2112 \times 302$ "*ref*" grid. The table indicates that for a grid of size $1507 \times 202$, more economical solutions could be obtained while retaining accuracy and solution precision. All computations below were then obtained with such grid, which is illustrated in Figure 9.17c.

### 9.2.4. Developing Flow

Developing profiles for $f$ and $Nu$ along the channel are shown in Figure 9.18 for solid baffles and in Figure 9.19 for porous fins. These numerical values for $\phi$ and $K$ used in

**Table 9.1.** Grid independence study.

| $x$ | $y$ | $\frac{f\,Re}{(f\,Re)_0}$ | $\frac{Nu}{Nu_0}$ | $\delta$ (%) | $\varepsilon$ (%) |
|---|---|---|---|---|---|
| 952 | 62 | 150.5 | 4.61 | 18.7 | −2.4 |
| 1207 | 62 | 158.0 | 4.74 | 14.7 | −5.3 |
| 1357 | 62 | 160.2 | 4.77 | 13.5 | −6.0 |
| 1507 | 82 | 167.6 | 4.69 | 9.5 | −4.2 |
| 1507 | 122 | 173.0 | 4.59 | 6.6 | −2.0 |
| 1507 | 202 | 178.7 | 4.51 | 3.5 | −0.2 |
| 1507 | 302 | 177.7 | 4.48 | 4.0 | 0.4 |
| 2112 | 302 | 185.2 | 4.50 | ref | ref |

Figure 9.19 correspond to those used in the literature (Lopez *et al.*, 1996; Hwang, 1997; Ko and Anand, 2003; Yang and Hwang, 2003). One can see in the figures that fully developed flow and heat transfer are obtained only after the 11th cell. Therefore, comparison of overall head losses and heat transfer characteristics will use values calculated past that particular axial position. It can also be seen that, although porous baffles were used for comparison with the solid cases, their values for $K$ and $\phi$ were such that no significant difference appears to exist between both situations. Therefore, shown below are the results past the 11th cell, which are compared with those by Kelkar and Patankar (1987) and Lopez *et al.* (1996).

## 9.2.5. Fully Developed Flow

### *9.2.5.1. Streamlines*

Figure 9.20 shows comparisons of calculated streamlines with those presented by Kelkar and Patankar (1987) for solid baffles. Also shown in the Figure are results for porous baffles. One can see that the present results (b) reproduce well the literature data for solid baffle cases (a). For porous baffles with $K = 1 \times 10^{-9}\,\mathrm{m}^2$ and $\phi = 0.4$, it is noticeable that the recirculation strength $\psi_0$ is slightly reduced (c). The numerical parameter $\psi_0$ was introduced by Kelkar and Patankar (1987) and represents a relation between the maximum recirculation rate within the cell and the mass flow rate through the channel. Also observed in the figure is the detachment of streamlines from the porous baffle since the no-slip condition, prevailing over a solid surface, is no longer applicable when a permeable baffle is considered.

Figure 9.21 presents the influence of the baffle height $h/H$ on the stream function pattern, indicating that, as expected, $\psi_0$ increases when $h/H$ rises, for both porous and solid baffles. Higher baffles, of either porous or solid type, retain more fluid within the recirculating bubbles.

Figure 9.22 shows the effect of permeability $K$ on the flow pattern for different values of $Re$. It can be seen that the cell recirculating strength $\psi_0$ decreases as $K$ increases, and

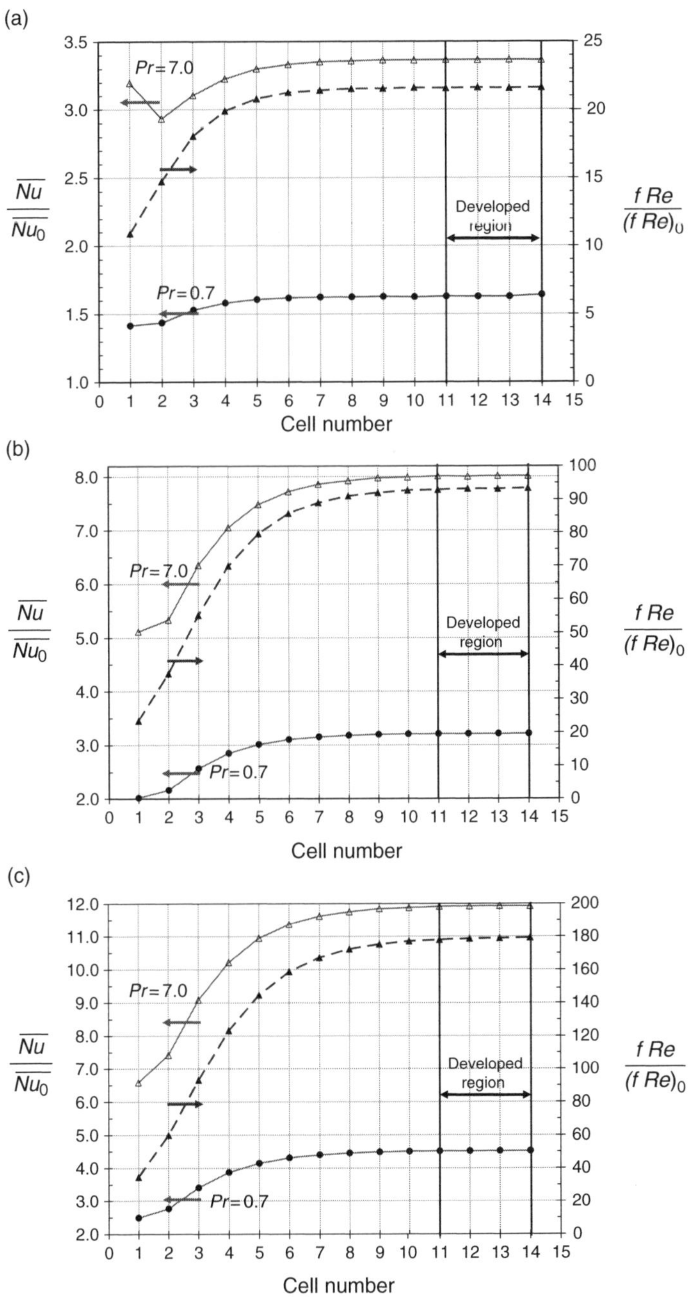


**Figure 9.18.** Developing $Nu$ and $f$ for a channel with solid baffles, $h/H = 0.5$, $\lambda = 0.4$: (a) $Re = 100$, (b) $Re = 300$, (c) $Re = 500$.

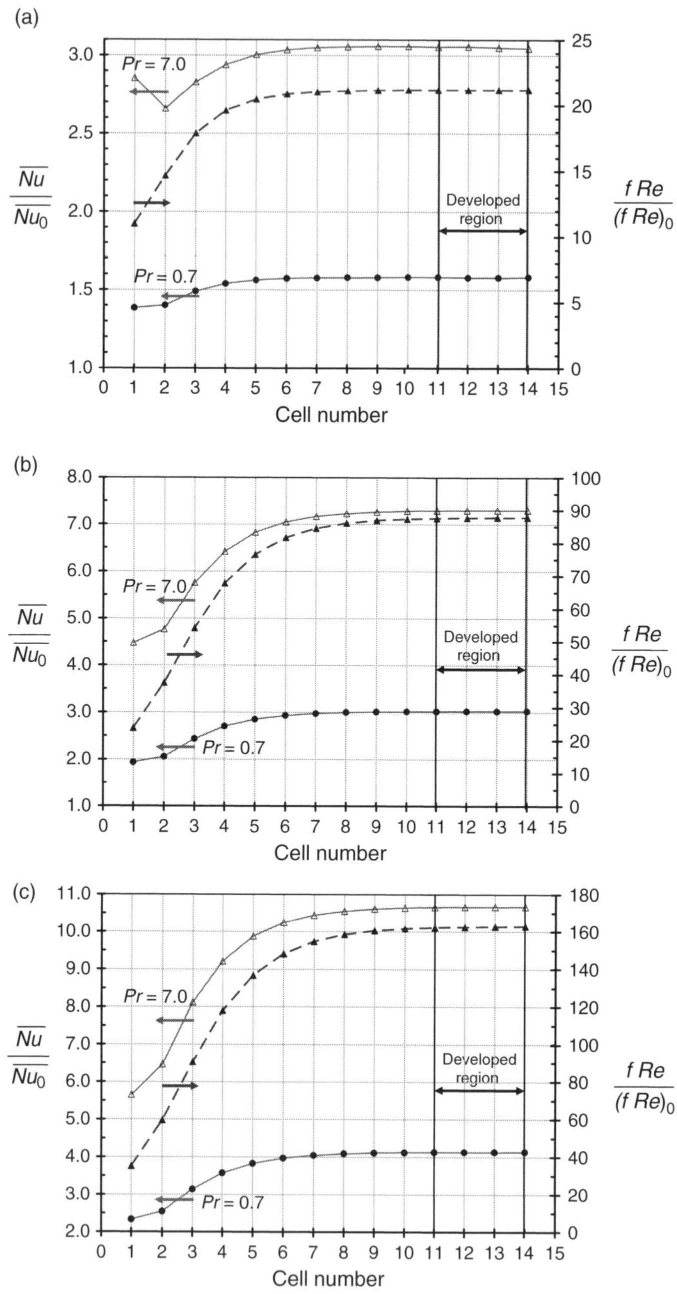


**Figure 9.19.** Developing $Nu$ and $f$ for a channel with porous fins, $h/H = 0.5$, $K = 1 \times 10^{-9}\,\mathrm{m}^2$, $\phi = 0.4$, $\lambda = 0.4$: (a) $Re = 100$, (b) $Re = 300$, (c) $Re = 500$.

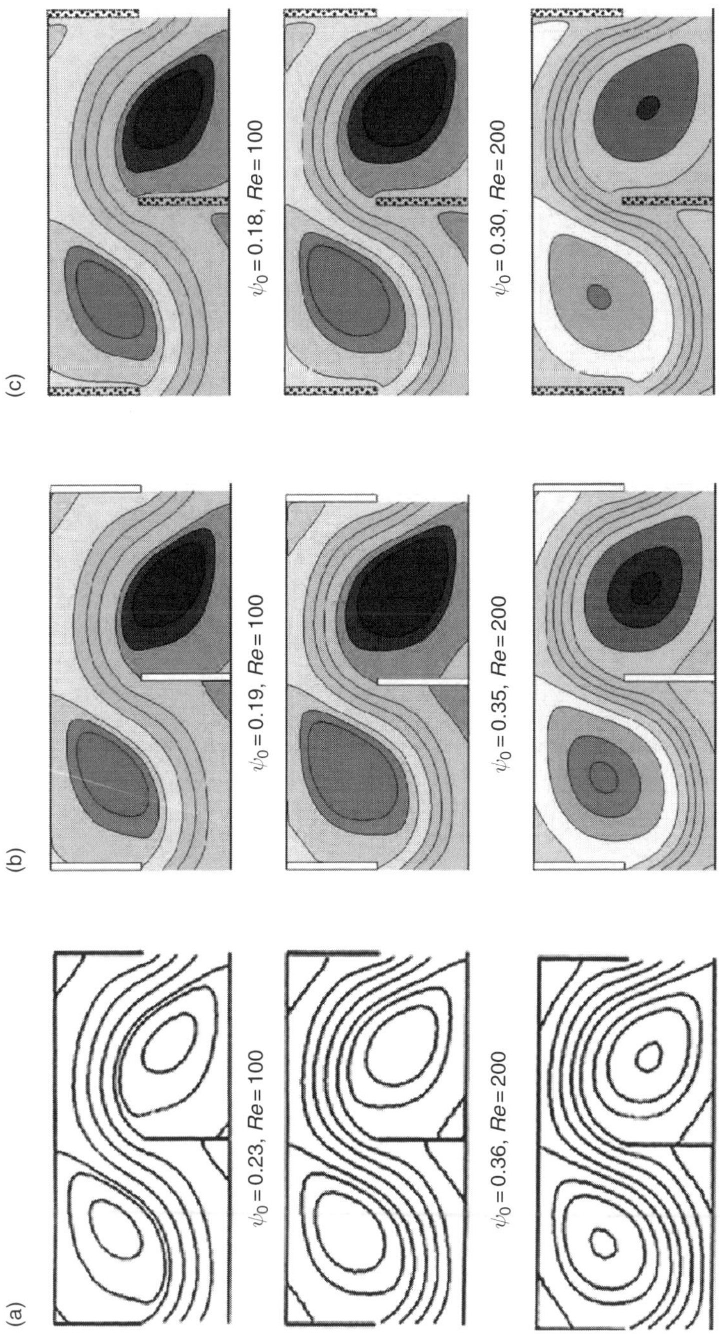


**Figure 9.20.** Streamlines pattern as a function of $Re$ for $h/H = 0.5$: (a) solid baffles (Kelkar and Patankar, 1987), (b) present work, solid baffles, (c) present work, porous baffles, $K = 1 \times 10^{-9}\,\mathrm{m}^2$, $\phi = 0.4$.

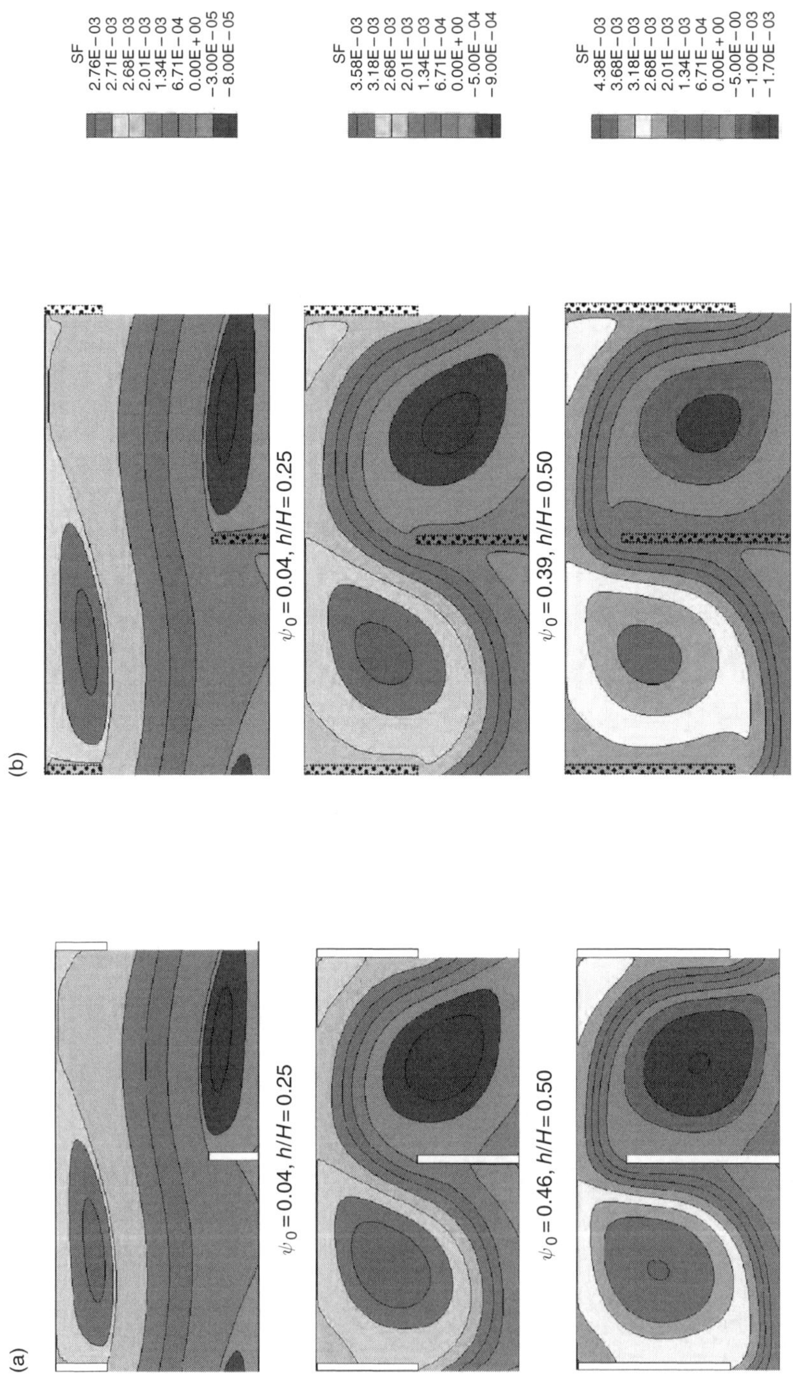


**Figure 9.21.** Streamlines pattern as a function of $h/H$, $Re = 300$: (a) solid baffles, (b) porous baffles, $K = 1 \times 10^{-9}\ \mathrm{m}^2$, $\phi = 0.4$.

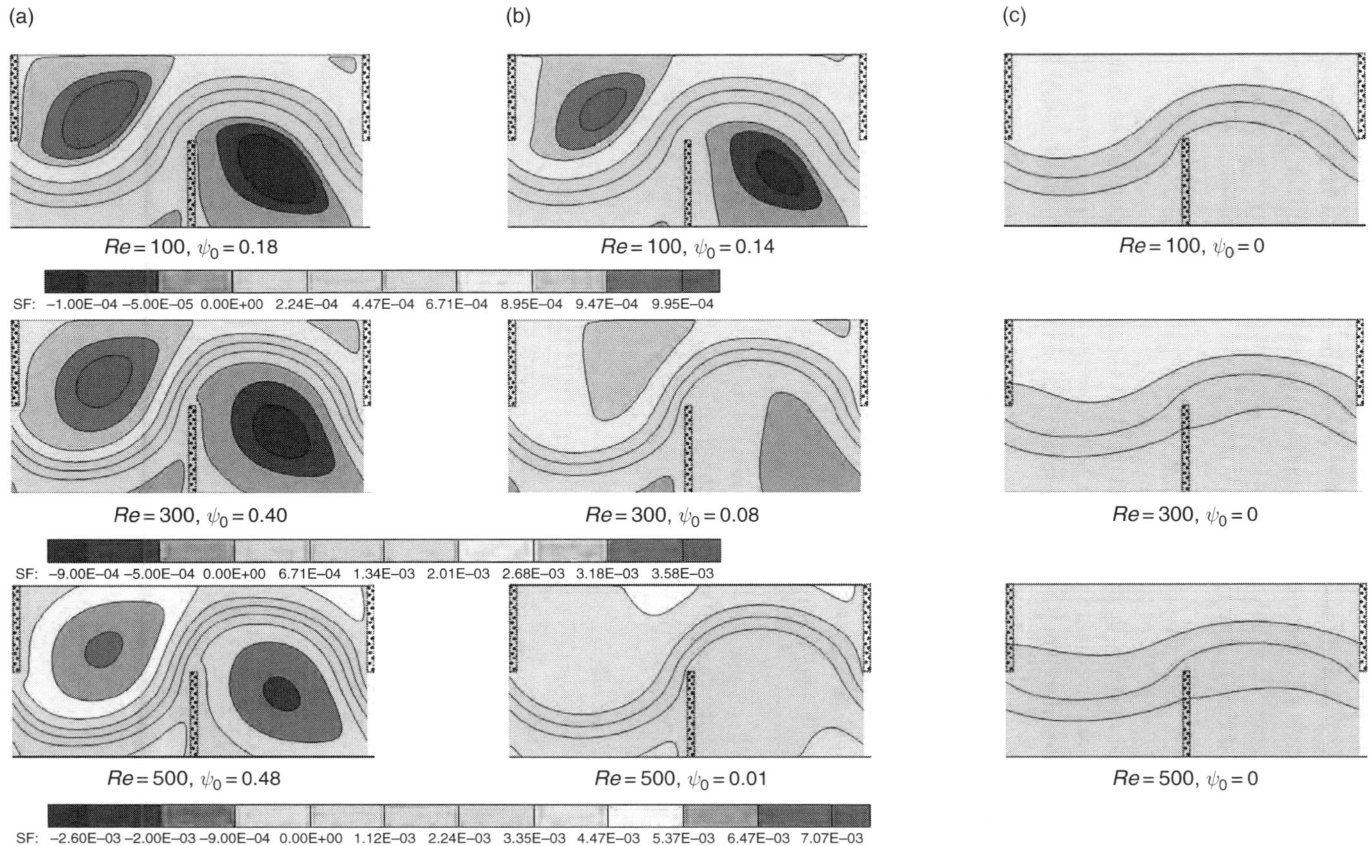


**Figure 9.22.** Streamlines pattern as a function of $K$ for $h/H = 0.5$ and $\phi = 0.9$: (a) $K = 1 \times 10^{-9}\ \mathrm{m}^2$, (b) $K = 1 \times 10^{-8}\ \mathrm{m}^2$, (c) $K = 1 \times 10^{-7}\ \mathrm{m}^2$.

essentially vanishes for $K = 10^{-7}\,\mathrm{m}^2$. It is interesting to observe the opposing effect of *Re* on the recirculating mass flow rate in the cell for different $K$ values. In Figure 9.22a for a less permeable baffle ($K = 10^{-9}\,\mathrm{m}^2$), an increase in *Re* increases $\psi_0$, whereas for a more permeable medium (Figure 9.22b, $K = 1 \times 10^{-8}\,\mathrm{m}^2$), more flow is "pushed" through the cell as the channel mass flow rate increases (lower $\psi_0$). The effect of porosity $\phi$ is presented in Figure 9.23, but as one can see, porosity does affect flow parameters as substantially as the permeability does.

### 9.2.5.2. *Friction Factor*

The effect of *Re* on fully developed values of $f$ is shown in Figure 9.24a, where the present results are compared with similar data from the literature. The curves reflect the rise in the $f$ value in relation to an unobstructed channel flow ("0" conditions). Results compare well with published data indicating an increase in $f$ as the mass flow rate through the channel is increased. The effect of permeability $K$ on friction factor $f$ is presented in Figure 9.24b. Results show that porous baffles of any type will always yield smaller friction factors and that a decrease in $f$ is observed for higher permeabilities. Also shown in Figure 9.24b is the slight reduction on $f$ for a more porous material (higher $\phi$). Less sensitivity of flow parameters on variations in $\phi$ is also seen in Figure 9.25a, where for a wide range of porosities no substantial change in $f$ is detected. The influence of baffle height is presented in Figure 9.25b, where it is noticed that the head loss increases significantly with greater $h/H$ values. Also, it is noted that when a porous baffle is employed, the higher the baffle height, the bigger the difference between the solid and the porous cases.

### 9.2.5.3. *Nusselt Number*

Figure 9.26a presents results for *Nu* as a function of *Re* and solid material properties. The coefficient $\lambda$ is defined as:

$$\lambda = \frac{k_{\mathrm{s}}}{k_{\mathrm{f}}}\frac{t}{L} \tag{9.55}$$

Results in the figure compare well with published data and reproduces the enhancement on the heat transfer characteristics as the solid conductivity or the fluid *Pr* number increases. Figure 9.26b shows the reduction on *Nu* for more permeable baffles, in coherence with reduction of recirculating strengths shown in Figure 9.22 and consequent lower drag in Figure 9.24. The effect on conductivity ratio is presented in Figure 9.26c where an increase in *Nu* with higher $\lambda$ is observed. Also, porosity affects heat transfer more efficiently for materials with high thermal conductivity. In Figure 9.26c, *Nu* values for $\phi = 0.4$ are higher than those of $\phi = 0.9$ when $\lambda$ is equal to 40.

Figure 9.27a shows a reduction on *Nu* in relation to the solid baffle case when a highly impermeable (compacted) material is used for manufacturing porous baffles. The

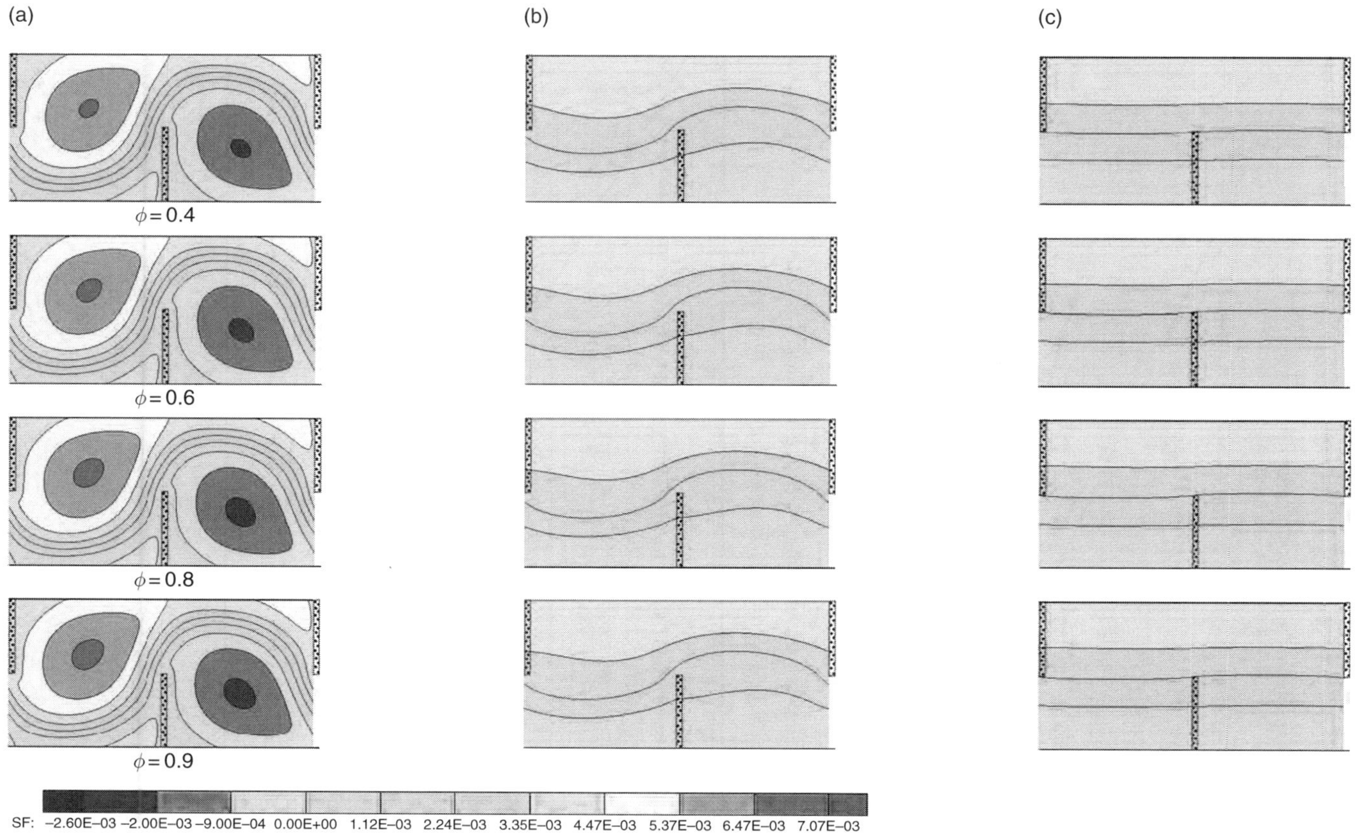


**Figure 9.23.** Streamlines as a function of $\phi$, $Re = 500$, $h/H = 0.5$: (a) $K = 1 \times 10^{-9}\,\text{m}^2$, (b) $K = 1 \times 10^{-7}\,\text{m}^2$, (c) $K = 1 \times 10^{-5}\,\text{m}^2$.

(a)

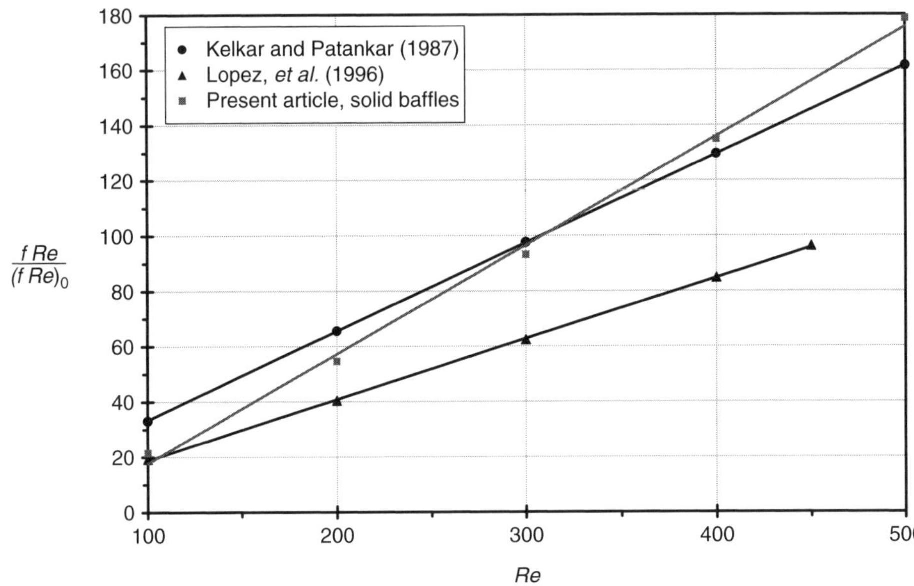


(b)

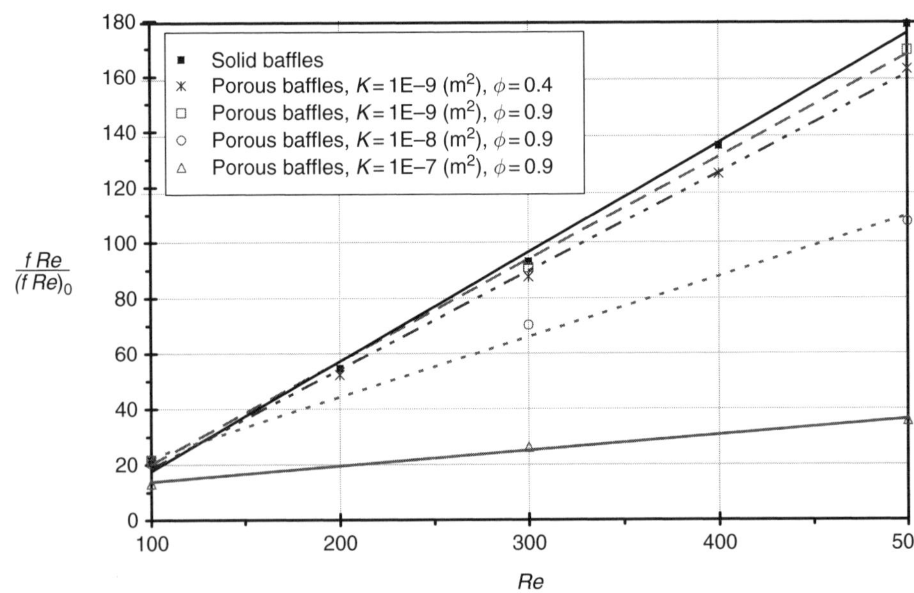


**Figure 9.24.** Friction factor for a channel as a function of $Re$, $h/H = 0.5$: (a) solid baffles, (b) porous baffles.

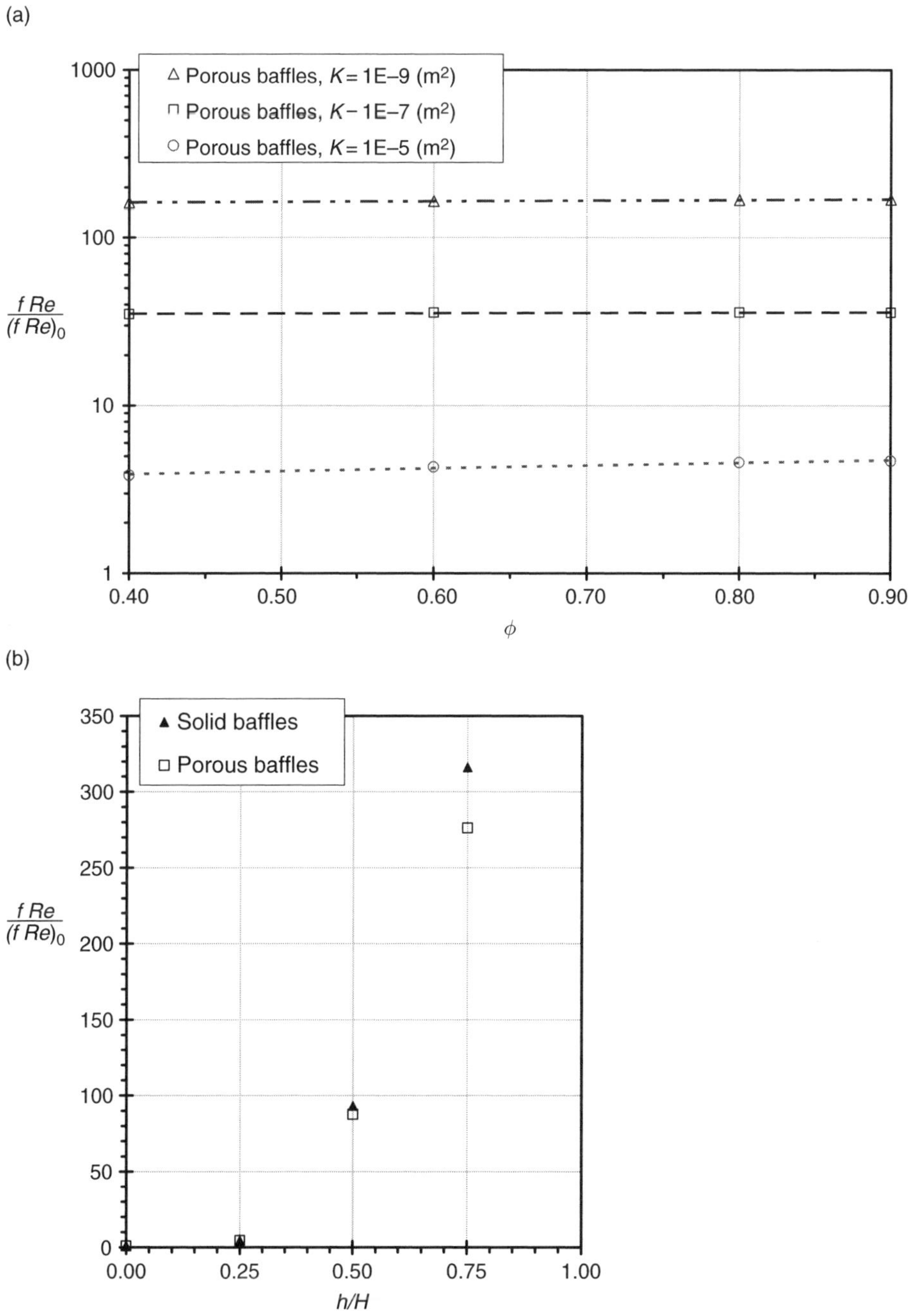


**Figure 9.25.** Friction factor for a channel with baffles: (a) effect of $\phi$, $h/H = 0.5$, (b) effect of $h/H$, $K = 1 \times 10^{-9}\,\mathrm{m}^2$ and $\phi = 0.4$.

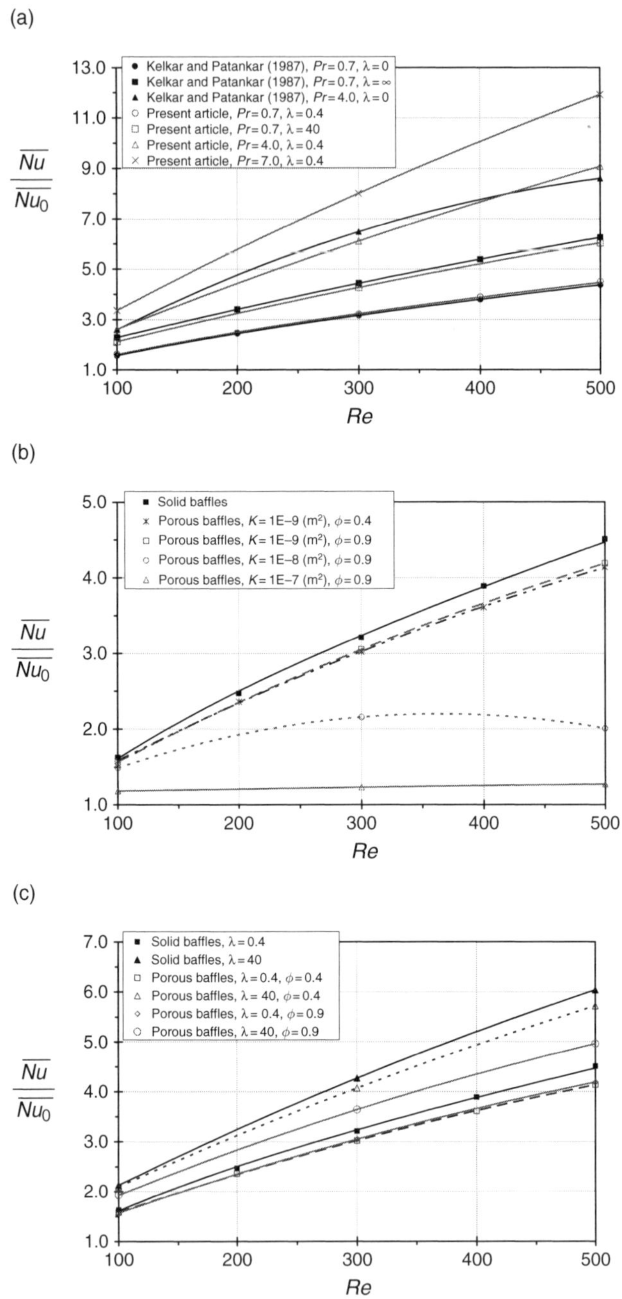


**Figure 9.26.** Nusselt number for a channel as a function of $Re$, $h/H = 0.5$: (a) solid baffles, (b) effect of $K$ and $\phi$, $\lambda = 0.4$, $Pr = 0.7$, (c) effect of $\lambda$ and $\phi$, $K = 1 \times 10^{-9}\,\mathrm{m}^2$, $Pr = 0.7$.

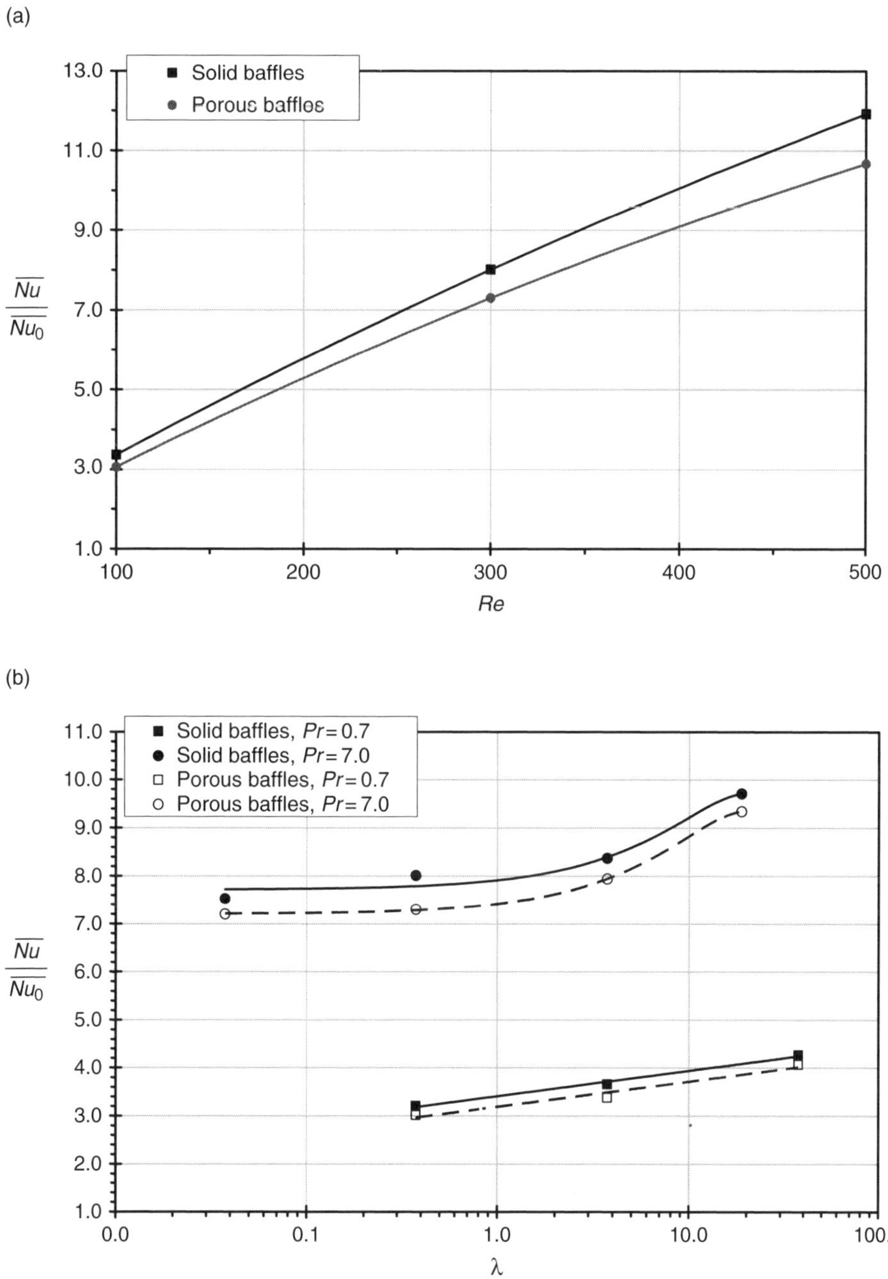


**Figure 9.27.** Nusselt number for a channel with solid and porous baffles, $h/H = 0.5$, $K = 1 \times 10^{-9}\,\text{m}^2$, $\phi = 0.4$: (a) effect of $Re$, $Pr = 7.0$ and $\lambda = 0.4$, (b) effect of $\lambda$.

influence of $\lambda$ on $Nu$ is presented in Figure 9.27b for two Prandtl numbers. For the conditions used, $Nu$ increases with both $Pr$ and $\lambda$ and is slightly higher for the solid baffle case. As in the case of $f$ (Figure 9.25a), $Nu$ is less sensitive to $\phi$, as seen in Figure 9.28a. Further, the channel height affects $Nu$ (Figure 9.28b) as it affects $f$ (Figure 9.25b), essentially by enhancing $\psi_0$ within the channel cell (see Figure 9.21).

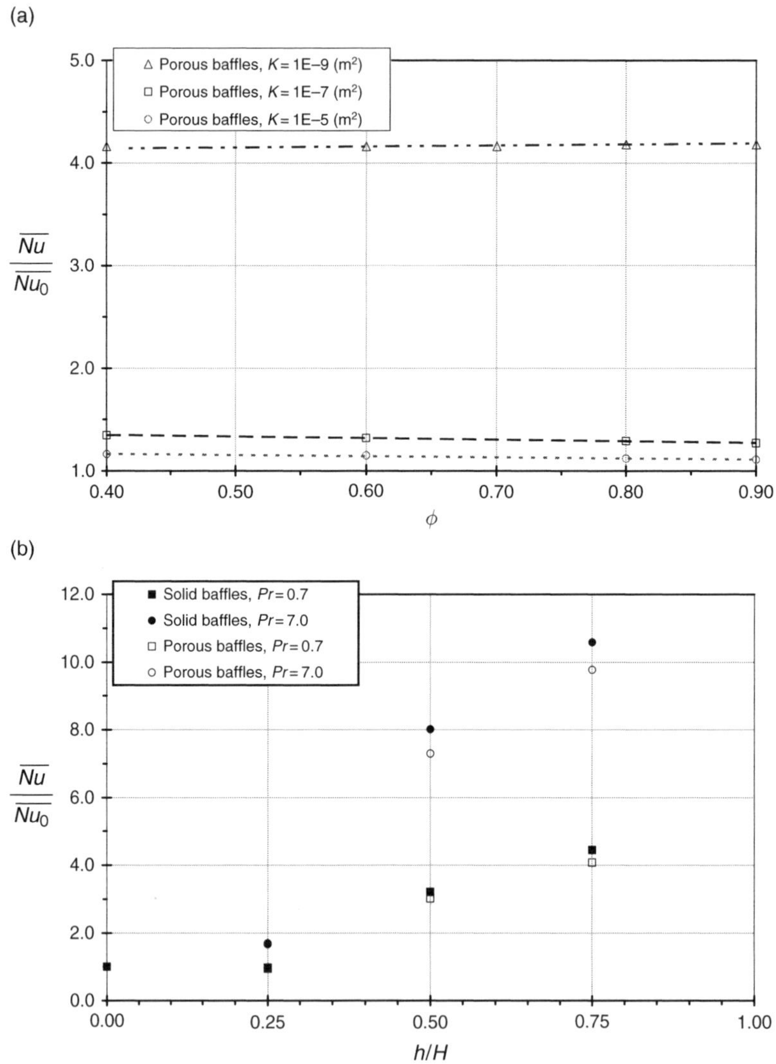


**Figure 9.28.** Nusselt number for a channel with baffles, $\lambda = 0.4$: (a) effect of $\phi$, $h/H = 0.5$, $Pr = 0.7$, (b) effect of $h/H$, $K = 1 \times 10^{-9}\,\text{m}^2$ and $\phi = 0.4$.

#### 9.2.5.4. *Effective Heat Transfer*

For analyzing the effective influence of the type of obstruction (solid or porous), a modified Nusselt number $\overline{Nu}^* = (\overline{Nu}/\overline{Nu_0})/(f/f_0)^{1/3}$ has been proposed in the literature (Ko and Anand, 2003). This modified Nusslet number is a measure of the head loss, when a porous medium is used, with the corresponding reduction in the heat transfer characteristics. Such analysis is useful when comparing performances of baffles of different types. Ultimately, if using porous baffles increases the value of $\overline{Nu}^*$, in relation to using solid material, then the system is considered to be more effective. In this case, although $Nu$ is reduced, the corresponding decrease in the necessary pumping power (reduction of $f$) makes it advantageous the use of permeable material for manufacturing the baffles (see Ko and Anand, 2003, for details).

Table 9.2 present values for $\overline{Nu}^*$ and compares, for different $Re$ and $Pr$, the effectiveness of both impermeable (solid) and porous baffled channels. The percent deviation, $\eta$, between these two values is also shown in the table and is calculated as:

$$\eta = \frac{\overline{Nu}^*|_{porous} - \overline{Nu}^*|_{solid}}{\overline{Nu}^*|_{solid}} \times 100 \tag{9.56}$$

A positive value for $\eta$ would then indicate an enhancement in the effective heat transfer when using porous materials. However, the table shows that, for the porous material here employed, and for the flow regime here simulated, no gain is obtained as far as heat transfer is of concern. Also noticed are more degraded conditions for the case of a higher Prandtl number.

Table 9.3 presents the influence of baffle height $h/H$ on $\overline{Nu}^*$ and $\eta$. The table shows that the use of shorter baffles is more advantageous for increasing heat transfer characteristics of this thermal equipment. This is related to a less significant head loss in porous material for low values of $h/H$, as indicated in Figure 9.25b.

Here, when comparing the corresponding performance of porous and solid baffles under the same conditions, i.e., same $Re$, $Pr$, $K$, $h/H$, the use of the parameter $\lambda$, previously

**Table 9.2.** Effects of $Re$ and $Pr$ on $\overline{Nu}^*$ for $h/H = 0.5$, $\lambda = 0.4$, $K = 1 \times 10^{-9}\,m^2$ and $\phi = 0.40$.

| | | $\overline{Nu}^*$ | | |
|---|---|---|---|---|
| $Re$ | $Pr$ | Porous | Solid | $\eta$ (%) |
| 100 | | 0.57 | 0.59 | −2.5 |
| 300 | 0.7 | 0.68 | 0.71 | −4.0 |
| 500 | | 0.76 | 0.80 | −5.3 |
| 100 | | 1.11 | 1.21 | −8.5 |
| 300 | 7.0 | 1.64 | 1.77 | −7.0 |
| 500 | | 1.95 | 2.12 | −7.6 |

**Table 9.3.** Effects of $h/H$ and $Pr$ on $\overline{Nu}^*$ for $Re = 300$, $\lambda = 0.4$, $K = 1 \times 10^{-9}\,\mathrm{m}^2$ and $\phi = 0.40$.

| $h/H$ | $Pr$ | $\overline{Nu}^*$ Porous | Solid | $\eta$ (%) |
|---|---|---|---|---|
| 0.25 | | 0.57 | 0.59 | −3.1 |
| 0.5 | 0.7 | 0.68 | 0.71 | −4.0 |
| 0.75 | | 0.63 | 0.65 | −4.1 |
| 0.25 | | 1.01 | 1.03 | −1.8 |
| 0.5 | 7.0 | 1.64 | 1.77 | −7.0 |
| 0.75 | | 1.50 | 1.55 | −3.4 |

proposed in Kelkar and Patankar (1987), was adopted. This parameter was defined in Kelkar and Patankar (1987) for analyzing the performance of solid baffles and did not consider the effective conductivity of the porous material due to the porosity. Accordingly, for a fixed $\lambda$, variation of $\phi$ indicates how the amount of solid material in the porous baffle affects heat transfer characteristics. On the other hand, calculations for a constant porosity $\phi$, using distinct values of $\lambda$, reveal the influence of thermal conductivity of the solid material.

In Table 9.4, the effect of the solid thermal conductivity $k_s$ (or $\lambda$) and Prandtl number on $\overline{Nu}^*$ is evaluated for baffles with $\phi = 0.40$ and $h/H = 0.5$. It is possible to verify that an increase in $\lambda$ (or $k_s$) results in a raise in $\overline{Nu}^*$, regardless of the $Pr$ used. The table also indicates that optimal values for $\eta$ will depend on a combination of the solid material ($\lambda$) and the working fluid ($Pr$) used. Table 9.5 shows the effect of porosity $\phi$ on $\overline{Nu}^*$ for different $Re$ and $\lambda$. Only for larger values of $\lambda$ a reduction on $\overline{Nu}^*$ is observed as $\phi$ increases. Porous baffles made of highly conductive material ($\lambda = 40$) and of relatively low porosity ($\phi = 0.4$) present the best advantages (higher $\eta$) over the range of $Re$ here investigated.

**Table 9.4.** Effects of $Pr$ and $\lambda$ on $\overline{Nu}^*$ for $h/H = 0.5$, $Re = 300$, $K = 1 \times 10^{-9}\,\mathrm{m}^2$ and $\phi = 0.40$.

| $Pr$ | $\lambda$ | $\overline{Nu}^*$ Porous | Solid | $\eta$ (%) |
|---|---|---|---|---|
| | 4.E−01 | 0.68 | 0.71 | −4.0 |
| 0.7 | 4.E+00 | 0.76 | 0.81 | −5.8 |
| | 4.E+01 | 0.92 | 0.94 | −2.7 |
| | 4.E−02 | 1.62 | 1.66 | −2.3 |
| 7.0 | 4.E−01 | 1.64 | 1.77 | −7.0 |
| | 4.E+00 | 1.79 | 1.85 | −3.2 |
| | 2.E+01 | 2.10 | 2.14 | −1.8 |

**Table 9.5.** Effect of $Re$, $\lambda$ and $\phi$ on $\overline{Nu}^*$ for $h/H = 0.5$, $Pr = 0.7$ and $K = 1 \times 10^{-9}\,m^2$.

| | | $\overline{Nu}^*$ $\phi$ | | | $\eta(\phi)$ (%) | |
|---|---|---|---|---|---|---|
| $Re$ | $\lambda$ | 0.4 | 0.9 | Solid | 0.4 | 0.9 |
| 100 | | 0.57 | 0.56 | 0.59 | −2.5 | −3.8 |
| 300 | 4.E−01 | 0.68 | 0.68 | 0.71 | −4.0 | −4.2 |
| 500 | | 0.76 | 0.76 | 0.80 | −5.3 | −5.4 |
| 100 | | 0.76 | 0.69 | 0.76 | 0.0 | −9.0 |
| 300 | 4.E+01 | 0.92 | 0.81 | 0.94 | −2.7 | −14.0 |
| 500 | | 1.05 | 0.90 | 1.07 | −2.3 | −16.4 |

**Table 9.6.** Effects of $K$ and $\phi$ on $\overline{Nu}^*$ for $h/H = 0.5$, $Re = 500$, $Pr = 0.7$ and $\lambda = 0.4$.

| | | $\overline{Nu}^*$ | | | |
|---|---|---|---|---|---|
| $\phi$ | $K$ ($m^2$) | 1.0E−09 | 1.0E−07 | 1.0E−05 | Solid |
| 0.4 | | 0.76 | 0.41 | 0.74 | |
| 0.6 | | 0.76 | 0.40 | 0.71 | 0.80 |
| 0.8 | | 0.76 | 0.39 | 0.67 | |
| 0.9 | | 0.76 | 0.39 | 0.66 | |

Table 9.6 shows the effects of $K$ and $\phi$ on $\overline{Nu}^*$, covering a wide range of permeabilities, from $K = 10^{-9}\,m^2$ to $10^{-5}\,m^2$. The table indicates that porosity has a very weak influence on heat transfer characteristics for less permeable material, a result already shown in Figures 9.25a and 9.28a. As the baffle becomes more permeable ($K = 10^{-5}\,m^2$), then porosity starts to play a certain role on heat transfer performance, a result also seen in Figures 9.25a and 9.28a. As expected, less permeable baffles ($K = 10^{-9}\,m^2$) tend to produce results for $\overline{Nu}^*$ close to those of using solid material. Finally, Table 9.7 indicates that for a raise in $Re$, the values of $\overline{Nu}^*$ also rises for highly flow resistant porous material

**Table 9.7.** Effects of $Re$, $K$ and $\phi$ on $\overline{Nu}^*$ for $h/H = 0.5$, $Pr = 0.7$ and $\lambda = 0.4$.

| | | $\overline{Nu}^*$ | | | | |
|---|---|---|---|---|---|---|
| | $K$ ($m^2$) | 1.0E−09 | 1.0E−09 | 1.0E−08 | 1.0E−07 | |
| $Re$ | $\phi$ | 0.4 | 0.9 | 0.9 | 0.9 | Solid |
| 100 | | 0.57 | 0.56 | 0.55 | 0.50 | 0.59 |
| 200 | | 0.63 | – | – | – | 0.65 |
| 300 | | 0.68 | 0.68 | 0.52 | 0.41 | 0.71 |
| 400 | | 0.72 | – | – | – | 0.76 |
| 500 | | 0.76 | 0.76 | 0.42 | 0.39 | 0.80 |

($K = 10^{-9}\,\mathrm{m}^2$) but, in turn, it decreases for lower values of $K$. High *Re* flow across nearly solid porous baffles gives rise to strong recirculation motion, ultimately increasing $\psi_0$ (see Figure 9.22a). On the other hand, easier-to-flow porous baffles ($K = 1 \times 10^{-5}\,\mathrm{m}^2$) will let the more fluid pass through the channel, as *Re* increases, causing no recirculation motion and keeping $\psi_0$ as zero (see Figure 9.22c).

In summary, none of the results above showed a situation where porous baffles presented a better performance than the solid ones ($\eta > 0$). This, however, is in agreement with results already presented in Da Silva Miranda and Anand (2004), Santos and de Lemos (2004a,b) and de Lemos and Santos (2004). Nevertheless, for turbulent flow cases, advantages in using porous baffles might be possible, as already indicated in recent research efforts in the literature (Hwang, 1997; Ko and Anand, 2003, Yang and Hwang, 2003).

### 9.2.6. Section Summary

This session reviewed the work in Santos and de Lemos (2006), who presented results for the numerical solution of laminar flow in a channel containing porous and solid (impermeable) obstructions. Discretization of the governing equations used the finite volume method and the set of algebraic equations was solved by the SIMPLE method. The presented numerical results for the friction factor $f$ and for the Nusselt number *Nu* were compared with data of Kelkar and Patankar (1987) and Lopez *et al.* (1996), indicating that results herein differ in less than 5% in relation to published results. Further simulations comparing the effectiveness of the porous material used showed that no advantages are obtained when using low porosity and low permeability baffles in the laminar flow regime. Finally, results herein are encouraging and motivate further analyses for simulating turbulent flow in porous baffled channels.

## 9.3. TURBULENT IMPINGING JET ONTO A POROUS LAYER

The problem considered in this section is schematically presented in Figures 9.29 and 9.30. A fluid jet enters a cylindrical chamber through an aperture in the center of an upper disk (Figure 9.29). An annular clearance between the cylinder lateral wall and the disc allows fluid to flow out the enclosure. The incoming jet diameter, $D_j$, is 0.019 m and the inner cylinder diameter, $D$, is 0.39 m. The clearance between the cylinder and the disc holding the jet has a width, $w$, equal to 0.005 m. At the bottom of the chamber, a layer of porous material covers the surface and is hit by the incoming jet (Figure 9.30). Three different heights of fluid column above the porous substrate, $H$, namely 0.05 m, 0.1 m and 0.15 m, were used in the simulations. Two thicknesses of the porous layer $h_p$, were considered, namely 0.05 m and 0.1 m. The average velocity of the incoming jet was 1.6 m/s, representing a Reynolds number of 30,000, which was based on jet exit

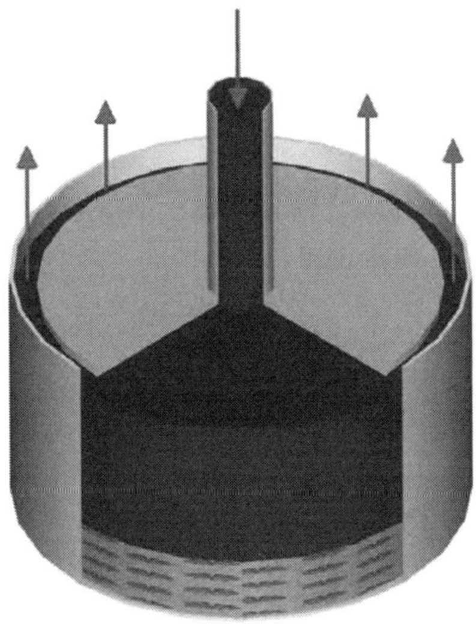

**Figure 9.29.** Physical model: jet impinging against a cylinder covered with a porous layer.

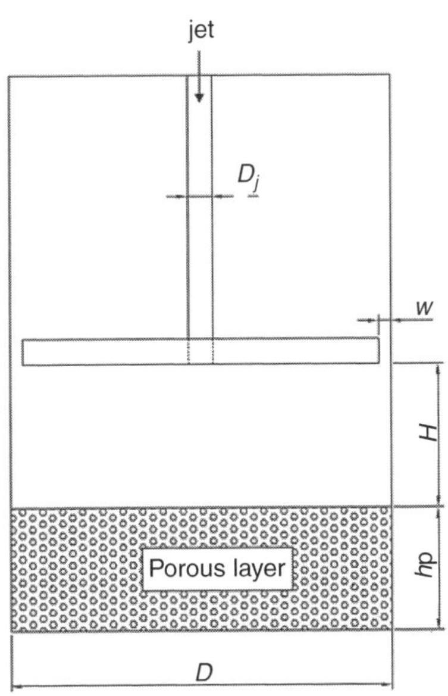


**Figure 9.30.** Geometry under consideration.

diameter. Different velocities were also used when evaluating the influence of Reynolds number on the main flow. These two other values were 1.0 and 2.5 m/s corresponding to $Re = 18{,}900$ and 47,000, respectively. Turbulent results for streamfunction, primary vortex center, velocity and turbulence kinetic energy profiles are here compared with those from Prakash *et al.* (2001a,b).

### 9.3.1. Numerical details

Due to the symmetry condition at the cylinder center, the computational domain adopted covered only half of the cylinder vertical cross section. Simulations were carried with a High Reynolds turbulence model. Grid independence studies were conducted and a computational grid of $100 \times 100$ nodes was found to suffice. Such mesh was used in all numerical runs.

The method employed to solve the flow equations was the finite volume method applied to a boundary-fitted coordinate system. Equations were discretized in a two-dimensional control volume involving both the clear and the porous media. For solving the algebraic equation set, the SIMPLE algorithm was used (Patankar, 1980), and residues for the transport equations were brought down to $10^{-6}$. The interface was positioned to coincide with the border between two adjacent control volumes, generating in such a way volumes that were either "totally porous" or "totally clear". Details of the numerical implementation can be seen in Pedras and de Lemos (2001b) and Silva and de Lemos (2003a).

### 9.3.2. Clear Medium

#### *9.3.2.1. Mean Flow*

All figures below are plotted for the symmetry axis in Figure 9.30 aligned with the horizontal direction, with the jet coming at the lower left corner. Figure 9.31 shows a comparison of the numerically simulated streamfunctions with experimental flow visualizations and CFD results from Prakash *et al.* (2001a) for $H = 0.15$ m. For the present case, the entire half plane of the cylinder is filled with one large recirculation cell. A small secondary recirculation bubble can be noticed in the top-right corner of the picture, but with negligible dimensions compared to the main one. Simulation seems to be close to literature results, with main flow streamlines following very closely the pattern given by the LDV experiments. The flow in the core region and close to the walls agrees well with literature CFD results, indicating that simulations in de Lemos and Graminho (2005) may well represent actual flow behavior within the entire chamber.

The effect of the fluid layer height, $H$, is given by Figure 9.32, showing the streamfunctions for $H = 0.15$, 0.10 and 0.05 m. By decreasing the liquid height, main flow behavior remains the same, with a central dominating recirculation. The peripheral recirculations get smaller with the decrease in $H$. For $H = 0.1$ m, a recirculation appears close to the jet

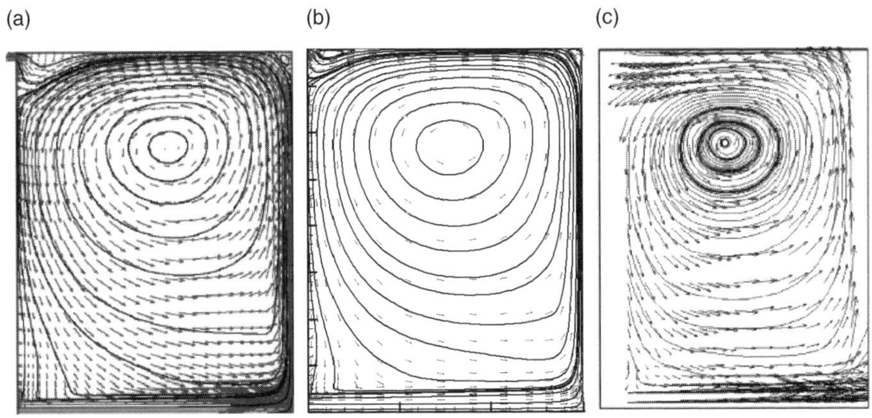


**Figure 9.31.** Comparison of streamfunctions for $H = 0.15\,\text{m}$, (a) CFD results – Prakash *et al.* (2001a), (b) present results, (c) LDV measurements – Prakash *et al.* (2001b).

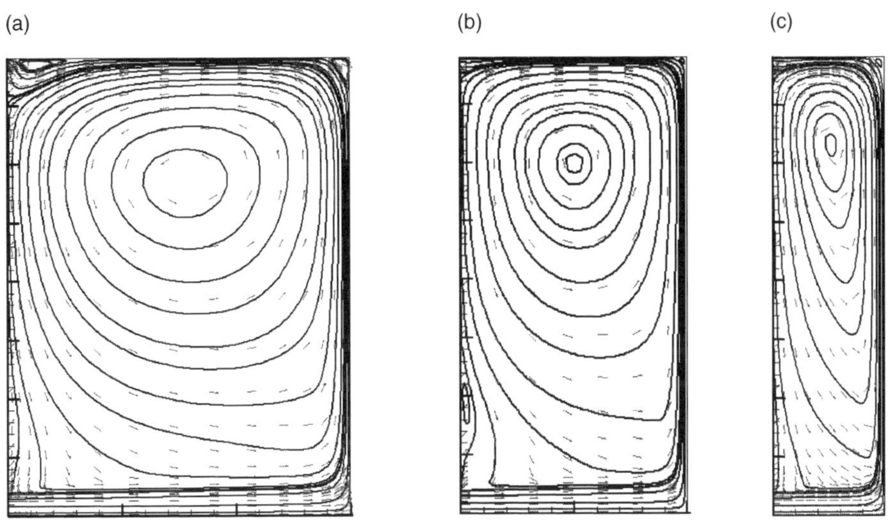


**Figure 9.32.** Effect of fluid layer height on streamlines for $Re = 30{,}000$: (a) $H = 0.15\,\text{m}$, (b) $H = 0.10\,\text{m}$, (c) $H = 0.05\,\text{m}$.

exit and both secondary recirculations close to the cylinder wall are reduced compared to the case with $H = 0.15\,\text{m}$. With a further decrease in $H$, the main recirculation becomes elongated and the peripherical ones nearly vanish.

Figure 9.33 shows the horizontal position of the main recirculation for different liquid heights. The $R_c/R$ axis represents the normalized radial position of the recirculation, where $R$ is the cylinder radius. From the picture it can be seen that the center of the

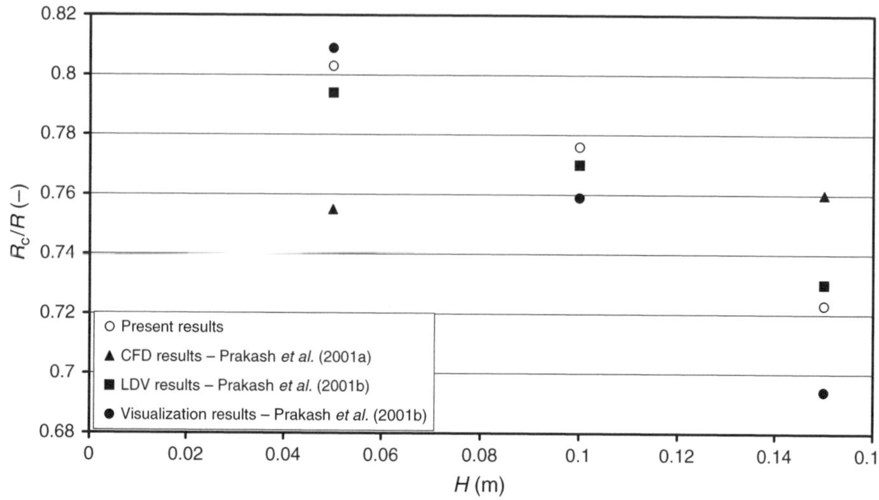


**Figure 9.33.** Radial position of center recirculation as function of fluid height.

recirculation moves closer to the wall as $H$ decreases. Present computations seem to follow the trend in both measurement and visualization results i.e., reduction of $R_c$ as $H$ increases.

Figure 9.34 gives the influence of Reynolds number on the main flow pattern. It can be seen that a change in $Re$ does not have a significant effect on the main flow pattern. The

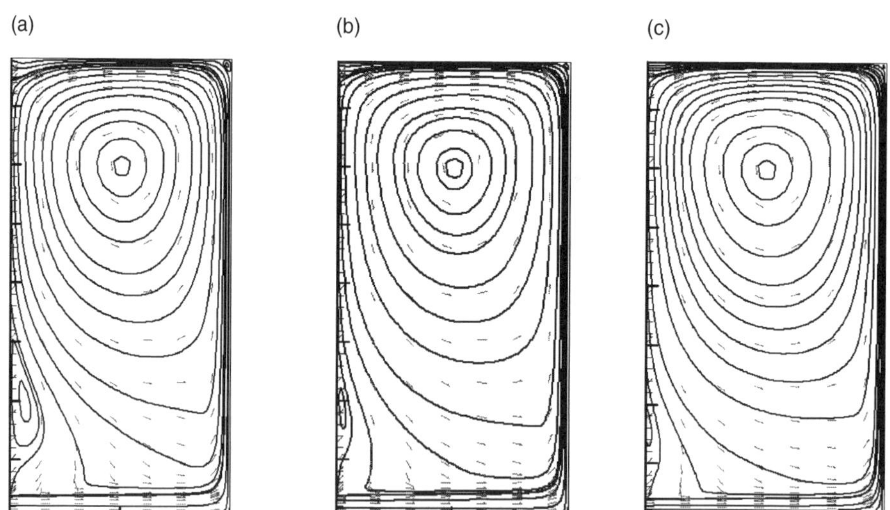


**Figure 9.34.** Effect of Reynolds number on streamfunctions for $H = 0.10\,\text{m}$: (a) $Re = 18{,}900$, (b) $Re = 30{,}000$, (c) $Re = 47{,}000$

most noticeable change refers to a secondary recirculation developed close to the jet exit (lower left corner). With an increase in Reynolds number up to 30,000, the recirculation tends to get smaller. Past this value, the effect of Reynolds number on the flow seems to be small.

Figure 9.35a,b,c shows the axial velocity profiles for fluid heights of 0.15, 0.10 and 0.05 m. In this figure, the profiles have been plotted for six different axial locations covering the entire cylindrical chamber. It can be seen that close to the jet centerline and near the surface at the bottom, present results seem to fall in between numerical and experimental data available in the literature. At higher axial distance close to the jet exit, present simulations align with CFD predictions in the literature, but both overpredict the axial velocity at the centerline. According to Prakash *et al.* (2001b), experimental conditions at inlet did not correspond to fully developed flow, whereas numerical simulations here reported assumed such condition. In general, the present simulations fail to predict the reduction in the centerline velocity as the bottom surface approaches. However, elsewhere in the chamber the main features of the flow were reproduced.

Radial velocity profiles for the three fluid layers investigated are presented in Figure 9.36 a,b,c. For $H = 0.15$, results follow very closely the experimental data for $z/H > 0.5$, but underpredicts measured velocities as one gets closer to the collision plate at the bottom. The same trend is observed in the reported CFD predictions. For $H = 0.10$ and $r/R$ less than 0.1, the present computations reproduce well experimental data, but in the range $0.1 < r/R < 0.6$, and for lower values of $z/H$, simulations herein fall above experimental values for velocities. At $H = 0.05$, present results seem to closely follow the experimental data trend for higher values of $z/H$. As the flow moves away from the symmetry axis and further down from the jet exit, the quality of predictions seems to deteriorate as one gets closer to the bottom surface. Nevertheless, the main features of the flow were captured by the simulations here reported.

#### 9.3.2.2. *Turbulent Field*

Finally, Figure 9.37a,b,c shows the turbulence kinetic energy profiles for the same conditions as above. For all three fluid layers, results reproduce very closely experimental literature data, including the peak value near the symmetry axis, which is caused by the strong shear layer formed around the incoming jet. Close to the impingement plate, a second elevation in turbulence kinetic energy appears. This second maximum is attributed to a laminar-turbulent boundary layer transition in the wall jet region as observed in Incropera and DeWitt (1990).

### 9.3.3. Porous Medium

As mentioned, two different thicknesses were considered for the porous layer placed at the bottom of the cylinder, namely $h_p = 0.10$ m and 0.05m. Porous medium properties are

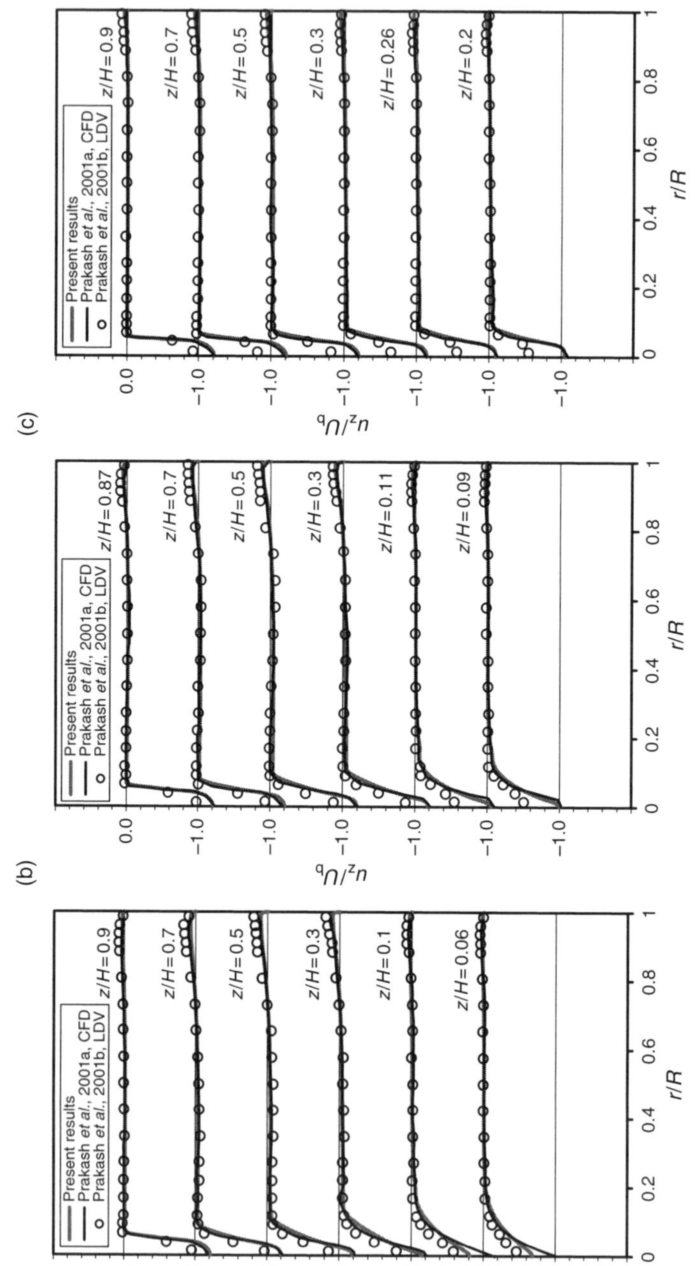


**Figure 9.35.** Axial velocity profiles for cases without foam: (a) $H = 0.15$ m, (b) $H = 0.10$ m, (c) $H = 0.05$ m.

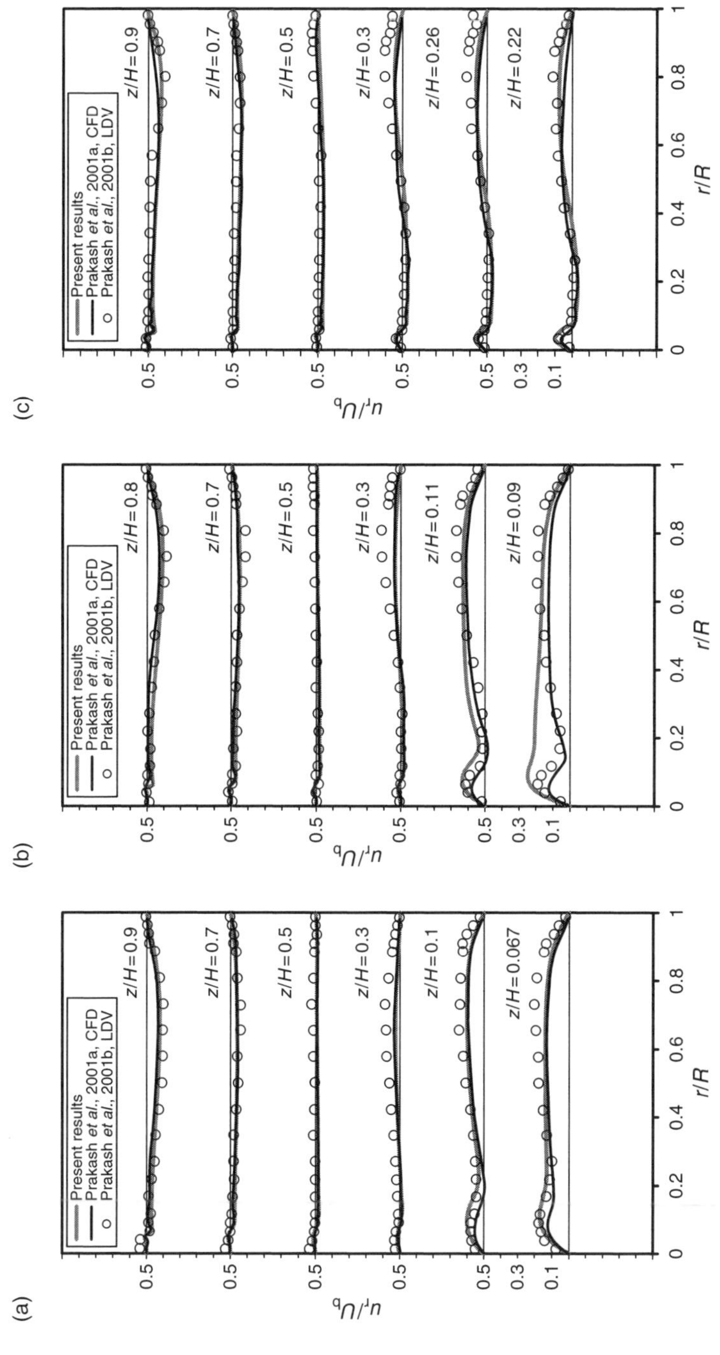


**Figure 9.36.** Radial velocity profiles for cases without foam: (a) $H = 0.15$ m, (b) $H = 0.10$ m, (c) $H = 0.05$ m.

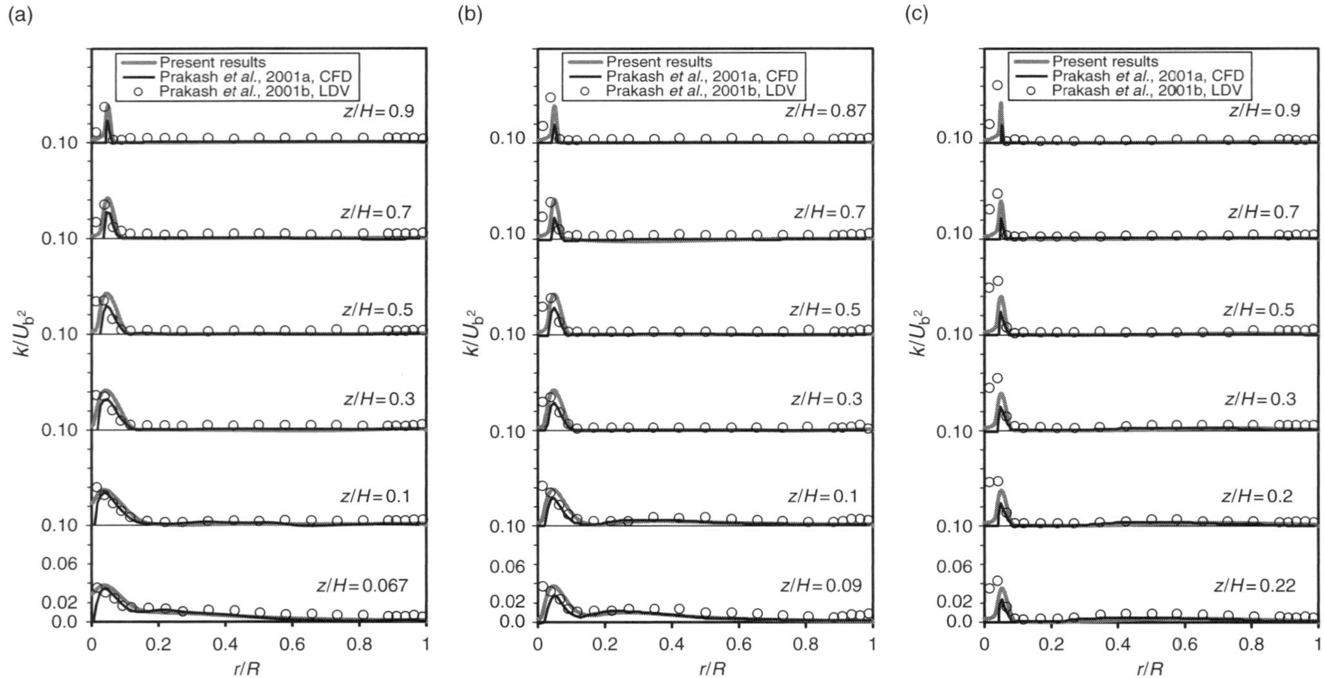


**Figure 9.37.** Turbulence kinetic energy profiles for cases without foam: (a) $H = 0.15\,\text{m}$, (b) $H = 0.10\,\text{m}$, (c) $H = 0.05\,\text{m}$.

**Table 9.8.** Porous medium properties for simulated foams.

| Foam | G10 | G30 | G45 | G60 |
|---|---|---|---|---|
| $\phi$ | 0.971 | 0.9755 | 0.978 | 0.976 |
| $K$ (m$^2$) | 2.84E−7 | 6.95E−8 | 1.6E−8 | 1.2E−8 |
| $c_F$ (m) | 0.1227 | 0.1218 | 0.1232 | 0.1161 |

given in Table 9.8, where these values are the same as those for metallic porous foams used by Prakash *et al.* (2001a,b). All porosities considered are nearly equal and close to unity, so that the changes on the porous material effectively represent a change in its permeability. The code used to label the metallic foams is associated with the number of pores per inch (ppi).

### 9.3.3.1. *Mean Flow*

Figure 9.38 shows a comparison of the numerically simulated streamfunctions with experimental flow visualizations and CFD data from Prakash *et al.* (2001a,b) for $H = 0.10\,\text{m}$, $h_p = 0.05\,\text{m}$ and $Re = 30{,}000$ on porous foam G10. Differently from the previous section where, for clear flow, we found a large central recirculation and small ones at the corners, it can be noticed in Figure 9.38 the presence of two recirculating zones dominating the entire flow. For comparisons, the recirculating bubble closer to cylinder centerline will be called *primary* whereas the one near the cylinder wall will be named *secondary* recirculation. For the present case, both primary and secondary recirculations seem to be slightly larger than those from flow visualizations with LDV, but in a general,

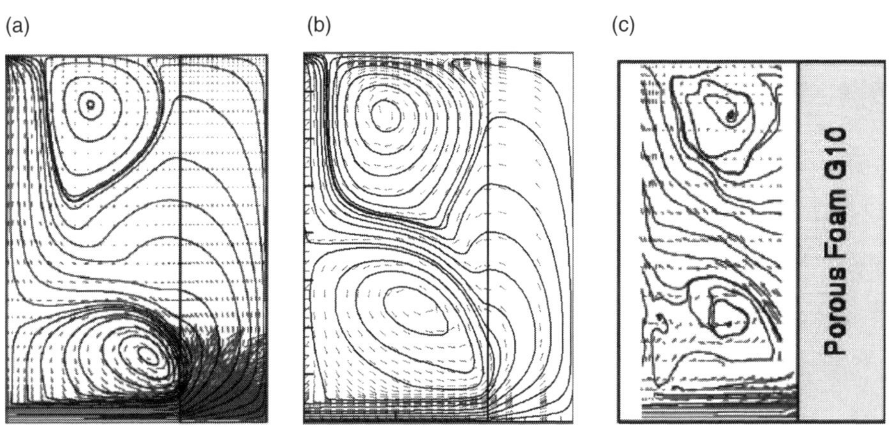


**Figure 9.38.** Comparison of streamfunctions for $H = 0.10\,\text{m}$, $h_p = 0.05\,\text{m}$ and porous foam G10: (a) CFD results – Prakash *et al.* (2001a), (b) present results, (c) LDV measurements – Prakash *et al.* (2001b).

results appear to be in good agreement with experimental data. The streamlines inside the porous medium, even though they cannot be seen by flow visualization methods due to the opaque characteristic of the material, can be well predicted from the flow behavior in the fluid layer above the foam.

The effect of the height of the fluid layer is presented in Figure 9.39, showing the streamfunction behavior for $H = 0.15$, 0.10 and 0.05 m. For $H = 0.15$ m, the primary recirculation is much stronger than the secondary one, indicating a smaller effect of $H$ on jet penetration into the porous layer. Under such conditions, the flow behavior tends to be similar to that obtained in a cylinder without the porous layer, as shown in the previous section. This is due to the spread of the jet before colliding into the porous foam. With a decrease in $H$, the primary recirculation tends to decrease in size as the secondary one grows, so that for $H = 0.10$ m both bubbles have almost the same dimensions filling completely the fluid layer and compressing the exit flow streamlines between both recirculations. For $H = 0.05$ m, both recirculations decrease in size, moving their centers apart, toward the cylinder centerline (*primary*) and in the direction of the cylinder wall (*secondary*). Streamlines between both recirculations are distributed rather regularly, with the flow nearly perpendicular to the interface as fluid leaves the porous foam to the fluid layer, a different behavior from that occurring for higher values of $H$.

The effect of the porous layer thickness, $h_p$, is presented in Figure 9.40 while maintaining the same fluid layer height, $H$. From the picture, it can be inferred that the thickness of the porous foam has less influence on the main flow behavior in comparison with the effect caused by varying $H$ (see Figure 9.39). With an increase in $h_p$, the primary recirculation shows only a small enlargement while the secondary one decreases a little. Other than that, no other significant effect can be detected.

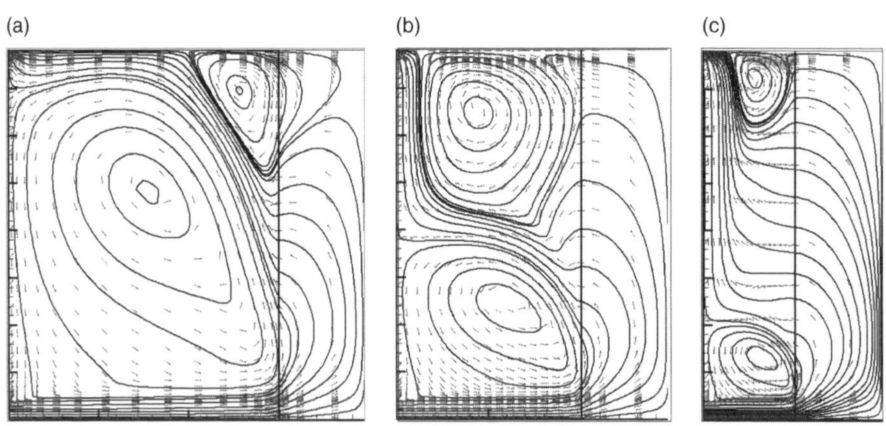


**Figure 9.39.** Effect of the fluid layer height on streamfunctions for $h_p = 0.05$ m, $Re = 30{,}000$ and porous foam G10: (a) $H = 0.15$ m, (b) $H = 0.10$ m, (c) $H = 0.05$ m.

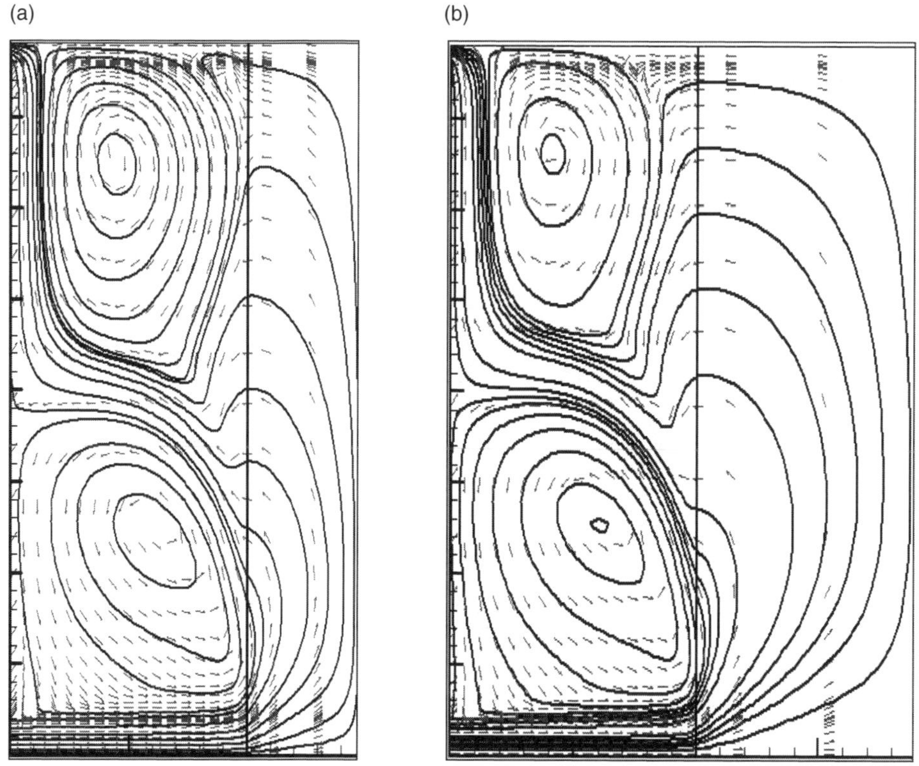


**Figure 9.40.** Effect of the porous layer thickness on streamfunctions for $H = 0.10\,\text{m}$, $Re = 30{,}000$ and porous foam G10: (a) $h_p = 0.05\,\text{m}$, (b) $h_p = 0.10\,\text{m}$.

Figures 9.41 and 9.42 show, respectively, radial and axial positions of central recirculation for porous foam G10 with $h_p = 0.05\,\text{m}$. From Figure 9.41 one can infer that the radial position of the primary recirculation, $R_c/R$, moves in the direction of the cylinder wall, for an increase in $H$, at a faster rate in present simulations than in the measurements found in the literature. Despite the fact that the results are reasonably good for smaller values of $H$, it seems that simulations herein fail to represent well the central recirculation position for higher values of $H$. This could be associated with the use of inadequate grid layout within the computational domain for $H = 0.15\,\text{m}$. For the axial position $Z_c/H$ (Figure 9.42), where $Z_c$ is measured in relation to porous/clear fluid interface, another pattern can be noticed. While LDV experiments indicates that the center of the main recirculation moves toward the porous plate as $H$ increases, the present simulation agrees with visualization results up to $H = 0.10\,\text{m}$, but indicates an opposite direction of that of LDV data for $H = 0.15\,\text{m}$. Also interesting to note is that other CFD simulations in the literature show degraded results for $H = 0.15\,\text{m}$ as well. In summary, experimental

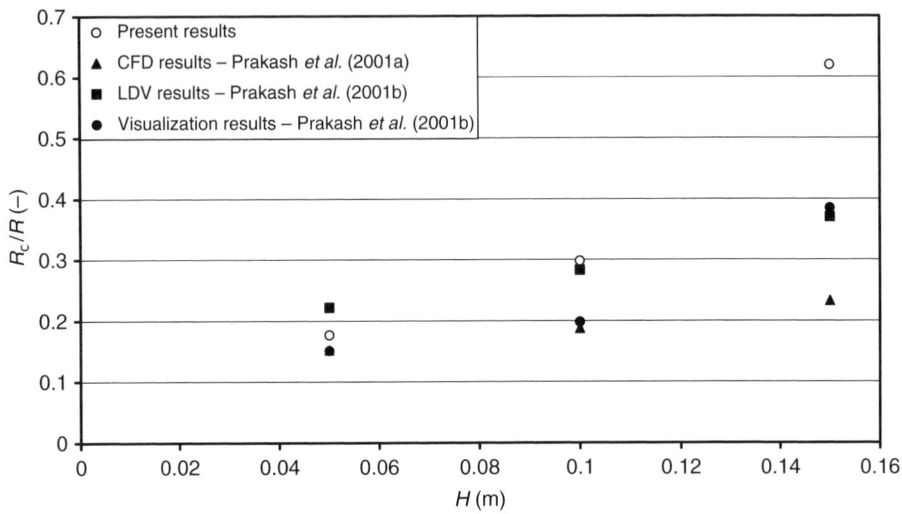


**Figure 9.41.** Radial position of center recirculation as function of fluid height for porous foam G10.

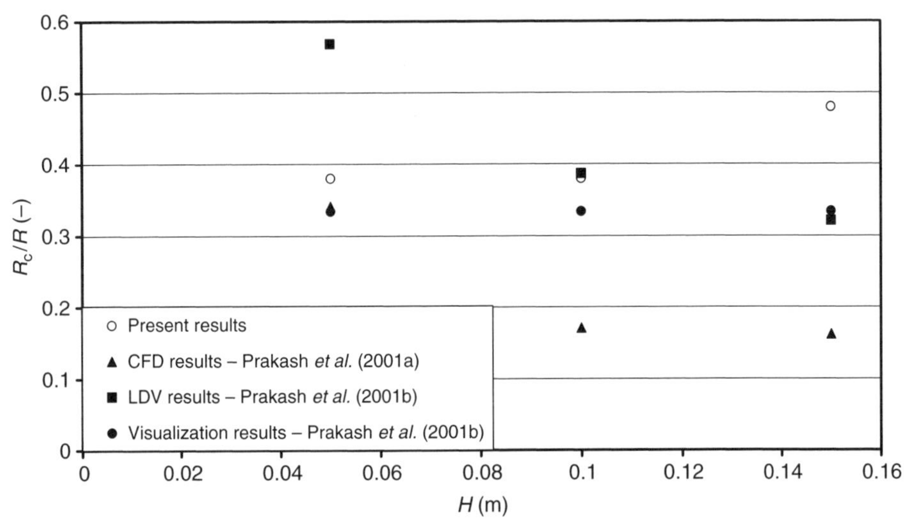


**Figure 9.42.** Axial position of center recirculation as function of fluid height for porous foam G10.

measurements and present simulations show good agreement for $H = 0.1$, but rather poor matches for $H = 0.05$ and $H = 0.15$. As mentioned, this could be related to the use of impropriated combination of grid layout and wall proximity model (High *Re* closure) for the particular cases of $H = 0.05$ and 0.15.

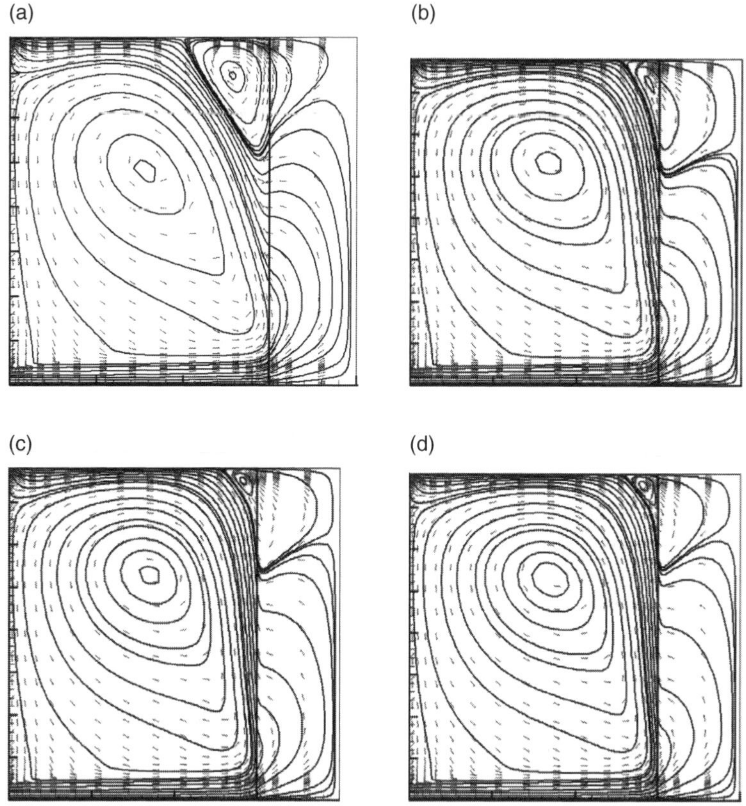


**Figure 9.43.** Effect of porous medium material on streamfunctions for $H = 0.15\,\text{m}$, $h_p = 0.05\,\text{m}$ on different porous foams: (a) porous foam G10, (b) porous foam G30, (c) porous foam G45, (d) porous foam G60.

The effect of the porous layer material, effectively representing a change in its permeability, can be well characterized by Figure 9.43. For the porous foam G10 with the highest permeability, a secondary recirculation develops with considerable size close to the cylinder wall. For the other foams G30, G45 and G60, this recirculation decreases with the reduction of the porous layer permeability, so that the porous layer tends to act as a solid obstacle being hit by a jet. For the less permeable foam, G60, the main flow pattern resembles those obtained for the cases shown in the Clear Medium section, as can be seen in Figure 9.31.

Figure 9.44 further shows the influence of the Reynolds number on the flow pattern for $H = 0.15\,\text{m}$. Apparently, a change in $Re$, under the conditions here investigated, does not cause a significant influence on the flow pattern for both the porous and the fluid layers. A similar conclusion was reached for clear flows as seen in Figure 9.34.

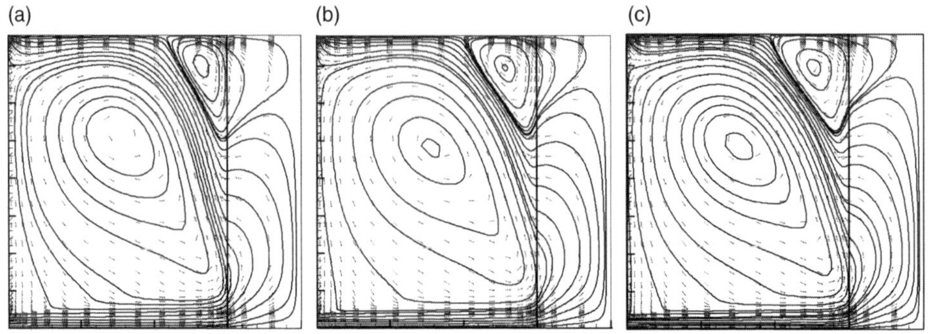


**Figure 9.44.** Effect of Reynolds number on streamfunctions for $H = 0.15\,\text{m}$, $h_p = 0.05\,\text{m}$ on porous foam G10: (a) $Re = 18{,}900$, (b) $Re = 30{,}000$, (c) $Re = 47{,}000$.

Results for the axial velocity profiles are shown in Figure 9.48a,b,c, for porous foam G10, $h_p = 0.05\,\text{m}$ and $Re = 30{,}000$. For all three fluid layer heights, in can be seen that both numerical simulations seem to slightly overpredict the actual flow behavior close to the jet centerline. For $H = 0.15\,\text{m}$, numerical predictions are closer to experimental data in comparison with $H = 0.10$ and $0.05\,\text{m}$. The general overprediction of the velocities profiles might be due to the fact that, in the experiments, the flow at the jet exit was not fully developed, a condition that was assumed in the present simulations. This same reasoning has been used when explaining discrepancies in the results presented in the Clear Medium section (Figure 9.35).

Figure 9.49a,b,c presents the radial profiles for the same parameters as in the previous figure. For higher values of $z/H$, simulations present difficulties in following experimental data, even though, closed to the interface region, the flow behavior is reasonably well predicted, particularly for $H = 0.10\,\text{m}$. For the second hump along the radial direction, while numerical simulations by others underpredict experimental values, the present simulations overpredicts them by a small amount.

### *9.3.3.2. Turbulent Field*

Figures 9.45, 9.46 and 9.47 show turbulent kinetic energy contours for $hp = 0.05\,\text{m}$, porous foam G10 and $Re = 30{,}000$. From the picture, it can be seen that turbulence penetrates into the porous medium, as can be noticed by the contour lines that go inside the porous bed. As the jet penetrates the foam, calculated turbulence intensities with the present model seem to be lower than those by Prakash *et al.* (2001a), who used only the Darcy term (dashed lines in Figures 9.45b, 9.46b and 9.47b). However, calculated levels of turbulence herein are higher than those computed by the same authors using both the Darcy and Forchheimer terms (solid lines in the same pictures). This second set of results by Prakash *et al.* (2001a,b) indicates that turbulence is damped almost completely at the interface. At the fluid layer, the present results are in agreement with published data.

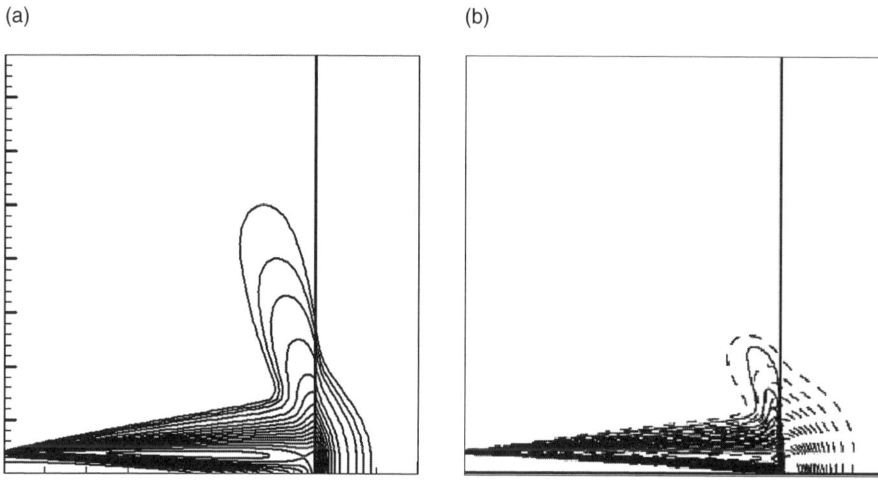

**Figure 9.45.** Comparison of turbulence kinetic energy contours for $H = 0.15\,\text{m}$, $h_p = 0.05\,\text{m}$ on porous foam G10: (a) present results, (b) CFD results – Prakash *et al.* (2001a).

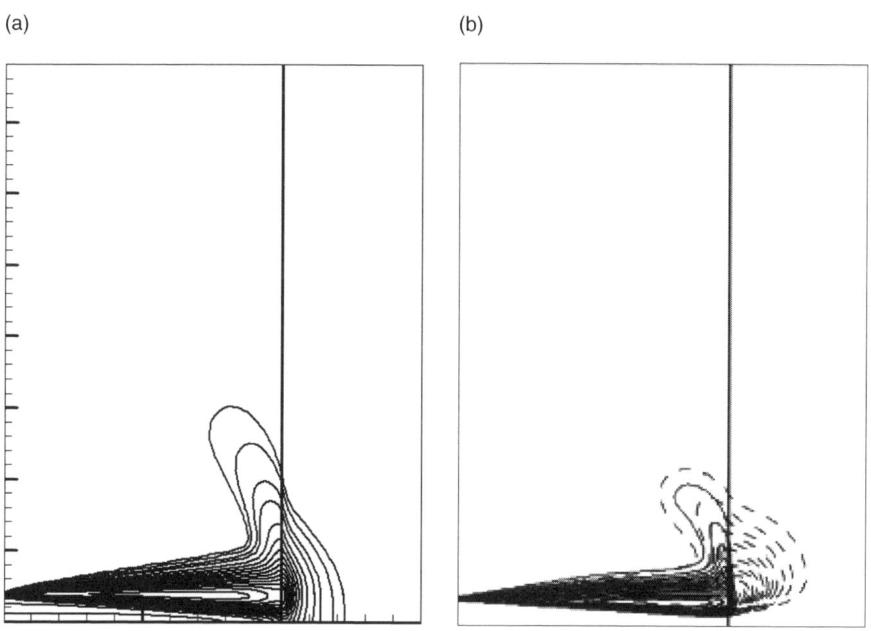

**Figure 9.46.** Comparison of turbulence kinetic energy contours for $H = 0.10\,\text{m}$, $h_p = 0.05\,\text{m}$ on porous foam G10: (a) present results, (b) CFD results – Prakash *et al.* (2001a).

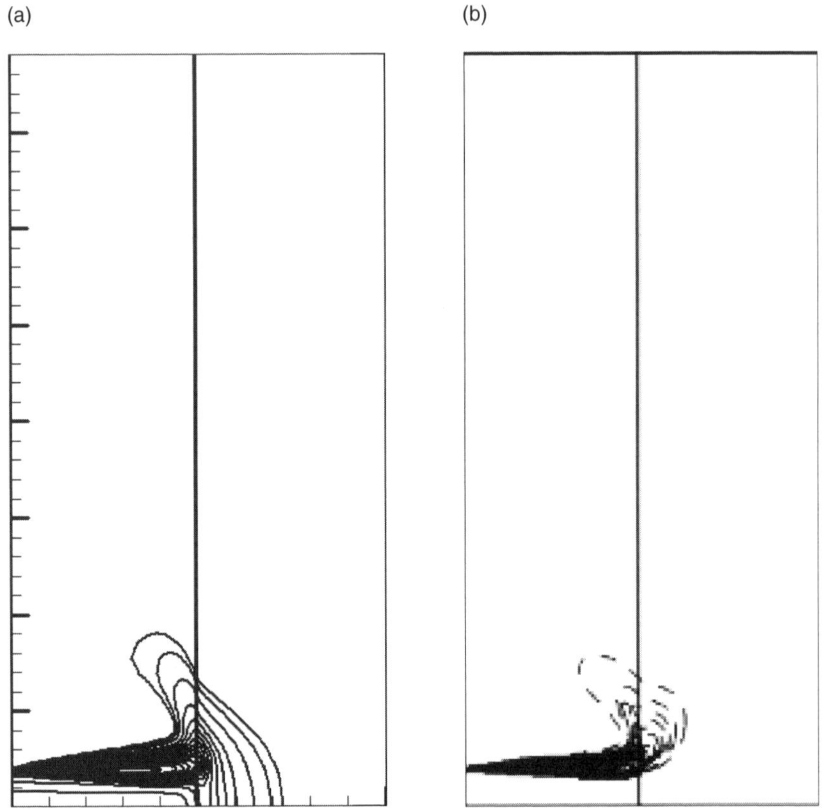

**Figure 9.47.** Comparison of turbulence kinetic energy contours for $H = 0.05\,\mathrm{m}$, $h_p = 0.05\,\mathrm{m}$ on porous foam G10: (a) present results, (b) CFD results – Prakash *et al.* (2001a).

Finally, turbulence kinetic energy profiles are shown in Figure 9.50a,b,c. Similarly to the case without foam (see Figure 9.37), this figure indicates that the present results predict the measured profiles with good accuracy, following closely the peak value of turbulent energy near the central axis.

## 9.4. BUOYANT FLOWS

The analysis of buoyancy-driven flows in an enclosed cavity provides useful comparisons for evaluating the robustness and performance of numerical methods dealing with viscous flow calculations. The importance of natural-convection phenomena in enclosures can best be appreciated by noting several of their application areas. Optimal design of furnaces and solar collectors, which contribute to energy losses minimization, nuclear reactor

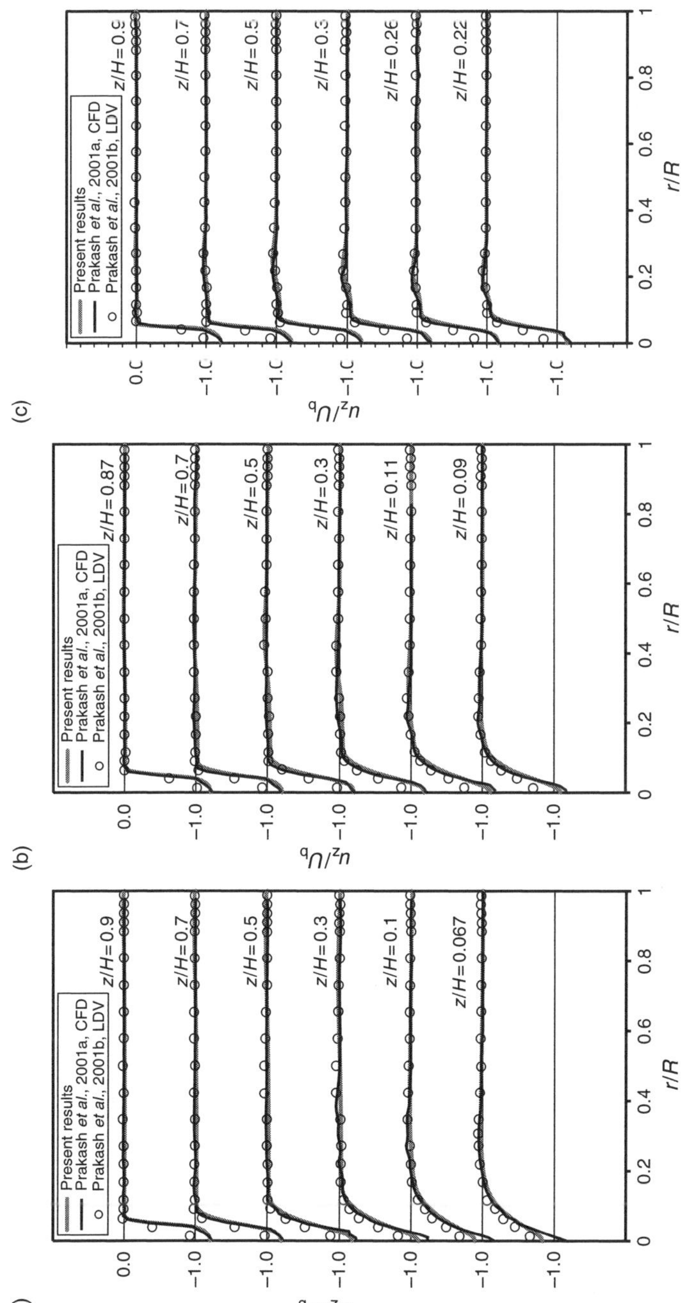


**Figure 9.48.** Axial velocity profiles for $h_p = 0.05\,\mathrm{m}$ on porous foam G10: (a) $H = 0.15\,\mathrm{m}$, (b) $H = 0.10\,\mathrm{m}$, (c) $H = 0.05\,\mathrm{m}$.

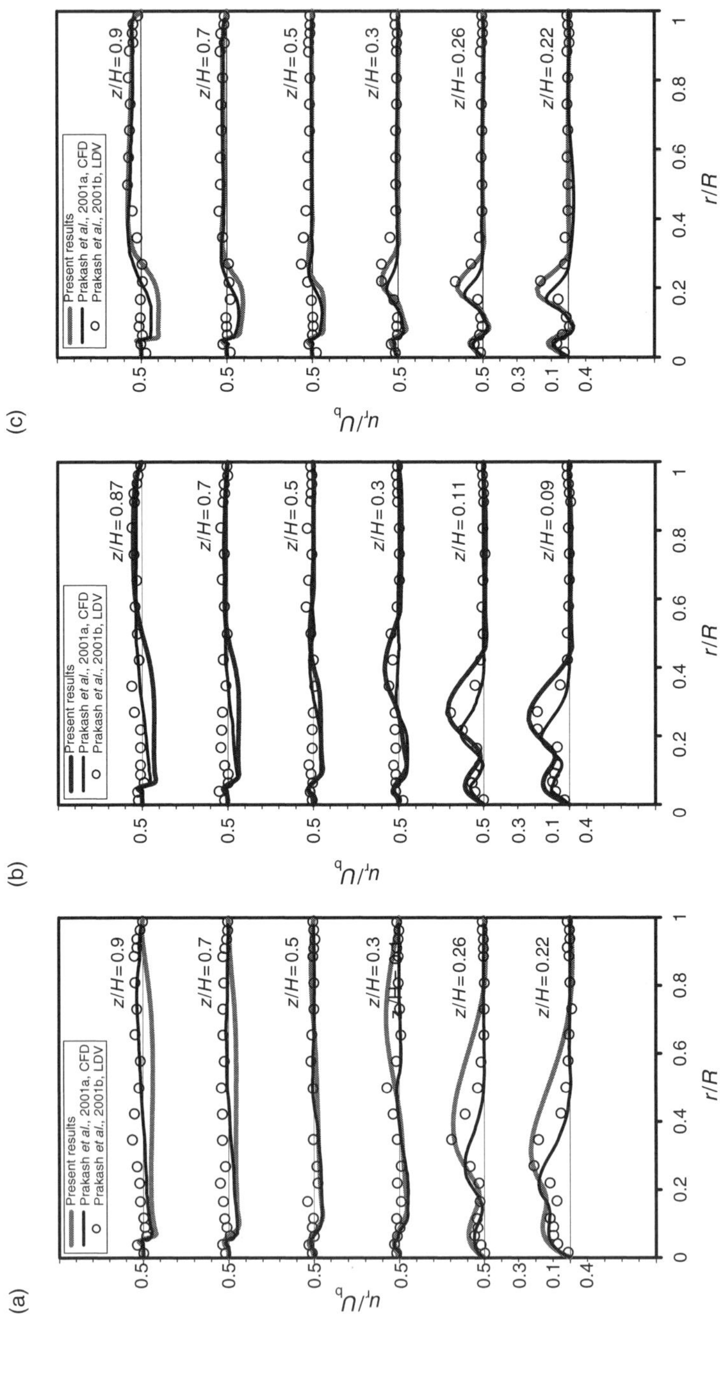


**Figure 9.49.** Radial velocity profiles for $h_p = 0.05\,\text{m}$ on porous foam G10: (a) $H = 0.15\,\text{m}$, (b) $H = 0.10\,\text{m}$, (c) $H = 0.05\,\text{m}$.

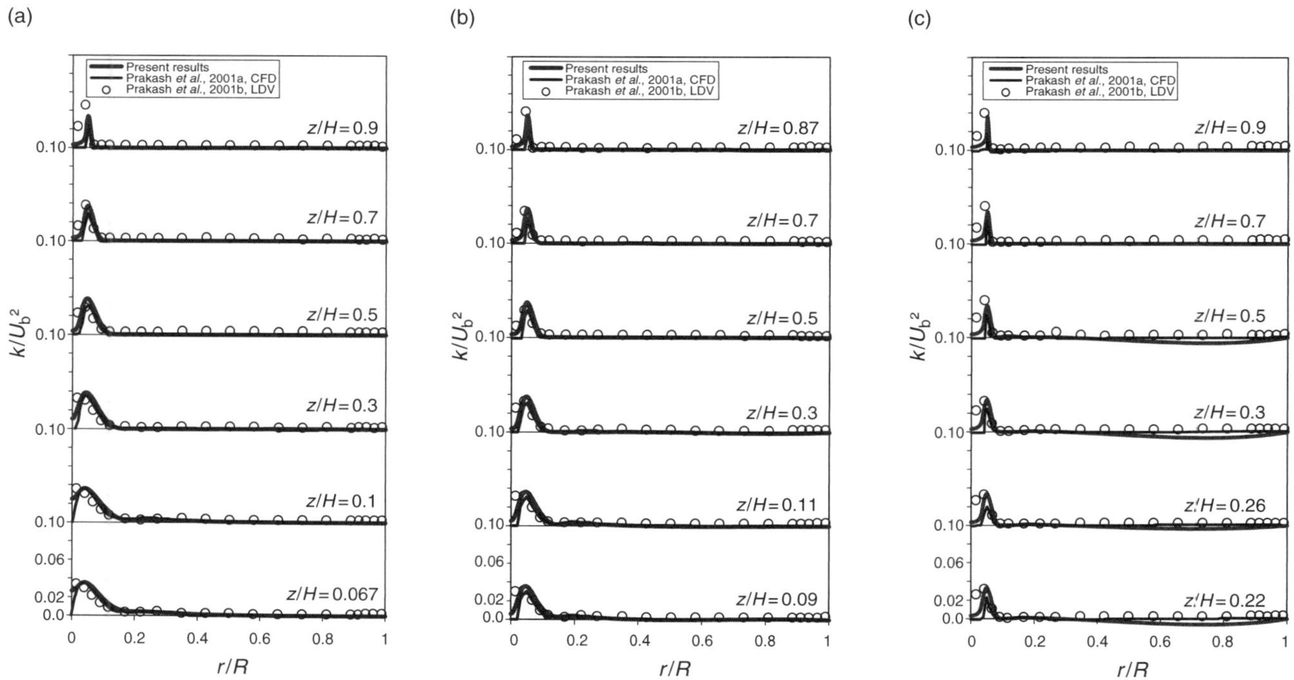


**Figure 9.50.** Turbulence kinetic energy profiles for $h_p = 0.05$ m on porous foam G10: (a) $H = 0.15$ m, (b) $H = 0.10$ m, (c) $H = 0.05$ m.

safety and insulation, ventilating systems, crystal growth in liquids, grain storage and food processing are some examples of applications of heat removal or addition by free convection mechanism. This section discusses some numerical results in cavities totally and partially filled with solid and porous material. The interested reader is referrred to the monograph of Nield and Bejan (1999) and edited book by Ingham and Pop (1998) for an in-dept documentation on natural convection in porous media.

### 9.4.1. Cavities Partially Filled with Vertical Layer of Porous material

Computations detailed in de Lemos and Magro (2003a) are reviewed in this section.

As mentioned in this book, recent works found in the literature propose a macroscopic treatment of the properties of interest, integrating these quantities in a representative elementary volume so that macroscopic equations for the flow arise (Antohe and Lage, 1997; Pedras and de Lemos, 2001a). With respect to cavity flows in clear and in porous media subjected to a temperature gradient across the layer, the literature is vast and a great number of solutions can be found. Examples for studies involving clear cavities can be found in de Vahl Davis (1983) and de Lemos (2000). For porous enclosures, the investigation by Charrier-Mojtabi (1997) shows analytical and numerical results. Also for porous cavities, the work of Braga and de Lemos (2002a) presents results for laminar convection in square cavities heated on the sides. Later, Braga and de Lemos (2002b) extended their results to horizontal annuli. Turbulent flow in eccentric and concentric annuli was also investigated (Braga and de Lemos, 2002c). In addition, a study on natural convection in cavities completely filled with porous material was presented in Braga and de Lemos (2002d). In the latter, the two geometries previously analyzed, namely square and annular region, were considered. More recently, Braga and de Lemos (2002e) presented results for laminar and turbulent flow in square cavity for clear and porous media. They employed the standard $k$–$\varepsilon$ turbulence model with wall function.

In all the results mentioned above, the cavity was either totally clear or totally filled with a porous substrate. In de Lemos and Magro (2003a), the considered problem is shown schematically in Figure 9.51. Figure 9.52 shows the corresponding computational grid. There, the situation investigated treats two-dimensional flow of an incompressible fluid in a square cavity of height $H$ and width $L$, partially filled with porous material. Further, for the cavity of the illustration one considers constant temperatures on the left face, $T_H$, and on the right, $T_C$, $T_H$ being greater than $T_C$. The other two walls are maintained isolated. Under such boundary conditions, the cavity is said to be of vertical type (in the present context, a vertical cavity refers to flow of heat in the horizontal direction).

In Magro and de Lemos (2002a), the heat transfer in the vertical cavity of Figure 9.51 was investigated. In that work, the effect of the Rayleigh number and the treatment of the interface, located at $x = L/2$, were the objective of the analysis. There, the interface treatment used was the one proposed in Ochoa-Tapia and Whitaker (1995a). Later, in

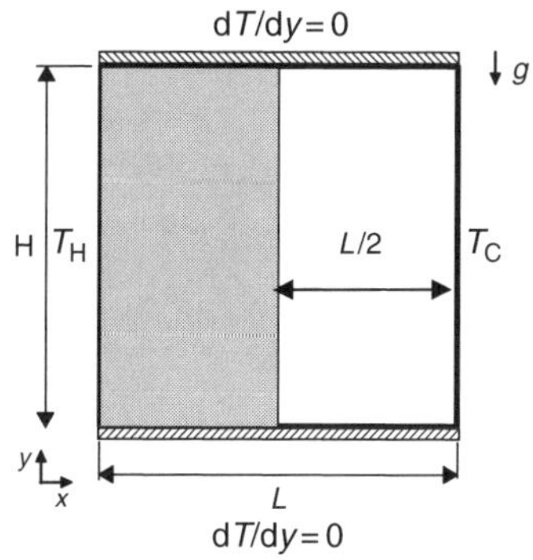


**Figure 9.51.** Vertical cavity partially filled with porous material.

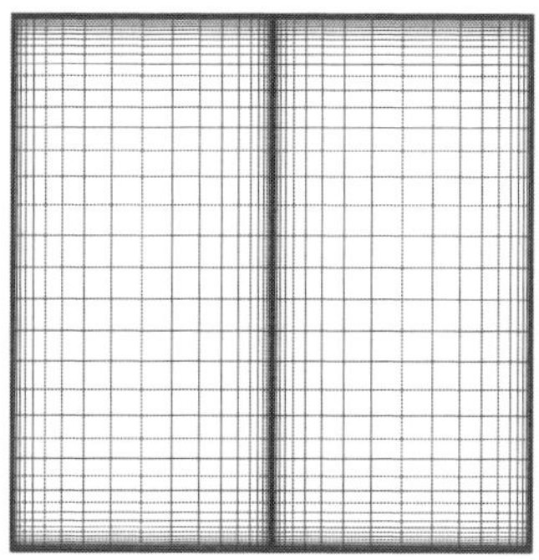

**Figure 9.52.** Computational grid.

Magro and de Lemos (2002b) the previous investigation was complemented, taken then into account the effects of porosity and permeability of the porous region. In both works, the analysis was made for flow in laminar regime.

As mentioned, the objective of this section is to dicuss results in de Lemos and Magro (2003a), which extend simulations in Magro and de Lemos (2002a,b) considering then turbulent flow in the cavity of Figure 9.51. Additional studies on turbulent flow in vertical cavities have also been documented (Magro *et al.*, 2003). Here, results are reviewed for the streamlines as well as for velocity and temperature fields in the geometry of Figure 9.51.

Figure 9.53 displays the effect of the Rayleigh number on the hydrodynamic field for both clear and porous areas. It is noticed that for low Rayleigh numbers, the low intensity of the buoyancy forces provokes flow just in the unobstructed regions, at the right of the cavity. Starting from $Ra > 10^6$ (Figure 9.53d), the flow begins to penetrate into the porous matrix, becoming more intense as $Ra$ increases. For $Ra > 10^7$, the existence of

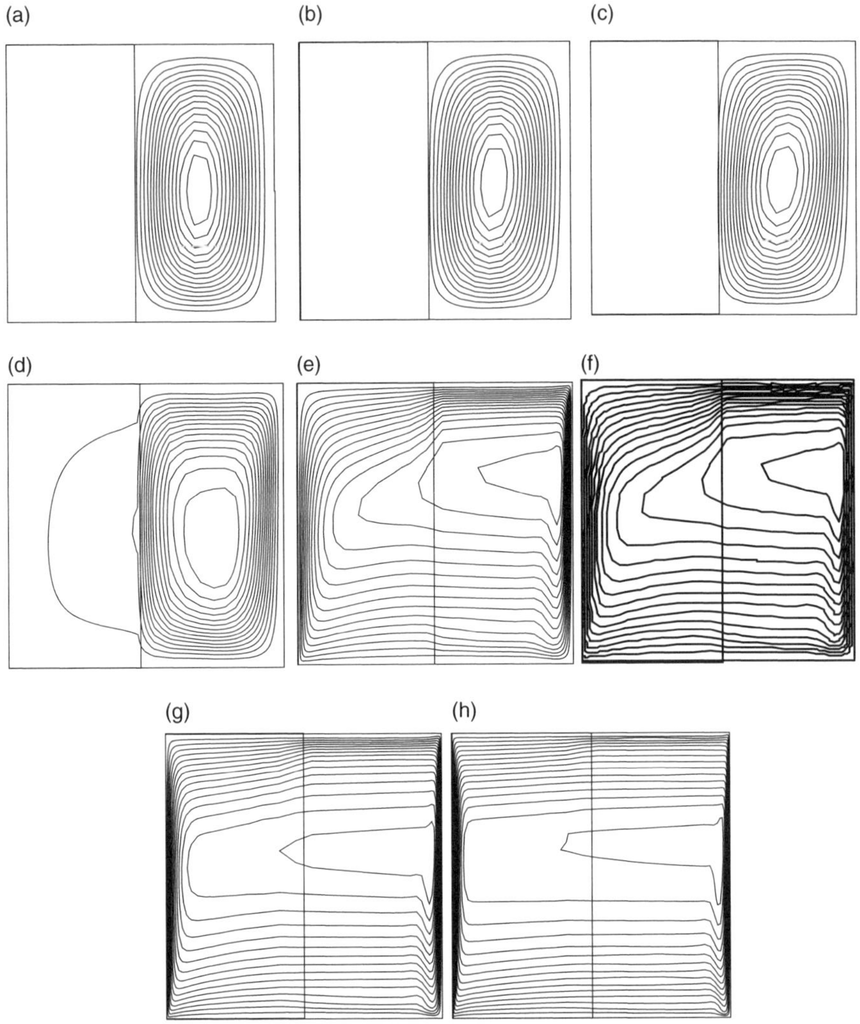


**Figure 9.53.** Effect of *Ra* on streamlines, $\phi = 0.8$, $K = 8.88 \times 10^{-6}$ m$^2$: (a) $Ra = 10^3$, (b) $Ra = 10^4$, (c) $Ra = 10^5$, (d) $Ra = 10^6$, (e) $Ra = 10^7$, (f) $Ra = 10^8$, (g) $Ra = 10^9$, (h) $Ra = 10^{10}$.

boundary layers close to the right face (clear medium) and at the left wall becomes clear. For $Ra = 10^{10}$, although the center of the recirculating flow is still in the clear region, there is an appreciable current convected through the porous matrix.

Corresponding temperature fields and vector plots are shown in Figure 9.54. Figure 9.54a indicates that for low *Ra* numbers, the predominant heat transfer mechanism across the cavity is conduction and that, for Rayleigh greater than about $10^7$, stratification

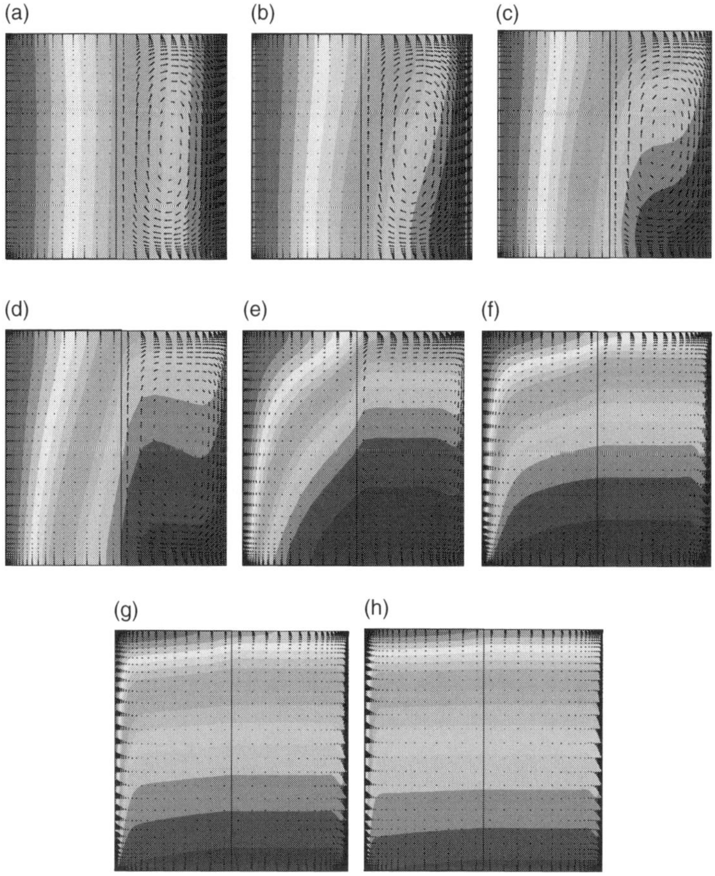


**Figure 9.54.** Effect of $Ra$ on temperature field for $\phi = 0.8$ and $K = 8.88 \times 10^{-6}\,\mathrm{m}^2$: (a) $Ra = 10^3$, (b) $Ra = 10^4$, (c) $Ra = 10^5$, (d) $Ra = 10^6$, (e) $Ra = 10^7$, (f) $Ra = 10^8$, (g) $Ra = 10^9$, (h) $Ra = 10^{10}$. Value range: Left wall: $T_{\mathrm{H}} = 1$; Right wall: $T_{\mathrm{C}} = 0$.

of the thermal field begins to take place also within the porous material. For $Ra = 10^{10}$, the thermal field presents the stratified behavior and a fine boundary layer appears along both lateral faces. The evolution of these boundary layers along the lateral walls can be better observed in Figure 9.54. Starting at $Ra = 10^6$, the convective currents penetrate more intensely into the porous bed and returns along the left wall. The center of the recirculating bubble moves to the left and the velocities in the central area reduced, a larger fraction of the mass flow being concentrated in the boundary layers along the vertical walls.

This can be better visualized in Figure 9.55, which presents horizontal profiles for the vertical velocity component taken at the cavity mid-height ($y = H/2$). For low $Ra$ cases

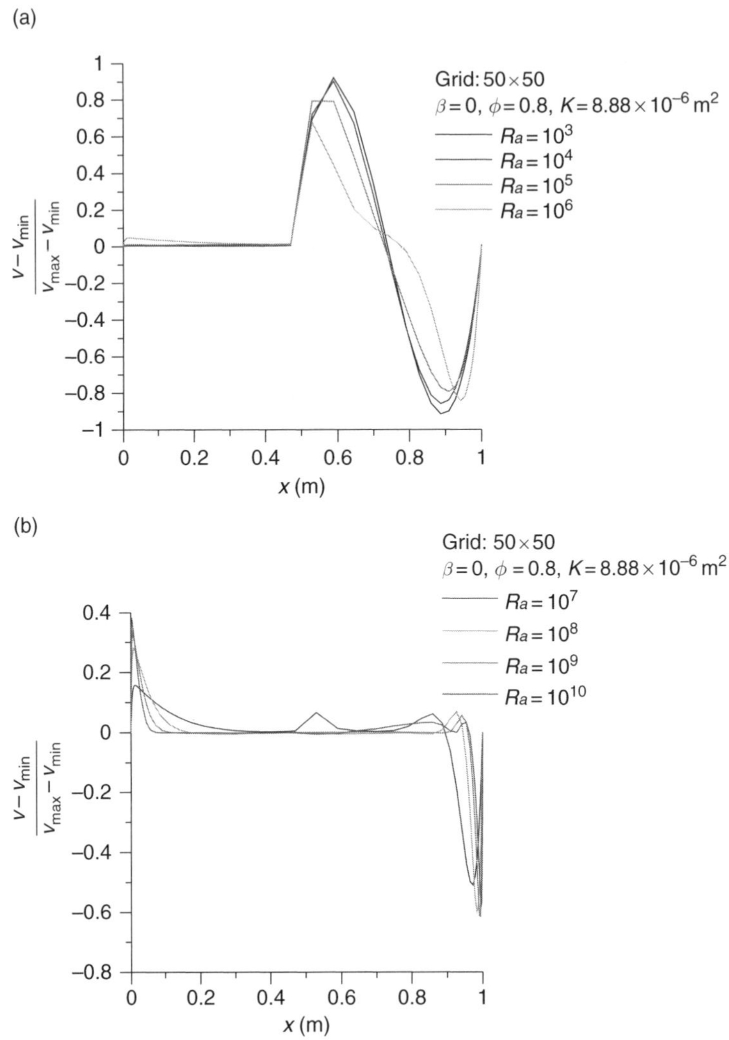


**Figure 9.55.** Effect of *Ra* on vertical velocity at $y = H/2$: (a) $10^3 < Ra < 10^6$, (b) $10^7 < Ra < 10^{10}$.

(Figure 9.55a) most of the flow is concentrated within the void space (right) of the cavity. As *Ra* increases (Figure 9.55b), the flow penetrates the porous matrix and the boundary layers along both vertical walls can be clearly seen.

Table 9.9 presents values for the Nusselt number calculated by:

$$\overline{Nu} = \frac{1}{H}\int_0^H Nu\,\mathrm{d}y \tag{9.57}$$

**Table 9.9.** Nusselt numbers for vertical cavities partially filled with porous material.

| $\phi$ \ $Ra$ | $10^7$ | $10^8$ | $10^9$ | $10^{10}$ |
|---|---|---|---|---|
| 0.5 | 1.69 | 2.901 | 14.11 | 53.97 |
| 0.8 | 5.39 | 20.18 | 51.10 | 113.31 |

where

$$Nu = \left.\frac{\partial T}{\partial x}\right|_{x=0} \frac{L}{T_\mathrm{H} - T_\mathrm{C}} \tag{9.58}$$

is taken along the wall where the higher temperature $T_\mathrm{H}$ prevails. For both porosities considered, there is an increase on $Nu$ as $Ra$ increases, reflecting the enhancement of the convective currents as $Ra$ attains higher values. For a given $Ra$, the same effect seems to occur on $Nu$ as the porosity $\phi$ increases.

### 9.4.2. Cavities Partially Filled with Horizontal Layer of Porous material

In this section numerical simulation by de Lemos and Magro (2003b) are reviewed. Additional results for buoyant flows in cavities containing a small horizontal layer of porous material can be found in Braga and de Lemos (2003a).

The considered cavity in de Lemos and Magro (2003b) is shown schematically in Figure 9.56, with corresponding computational grid in Figure 9.57. There, the authors simulated two-dimensional flow of an incompressible fluid in a square cavity of height $H$ and width $L$, partially filled with porous material. Differently from the previous section,

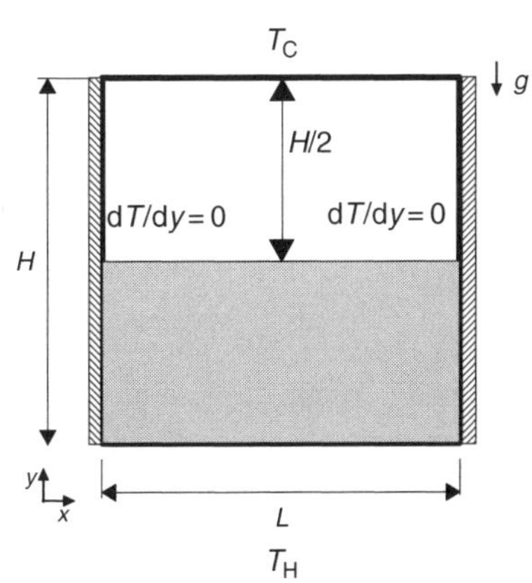


**Figure 9.56.** Horizontal cavity with a layer of porous material at the bottom.

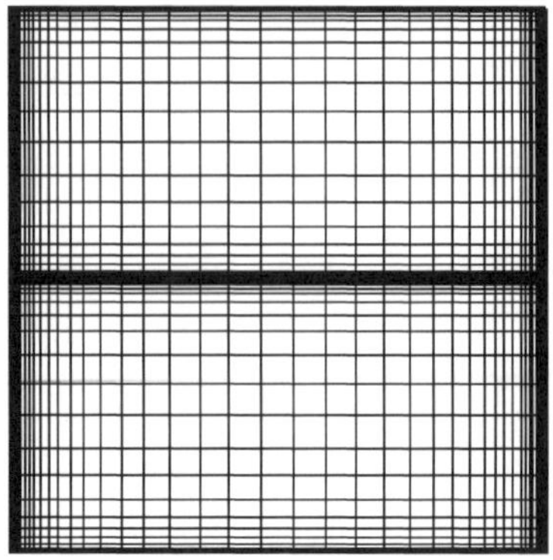

**Figure 9.57.** Computational grid.

the porous substrate is now positioned horizontally, at the bottom of the cavity, which is kept with constant temperatures on the bottom face, $T_H$, and on the top, $T_C$, being $T_H > T_C$. The vertical walls are maintained isolated and, according to our nomenclature, such configuration yields a *horizontal* cavity problem (vertical heat flux).

### *9.4.2.1. Numerical Method*

Governing equations seen on previous Chapters were discretized in the computational mesh shown in Figure 9.57. This mesh concentrates nodal points close to the four walls and around the interface. Grid-independence studies were previously conducted for checking the correctness of the solution (Braga and de Lemos, 2002a–e).

The numerical method used in the resolution of the equations above was the Finite Volumes technique and the SIMPLE algorithm of Patankar (1980). The interface is positioned to coincide with the border between two control volumes, generating, in such a way, only volumes of the types "totally porous" or "totally clear". The flow and energy equations are resolved on both sides of the interface, namely in the porous and in the clear domain, along with the interface conditions given by Equations (9.37) through (9.38).

### *9.4.2.2. Numerical Results*

In this section, results are presented for the streamlines as well as for velocity and temperature fields in the geometry of Figure 9.56. Only numerical simulations are presented since no experimental results could be found in the literature.

Figure 9.58 shows the effect of porosity on the hydrodynamic field for both porous and clear regions. One can see that for low porosity, the numerical solution indicates four recirculation bubbles above the porous layer. For porosity between 0.3 and 0.5, the

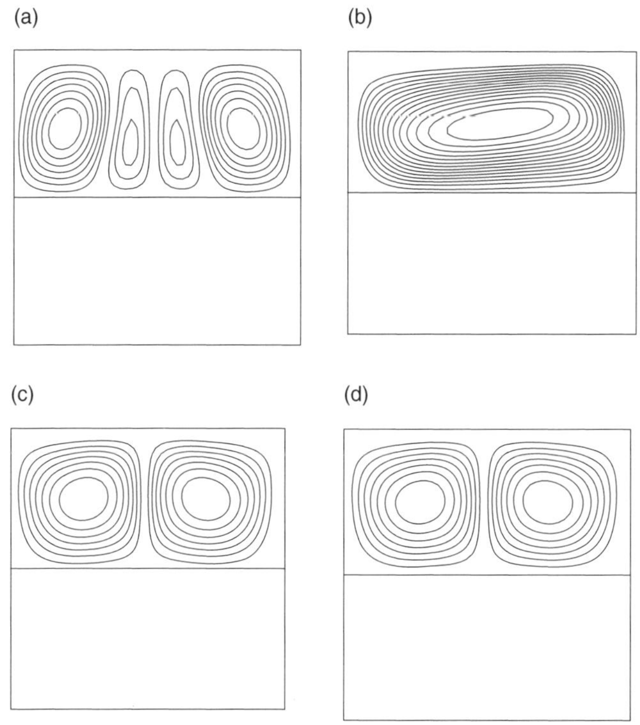


**Figure 9.58.** Effect of porosity $\phi$ on streamlines, grid $50 \times 50$, $Ra = 10^6$, $\beta_i = 0$, $K = 4 \times 10^{-5}\,\mathrm{m}^2$: (a) $\phi = 0.3$, (b) $\phi = 0.4$, (c) $\phi = 0.5$, (d) $\phi = 0.8$.

flow pattern changes drastically, going through a single recirculation structure to two vortices. In order to make this transition in the flow structure more clear, Figures 9.59, 9.60 and 9.61 show details of the vertical velocity component located horizontally at the interface. A sensitivity effect on the values of $K$ can be noted when comparing Figures 9.59 and 9.60.

Corresponding temperature fields are shown in Figure 9.62 for $Ra = 10^6$. The figure indicates that as $\phi$ increases, the flow develops from a ascending plume emerging into the clear region at the middle of the cavity (Figure 9.62a), changes its main pattern by moving upward along the walls (Figure 9.62b,c), turns into a single recirculation bubble (Figure 9.62d) and, for high porosities, reverts back to the initial flow pattern of an ascending flow at $x/L = 0.5$.

Table 9.10 presents values for the Nusselt number calculated by Equations (9.57) and (9.58) taken along the wall where the higher temperature $T_H$ prevails. For both porosities considered, there is an increase on $Nu$ as $Ra$ increases, reflecting the enhancement of the convective currents as $Ra$ attains higher values. For a given $Ra$, the same effect seems to occur on $Nu$ as the porosity $\phi$ increases.

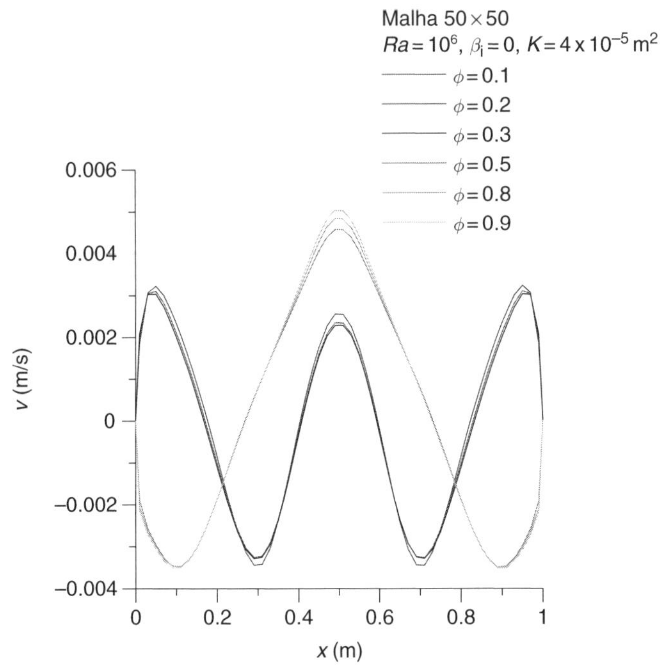


**Figure 9.59.** Effect of porosity $\phi$ on vertical velocity along the interface, grid $50 \times 50$, $Ra = 10^6$, $\beta_i = 0$, $K = 4 \times 10^{-5}\,\text{m}^2$.

#### 9.4.2.3. Section Summary

In this section, numerical results were presented for laminar flows in hybrid domains with heat transfer, which involved an interface between the porous bed and the clear medium. The used numerical method made possible the simultaneous treatment of the porous matrix and of the unobstructed region as a single calculation domain, naturally considering the interface conditions between the two media. It was observed that as the porosity increases, interesting changes in the flow and heat transfer pattern were calculated. The increment on the heat transferred was also observed with the increase of the porosity or the permeability of the material.

### 9.4.3. Fluid–Porous–Solid Systems

When a substance that is a mixture of two or more constituents undergoes a phase change from liquid to solid, or vice-versa, a partially solidified region often forms. In many circumstances, the solid forms a rigid framework with respect to which the liquid may move and that region is called "mushy-zone". An interesting manner to study that region is to consider it as a porous medium. Here, the problem considered is a composite

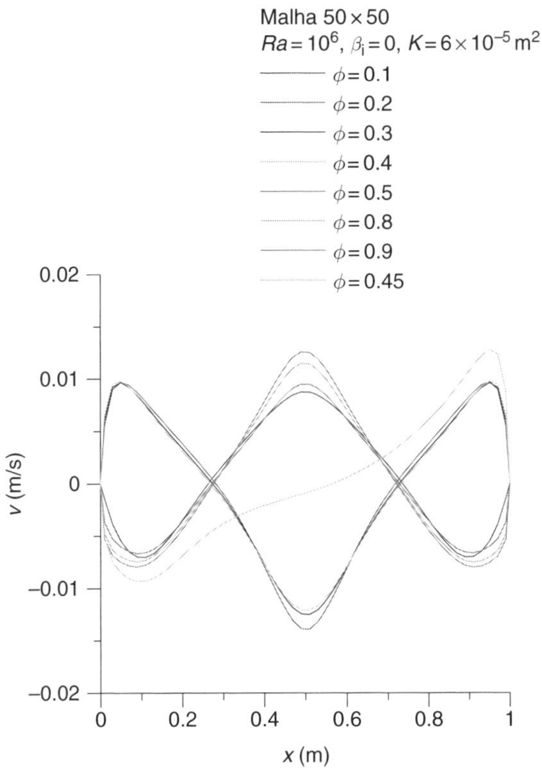


**Figure 9.60.** Effect of porosity $\phi$ on vertical velocity along the interface, grid $50 \times 50$, $Ra = 10^6$, $\beta_i = 0$, $K = 6 \times 10^{-5}\ m^2$.

square cavity formed by three distinct regions, namely, clear, porous and solid region (see Figure 9.63a and Braga and de Lemos, 2003b).

It is important to emphasize that no phase change is considered and the interface between such regions remains immovable. Accordingly, the development of a numerical tool able to treat all these regions as one computational domain is of advantage for engineering design of thermal systems.

The composite cavity is isothermally heated from the left side and cooled from the opposite side. The other two walls are insulated. The porous medium is considered to be rigid and saturated with an incompressible fluid.

Calculations for turbulent flow were performed for all cases using a $120 \times 80$ grid with several points inside the boundary layer as shown in Figure 9.63b. A turbulent case with $160 \times 80$ grid with several points near the walls was performed for $Ra = 10^{10}$. The percent difference for the average Nusselt number on the heated wall between these two meshes is less than 3%.

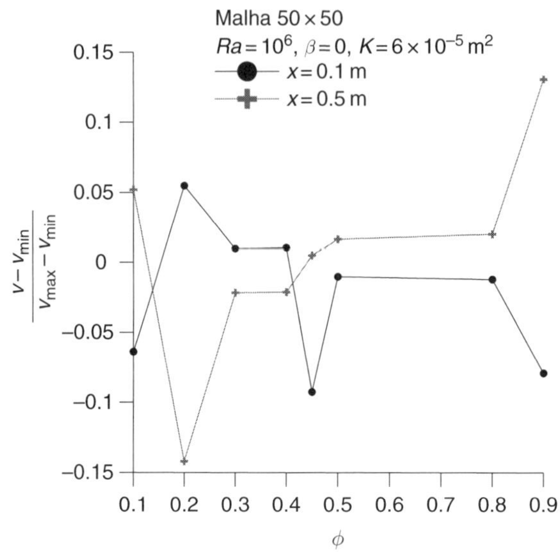


**Figure 9.61.** Effect of porosity on vertical velocity at the interface, grid $50 \times 50$, $K = 6 \times 10^{-5}\ \mathrm{m}^2$.

Figure 9.64 shows streamlines (left) and isotherms (right) for a composite square cavity with *Ra* number ranging from $10^4$ to $10^{10}$ with $\phi = 0.95$, $K = 0.2382 \times 10^{-5}\ \mathrm{m}^2$, $k_s/k_f = 1$, and $Pr = 1$. For the sake of simplicity, all calculations were performed with $\mathbf{K}_{\mathrm{disp}} = 0$.

For $Ra = 10^4$, the streamlines (Figure 9.64a) from a single vortex confined only in the clear region and the flow circulation in the porous medium is almost nonexistent. The isotherms (Figure 9.64b) are almost parallel to the heated wall indicating that the main mechanism of heat transfer is conduction.

Increasing the *Ra* number to $10^6$, the streamlines are now stronger than those for $Ra = 10^4$, but the flow structure still remains mainly in the clear region, Figure 9.64c. The isotherms in the clear region start to be distorted due to the increasing of the natural convection strength. However, in the porous region, the main mechanism of heat transfer is still conduction (Figure 9.64d).

Further increasing *Ra* to $10^8$, the fluid motion in the clear region is now very intense and the center of the single vortex is moved toward the heated wall. In the porous region, the fluid starts to permeate the porous matrix driven by natural convection in that region, Figure 9.64e. The isotherms in the clear region are stratified and the convection mechanism is fully developed in such region. In the porous region, the isotherms become distorted due to the fluid motion pumped by natural convection, Figure 9.64f.

Finally for $Ra = 10^{10}$, the fluid movement is strong in both clear and porous regions where natural convection is developed, Figure 9.64g. The isotherms in both such regions are stratified and heat transfer in the solid occurs under a strong temperature gradient

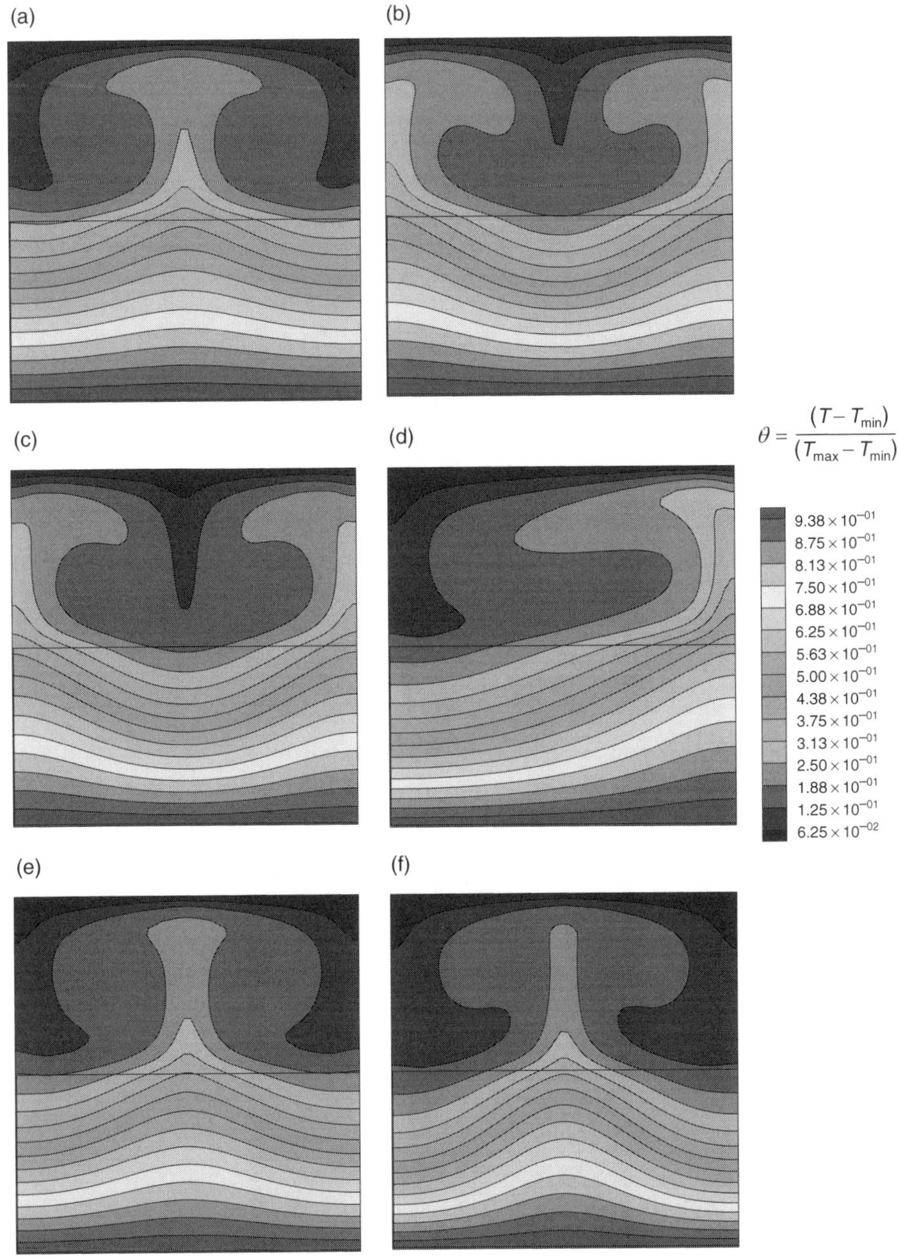


**Figure 9.62.** Effect of porosity on non-dimensional temperature field, grid $50 \times 50$, $\beta_i = 0$, $Ra = 10^6$: (a) $\phi = 0.1$, (b) $\phi = 0.2$, (c) $\phi = 0.4$, (d) $\phi = 0.45$, (e) $\phi = 0.5$, (f) $\phi = 0.9$.

**Table 9.10.** Nusselt number for horizontal cavity partially filled with porous material, grid $50 \times 50$, $\beta_i = 0$, $Ra = 10^6$.

| | $\phi$ | | | | | | |
|---|---|---|---|---|---|---|---|
| $K$ | 0.1 | 0.2 | 0.3 | 0.4 | 0.5 | 0.8 | 0.9 |
| $2 \times 10^{-5}\,m^2$ | 1.28 | 1.31 | 1.34 | 1.37 | 1.39 | 1.53 | 1.56 |
| $4 \times 10^{-5}\,m^2$ | 1.4 | 1.33 | 1.35 | 1.38 | 1.41 | 1.61 | 1.65 |
| $6 \times 10^{-5}\,m^2$ | 1.41 | 1.54 | 1.58 | 1.62 | 1.68 | 1.84 | 1.96 |

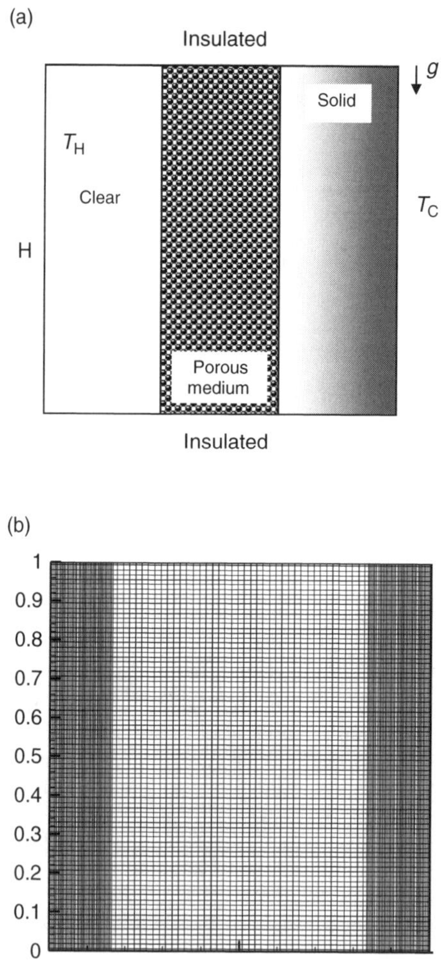


**Figure 9.63.** (a) Geometry under consideration, (b) $120 \times 80$ grid used in calculations for turbulent flow.

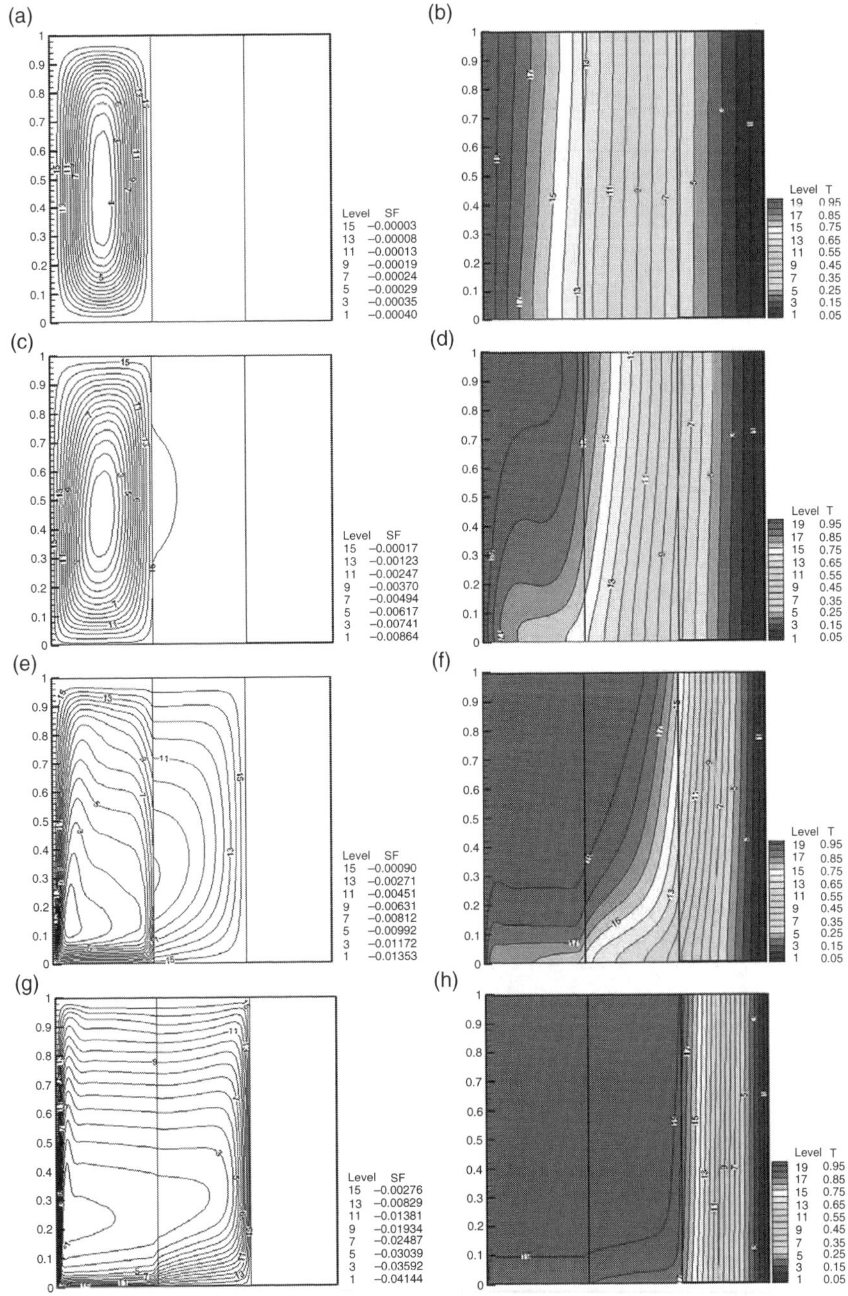


**Figure 9.64.** Turbulent streamlines (left) and isotherms (right) of a composite square cavity for $Ra$ ranging from $10^4$ to $10^{10}$ with $\phi = 0.95$, $K = 0.2382 \times 10^{-5}$ m$^2$, $k_s/k_f = 1$ and $Pr = 1$: $Ra = 10^4$ (a,b), $Ra = 10^6$ (c,d), $Ra = 10^8$ (e,f), $Ra = 10^{10}$ (g,h).

**Table 9.11.** Average Nusselt numbers for 104 < *Ra* < 1010 with $\phi = 0.95$, $K = 0.2382 \times 10^{-5}$, $k_s/k_f = 1$ and $Pr = 1$.

| | *Ra* | | | |
|---|---|---|---|---|
| Model applied | $10^4$ | $10^6$ | $10^8$ | $10^{10}$ |
| Laminar solution | 1.0089 | 1.2752 | 2.1292 | 2.8754 |
| Turbulent solution | 1.0107 | 1.2843 | 2.1249 | 2.8703 |

(Figure 9.64h), indicating that in this case the solid material poses the most heat resistance across the entire cavity.

Table 9.11 shows the average Nusselt number in the heated wall for *Ra* ranging from $10^4$ to $10^{10}$. Table 9.11 shows that for the range of *Ra* analyzed there are no significant variation between the laminar and turbulent model solution.

Figure 9.65 shows the isolines of $\langle k \rangle^i$ for $Ra = 10^{10}$ with $\phi = 0.95$, $K = 0.2382 \times 10^{-5}\,\text{m}^2$, $k_s/k_f = 1$ and $Pr = 1$. The figure indicates two regions where high generation rates of turbulent kinetic energy occur. One of those regions is located between the heated wall and the recirculation vortex that appears in the clear region (see Figure 9.64g). There, velocity gradients close to the wall generate turbulence. The other region is located in the vicinity of the interface between the porous medium and the solid obstacle. In this

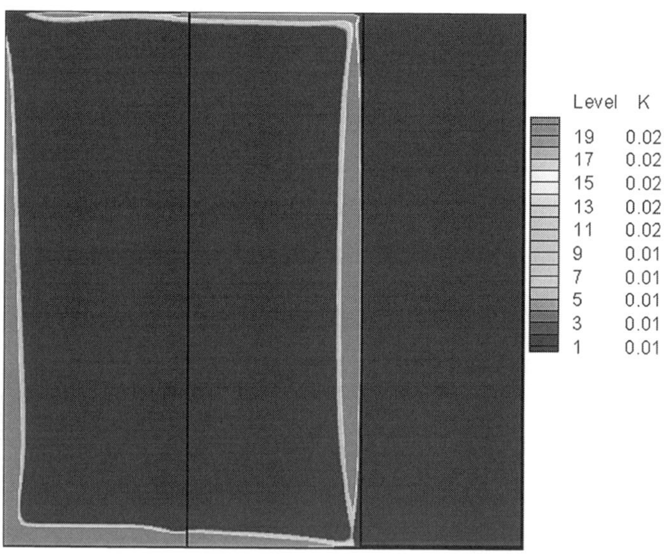


**Figure 9.65.** Turbulent isolines of $\langle k \rangle^i$ of a composite square cavity for $Ra = 10^{10}$ with $\phi = 0.95$, $K = 0.2382 \times 10^{-5}$, $k_s/k_f = 1$ and $Pr = 1$.

second region, generation of $\langle k \rangle^{\mathrm{i}}$ occurs due to velocity gradients and the presence of the porous matrix (Pedras and de Lemos, 2001a).

In fact, as proposed by Pedras and de Lemos (2001a), the porous matrix contributes with the generation of turbulent kinetic energy such that a new term in the $\langle k \rangle^{\mathrm{i}}$ transport equation was introduced. For a fixed value of the Darcy velocity through a porous bed, the amount of mechanical energy converted into turbulence should depend on the medium properties. For the limiting case of high porosity and permeability media ($\phi \to 1 \Rightarrow K \to \infty$) no fraction of this available mechanical energy is expected to generate turbulence. The flow, in this situation, behaves like clear fluid flow. As the flow resistance increases, by increasing $\phi/\sqrt{K}$, gradients of local $\mathbf{u}$ within the pore will contribute to increasing $\langle k \rangle^{\mathrm{i}}$ (see Chapter 4).

### 9.4.4. Cavities Totally Filled with a Porous material

The case of free convection in a rectangular cavity heated on a side and cooled at the opposing side is an important problem in thermal convection in porous media. Walker and Homsy (1978), Bejan (1979), Prasad and Kulacki (1984), Beckermann *et al.* (1986), Gross *et al.* (1986) and Manole and Lage (1992) have contributed with some important results to this problem. The recent work of Baytas and Pop (1999) concerned a numerical study of the steady free-convection flow in rectangular and oblique cavities filled with homogeneous porous media using a non-linear axis transformation. The Darcy momentum and energy equations are solved numerically using the ADI method.

To illustrate some results on cavities totally filled with a permeable medium, simulations by Braga and de Lemos (2004) are reviewed. There, in order to guarantee grid-independent solutions, runs were performed in grids up to $110 \times 110$ control volumes, using stretched meshes for turbulent flow with $Ra_{\mathrm{m}} = 10^6$. The percent difference of the averaged Nusselt number at the hot wall, compared with results obtained with the $80 \times 80$ grid, is 1.15%. Therefore, the $80 \times 80$ stretched mesh seems to be refined enough near to the walls to capture the thin boundary layers that appear along the vertical surfaces.

#### *9.4.4.1. Laminar Model Solution*

Runs for laminar model solution were performed with an $80 \times 80$ control volume in a stretched grid as shown in Figure 9.66b. The present results were performed with $\phi = 0.8$ and the Prandtl number and the conductivity ratio between the solid and fluid phases are assumed to be a unit. The available literature shows that for the non-Darcy region (Merrikh and Mohamad, 2002), the fluid flow and the heat transfer depend on the fluid Rayleigh number, $Ra_{\mathrm{f}}$, and the Darcy number, $Da$, when other parameters (e.g., Porosity, Prandtl number, conductivity ratio between the fluid and solid matrix) are fixed.

Hence, herein, porosity, Prandtl number and conductivity ratio were kept fixed. It is also important to emphasize that all runs were performed without the contribution of

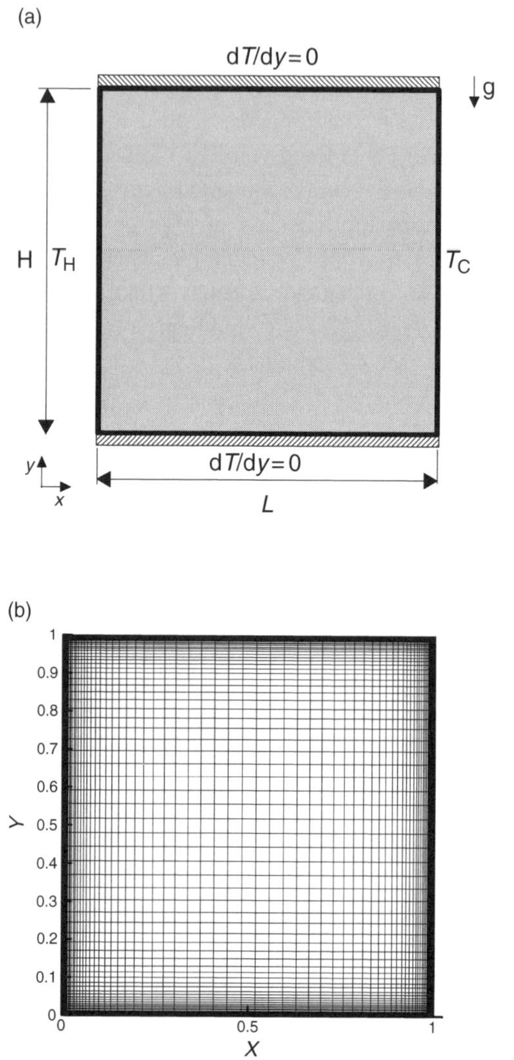


**Figure 9.66.** (a) Geometry under consideration, (b) computational grid.

the thermal dispersion, $\mathbf{K}_{disp}$. However, a few cases considering the effect of thermal dispersion on the Nusselt number were also computed in order to show its influence on the overall heat transport.

Table 9.12 shows some previous laminar numerical results for $Ra_m$ ranging from 10 to $10^4$. Table 9.13 shows the average Nusselt number for different Darcy numbers for $Ra_m$ ranging from 10 to $10^4$. It is clearly seen from the Table 9.13 that for a fixed $Ra_m$ the lower the permeability, the higher the average Nusselt number at the hot wall. It is

**Table 9.12.** Some previous laminar numerical results for average Nusselt number for $Ra_m$ ranging from 10 to $10^4$.

| | $Ra_m$ | | | |
|---|---|---|---|---|
| | 10 | $10^2$ | $10^3$ | $10^4$ |
| Walker and Homsy (1978) | – | 3.097 | 12.96 | 51.0 |
| Bejan (1979) | – | 4.2 | 15.8 | 50.8 |
| Beckermann *et al.* (1986) | – | 3.113 | – | 48.9 |
| Gross *et al.* (1986) | – | 3.141 | 13.448 | 42.583 |
| Manole and Lage (1992) | – | 3.118 | 13.637 | 48.117 |
| Moya *et al.* (1987) | 1.065 | 2.801 | – | – |
| Baytas and Pop (1999) | 1.079 | 3.16 | 14.06 | 48.33 |

**Table 9.13.** Behavior of the average Nusselt number for different values of *Da* for $Ra_m$ ranging from 10 to $10^4$.

| | | $Ra_m$ | | | |
|---|---|---|---|---|---|
| *Da* | $\mathbf{K}_{disp}$ | 10 | $10^2$ | $10^3$ | $10^4$ |
| $10^{-7}$ | $\mathbf{K}_{disp}$ given by (10) | 1.0907 | 3.0866 | 12.9641 | 41.7693 |
| $10^{-7}$ | $\mathbf{K}_{disp}=0$ | 1.0902 | 3.0831 | 12.8930 | 38.6494 |
| $10^{-8}$ | $\mathbf{K}_{disp}=0$ | 1.0908 | 3.0979 | 13.2751 | 43.5799 |
| $10^{-9}$ | $\mathbf{K}_{disp}=0$ | 1.0910 | 3.0985 | 13.3848 | 46.1659 |
| $10^{-10}$ | $\mathbf{K}_{disp}=0$ | 1.0912 | 3.1016 | 13.4289 | 47.2653 |

evident that different combinations of $Ra_f$ and *Da* yield different heat transfer results. The increasing of the fluid Rayleigh number increases the natural convection inside the enclosure. Since $Ra_m$ is fixed, a higher fluid Rayleigh number is associated with a less permeable media (i.e. lower Darcy number). Its also clearly seen from the Table 9.13 that the Nusselt numbers computed with the thermal dispersion are higher than those computed without it for $Da = 10^{-7}$. It seems evident that this additional mechanism increases heat transfer. Table 9.13 also shows that for higher values of $Ra_m$, the effect of the thermal dispersion on the Nusselt number are more pronounced. However, the computational cost due to the inclusion of this mechanism increases significantly (not shown here). In comparison with results of Table 9.12, more accurate simulations were obtained for lower permeability media.

The local Nusselt number on the hot wall for the square cavity at $x = 0$ is defined as:

$$Nu = \frac{hL}{k_{eff}} \quad \therefore \quad Nu = \left(\frac{\partial \langle T \rangle^{v}}{\partial x}\right)_{x=0} \frac{L}{T_H - T_C} \tag{9.59}$$

and the average Nusselt number is given by:

$$\overline{Nu} = \frac{1}{H}\int_0^H Nu\,\mathrm{d}y \tag{9.60}$$

Figure 9.67 shows streamlines and isotherms for a laminar model solution in a square cavity filled with porous medium for $Ra_m$ ranging from $10^3$ to $10^6$. The cavity is heated from the left side and cooled from the opposing side. The other two walls are kept insulated.

For lower Rayleigh number values, $Ra_m \leq 10^2$, not shown here, the isotherms are almost parallel to the heated walls, indicating that the most part of heat transfer is by conduction mechanism, while the streamlines are a single vortex with its center in the center of the square cavity.

At $Ra_m = 10^3$, Figure 9.67b, the streamlines are an elliptic flattened vortex. In contrast with the clear cavity case, the porous matrix makes the flow to be more intense near the heated and cooled walls and damped in the center due to the presence of the porous matrix. Corresponding isotherms are shown in Figure 9.67a. The enhancing of the natural convection begins to distort the isotherms. The vortex is generated due the horizontal temperature gradient across the section. This gradient, $\delta T/\delta y$, is negative everywhere, giving a clockwise vertical rotation.

Increasing $Ra_m$ to $10^4$, Figure 9.67d, the central vortex becomes rectangular and the effect of convection is now more pronounced in the isotherms, as can be seen in Figure 9.67c. The flow pattern comprises a primary cell of relatively high velocity, circulating around the entire cavity. Temperature gradients are stronger near the vertical walls, but decrease in the center.

For higher values of Rayleigh numbers, $Ra_m = 10^5$ and $10^6$, the flow moves faster close to heated walls, Figure 9.67f,h, and the isotherms tend to stratification, Figure 9.67e,g, respectively.

#### 9.4.4.2. *Turbulent Model Solution*

It is important to emphasize that the main objective of this work is not to simulate the transition mechanism from laminar regime to fully turbulent flow, which involves modeling of complex physical processes and hydrodynamic instabilities. The aim of this work is to establish an $Ra_{cr}$ below which both turbulent and laminar models did not differ substantially as far as predictions of overall $Nu$ are of concern. Therefore, a strategy for determining the range of validity of a laminar flow solution was to simulate the *laminarization* of the flow when the Raleigh number is reduced.

For clear flows, when $Ra_f$ is varied, the literature often refers to laminar and turbulent "branches" of solutions as $Ra_f$ passes a critical value. When a turbulence model is included, the turbulent solution can deviate from the laminar branch for $Ra_f > Ra_{cr}$ and follows its

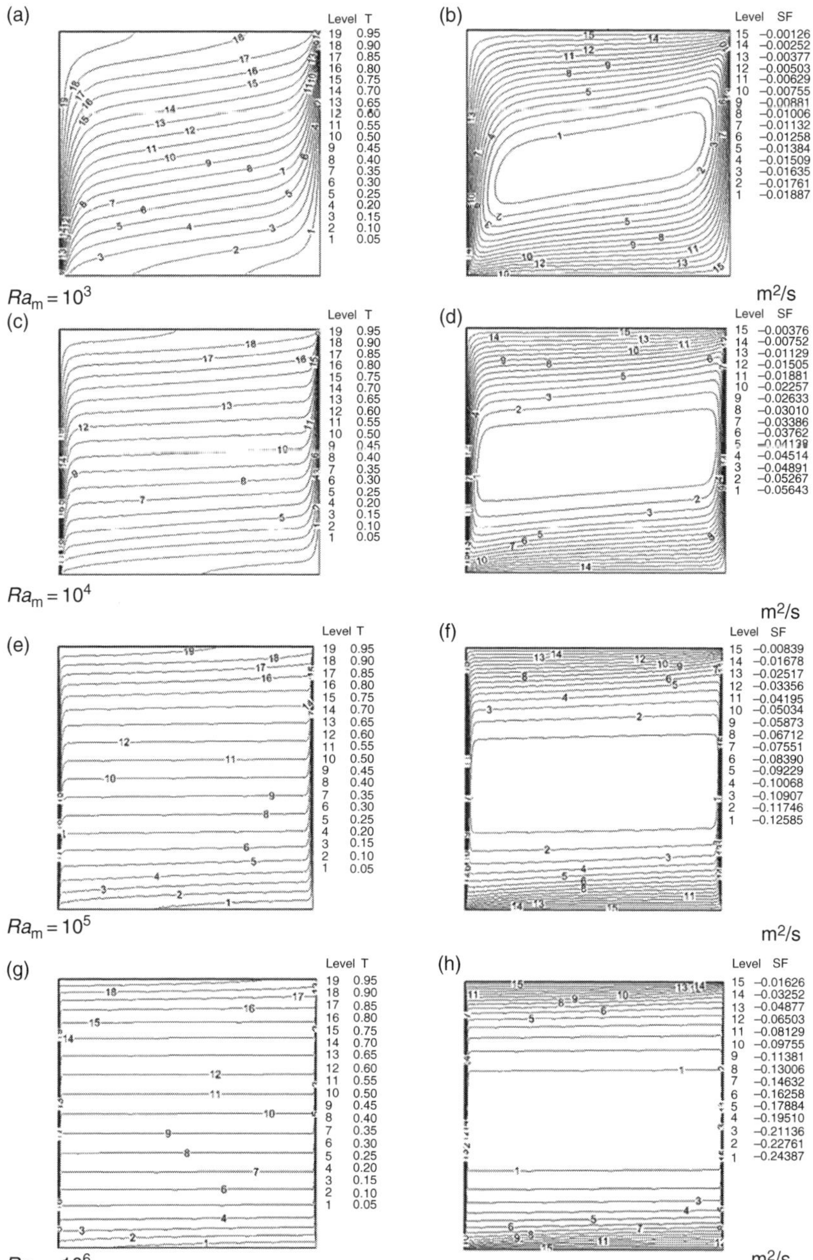


**Figure 9.67.** Isotherms and streamlines for laminar model solution for a square cavity filled with porous material with $\phi = 0.8$, $Da = 10^{-7}$ and $\mathbf{K}_{\text{disp}} = 0$.

turbulent branch. According to Henkes *et al.* (1991), the deviation of the averaged wall-heat transfer between laminar and turbulent fields depends on the turbulence model used.

When the standard $k$–$\varepsilon$ model is used, the laminar solution is not a solution of the equation system, because it does not satisfy the boundary condition, namely wall function, for the kinetic energy at the first inner grid point close to the wall. Below a critical $Ra_f$ number, the standard $k$–$\varepsilon$ model gives a turbulent viscosity close to zero everywhere. This reduction of turbulent transfer can be interpreted as an indication of the laminarization process. However, above this critical value, the turbulent viscosity suddenly increases and a turbulent solution is obtained.

With these ideas in mind, this part of the work tries to find, for flow in porous media, critical Rayleigh number, $Ra_{cr}$, for which simulations with the turbulence model deviates from those considering laminar flow, as done in the literature for the clear fluid case, as in Henkes *et al.* (1991) and Barakos *et al.* (1994). In order to achieve this goal, the turbulence model is first "switched off" and the laminar branch of the solution is found when increasing the Rayleigh number, $Ra_m$. Subsequently, the turbulence model is included so that the solution merges to the laminar branch for a reducing $Ra_m$ and for $Ra_m < Ra_{cr}$. This convergence of results as $Ra_m$ decreases can be seen to characterize the so-called *laminarization* phenomenon.

Calculations for turbulent model solution were performed with the same grid used for the laminar model solution and the parameters (porosity, Prandtl number and conductivity ratio between the fluid and the solid matrix) are fixed.

Figure 9.68 shows the isotherms and streamlines for turbulent model solution for $Ra_m$ ranging from $10^3$ to $10^6$. For $Ra_m \leq 10^2$, not shown here, the solution with the turbulence model gives nearly the same values as those obtained with laminar flow computations. Even for $Ra_m$ up to $10^6$ the flow patterns resemble those from the laminar model solution, but the values of the streamlines and the average Nusselt numbers at the hot wall are significantly increased.

Table 9.14 shows the average Nusselt number at the hot wall for the two types of regime, namely laminar and turbulent for two distinct Darcy numbers. Table 9.14 shows that the turbulent solution deviates from the laminar one for $Ra_m$ greater than around $10^4$. Consequently, the calculations herein suggest that a critical value for Rayleigh is of the order of $10^4$ and from that value on simulations considering a turbulence model are higher than their laminar counterpart.

Figure 9.69 shows the behavior of the average Nusselt number versus the Rayleigh number for the two models here considered, namely the laminar and the turbulence models for $Da = 10^{-7}$ and $10^{-8}$ illustrating the two regions mentioned above. It is clearly seen from Figure 9.69 that $Ra_{cr}$ is not affected due to small variations on the Darcy number.

In the first region for $Ra_m < Ra_{cr} \sim 10^4$, both laminar and turbulent flow simulations give nearly the same results. After this point, Nusselt numbers calculated with a full turbulence model give higher values for $\overline{Nu}$.

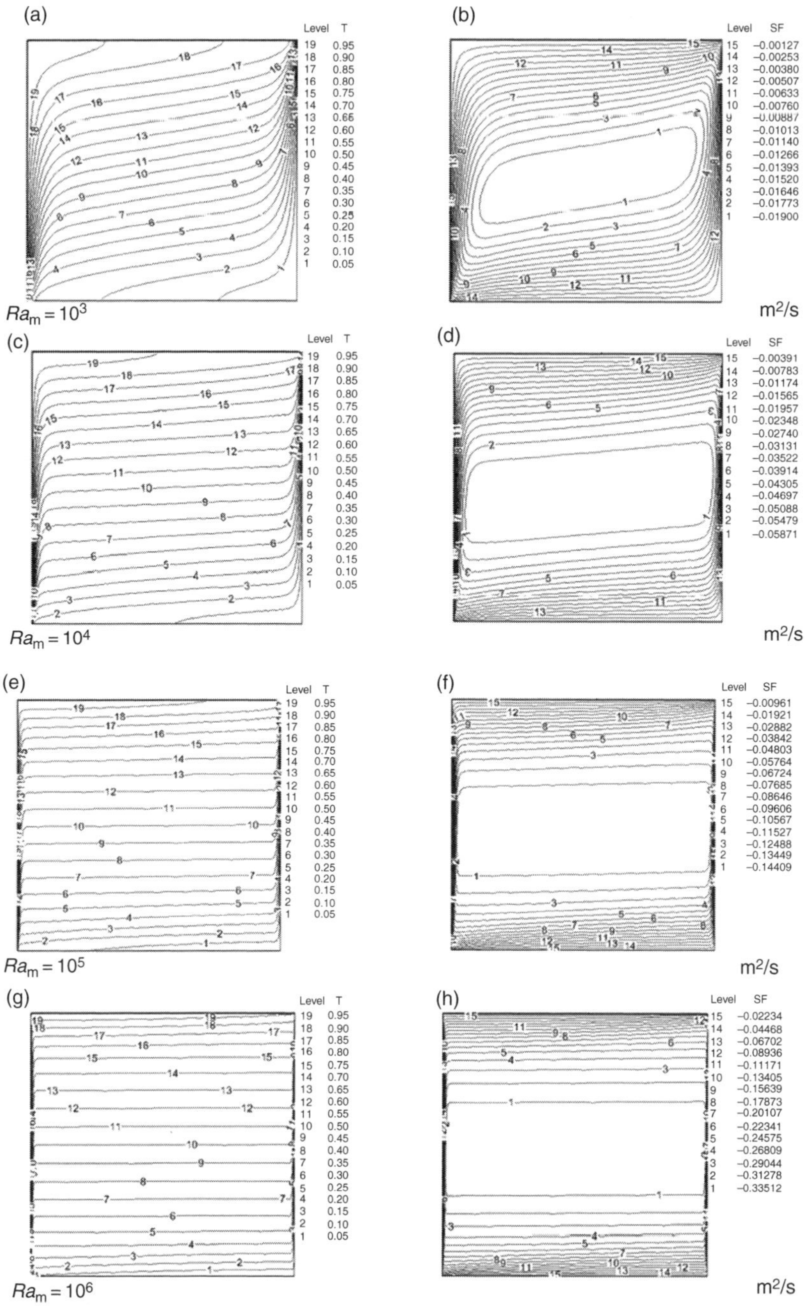


**Figure 9.68.** Isotherms and streamlines for turbulent model solution for a square cavity filled with porous material with $\phi = 0.8$, $Da = 10^{-7}$ and $\mathbf{K}_{\text{disp}} = 0$.

**Table 9.14.** Comparison between laminar and turbulent model solutions for the average Nusselt number at the hot wall for $Da = 10^{-7}$ and $10^{-8}$ and $Ra_m$ ranging from 10 to $10^6$.

| | $Ra_m$ | | | | | |
|---|---|---|---|---|---|---|
| Model solution | 10 | $10^2$ | $10^3$ | $10^4$ | $10^5$ | $10^6$ |
| $Da = 10^{-7}$ with $\mathbf{K}_{disp}$ given by (10) | | | | | | |
| Laminar | 1.0907 | 3.0866 | 12.9641 | 41.7693 | 110.1664 | – |
| Turbulent | 1.0910 | 3.0896 | 13.0652 | 43.2809 | 120.4175 | – |
| $Da = 10^{-7}$ with $\mathbf{K}_{disp} = 0$ | | | | | | |
| Laminar | 1.0902 | 3.0831 | 12.8930 | 38.6494 | 87.3268 | 169.9404 |
| Turbulent | 1.0907 | 3.0860 | 12.9956 | 40.3077 | 100.6035 | 235.5515 |
| $Da = 10^{-8}$ with $\mathbf{K}_{disp} = 0$ | | | | | | |
| Laminar | 1.0908 | 3.0979 | 13.2751 | 43.5799 | 109.1877 | 222.5915 |
| Turbulent | 1.0910 | 3.1006 | 13.3525 | 44.7605 | 119.1966 | 277.0930 |

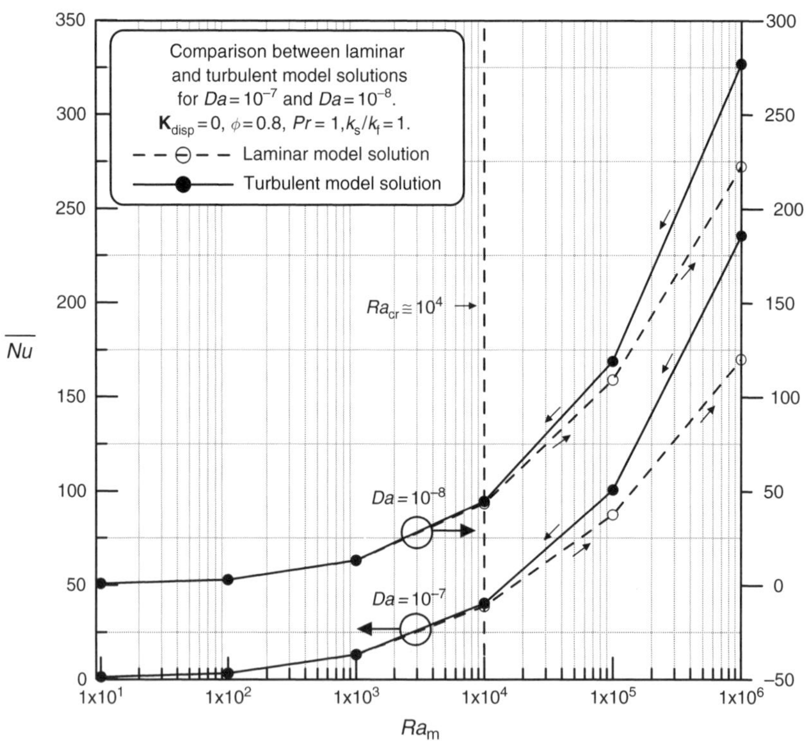


**Figure 9.69.** Comparison between the laminar and turbulent model solutions with the averaged Nusselt number at the hot wall.

#### *9.4.4.3. Section Summary*

This session reviewed computations for laminar and turbulent flows with the macroscopic $k$–$\varepsilon$ model with a wall function for natural convection in a square cavity totally filled with porous material were performed. The cavity was heated from the left and cooled from the opposing side. The numerical values yielded generally satisfactory agreement with similar data available in the literature. This agreement was also found when comparing average Nusselt numbers along the hot wall.

In general, when fluid and medium properties (Prandtl number, porosity and conductivity ratio between the fluid and the solid matrix) are kept fixed and $Ra_{\mathrm{m}}$ is constant, the lower the Darcy number (or media permeability), the higher the average Nusselt number at the hot wall. The increasing of the fluid Rayleigh number increases the natural convection inside the enclosure. Since the $Ra_{\mathrm{m}}$ is fixed, a higher fluid Rayleigh number is associated with a less permeable media (i.e. lower Darcy number). In the end, for $Ra_{\mathrm{m}}$ values greater than around $10^4$, both laminar and turbulent flow solutions deviate from each other, indicating that such critical value for $Ra_{\mathrm{m}}$ was reached. Accordingly, in order to observe that, the turbulence model was first switched off and the laminar branch of the solution was found when increasing the Rayleigh number, $Ra_{\mathrm{m}}$. Subsequently, the turbulence model was included so that the solution merged to the laminar branch for $Ra_{\mathrm{m}} < Ra_{\mathrm{cr}}$. This convergence of results as $Ra_{\mathrm{m}}$ decreases can be seen as an estimate of the so-called *laminarization* phenomenon.

Ultimately, the inclusion of thermal dispersion increases the Nusselt number on the hot wall by a fair amount for higher $Ra_{\mathrm{m}}$, since it represents an additional mechanism of mixing. However, the inclusion of this effect also significantly reduces convergence rates and associated computational cost. Further, $Ra_{\mathrm{cr}}$ is not affected due to the inclusion of the thermal dispersion mechanism, $\mathbf{K}_{\mathrm{disp}}$, or due to small variations on the Darcy number.

### 9.4.5. Heterogeneous vs. Homogenous Systems

#### *9.4.5.1. The Problem Considered*

This section reviews some heat transfer calculations in cavities containing a fixed amount of solid material, which were detailed in Braga and de Lemos (2005). While maintaining the same overall volume, the morphology, shape and distribution of the solid phase within the cavity may differ from case to case. If one associates a permeability to such systems, its value will be different depending on how easy the fluid is able to flow through the solid matrix within the cavity. Also, different fluids are characterized by distinct properties such as the thermal expansion coefficient $\beta$. Figure 9.70a illustrates situations considering cavities saturated with distinct fluids (different $\beta$) with the solid material of different morphology, size and distribution (distinct permeability).

To analyze such an arrangement, the *continuum* model is here employed, in which the flow equations are solved within the void (fluid) space. More specifically, the problem here

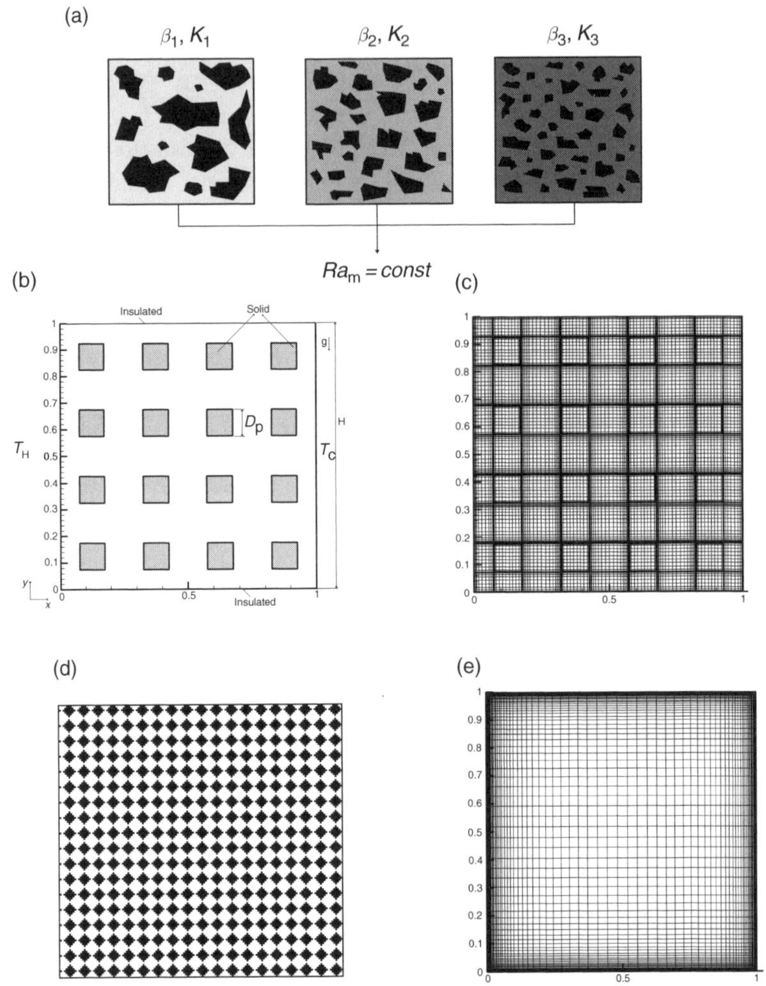


**Figure 9.70.** Physical systems: cavities with different fluids in distinct media (a); continuum model: cavities with distributed solid material (b) and corresponding grid (c); porous continuum model: porous cavity (d) and corresponding grid (e).

investigated is schematically presented in Figure 9.70b. The corresponding computational grid is presented in Figure 9.70c and refers to a square cavity of side $H = 1\,\text{m}$. In the continuum model, the cavity is partially filled with a fixed amount of solid conducting material, in the form of square obstacles of size $D_p$ that is equally distributed within the cavity. Also, the cavity is isothermally heated from the left, with temperature $T_H$ prevailing over that side, and cooled from the opposing surface, where a constant temperature $T_C$ is maintained. The horizontal walls are kept insulated.

On the other hand, the same physical system of Figure 9.70a was also treated as a permeable structure (Figure 9.70d), having the same void to void-plus-solid ratio, or porosity $\phi$. In this case, the *porous-continuum* model was applied, in which the flow properties are integrated over an REV. The corresponding computational grid is presented in Figure 9.70e.

Here, the coefficient $\beta_\phi$ is an effective macroscopic or volume averaged thermal expansion coefficient defined by de Lemos and Braga (2003) and Braga and de Lemos (2004). Note that $\beta_\phi$ is not necessarily equal to the thermal expansion coefficient $\beta$. Assuming that gravity is constant over the REV, an expression for it based on (5.90) is given in de Lemos and Braga (2003) as:

$$\beta_\phi = \frac{\langle \rho\beta(T - T_{\text{ref}})\rangle^{\text{v}}}{\rho\phi(\langle T\rangle^{\text{i}} - T_{\text{ref}})} \tag{9.61}$$

It is important to emphasize that when $K \to \infty$ and $\phi = 1$ the equation set above resembles the one applied in the cases of clear (unobstructed) flows.

### *9.4.5.2. Nondimensional Parameter*

For the systems in Figure 9.70a, here also modeled with the geometry and grid of Figures 9.17d,e, the literature has defined a modified Rayleigh number $Ra_{\text{m}}$ in the form (Walker and Homsy, 1978; Bejan, 1979; Beckermann *et al.*, 1986; Gross *et al.*, 1986; Moya *et al.*, 1987; Manole and Lage, 1992; Baytas and Pop, 1999; Merrikh and Mohamad, 2001, 2002; Merrikh *et al.*, 2002; Massarotti *et al.*, 2003; Merrikh and Lage, 2004, 2005; Merrikh *et al.*, 2005):

$$Ra_{\text{m}} = RaDa \tag{9.62}$$

with

$$Ra = \frac{g\beta H^3 \Delta T}{\nu_{\text{f}}\alpha_{\text{eff}}} \tag{9.63}$$

$$Da = \frac{K}{H^2} \tag{9.64}$$

where $\alpha_{\text{eff}} = k_{\text{eff}}/(\rho c_{\text{p}})_{\text{f}}$ and $H$ is the size of the square cavity. One can note that $Ra$, as defined by Equation (9.63), is similar to the parameter used in clear (unobstructed) heated cavities when $k_{\text{eff}}$ is equal to the fluid thermal conductivity. On the other hand, in Equation (9.64) the attribution of a permeability $K$ to a porous structure is common path followed in porous media analysis. Equation (9.62) then involves *continuum* properties such as the thermal expansion coefficient for fluid phase $\beta$ and *porous-continuum* parameters such as permeability $K$. Here, a different Rayleigh number is associated with heated porous cavities, being defined as:

$$Ra_{\phi} = \frac{g\beta_{\phi} H^3 \Delta T}{\nu_f \alpha_{eff}} \tag{9.65}$$

where $\beta_{\phi}$ is defined in Equation (9.61).

Furthermore, in order to associate a value for an "equivalent" permeability of the arrangement in Figure 9.17b, for comparisons with the *porous-continuum* model, the correlation of Nakayama and Kuwahara (1999) was applied. That correlation is based on the work of Ergun (1952) and reads:

$$K_{eq} = \frac{D_p^2 \phi^3}{120(1-\phi)^2} \tag{9.66}$$

where $D_p$, as seen, is the size of the square rods. It is important to note that $K_{eq}$ given by Equation (1.9) was proposed for forced flows thorough permeable media and that use of such correlation in buoyancy-driven flows might be questionable. Nevertheless, in the absence of better information, this work associates a permeability $K_{eq}$ to the system in Figure 9.17b using Equation (1.9). This equivalent permeability is used to form an "equivalent" Darcy number:

$$Da_{eq} = \frac{K_{eq}}{H^2} \tag{9.67}$$

As such, this work is based on the hypothesis that both systems in Figure 9.17b,d can be compared if the same $Ra_m$ is applied, or say, $Ra_m = RaDa_{eq}$ characterizing Figure 9.17b was equal to $Ra_m = Ra_{\phi}Da$ describing Figure 9.17d. This imposed condition is therefore:

$$\begin{aligned} Ra_m = Ra \cdot Da_{eq} &= \left.\frac{g\beta H \Delta T K_{eq}}{\nu_f \alpha}\right\} \text{Continuum model – Figure 9.70b} \\ = Ra_{\phi} \cdot Da &= \left.\frac{g\beta_{\phi} H \Delta T K}{\nu_f \alpha_{eff}}\right\} \text{Porous-continuum model – Figure 9.70d} \end{aligned} \tag{9.68}$$

Also, if values for $Ra$ or $Ra_{\phi}$ and $Da$ or $Da_{eq}$ are selected while $Ra_m$ is kept constant, a family of curves is obtained as schematically shown in Figure 9.71. Each curve in the figure represents distinct systems consisting of possibly different fluids and solid distribution, but all having the same modified Rayleigh number $Ra_m$ (see illustration in Figure 9.70a). Considering such premise, the present work intends to study a family of cases with different $Ra$ in distinct media (different $Da$), having all of them the same $Ra_m = 10^4$.

The local Nusselt number on the hot wall for the square cavity at $x = 0$ is defined as:

$$Nu_y = hH/k_{eff} \quad \therefore \quad Nu_y = \left(\frac{\partial \langle T \rangle^v}{\partial x}\right)_{x=0} \frac{H}{T_H - T_C} \tag{9.69}$$

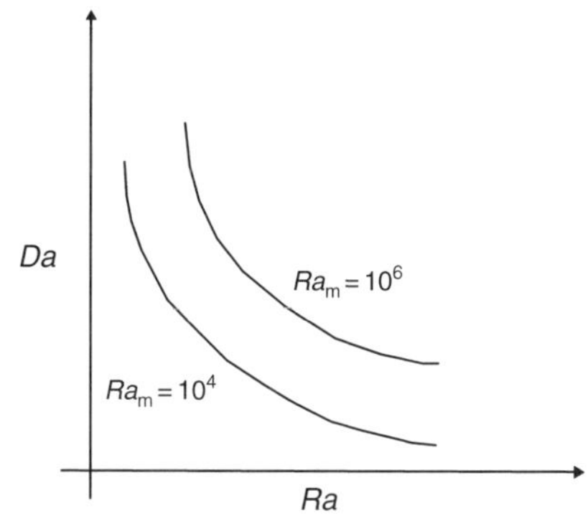


**Figure 9.71.** Family of curves for cavities of Figure 9.70a with different fluids in media having the solid phase distributed in different forms and $Ra_m = \text{const}$.

and the average Nusselt number for the *continuum* model (see Figure 9.70b,c) is given by,

$$Nu = \frac{1}{H}\int_0^H Nu_y \, dy \tag{9.70}$$

For porous cavities computed with the *porous-continuum* model (see Figure 9.70d,e), the Nusselt number is here given the symbol $Nu_\phi$.

### 9.4.5.3. Numerical Method and Solution Procedure

The numerical method employed for discretizing the governing equations is the control-volume approach with a generalized grid. The flux blended deferred correction, which combines linearly the Upwind Differencing Scheme (UDS) and Central Differencing Scheme (CDS), was used for interpolating the convective fluxes. The well-established SIMPLE algorithm (Patankar, 1980) is followed for handling the pressure-velocity coupling. Individual algebraic equation sets were solved by the SIP procedure (see Stone, 1968, for details). Further, concentration of nodal points to walls reduces eventual errors due to numerical diffusion which, in turn, are further annihilated due to the hybrid scheme here adopted.

### 9.4.5.4. Results and Discussion

*Grid-Independence Studies* In order to guarantee grid-independent solutions, runs for *porous-continuum* model were also performed with $110 \times 110$ control volumes, in addition to $80 \times 80$ mesh, in a stretched grid for $Ra_m = 10^4$. The difference of the average Nusselt number at the hot wall between these two meshes was smaller than 1%. Therefore, the $80 \times 80$ mesh was found to be refined enough near the walls to capture the thin boundary layers that appear along the vertical surfaces. The use of a stretched grid rather than a uniform one is due to the fact that a uniform grid will require several grid points in order to guarantee a first grid point closer enough to the wall to capture the thin boundary layer that appears along the heated walls, mainly for turbulent flows.

For the *continuum* model and for the most stringent case simulated, i.e., $N = 256$ obstacles, a stretched grid with $151 \times 151$ nodes yielded average Nusselt numbers less than 1% different from those obtained when using a grid size of $117 \times 117$ stretched nodes. In spite of such small difference, all results herein were computed with the finer $151 \times 151$ stretched grid for conservativeness and accuracy enhancement.

*Velocity and Temperature Fields* First, to validate the *continuum* approach, runs were performed for a clear (of obstacles) cavity and compared with other values from the open literature. Present results were summarized in Table 9.15 and show good agreement with those obtained from other sources. A case with a single conducting square solid located at the center of the cavity was also performed showing good agreement with those summarized in Table 9.16.

As said, the main idea of this work is to compare heat transfer simulations in a porous square cavity using the *porous-continuum* and the *continuum* models with several obstacles. Comparisons are based on similar conditions in order to verify if the two models yield equivalent values for the overall Nusselt numbers. As mentioned, runs

**Table 9.15.** Average Nusselt numbers for buoyancy-driven laminar flow in clear cavities; $104 < Ra < 108$, $Pr = 0.71$ (unless otherwise noted).

| | *Ra* | | | | |
|---|---|---|---|---|---|
| | $10^4$ | $10^5$ | $10^6$ | $10^7$ | $10^8$ |
| de Vahl Davis (1983) | 2.243 | 4.519 | 8.800 | – | – |
| House *et al.* (1990) | 2.254 | 4.561 | 8.923 | – | – |
| Merrikh and Lage (2004) | 2.244 | 4.536 | 8.860 | 16.625 | 31.200 |
| Kalita *et al.* (2001) | 2.245 | 4.522 | 8.829 | 16.52 | – |
| Lage and Bejan (1991), $Pr = 1$ | – | 4.9 | 9.2 | 17.9 | 31.8 |
| Braga and de Lemos (2005), $Pr = 1$ | 2.249 | 4.575 | 8.918 | 16.725 | 30.642 |

**Table 9.16.** Average Nusselt number for cavity with a single conducting solid at the center; $Ra = 105$, $Pr = 0.71$ (unless otherwise noted).

| *Ra* | $D_p$ | $k_s/k_f$ | House *et al.* (1990) | Merrikh and Lage (2004) | Braga and de Lemos (2005) $Pr = 1$ |
|---|---|---|---|---|---|
| $10^5$ | 0.5 | 0.2 | 4.624 | 4.605 | 4.667 |
| $10^5$ | 0.5 | 5.0 | 4.324 | 4.280 | 4.375 |

for *continuum* model solution were performed with a $151 \times 151$ control volumes in a stretched grid shown in Figure 9.17c. For the sake of comparison of the two models, all calculations were made with $Ra_m = RaDa = 10^4$. For the continuum model, a Darcy number is associated with the flow in the arrangement of Figure 9.17b with permeability $K$ calculated by expression (1.9), using for $D_p$ the size of the square rod.

As such, when the number of blocks $N$ in the cavity is increased while keeping the overall solid-to-void ratio (equivalent to a constant porosity cavity case), a reduced value of the square rod size $D_p$ yields different $K_{eq}$ values, according to (1.9), implying in distinct $Da_{eq}$ numbers (see Equation (9.64)). However, looking back at Figure 9.71 and remembering the definition of $Ra$ (Equation (9.63)), for different Darcy numbers one has to modify $Ra$ in order to keep $Ra_m$ fixed at $10^4$. One way to accomplish this is to modify the numerical value of coefficient $\beta$ in Equation (9.63), assuming that a different fluid is being computed for all point lying in the same curve in Figure 9.71. One could also modify $H$, $\rho$, $\Delta T$ or another variable composing $Ra$. This work used several pairs of values for $K$ and $\beta$ so that the corresponding $Ra$ and $Da$ were such that their product would yield always $Ra_m = 10^4$. Therefore, coefficients $\beta$ and $K$ are the variables to be modified in order to maintain $Ra_m$ constant and all cases here analyzed.

$$\Delta \cdot \mathbf{u}_D = 0 \tag{9.71}$$

$$\rho\left[\nabla\cdot\left(\frac{\mathbf{u}_D\mathbf{u}_D}{\phi}\right)\right] = -\nabla\left(\phi\langle p\rangle^i\right) + \mu\nabla^2\mathbf{u}_D - \left[\frac{\mu\phi}{K}\mathbf{u}_D + \frac{c_F\phi\rho|\mathbf{u}_D|\mathbf{u}_D}{\sqrt{K}}\right] - \rho\beta_\phi\mathbf{g}\phi\left(\langle T\rangle^i - T_{ref}\right) \tag{9.72}$$

It is also interesting to note that when $Da = 1$, $Ra_m$ calculated for the *continuum* model reduces to $Ra$, similar to the one used for clear (unobstructed) fluid cavities. For that, also shown with this set of results are calculations with the *porous-continuum* model with $Da = 1$ (high permeability $K$) and $\phi = 0.998$ (high porosity) in order to simulate a clear fluid cavity, which, in turn, would correspond to the case of having no obstacle at all ($N = 0$). In fact, a cavity filled with a porous material with $Da = 1$ and $\phi = 0.998$ is analogous to a continuum model solution with a very small obstacle in its center, which does not contribute effectively to the heat transfer process (Merrikh and Lage, 2004). Indeed, the difference between the average Nusselt number calculated with the two models, for clear cavity ($N = 0$) and porous cavity with $Da = 1$ and $\phi = 0.998$, is less than 3%, indicating that a porous-continuum model will reproduce a clear fluid cavity solution when appropriate parameters are set in Equations (9.71), (9.72) and (5.33). Also, solid obstacles in the cavity yield an overall cavity porosity $\phi = 0.84$ (unless otherwise noted). The fluid Prandtl number and the conductivity ratio between the solid and fluid phases were assumed to be equal to one. Table 9.17 summarizes the parameters used in the calculations.

**Table 9.17.** Parameters used in the arrangement of Figure 9.70b for the continuum model and $Pr = 1$, equivalent $\phi = 0.84$, $Ra_m = Ra\ Da = 10^4$, $k_s/k_f = 1$.

| $Da = Da_{eq}$ ($K_{eq}$ from (1.9)) | $Ra$ | $D_p$ (m) | $N$ = number of obstacles |
|---|---|---|---|
| Porous cavity with $Da = 1$ and $\phi = 0.998$ | $Ra_\phi = Ra = 10^4 (\beta_\phi = \beta = 0.001)$ | 0.022 | Equivalent to 0 |
| Clear cavity | $10^4$ | 0 | 0 |
| $0.3087 \times 10^{-1}$ | $0.0324 \times 10^7$ | 0.400 | 1 |
| $0.7717 \times 10^{-2}$ | $0.1295 \times 10^7$ | 0.200 | 4 |
| $1.9290 \times 10^{-3}$ | $0.5000 \times 10^7$ | 0.100 | 16 |
| $0.4823 \times 10^{-3}$ | $2.0734 \times 10^7$ | 0.050 | 64 |
| $1.2060 \times 10^{-4}$ | $8.2918 \times 10^7$ | 0.025 | 256 |

Figures 9.72 and 9.73 show the streamlines and isotherms, respectively, for the heterogeneous system of Figure 9.70a with several obstacles and equivalent $Da$ ranging from 1 to $1.2060 \times 10^{-4}$, $Ra_m = 10^4$, $\phi = 0.84$ and $k_s/k_f = 1$. Figures 9.74 and 9.75 show corresponding results, i.e., same $Ra_m$, $\phi$, $Pr$ and ratio $k_s/k_f$, for a square cavity completely filled by a porous material. Results in Figures 9.72 and 9.73 are computed with the *continuum* approach whereas Figures 9.74 and 9.75 made use of the *porous-continuum* described earlier.

Comparing Figures 9.72a and 9.74a one can see that clear cavity flow is reproduced in both models if appropriated parameters are set, or say, $N = 0$ (continuum model with no obstacle or clear cavity flow) or $Da = 1$ and $\phi = 0.998$ (highly permeable and highly porous cavity). The same comparisons hold for the thermal field (see Figures 9.73a and 9.75a).

Figure 9.72 shows that, in comparison with corresponding cases run with the porous-continuum model (Figure 9.74), the higher the number of obstacles inside the clear cavity, the higher the similarity of the flow pattern between the two approaches, i.e., the porous-continuum and the continuum solutions resemble each other for greater values of $N$ (see Figures 9.72f and 9.74f). In other words, the porous-continuum model seems to be more representative of reality when the number of obstacles inside the cavity is higher, which, in turn, correspond to lower permeability cases. Further, Figure 9.73 shows that, the higher the number of obstacles, the higher the stratification of the thermal field. This characteristic is also observed for the isotherms of the porous-continuum model (Figure 9.75).

Figure 9.74 also indicates that the recirculation intensity increases as the medium permeability decreases and the flow patterns comprise primarily cells of relatively high velocity, which circulate around of the entire cavity. However, the secondary recirculation that appears in the center of the cavity, for the higher Darcy numbers analyzed, tends to

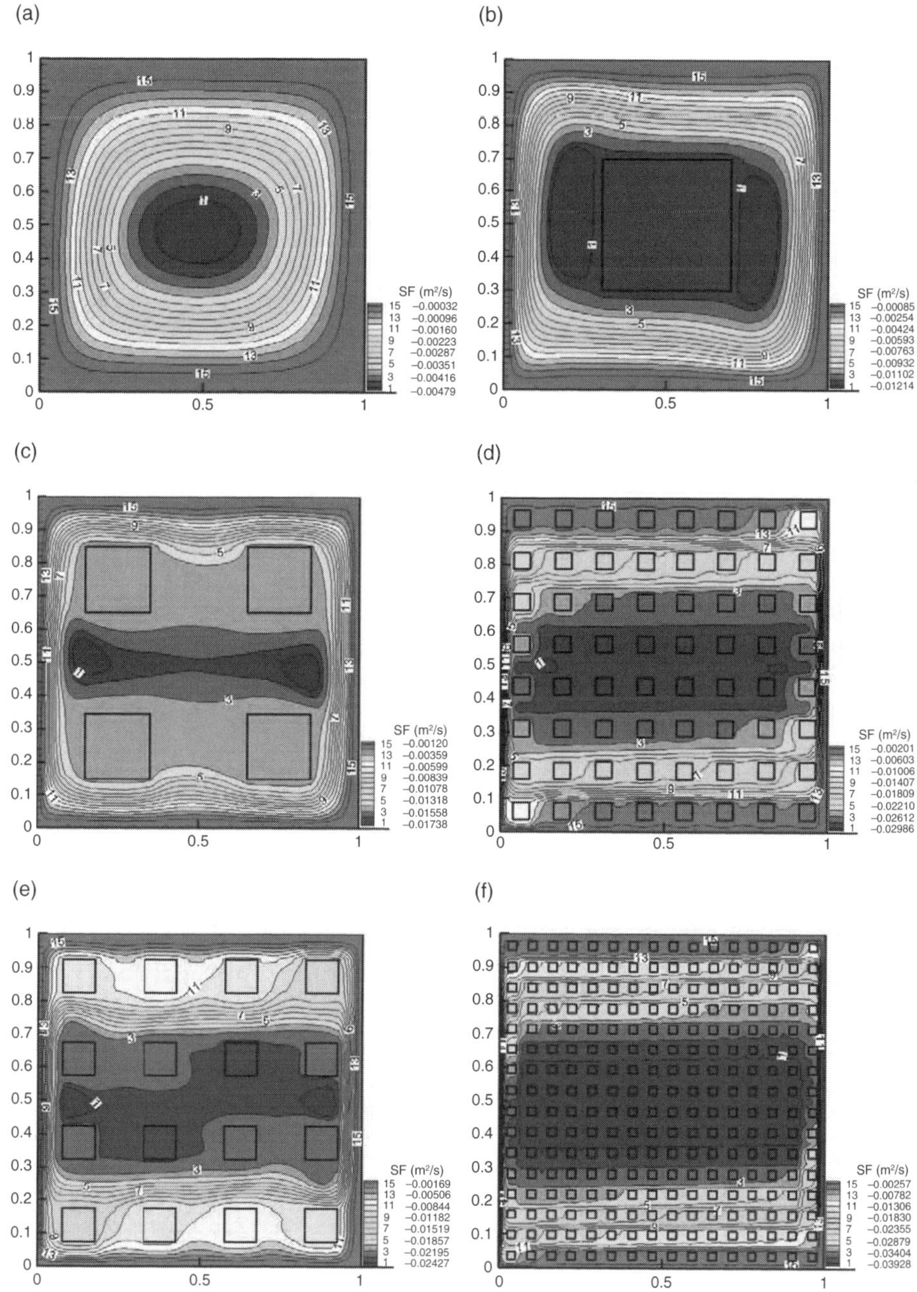


**Figure 9.72.** Streamlines for continuum model solution, $\phi = 0.84$, $Pr = 1$, $k_s/k_f = 1$, $Ra_m = 104$: (a) $Da = 1$, $N = 0$, (b) $Da = 0.3087 \times 10^{-1}$, $N = 1$, (c) $Da = 0.7717 \times 10^{-2}$, $N = 4$, (d) $Da = 1.929 \times 10^{-3}$, $N = 16$, (e) $Da = 0.4823 \times 10^{-3}$, $N = 64$, (f) $Da = 1.206 \times 10^{-4}$, $N = 256$.

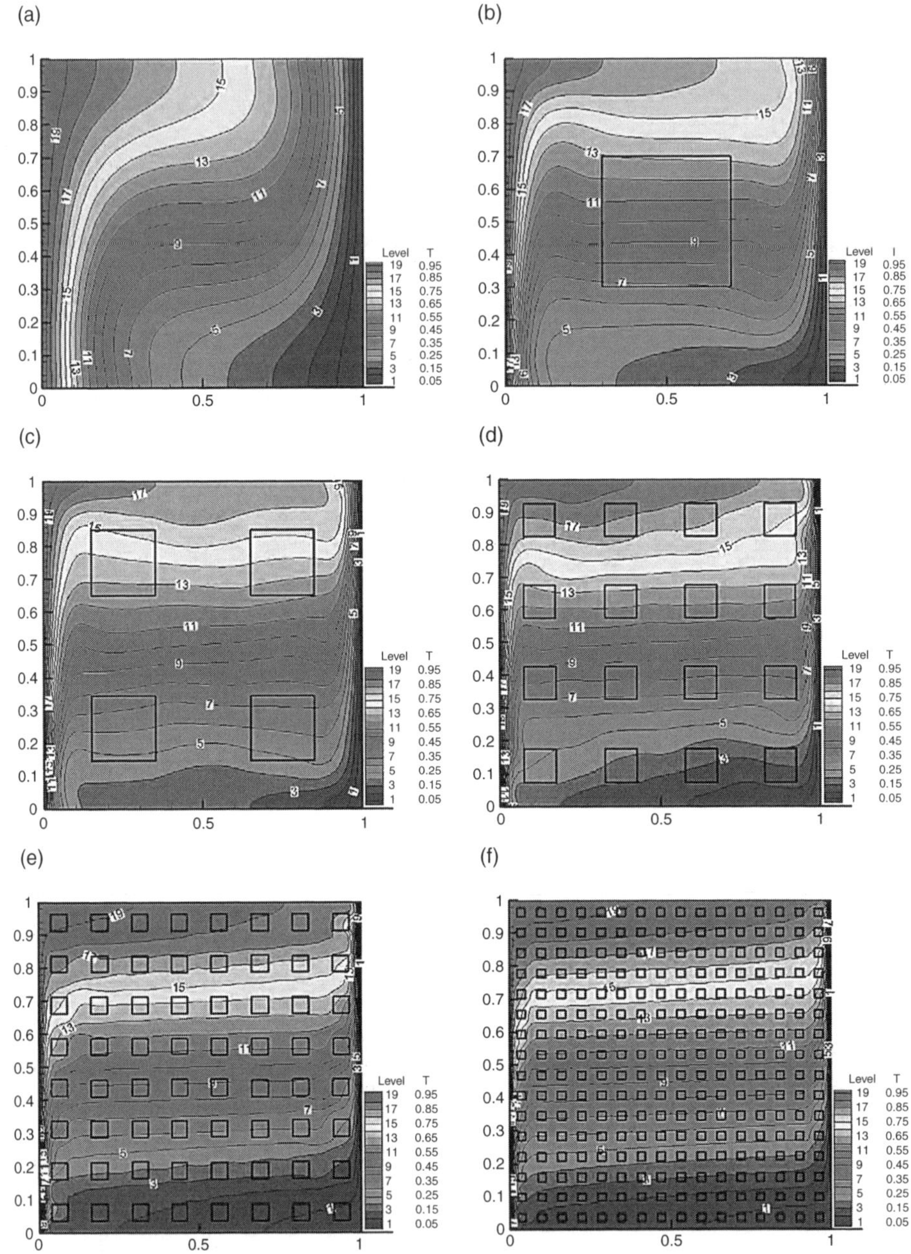


**Figure 9.73.** Isotherms for continuum model solution, $\phi = 0.84$, $Pr = 1$, $k_s/k_f = 1$, $Ra_m = 104$: (a) $Da = 1$, $N = 0$, (b) $Da = 0.3087 \times 10^{-1}$, $N = 1$, (c) $Da = 0.7717 \times 10^{-2}$, $N = 4$, (d) $Da = 1.929 \times 10^{-3}$, $N = 16$, (e) $Da = 0.4823 \times 10^{-3}$, $N = 64$, (f) $Da = 1.206 \times 10^{-4}$, $N = 256$.

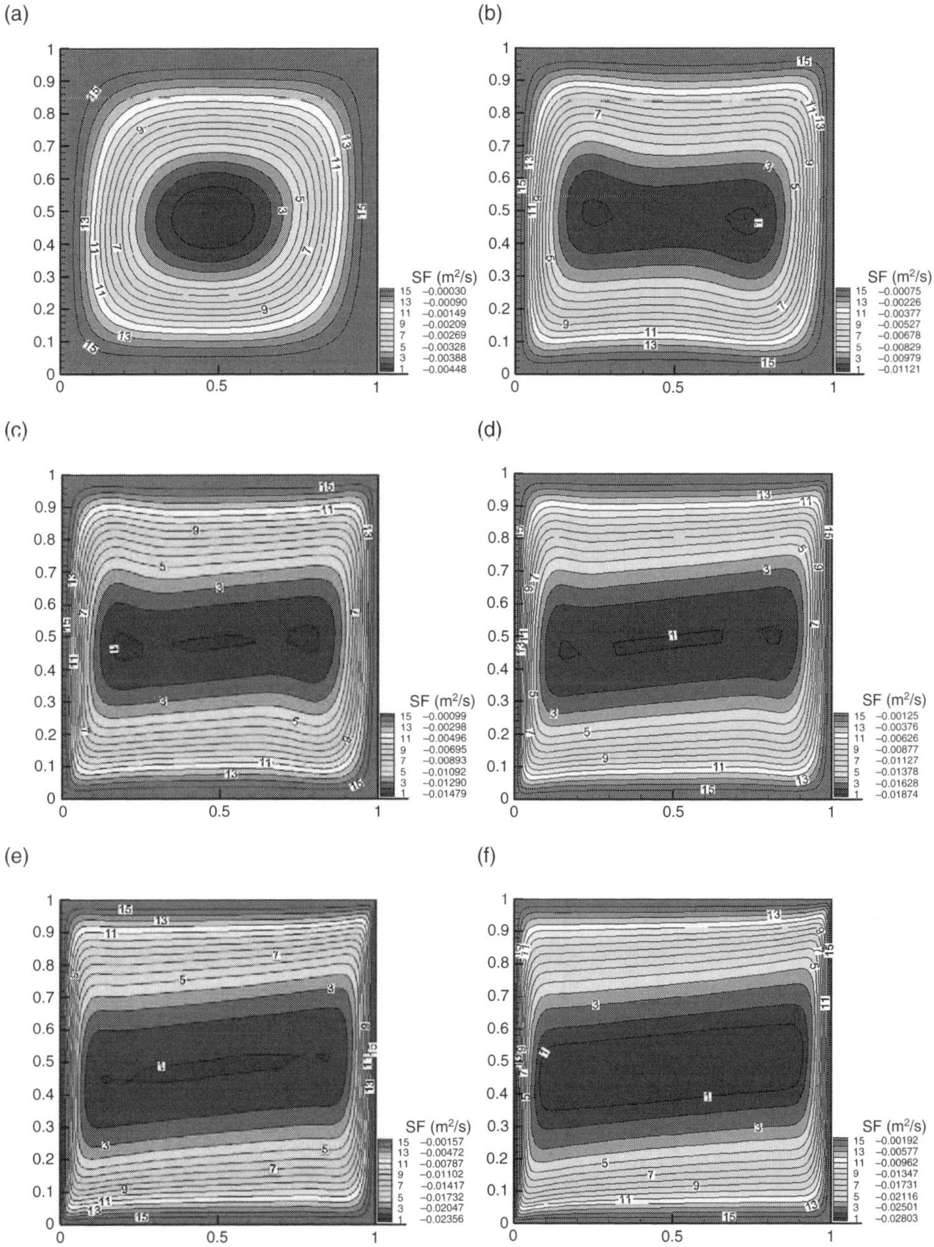


**Figure 9.74.** Streamlines for a porous cavity with $\phi = 0.84$, $Pr = 1$, $k_s/k_f = 1$, $Ra_m = 10^4$; (a) $Da = 1$, $\phi = 0.998$, (b) $Da = 0.3087 \times 10^{-1}$, (c) $Da = 0.7717 \times 10^{-2}$, (d) $Da = 1.929 \times 10^{-3}$, (e) $Da = 0.4823 \times 10^{-3}$, (f) $Da = 1.206 \times 10^{-4}$.

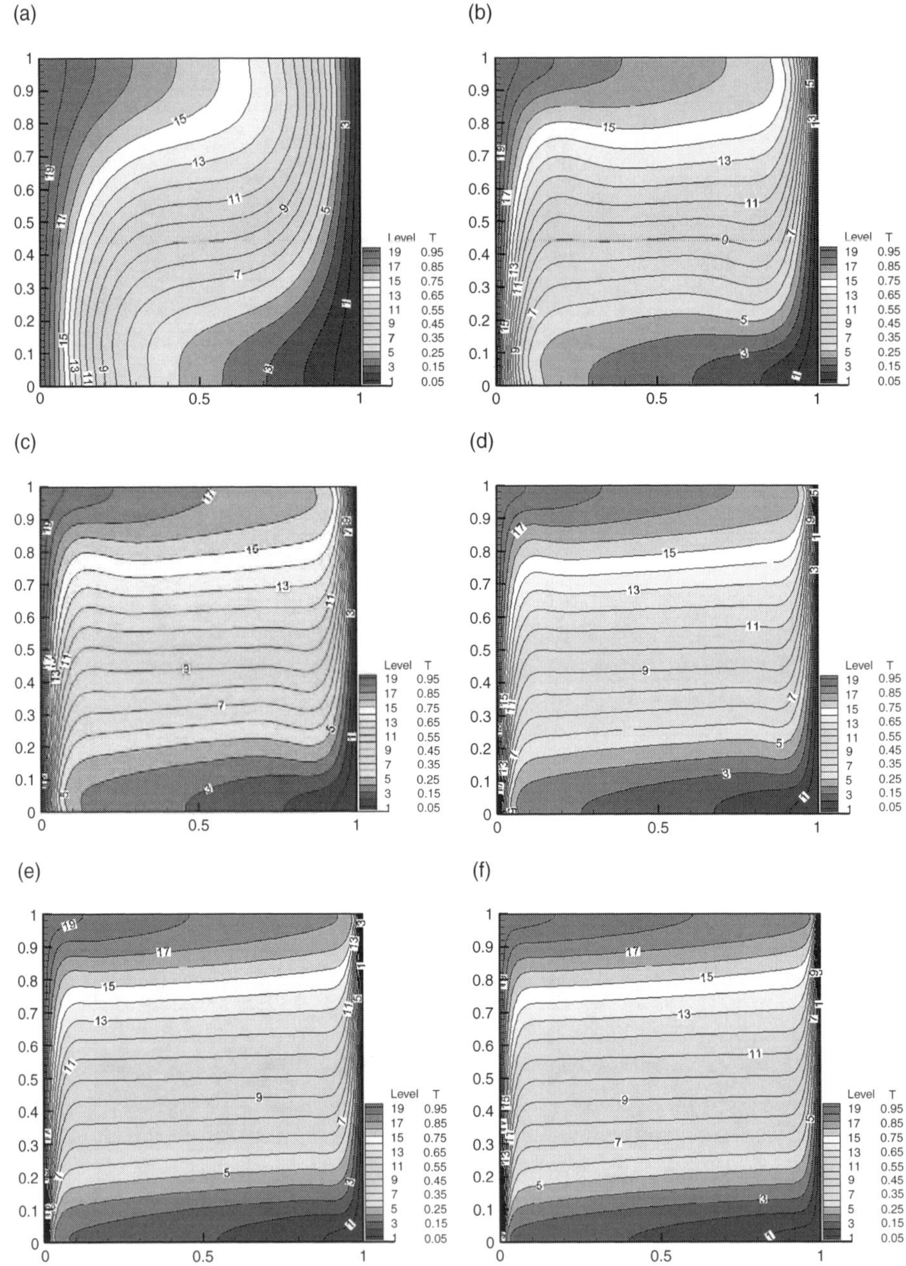


**Figure 9.75.** Isotherms for a porous cavity with $\phi = 0.84$, $Pr = 1$, $k_s/k_f = 1$, $Ra_m = 10^4$; (a) $Da = 1$, $\phi = 0.998$, (b) $Da = 0.3087 \times 10^{-1}$, (c) $Da = 0.7717 \times 10^{-2}$, (d) $Da = 1.929 \times 10^{-3}$, (e) $Da = 0.4823 \times 10^{-3}$, (f) $Da = 1.206 \times 10^{-4}$.

disappear as the permeability decreases. In a similar way, the temperature gradients are stronger near the vertical walls, but decrease at the center. Figure 9.75 shows that the isotherms tend to stratification as *Da* is decreased, i.e., as the medium permeability is decreased.

According to Merrikh and Lage (2004), as the number of square rods increases, and their size becomes reduced, the flow tends to migrate away from the wall towards the center of the cavity. This phenomenon is seen in Merrikh and Lage (2004) as a response of the system due to the increasing flow resistance closer to the solid wall, as the obstacles get closer to the solid surface.

Further, the available literature shows that for the non-Darcy region in a porous cavity (Merrikh and Mohamad, 2002; Braga and de Lemos, 2004), fluid flow and heat transfer depend on the fluid Rayleigh number, $Ra_f$, and on the Darcy number, *Da*, when other parameters, such as porosity, Prandtl number, and conductivity ratio between the fluid and solid matrix, are held constant. In Braga and de Lemos (2004) it was shown that for a fixed $Ra_m$, the lower the permeability (lower *Da*), the higher the average Nusselt number at the hot wall. It then looks evident that different combinations of $Ra_f$ and *Da* yields different heat transfer results, even when $Ra_m$ is the same. The increasing of the fluid Rayleigh number increases the natural convection inside the enclosure. For a fixed $Ra_m$, a higher fluid Rayleigh number is associated with a less permeable media (i.e. lower Darcy number).

To further validate the present porous-continuum approach, runs were performed for a square cavity totally filled with porous material and results were compared with other values from the open literature. Calculations in Braga and de Lemos (2004) made use of the same numerical procedure here exploited. Their results are summarized in Table 9.18, showing good agreement with several other authors.

*Turbulent Field* The set of macroscopic equations used to perform turbulent model solutions, fully documented in Pedras and de Lemos (2000a–2003), was extended to

**Table 9.18.** Average Nusselt number for buoyancy-driven laminar flow in porous cavities.

| | $Ra_m$ | | | |
|---|---|---|---|---|
| | 10 | $10^2$ | $10^3$ | $10^4$ |
| Walker and Homsy (1978) | – | 3.097 | 12.96 | 51.0 |
| Bejan (1979) | – | 4.2 | 15.8 | 50.8 |
| Beckermann *et al.* (1986) | – | 3.113 | – | 48.9 |
| Gross *et al.* (1986) | – | 3.141 | 13.448 | 42.583 |
| Manole and Lage (1992) | – | 3.118 | 13.637 | 48.117 |
| Moya *et al.* (1987) | 1.065 | 2.801 | – | – |
| Baytas and Pop (1999) | 1.079 | 3.16 | 14.06 | 48.33 |
| Braga and de Lemos (2004), $Da = 10^{-8}$, $Pr = 1$ | 1.0908 | 3.0979 | 13.2751 | 43.5799 |

natural convection in de Lemos and Braga (2003) and Braga and de Lemos (2004) and for that they need not be repeated here. Turbulent model solutions were performed in the same grids used for laminar model solutions. The stretched grid here adopted is refined enough to capture the thin boundary layer that appears along the heated walls.

Table 9.19 shows laminar and turbulent average Nusselt numbers for the continuum and porous-continuum models for $0.7717 \times 10^{-2} < Da < 0.4823 \times 10^{-3}$, $Pr = 1$, $\phi = 0.84$, $Ra_m = 10^6$ and $k_s/k_f = 1$. According to Table 9.19, in both laminar and turbulent solutions a macroscopic model ($Nu_\phi$) underestimates the value of $Nu$ and such discrepancy increases as the number of blocks ($N$) increases. For laminar solution and $N = 4$ (see Table 9.19), $Nu_\phi$ is $(27.8 - 32.9)/32.9. \times 100 = -15.5\%$ lower than $Nu$ and for $N = 64$ such decrease is $(48.3 - 66.7)/66.7 \times 100 = -27.6\%$. The same trend can be observed when turbulence is included: $Nu_\phi/Nu = (30.73/32.9) = 0.93$ for $N = 4$ and $(62.2/70.8) = 0.88$ when $N = 64$.

Also, both the continuum and the porous-continuum models give higher values for the Nusselt number when turbulence is considered. Further, *macroscopic* solutions are seen to be more sensitive to the inclusion of a turbulence model (see Braga and de Lemos, 2004, for details), and this sensitivity increases with the number of blocks $N$. When $N = 4$, inclusion of turbulence in the calculations rises $Nu_\phi$ by $(30.7 - 27.8)/27.8. \times 100 = +\mathbf{10.4}\%$ and for $N = 64$ Nusselt will be elevated by $(62.2 - 48.3)/48.3 \times 100 = +\mathbf{28.6}\%$. On the other hand, turbulent solution using the *continuum* approach raises $Nu$ by $(32.94 - 32.87)/32.87 \times 100 = +\mathbf{0.2}\%$ for $N = 4$ and $(70.8 - 66.7)/66.7 \times 100 = +\mathbf{6}\%$ for $N = 64$, which represents a much weaker influence of turbulence than in the case of $Nu_\phi$. Therefore, one can infer that for $Ra_m = 10^6$ the flow is already fully turbulent because results using only laminar model solution gives average Nusselt numbers lower than those obtained with the turbulent model solution.

Figure 9.76 further shows results using the turbulent model solution for: (a,b) streamlines, (c,d) isotherms, (e,f) isolines of turbulent kinetic energy, $k$, for $\phi = 0.84$, $Ra_m = 10^6$, $Da = 0.4823 \times 10^{-3}$, $k_s/k_f = 1$, $Pr = 1$ for continuum and porous-continuum model respectively. Figure 9.76a presents some recirculations between the solid obstacles and that recirculations are turbulent kinetic energy generators in such regions, see Figure 9.76e. Nevertheless, the flow pattern for the two models considered show satisfactory agreement

**Table 9.19.** Laminar and turbulent average Nusselt number for the continuum and porous-continuum models for $0.7717 \times 10^{-2} < Da < 0.4823 \times 10^{-3}$ and $Pr = 1$, $\phi = 0.84$, $Ra_m = 10^6$ and $k_s/k_f = 1$.

| | Laminar model solution | | Turbulent model solution | | Number of rods, $N$ |
|---|---|---|---|---|---|
| $Da = Da_{eq}$ | $Nu$ | $Nu_\phi$ | $Nu$ | $Nu_\phi$ | |
| $0.7717 \times 10^{-2}$ | 32.8734 | 27.7820 | 32.9399 | 30.7346 | 4 |
| $1.9290 \times 10^{-3}$ | 46.4907 | 36.7713 | 46.6804 | 43.7739 | 16 |
| $0.4823 \times 10^{-3}$ | 66.6977 | 48.3144 | 70.7901 | 62.1561 | 64 |

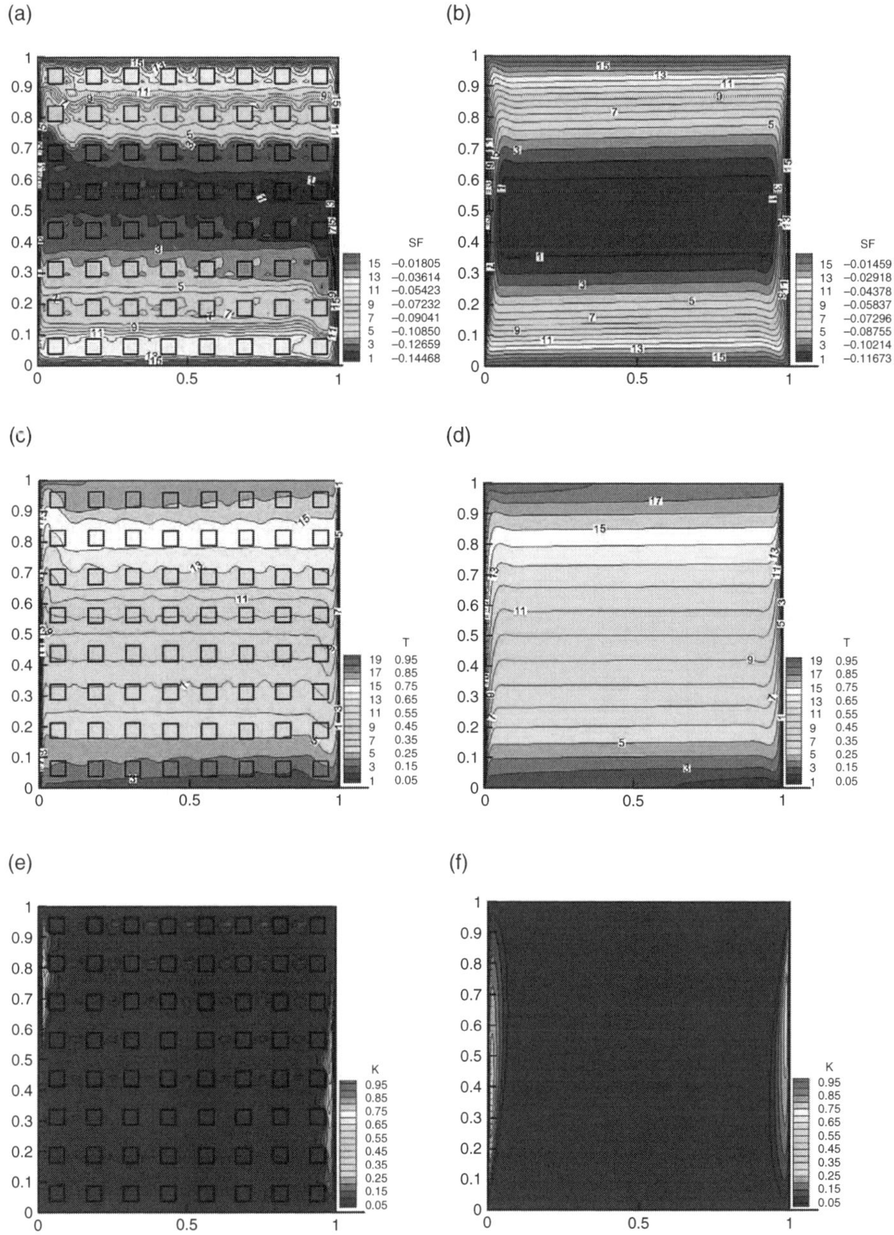


**Figure 9.76.** Turbulent model solution using the continuum (left) and porous-continuum (right) models for $\phi = 0.84, Ra_{\mathrm{m}} = 10^6, Da = 0.4823 \times 10^{-3}, k_{\mathrm{s}}/k_{\mathrm{f}} = 1, Pr = 1$: (a,b) streamlines, (c,d) isotherms, (e,f) isolines of turbulent kinetic energy, $k$.

when qualitatively compared, see Figure 9.76a,b. As mentioned, the porous-continuum model seems to be more representative of reality when the number of obstacles inside the cavity is large, which, in turn, correspond to lower permeability cases.

Figure 9.76c shows some peeks in the isotherms when compared with the isotherms from the Figure 9.76d. This distortion on the thermal field is a response of the system due to the non-homogeneity of the heat mixing process. However, both models show a stratificated thermal field, see Figure 9.76c,d.

Figure 9.76e shows the isolines of $k$ for the continuum model solution and one can note that the steepest velocity gradients are found near to the heated walls and between the solid obstacles. For that reason the generation of turbulent kinetic energy is considerably higher in such regions. On the other hand, the porous-continuum approach, Figure 9.76f, shows a high turbulent kinetic energy generation only close to the heated wall while the center of the cavity does not contribute to the $k$ generation.

*Nusselt Number* Finally, Figure 9.77 compares the behavior of the average Nusselt number for the two models here investigated, namely the porous-continuum and continuum models. Both calculation methods were based on the same numerical values for both

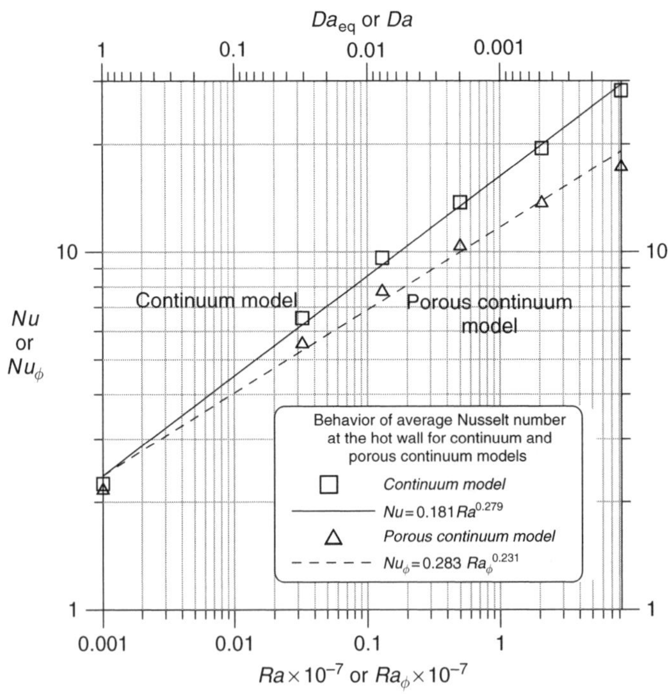


**Figure 9.77.** Comparison between the continuum and porous-continuum models with respect to the average Nusselt number at the hot wall.

the Rayleigh and the Darcy numbers. It is clearly seen in the figure that, if $Ra = Ra_\phi$ and $Da_{eq} = Da$ for both models, the overall values of average Nusselt number for the porous-continuum, $Nu_\phi$, model are lower than those obtained with the continuum model, $Nu$. This difference increases, with increasing number of rods $N$. Therefore, the porous-continuum model fails in predicting the average Nusselt number, when compared with those obtained from the continuum model with several obstacles.

A possible explanation for such discrepancy is twofold: First, similar cases for the porous and porous-continuum models were compared under the condition $Ra_\phi = Ra$ (or $\beta_\phi = \beta$). However, it has already been pointed out in de Lemos and Braga (2003) that these two thermal expansion coefficients do not, necessarily, have equal values. In fact, expression (9.61), derived in de Lemos and Braga (2003), shows the relationship between these two parameters. The first one, $\beta$, is a fluid property and for an ideal gas it is given by $1/T$, where $T$ is the absolute gas temperature. On the other hand, $\beta_\phi$ is a macroscopic quantity, defined in Equation (9.61), and by no means represents a local fluid property. Consequently, comparisons with the two models here investigated on the basis $\beta_\phi = \beta$ are strictly not quite correct. For having the same $Nu$, an inspection on Figure 9.77 indicates that one should have $Ra_\phi > Ra$ when comparing the two models.

Secondly, both models were compared in Figure 9.77 using the same value for parameter Darcy, or $Da_{\mathrm{eq}} = Da$. Note that $Da_{\mathrm{eq}}$ is an "equivalent" Darcy number, whose associated $K_{\mathrm{eq}}$ in the continuum model was estimated by (1.9), $D_{\mathrm{p}}$ being the rod size. On the other hand, $Da$ is the Darcy number for the porous cavity of Figure 9.70d formed with the porous medium permeability $K$. In the simulation herein, the same $K_{\mathrm{eq}}$ value was prescribed in the porous-continuum model for use in Equation (9.72), or say, $K_{\mathrm{eq}} = K$. However, Equation (1.9) was derived for forced convection flow over a porous bed so that its application to the natural convection problem under analysis here is questionable. Also, real porous systems like the ones in Figure 9.70a may have lower permeability than those associated with their counterpart continuum models of Figure 9.70b. As the number of rods, $N$, increases, $K_{\mathrm{eq}}$ associated with Figure 9.70b might be reduced at a rate faster than that given by Equation (1.9). So, when calculating a counterpart porous-continuum model for the system in Figure 9.70b, a lower value for $K$ should be used instead of $K_{\mathrm{eq}}$ given by Equation (1.9), or say, $Da < Da_{\mathrm{eq}}$.

Therefore, in order to try to match Nusselt numbers calculated with both models here described, a correction is applied to the values given by the porous-continuum model. This correction is set by curve fitting the values of $Nu$ in Figure 9.77. For the two models applied, the following curves result:

$$Nu = 0.181 Ra^{0.279} \tag{9.73}$$

$$Nu_\phi = 0.283 Ra_\phi^{0.231} \tag{9.74}$$

where Equations (9.73) and (9.74) refer to fitting curves obtained from the points of the average $Nu$ (symbols) calculated with continuum and porous-continuum models, respectively. Now, by combining these two fittings, namely Equations (9.73) and (9.74), one yields the following expression:

$$Ra_\phi = 0.144 Ra^{1.208} \tag{9.75}$$

As such, for matching $Nu$ calculated from the two models, namely continuum and porous-continuum models, $Ra_\phi$ given by (9.75) should be used instead of $Ra_\phi = Ra$. Also, keeping all parameters the same in Equations (9.63) and (9.65), correction (9.75) is equivalent to setting $\beta_\phi = 4.128\beta^{1.208}$. This correction means that for calculating equivalent $Nu$ with both models, one has to have $Ra_\phi > Ra$.

Further, in order to keep

$$Ra_{\rm m} = \overbrace{Ra_\phi Da}^{\textit{Porous-continuum model}} = \underbrace{Ra\ Da_{\rm eq}}_{\textit{Continuum model}} = 10^4 \tag{9.76}$$

a relationship between the two Darcy numbers was also considered. Using Equation (9.75) under the constraint (9.76) yields for such connection:

$$Da = 1.005 Da_{\rm eq}^{1.208} \tag{9.77}$$

which is equivalent to having $K = 1.005 K_{\rm eq}^{1.208}$ since $Da = K/H^2$ and $Da_{\rm eq}$ is defined in Equation (9.67).

Therefore, in addition to using (9.75) for matching the Nusselt numbers given by the two models, which are represented by the correlations (9.73) and (9.74), one has also to use (9.77) instead of applying $Da = Da_{\rm eq}$.

After that, new runs were performed for the porous-continuum model using now a macroscopic Rayleigh number given by (9.75), instead of using $Ra_\phi = Ra$, with corresponding Darcy numbers expressed by (9.77), used in place of $Da = Da_{\rm eq}$. These new runs were performed while $Ra_{\rm m}$ was kept constant in such a way that

$$Ra_{\rm m} = \underbrace{Ra_\phi Da}_{\textit{Using } Ra_\phi = Ra,\ Da = Da_{\rm eq}} = \overbrace{Ra_\phi Da}^{\textit{Using Equations (9.75) and (9.77)}} = 10^4 \tag{9.78}$$

As a result, the use of (9.75) and (9.77) yielded corrected macroscopic Nusselt numbers, $Nu_{\phi,\rm corr}$. Table 9.20 shows average values of Nusselt numbers for the continuum ($Nu$), porous-continuum with $Ra_\phi = Ra$ and $Da = Da_{\rm eq}$ ($Nu_\phi$) and corrected porous-continuum using (9.75) and (9.77) ($Nu_{\phi,\rm corr}$). The percent errors shown in Table 9.20 are calculates as:

$$\varepsilon = \frac{Nu_\phi - Nu}{Nu} \times 100; \quad \eta = \frac{Nu_{\phi,\rm corr} - Nu}{Nu} \times 100 \tag{9.79}$$

**Table 9.20.** Average Nusselt number for the continuum, porous-continuum and corrected porous-continuum models for $Da$ ranging from 1 to $1.206 \times 10^{-4}$ and $Pr = 1$, $\phi = 0.84$, $Ra_m = 10^4$, $k_s/k_f = 1$.

| $N$ | $Da$ | $Nu$ | $Nu_\phi$ | $Nu_{\phi,corr}$ | $\varepsilon$ (%) | $\eta$ (%) |
|---|---|---|---|---|---|---|
| 0 | 1 | 2.2493 | 2.1854 | 2.1834 | −2.84 | −2.93 |
| 1 | $0.3087 \times 10^{-1}$ | 6.5254 | 5.5960 | 6.6804 | −14.24 | 2.37 |
| 4 | $0.7717 \times 10^{-2}$ | 9.6204 | 7.8484 | 9.7677 | −18.42 | 1.53 |
| 16 | $1.9290 \times 10^{-3}$ | 13.7276 | 10.4911 | 13.3775 | −23.57 | −2.55 |
| 64 | $0.4823 \times 10^{-3}$ | 19.4821 | 13.8199 | 17.9348 | −29.06 | −7.94 |
| 256 | $1.2060 \times 10^{-4}$ | 28.2085 | 17.4460 | 22.8239 | −38.15 | −19.08 |

According to Table 9.20, the overall values of the corrected average Nusselt numbers $Nu_{\phi,corr}$ are still lower than those obtained from the continuum model for lower $Da$ numbers or for cases with $N$ large. Nevertheless, as can be seen in the table, $\eta$ values are substantially lower in comparison with $\varepsilon$. One should point out that thermal dispersion was not considered in this work and that this additional mixing mechanism might be important for cases with a large number of blocks, which correspond to cases in Table 9.20 presenting the largest discrepancies between the continuum and porous-continuum solutions, even with the corrections on $Nu$ obtained with the use of Equations (9.75) and (9.77).

In the end, when results herein are seen along with those presented in Braga and de Lemos (2004), which used the porous-continuum model for simulating macroscopic turbulent flow in heated cavities with dispersion, there seems to be an indication that inclusion of turbulence and dispersion as additional exchange mechanisms is mandatory to reflect the actual heat transfer rate across the cavity, particularly for cases with large $N$. Ultimately, a more realistic macroscopic model, accounting for all possible mechanisms, will result in macroscopic values for $Nu_\phi$ that will be closer to microscopic computations of $Nu$, making cheaper and easy-to-implement macroscopic models a suitable engineering tool for analyzing complex physical systems as the ones in Figure 9.70a.

#### 9.4.5.5. *Conclusions*

This work presented numerical solutions for steady laminar and turbulent natural convection within a square cavity filled by a fixed amount of conducting solid material using a continuum model. The solid phase was composed by square obstacles, equally spaced within the cavity. In addition, an isotropic and homogeneous porous-continuum model was used for simulating the flow and heat transfer across the cavity, assuming the enclosure as totally filled with a porous material. The main conclusions of this work are:

(1) The porous-continuum model failed to correctly predict the average Nusselt number when compared with those obtained from the continuum model with several obstacles.

This same conclusion has also been reached in the recent literature (Massarotti *et al.*, 2003).

(2) An adjustment for the Rayleigh and Darcy numbers associated with the macroscopic model was proposed in order to match the average Nusselt number computed by the two approaches for a fixed $Ra_m = 10^4$. After such correction, the overall values of the corrected average Nusselt numbers were still lower than those obtained from the continuum model, mostly for the lower *Da* numbers cases, which correspond to cases with greater number of blocks *N*.

(3) Both the continuum and the porous-continuum models give higher values for the Nusselt number when turbulence is considered, being *macroscopic* solutions more sensitive to the inclusion of a turbulence model. This sensitivity increases with the number of blocks *N*.

Finally, results herein indicate that inclusion of turbulence and dispersion construct more realistic macroscopic models, making them a suitable engineering tool for analyzing complex physical systems.

## 9.5. FLOW AND HEAT TRANSFER IN A BACK-STEP

The results to follow are detailed in references Assato and de Lemos (2000, 2001) and Assato *et al.* (2005). They considered the simulation of turbulent flow in a back-step, in which a porous insert is positioned past the sudden expansion of the flow. A schematic is presented in Figure 9.78. The objective is to see the influence of the porous medium on the pressure and velocity fields. Next are the equations resolved and the results obtained.

### 9.5.1. Macroscopic Mean Equations

$$\nabla \cdot \bar{\mathbf{u}}_D = 0 \tag{9.80}$$

where $\bar{\mathbf{u}}_D$ is the average surface velocity ("seepage" or Darcy velocity). The Equation (9.80) represents the macroscopic continuity equation for an incompressible fluid.

$$\left[\nabla \cdot \left(\rho \frac{\bar{\mathbf{u}}_D \bar{\mathbf{u}}_D}{\phi}\right)\right] = -\nabla\left(\phi \langle \bar{p} \rangle^i\right) + \mu \nabla^2 \bar{\mathbf{u}}_D + \nabla \cdot \left(-\rho\phi \langle \overline{\mathbf{u}'\mathbf{u}'} \rangle^i\right) - \left[\frac{\mu\phi}{K}\bar{\mathbf{u}}_D + \frac{c_F \phi \rho |\bar{\mathbf{u}}_D| \bar{\mathbf{u}}_D}{\sqrt{K}}\right] \tag{9.81}$$

where $-\rho\phi\langle \overline{\mathbf{u}'\mathbf{u}'} \rangle^i$ is the macroscopic Reynolds stress and the last two terms represent the Darcy-Forchheimer contributions. As before, the symbol $K$ is the porous medium

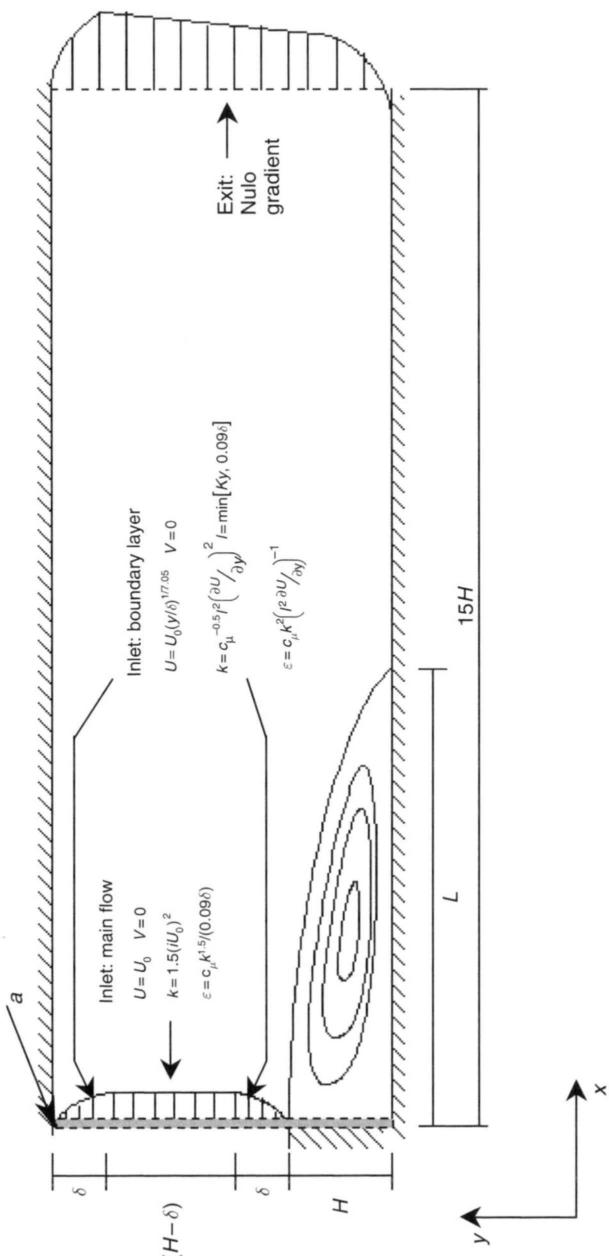


**Figure 9.78.** Boundary conditions for turbulent flow past a backward facing step with porous insert.

permeability, $c_F$ is the form drag coefficient (Forchheimer coefficient), $\langle \bar{p} \rangle^i$ is the intrinsic average pressure of the fluid, $\rho$ is the fluid density, $\mu$ represents the fluid viscosity and $\phi$ is the porosity. Here, it is also important to acknowledge a possible influence of the medium morphology on macroscopic models that, in principle, do not explicitly account for any effect of turbulence such as the inclusion of the macroscopic Reynolds Stress tensor of Equation (9.81). In fact, recent literature results by Bhattacharya *et al.* (2002) propose correlations for the inertia coefficient $c_F$ as a function of medium and flow properties. In the path here followed, however, one unique value for the inertia coefficient will be used when presenting macroscopic results later. Here, the explicit accounting for turbulent transport, while keeping a unique macroscopic inertia coefficient, can be seen as an alternative path on adjusting the Forchheimer coefficient for large values of *Re*. A model for the macroscopic Reynolds Stresses $-\rho\phi\langle \overline{\mathbf{u}'\mathbf{u}'} \rangle^i$, required in the present formulation, is given below.

A macroscopic linear stress–strain rate relationship is given by Pedras and de Lemos (2001a) as:

$$-\rho\phi\langle \overline{\mathbf{u}'\mathbf{u}'} \rangle^i = \mu_{t_\phi} \langle \overline{\mathbf{D}} \rangle^v - \frac{2}{3}\phi\rho\langle k \rangle^i \mathbf{I} \tag{9.82}$$

in analogy with clear flow cases. In (9.82) the term

$$\langle \overline{\mathbf{D}} \rangle^v = \left[ \nabla\bar{\mathbf{u}}_D + [\nabla\bar{\mathbf{u}}_D]^T \right], \tag{9.83}$$

represents the mean deformation tensor and $\mathbf{I}$ is the unity tensor.

$$\mu_{t_\phi} = \rho c_\mu \frac{\langle k \rangle^{i^2}}{\langle \varepsilon \rangle^i} \tag{9.84}$$

where $c_\mu = 0.09$ and $\langle k \rangle^i$ and $\langle \varepsilon \rangle^i$ are the intrinsic averages of the turbulent kinetic energy and its dissipation rate, respectively.

$$\rho\nabla\cdot(\bar{\mathbf{u}}_D\langle k \rangle^i) = \nabla\cdot\left[\left(\mu + \frac{\mu_{t_\phi}}{\sigma_k}\right)\nabla(\phi\langle k \rangle^i)\right] - \rho\langle \overline{\mathbf{u}'\mathbf{u}'} \rangle^i : \nabla\bar{\mathbf{u}}_D + c_k\rho\frac{\phi\langle k \rangle^i|\bar{\mathbf{u}}_D|}{\sqrt{K}} - \rho\phi\langle \varepsilon \rangle^i \tag{9.85}$$

where $-\rho\langle \overline{\mathbf{u}'\mathbf{u}'} \rangle^i$ is defined by Equation (9.82) and $\sigma_k = 1.0$.

$$\rho\nabla\cdot(\bar{\mathbf{u}}_D\langle \varepsilon \rangle^i) = \nabla\cdot\left[\left(\mu + \frac{\mu_{t_\phi}}{\sigma_\varepsilon}\right)\nabla(\phi\langle \varepsilon \rangle^i)\right] + c_{1\varepsilon}(-\rho\langle \overline{\mathbf{u}'\mathbf{u}'} \rangle^i : \nabla\bar{\mathbf{u}}_D)\frac{\langle \varepsilon \rangle^i}{\langle k \rangle^i} + c_{2\varepsilon}c_k\rho\frac{\phi\varepsilon_\phi|\bar{\mathbf{u}}_D|}{\sqrt{K}} - c_{2\varepsilon}\rho\phi\frac{\langle \varepsilon \rangle^{i^2}}{\langle k \rangle^i} \tag{9.86}$$

where $\sigma_\varepsilon = 1.33$, $c_{1\varepsilon} = 1.44$, $c_{2\varepsilon} = 1.92$ and $c_k$ assumes a value equal to 0.28 found by Pedras and de Lemos (2001b,c, 2003).

It is worthwhile to mention that the surface volume average quantities are related to the intrinsic average quantities through the porosity $\phi$ as:

$$\langle \varphi \rangle^{\mathrm{v}} = \phi \langle \varphi \rangle^{\mathrm{i}} \tag{9.87}$$

Also, the equations given above are valid for the clear medium as well, setting $\phi = 1$ ($K \to \infty$) and discarding the last two terms in Equation (9.81).

### 9.5.2. Macroscopic Non-Linear Model

In this work, results produced by non-linear eddy-viscosity models (NLEVM) are investigated. Different from the linear stress–strain relationship (9.82), a more general non-linear constitutive equation will be employed. These models originated in a general proposal done by Pope (1975). However, only in the 1980s such closures had greater progress, particularly due to the works of Speziale (1987), Nisizima and Yoshizawa (1987), Rubinstein and Barton (1990) and Shih *et al.* (1993), among others. In these works, quadratic products were introduced involving the strain rate and vorticity tensors with different derivations and calibrations for each model. These quadratic forms produce a certain anisotropy degree among the normal stresses, which make possible to predict, among other processes, the presence of secondary motion in non-circular ducts.

The macroscopic non-linear turbulence model here proposed is constituted by the same system of Equations (9.80)–(9.86) formerly given by Pedras and de Lemos (2001a). The sole difference between both macroscopic models (linear and non-linear) lies in the expression for the macroscopic Reynolds stress. Using indexed notation now for clarity and keeping terms to second order, this new macroscopic non-linear stress–strain rate equation can be rewritten in the form:

$$\begin{aligned}
-\rho\varphi\langle\overline{u_i' u_j'}\rangle^{\mathrm{i}} &= \left(\mu_{t_\varphi}\langle\overline{D}_{ij}\rangle^{\mathrm{v}}\right)^{\mathrm{L}} - \left(c_{1\mathrm{NL}}\mu_{t_\varphi}\frac{\langle k\rangle^{\mathrm{i}}}{\langle\varepsilon\rangle^{\mathrm{i}}}\left[\langle\overline{D}_{ik}\rangle^{\mathrm{v}}\langle\overline{D}_{kj}\rangle^{\mathrm{v}} - \frac{1}{3}\langle\overline{D}_{kl}\rangle^{\mathrm{v}}\langle\overline{D}_{kl}\rangle^{\mathrm{v}}\delta_{ij}\right]\right)^{\mathrm{NL1}} \\
&\quad - \left(c_{2\mathrm{NL}}\mu_{t_\varphi}\frac{\langle k\rangle^{\mathrm{i}}}{\langle\varepsilon\rangle^{\mathrm{i}}}\left[\langle\overline{\Omega}_{ik}\rangle^{\mathrm{v}}\langle\overline{S}_{kj}\rangle^{\mathrm{v}} + \langle\overline{\Omega}_{jk}\rangle^{\mathrm{v}}\langle\overline{S}_{ki}\rangle^{\mathrm{v}}\right]\right)^{\mathrm{NL2}} \\
&\quad - \left(c_{3\mathrm{NL}}\mu_{t_\varphi}\frac{\langle k\rangle^{\mathrm{i}}}{\langle\varepsilon\rangle^{\mathrm{i}}}\left[\langle\overline{\Omega}_{ik}\rangle^{\mathrm{v}}\langle\overline{\Omega}_{jk}\rangle^{\mathrm{v}} - \frac{1}{3}\langle\overline{\Omega}_{lk}\rangle^{\mathrm{v}}\langle\overline{\Omega}_{lk}\rangle^{\mathrm{v}}\delta_{ij}\right]\right)^{\mathrm{NL3}} - \frac{2}{3}\phi\delta_{ij}\rho\langle k\rangle^{\mathrm{i}}
\end{aligned} \tag{9.88}$$

where $\delta_{\mathrm{ij}}$ is the Kronecker delta; the superscripts (L and NL) indicate Linear and Non-Linear contributions, $\mu_{t_\phi}$ is again the macroscopic turbulent viscosity given by

Equation (9.84), $\langle \overline{D}_{ij}\rangle^{v}$ and $\langle \overline{\Omega}_{ij}\rangle^{v}$ are the deformation and vorticity tensors, written in the indexed form, respectively, as:

$$\langle \overline{D}_{ij}\rangle^{v} = \left( \frac{\partial \overline{u}_{D_i}}{\partial x_j} + \frac{\partial \overline{u}_{D_j}}{\partial x_i} \right), \quad \langle \overline{\Omega}_{ij}\rangle^{v} = \left( \frac{\partial \overline{u}_{D_i}}{\partial x_j} - \frac{\partial \overline{u}_{D_j}}{\partial x_i} \right) \tag{9.89}$$

Table 9.21 shows the different values of $c_\mu$, $c_{1NL}$, $c_{2NL}$ and $c_{3NL}$ proposed in the literature. Note that Equation (9.82) is recovered if constants $c_{1NL}$, $c_{2NL}$ and $c_{3NL}$ in (9.88) are set to zero. It is important to emphasize here that values in Table 9.21 were obtained in the literature for models developed for clear medium although Equation (9.81) will be applied to both porous and clear domains. Here, in the absence of better information, no modifications are introduced in the constants or model parameters when Equation (9.81) is used within the porous structure.

### 9.5.2.1. *Interface Conditions*

The interface between the porous medium and the clear fluid is treated following Ochoa-Tapia and Whitaker (1995a), and the interface conditions can be expressed as:

$$\overline{\mathbf{u}}_{\mathrm{D,clear\ medium}} = \overline{\mathbf{u}}_{\mathrm{D,porous\ medium}} \tag{9.90}$$

$$\langle \overline{p}\rangle^{i}_{\mathrm{clear\ medium}} = \langle \overline{p}\rangle^{i}_{\mathrm{porous\ medium}} \tag{9.91}$$

$$\frac{1}{\phi}\frac{\partial U_{\mathrm{D}\|,\mathrm{porous\ med.}}}{\partial n} - \frac{\partial U_{\mathrm{D}\|,\mathrm{clear\ med.}}}{\partial n} = \frac{\beta}{\sqrt{K}} U_{\mathrm{D}\|,\mathrm{porous\ med.}} \tag{9.92}$$

where $U_{\mathrm{D}\|}$ is the component of the average surface velocity parallel to the interface, $n$ is the coordinate normal to the interface from the porous medium to the clear medium,

**Table 9.21.** Non-linear turbulence models.

| Models | $c_\mu$ | $c_{1NL}$ | $c_{2NL}$ | $c_{3NL}$ | Extra terms |
|---|---|---|---|---|---|
| Nisizima and Yoshizawa (1987) | 0.09 | −0.76 | 0.18 | 1.04 | |
| Speziale (1987) | 0.09 | −0.1512 | 0.0 | 0.0 | $T_{SPE}$ |
| Rubinstein and Barton (1990) | 0.0845 | 0.68 | 0.14 | −0.56 | |
| Shih *et al.* (1993) | $\frac{2/3}{1.25+s+0.9\Omega}$ | $\frac{0.75}{c_\mu(1000+s^3)}$ | $\frac{3.8}{c_\mu(1000+s^3)}$ | $\frac{4.8}{c_\mu(1000+s^3)}$ | |
| Park and Sung (1995) | 0.09 | 0.6 | 0.4 | 0.005 | |

$T_{SPE} = -0.3024\mu_{t_\phi}\frac{\langle k\rangle^i}{\langle\varepsilon\rangle^i}\left\{\tilde{D}_{ij} - \frac{1}{3}\tilde{D}_{mm}\delta_{ij}\right\}$, where: $\tilde{D}_{ij} = \frac{\partial\langle\overline{D}_{ij}\rangle^v}{\partial t} + \vec{V}_D\cdot\langle\overline{D}_{ij}\rangle^v - \frac{\partial V_{iD}}{\partial x_k}\langle\overline{D}_{kj}\rangle^v - \frac{\partial V_{jD}}{\partial x_k}\langle\overline{D}_{ki}\rangle^v$

and, $s = \frac{\langle k\rangle^i}{\langle\varepsilon\rangle^i}\sqrt{\frac{1}{2}\langle\overline{D}_{ij}\rangle^v\langle\overline{D}_{ij}\rangle^v}$, $\Omega = \frac{\langle k\rangle^i}{\langle\varepsilon\rangle^i}\sqrt{\frac{1}{2}\langle\overline{\Omega}_{ij}\rangle^v\langle\overline{\Omega}_{ij}\rangle^v}$.

and $\beta$ is a coefficient which expresses the stress jump condition at the interface. For all the cases treated in this article the coefficient $\beta$ was assumed to be null, i.e., $\beta = 0$. The effect of using a different value of such coefficient is shown below.

### 9.5.3. Results and discussion

Preliminary results for unobstructed flow past a back step were obtained in order to assess the performance of the linear and non-linear turbulence models in clear domains. Numerical parameters for this cases were $a = 0$, $c_F = 0$, $\phi = 1$ and $K \to \infty$. Figures 9.79, 9.80 and 9.81 show calculations for the axial velocity, turbulence intensity and shear stress, respectively, compared with experimental data reported by Kim *et al.* (1980). The figures indicate that overall behavior of the mean and statistical flow is reproduced by both linear and non-linear theories, the latter method being of superior quality when comparing results for the mean and statistical quantities.

Grid-independence studies were also conducted with the aim of checking the behavior of the solution as the grid size was varied. Table 9.22 presents calculations for the reattachment length $x_R/H$ for different grids. One can see that for grids greater than $200 \times 60$ grid nodes there is no detectable change in the calculated reattachment length. For that, in this work all results are shown for a mesh of size $200 \times 60$. With unobstructed results validated, calculations involving the mentioned porous insert can be better assessed.

As such, in the following figures, the effect of coefficient $\beta$, channel length $L/H$, porosity, thickness and permeability of porous insert on the flow pattern will be shown, for turbulent flow, using both the linear and non-linear models. In each figure the streamlines

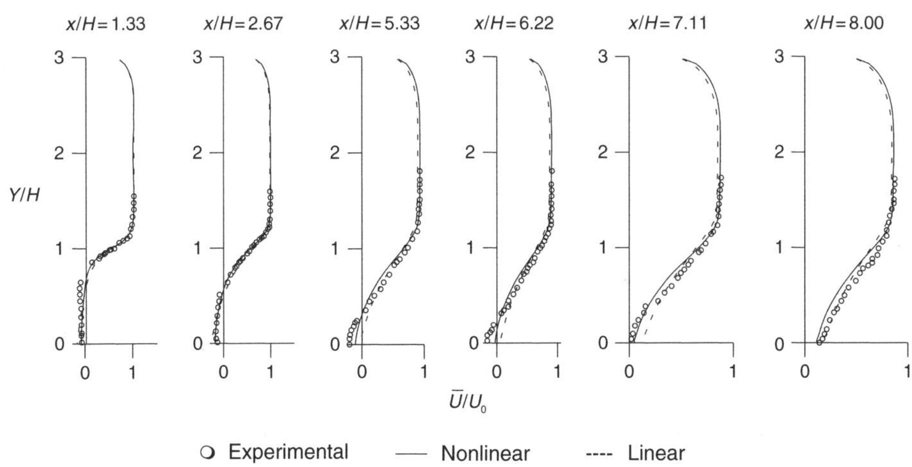


**Figure 9.79.** Axial mean velocity profiles along axial coordinate.

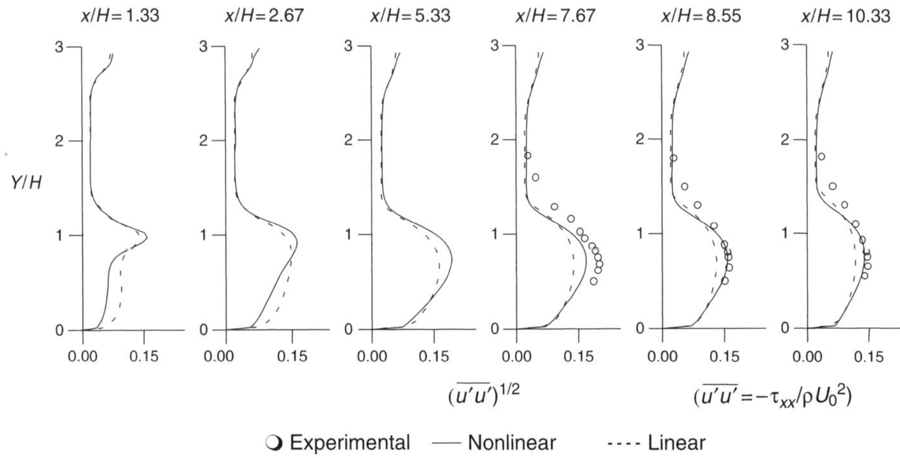


**Figure 9.80.** Non-dimensional turbulence intensity along axial coordinate compared with experiments by Kim *et al.* (1980).

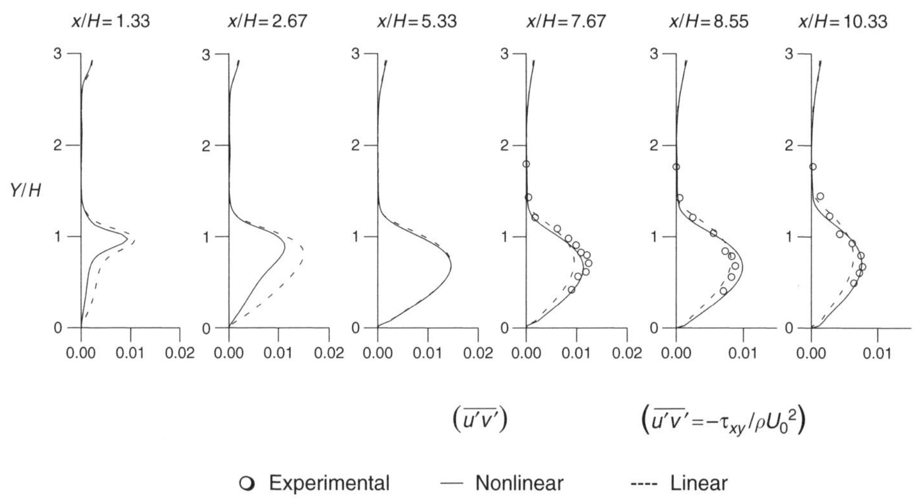


**Figure 9.81.** Non-dimensional turbulent shear stress compared with experiments by Kim *et al.* (1980).

are analyzed without the porous insert and with the porous material for the following thickness: $a = 0.15H$, $a = 0.30H$ and $a = 0.45H$, where $H$ is the step height.

The effect of using a different coefficient $\beta$ is shown in Figure 9.82. One can see that the flow pattern presents nearly the same behavior regardless of the value used for the

**Table 9.22.** Separation length as a function of grid size.

| Model | Dimensions (m) | Grid size | Reattachment length: $x_R/H$ | Percent deviation from experimental value of $x_R/H = 7.0$ |
|---|---|---|---|---|
| L_HRN | $1.5 \times 0.3$ | $150 \times 45$ | 5.50 | −21.43 |
| NL_HRN | | | 6.50 | −7.14 |
| L_HRN | $1.5 \times 0.3$ | $200 \times 60$ | 5.55 | −20.71 |
| NL_HRN | | | 6.45 | −7.86 |
| L_HRN | $1.5 \times 0.3$ | $200 \times 75$ | 5.55 | −20.71 |
| NL_HRN | | | 6.45 | −7.86 |
| L_HRN | $1.8 \times 0.3$ | $240 \times 60$ | 5.55 | −20.71 |
| NL_HRN | | | 6.45 | −7.86 |

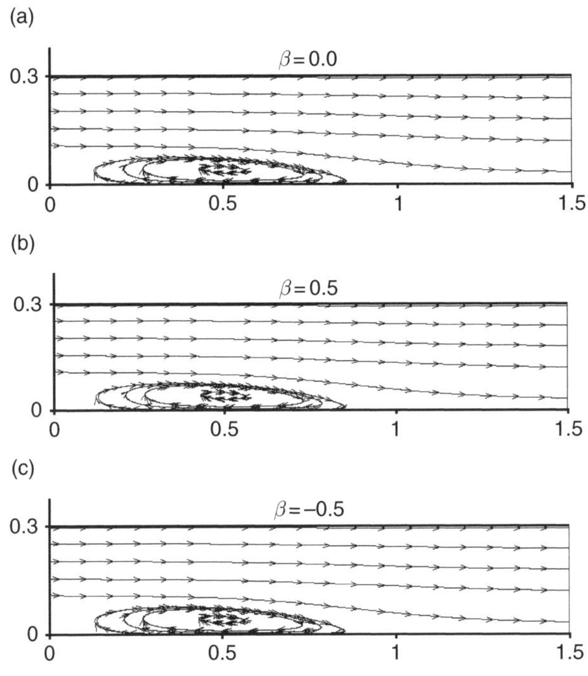


**Figure 9.82.** Calculated flow pattern using a linear model with $a = 0.15H$ m, $K = 10^{-6}$ m$^2$, $\phi = 0.85$: (a) $\beta = 0.0$, (b) $\beta = 0.5$; (c) $\beta = -0.5$.

jump coefficient in Equation (9.92). Further calculations for the friction coefficient along the bottom wall defined as

$$C_f = \frac{\tau_w}{\rho U_0^2/2} \tag{9.93}$$

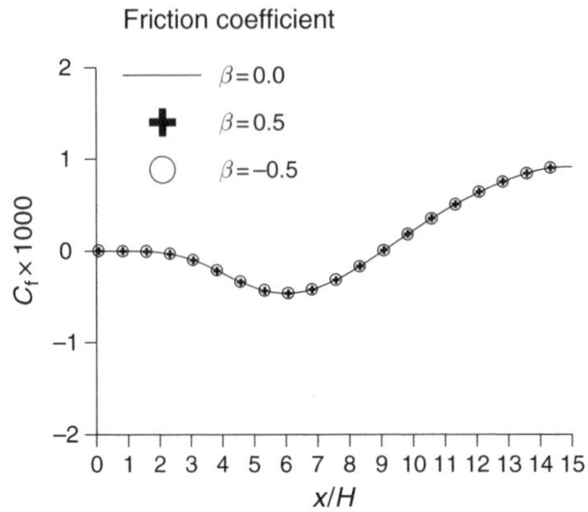


**Figure 9.83.** Friction coefficient at bottom surface for $\beta$ equal to $-0.5$, 0.0 and 0.5.

are presented in Figure 9.83, where a similar behavior can be noted. Since the flow in the geometry of Figure 9.78 is mostly *perpendicular* to the interface, the shear stress caused by the fluid at that location is negligible. In fact, previous work on the stress jump condition across an interface for laminar (Silva and de Lemos, 2003a) and turbulent flow (Silva and de Lemos, 2003b) *parallel* to a layer of porous material in a channel indicated a substantial modification of the velocity pattern depending on the mentioned $\beta$ parameter. Therein, however, the streamlines were aligned to the interface position, corresponding to a different flow configuration than the one explored in the present work.

An indication of the appropriateness of the channel length value used, $L/H = 15$, can be drawn from analyzing Figure 9.84 where one can see that the size and shape of the recirculating bubble is nearly the same if the channel is increased. Figure 9.85 further indicates that no influence is also detected on the friction coefficient, calculated by Equation (9.93), if the length of the computational domain is increased. Preliminary results shown so far support the used of $200 \times 60$ grid nodes, $L/H = 15$ and $\beta = 0$ in all computations to be presented.

Therefore, Figures 9.86, 9.87 and 9.88 show comparisons of streamlines between the linear and non-linear closures considering the following permeability and porosity combinations: Figure 9.86 – $K = 10^{-6}\,\mathrm{m}^2$, $\phi = 0.65$; Figure 9.87 – $K = 10^{-6}\,\mathrm{m}^2$, $\phi = 0.85$ and Figure 9.88 – $K = 10^{-7}\,\mathrm{m}^2$, $\phi = 0.85$. It can be seen that the size of the recirculation bubble simulated by the linear model, in all cases with porous inserts, is shorter than the one calculated by non-linear theories. Unfortunately, no experimental data seems to be available in the literature documenting measurements of flow properties in a back step with a porous insertion. Also, as the thickness of insert is increased, the recirculation bubble decreases, and for $a = 0.45H$, the recirculation bubble is nearly suppressed, independently of the turbulence model used. A possible explanation for this behavior is

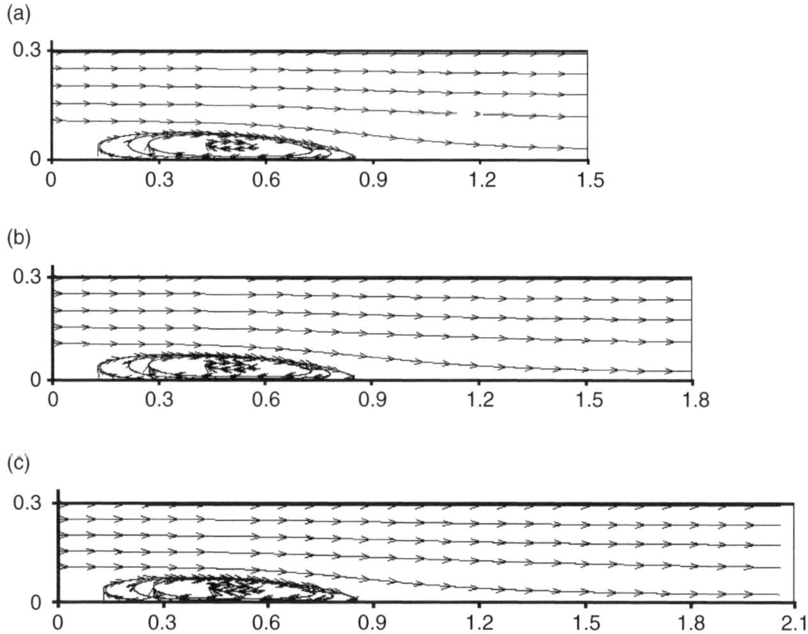


**Figure 9.84.** Calculated flow pattern using a linear model with $a = 0.15H$ m, $K = 10^{-6}$ m$^2$, $\phi = 0.85$: (a) $L/H = 15$, grid: $200 \times 60$; (b) $L/H = 18$, grid: $240 \times 60$; (c) $L/H = 21$, grid: $280 \times 60$.

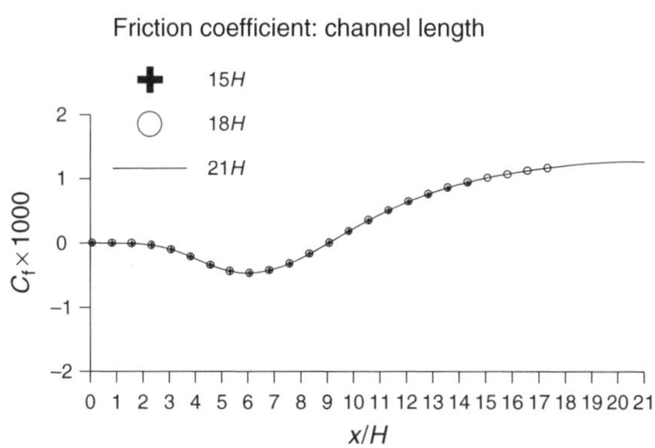


**Figure 9.85.** Friction coefficient at bottom surface for different values of $L/H$.

that the presence of the porous substrate tends to flatten the Darcy velocity profile due to the strong action of the two additional flow resistance terms caused by the solid phase and modeled by the last two terms on the rhs of Equation (9.81). As such, after a certain developing thickness within the porous matrix, the profile is sufficiently "flat" so that the

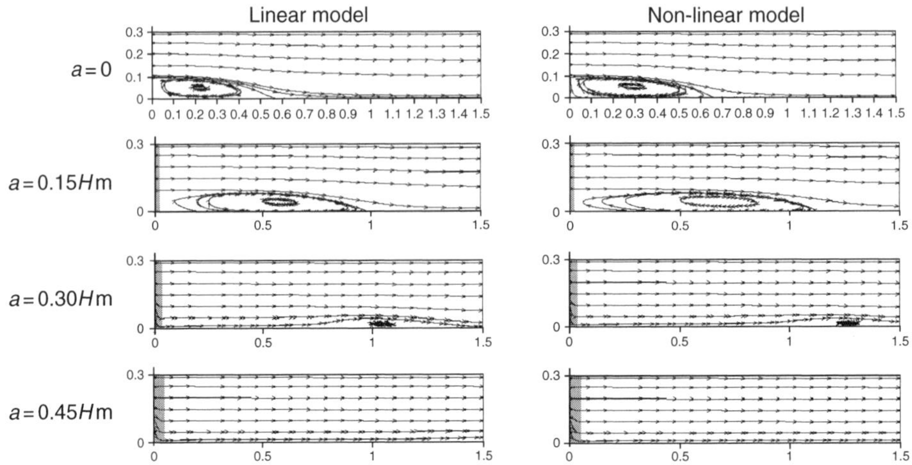


**Figure 9.86.** Comparison of streamlines between the linear and non linear models for back step flow with porous insert, $K = 10^{-6}\,\mathrm{m}^2$, $\phi = 0.65$.

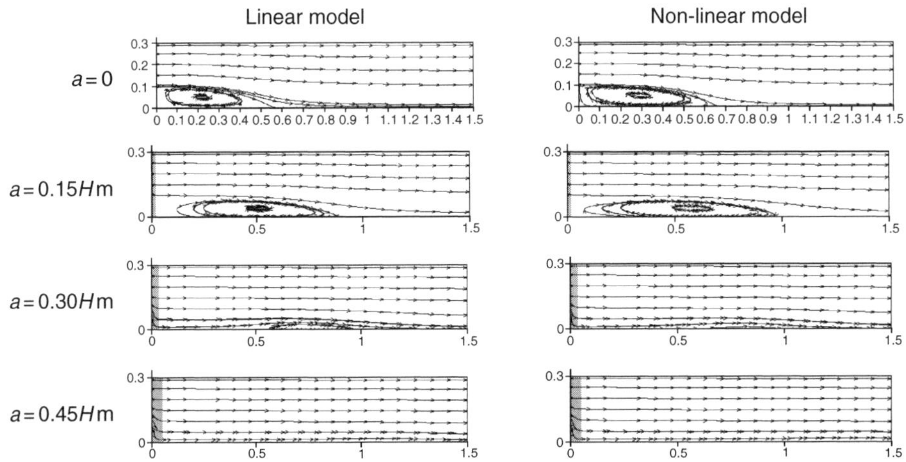


**Figure 9.87.** Comparison of streamlines between the linear and non linear models for back step flow with porous insert, $K = 10^{-6}\,\mathrm{m}^2$, $\phi = 0.85$.

fluid is "delivered" to the channel with a uniform pressure at each cross-station along the longitudinal $x$ direction.

From the figures one can further observe that the permeability $K$ and the porosity $\phi$ of the porous insert also plays a role in determining the flow pattern. However, their influence on the flow distribution past the obstacle seems to be not as intense as the

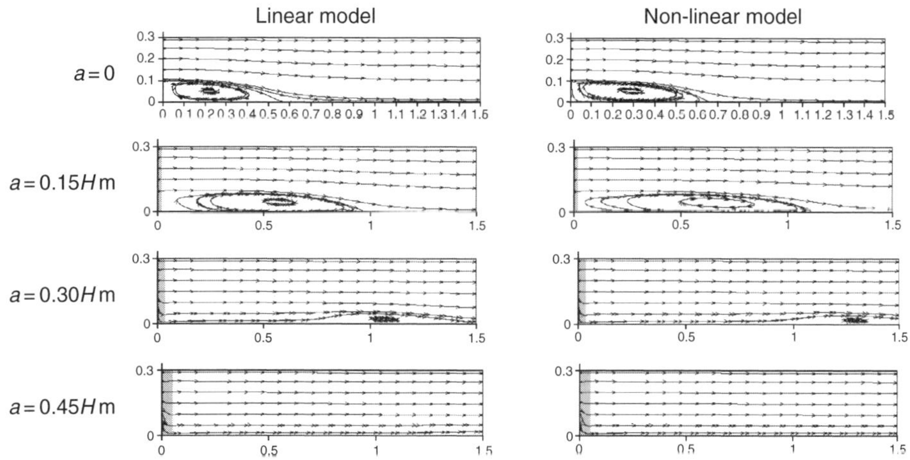


**Figure 9.88.** Comparison of streamlines between the linear and non linear models for back step flow with porous insert, $K = 10^{-7}\,\mathrm{m}^2$, $\phi = 0.85$.

effect of the thickness $a$. Or say, by just increasing the value of $a$, one can smooth the flow past the expansion, damping any existing recirculating stream.

Finally, Figures 9.89, 9.90 and 9.91 show the mean velocity field $\overline{U}/U_0$ at some stations along the channel with the porous insert. It can be noted that the deviations between the results produced with linear and non-linear theory decrease as the thickness of insert increases. Further, for $a = 0.45H$, both models produce nearly the same results.

### 9.5.4. Section Summary

In this work, two turbulence models (linear and non-linear), using wall functions, have been used to simulate turbulent flow past a backward-facing step with a porous insert. Parameters such as porosity $\phi$, permeability $K$, and thickness $a$ of the porous material were varied in order to analyze their effects on the flow pattern.

For validation, results without the insert were compared with experimental data of Kim *et al.* (1980). The experimental value for the separation length given in the literature is $x_R/H = 7.0$. The linear and non-linear models resulted in $x_R/H = 5.55$ and $x_R/H = 6.45$, respectively, indicating an advantage of non-linear closures in predicting more realistic results.

Figures 9.86, 9.87 and 9.88 showed that the recirculating bubble simulated with the linear model was always shorter than the one calculated with non-linear theories. Also, results indicate that the thickness of the insert had a more pronounced effect in suppressing the recirculation bubble than other parameters such as the permeability or the porosity. It has also been observed that the total damping of the recirculation bubble occurred for $a = 0.45H$, independently of the turbulence model employed.

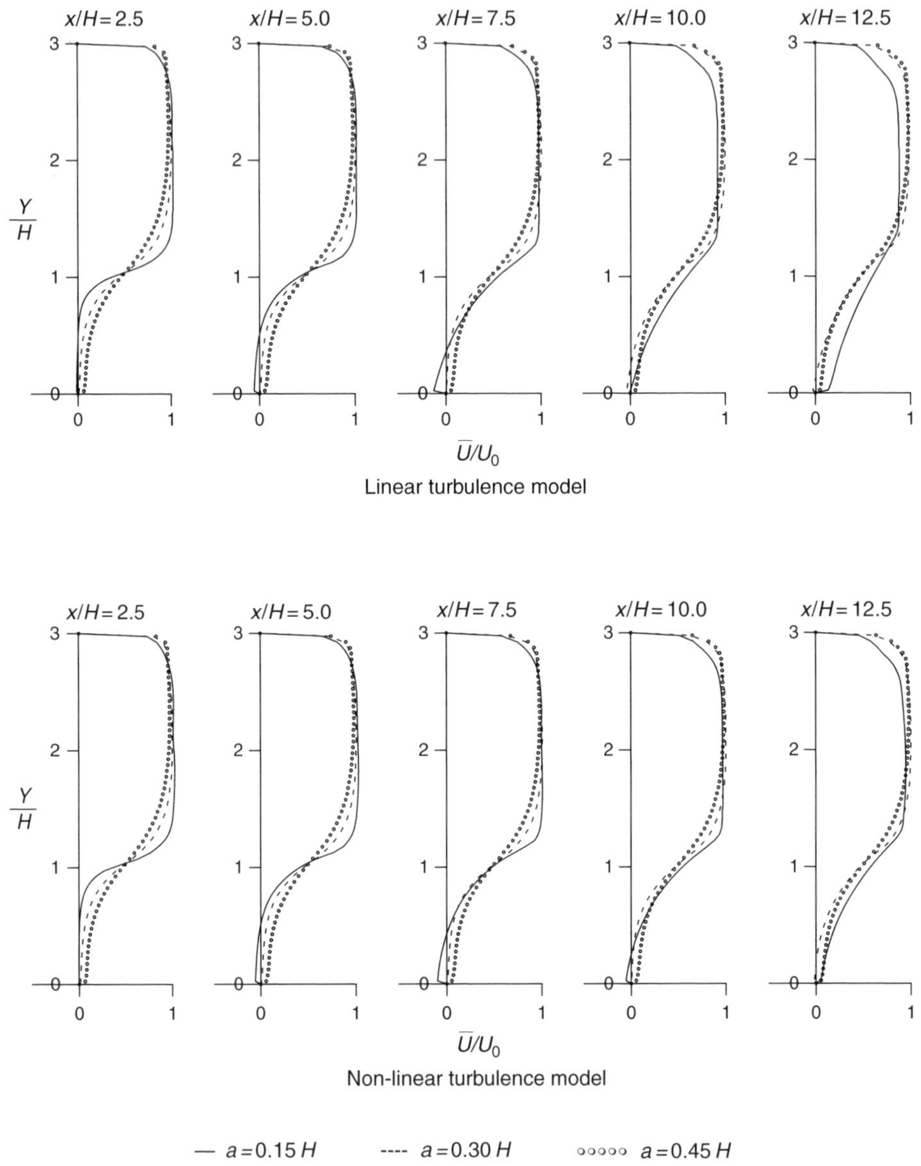


**Figure 9.89.** Mean velocity field simulated by linear and non-linear models. Porous insert with $K = 10^{-6}\,\text{m}^2$, $\phi = 0.65$.

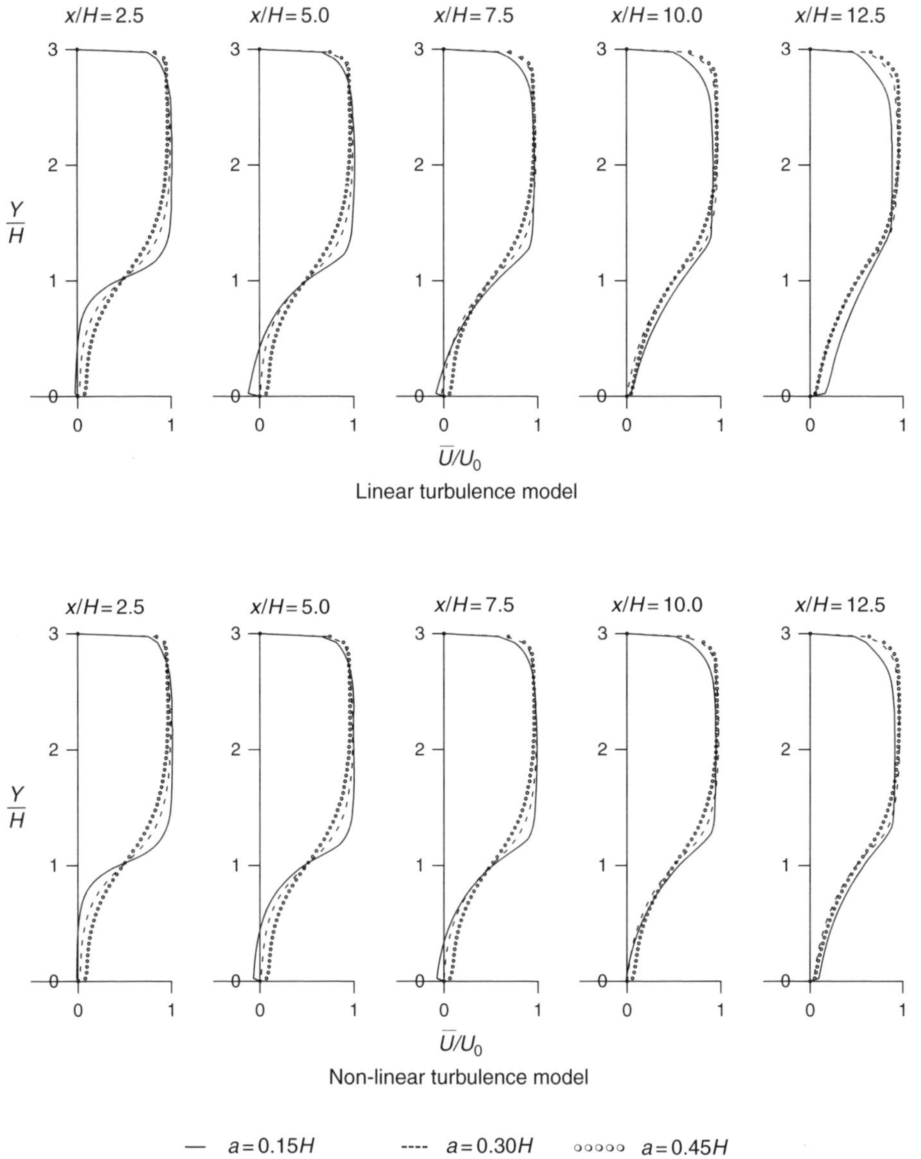


**Figure 9.90.** Mean velocity field simulated by linear and non-linear models. Porous insert $K = 10^{-6}\,\text{m}^2$, $\phi = 0.85$.

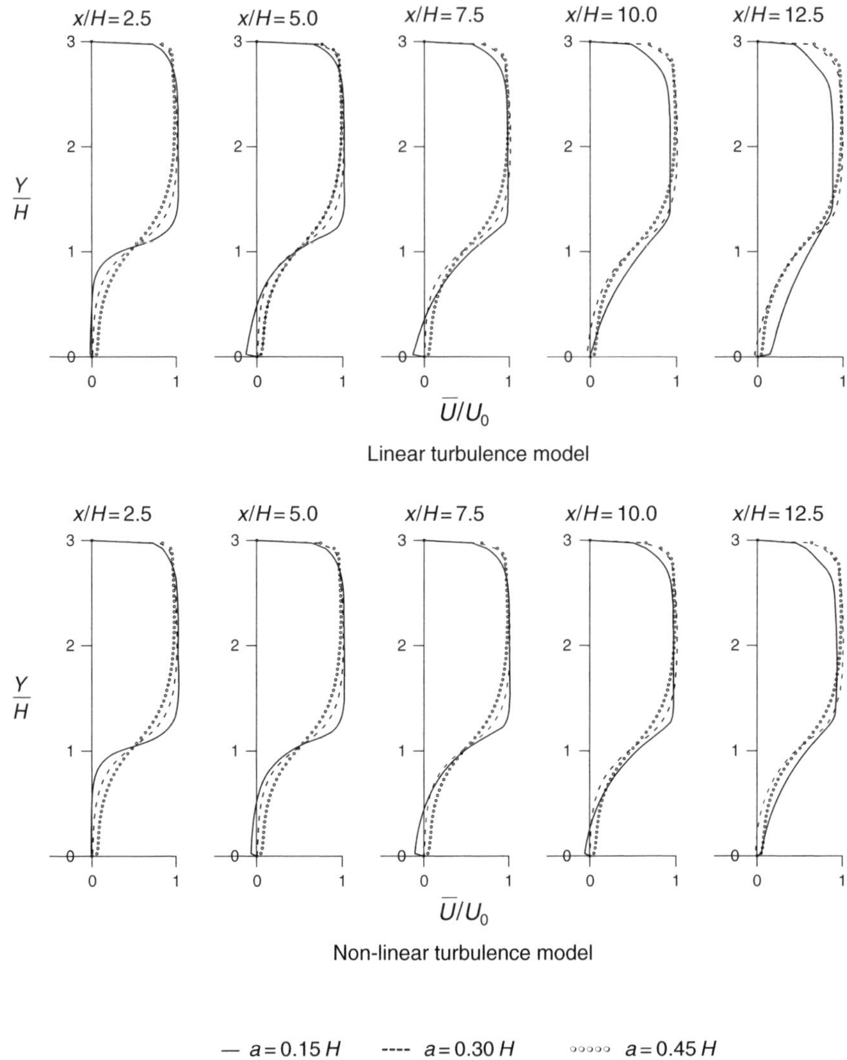


**Figure 9.91.** Mean velocity field simulated by linear and non-linear models. Porous insert with $K = 10^{-7}\,\text{m}^2$, $\phi = 0.85$.

In summary, the following conclusion can then be drawn from this work:

(1) The developed code and numerical methodology used are in agreement with findings in the literature as far as simulating unobstructed sudden expanding flows in channels.
(2) For the configuration in question, which includes a porous insert of thickness $a$ at the sudden expansion, the recirculating bubble calculated with linear models gave

shorter reattachment lengths than those simulated via a non-linear stress–strain rate relationship.

(3) For the cases analyzed here, the thicker the insert, the lower the differences in the value of $x_R$ calculated with the two models. This behavior might be explained by the fact that inside the porous material additional forces exerted by the solid on the fluid tend to flatten the Darcy velocity profiles. As such, as the porous matrix gets thicker, recirculating bubbles tend to disappear regardless of the turbulence model used.

# References


Adnani, P., Catton, I. and Abdou, M.A., 1995, Non-Darcian Forced Convection in Porous Media with Anisotropic Dispersion, *J. Heat Transfer*, 117, 447–451.

AGARD-AR-291, 1991, Technical Status Review – Appraisal of the Suitability of Turbulence Models in Flow Calculation.

Anderson, D.A., Tannehill, J.C. and Pletcher, R.H., 1984, *Computational Fluid Mechanics and Heat Transfer*, Hemisphere, New York.

Antohe, B.V. and Lage, J.L., 1997, A General Two-equation Macroscopic Turbulence Model for Incompressible Flow in Porous Media, *Int. J. Heat Mass Tran.*, 40, 3013–3024.

Assato, M. and de Lemos, M.J.S., 2000, Numerical Treatment and Applications of a Non-Linear Eddy Viscosity Model for High and Low Reynolds Formulations (In Portuguese), *Proc. of 2nd ETT – Brazilian School on Transition and Turbulence*, Uberlândia, MG, Brazil, December 11–15, pp. 459–483.

Assato, M. and de Lemos, M.J.S., 2001, Numerical Simulation of Turbulent Flow through Axisymmetric Stenosis Using Linear and Non-Linear Eddy-Viscosity Models, *Proc. of 16th COBEM – Braz. Cong. Mech Eng.* (on CD-ROM), Uberlândia, MG, Brazil.

Assato, M., Pedras, M.H.J. and de Lemos, M.J.S., 2005, Numerical Solution of Turbulent Channel Flow Past a Backward-Facing Step with a Porous Insert Using Linear and Nonlinear k-$\varepsilon$ Models, *J. Porous Media*, 8(1), 13–29.

Aziz, K. and Settari, A., 1979, *Petroleum Reservoir Simulation*, Applied Science Publishers, New York.

Baliga, B.H. and Patankar, S.V., 1980, A Novel Finite Element Formulation for Convection Diffusion Problems, *Num. Heat Transfer*, 3, 393–409.

Baliga, B.H., Phan, T.T. and Patankar, S.V., 1983, Solution of some Two-dimensional Incompressible Flow and Heat Transfer Problems using a Control-Volume Finite-Element, *Num. Heat Transfer*, 6, 263–282.

Barakos, G., Mitsoulis, E. and Assimacopoulos, D., 1994, Natural Convection Flow in a Square Cavity Revited: Laminar and Turbulent Models with Wall Function, *Int. J. Num. Meth. Fl.*, 18, 695–719.

Baytas, A.C. and Pop, I., 1999, Free Convection in Oblique Enclosures Filled with a Porous Medium, *Int. J. Heat Mass Tran.*, 42, 1047–1057.

Beam, R.M. and Warming, R.F., 1978, An Implicit Factored Scheme for the Compressible Navier-Stoes Equation, *AIAA J.*, 16, 393–402.

Bear, J., 1972, *Dynamics of Fluids in Porous Media*, American Elsevier Pub. Co., New York.

Bear, J. and Bachmat, Y., 1967, A Generalized Theory on Hydrodynamic Dispersion in Porous Media, *I.A.S.H. Symposium on Artificial Recharge and Management of Aquifers*, Haifa, Israel, Pub. No 72, I.A.S.H 7–16.

Bear, J. and Bachmat, Y., 1990, *Introduction to Modeling of Transport Phenomena in Porous Media*, Kluwer Academic Publishers, Dordrecht.

Beckermann, C., Viskanta, R. and Ramadhyani, S., 1986, A Numerical Study of Non-Darcian Natural Convection in a Vertical Enclosure Filled with a Porous Medium, *Num. Heat Transfer*, 10, 557–570.

Behie, A. and Vinsome, P.K.W., 1982, Block Iterative Methods for Fully Implicit Reservoir Simulation, *Soc. Pet. Eng. J.*, 22.

Behie, G.A. and Forsyth, Jr., P.A., 1984, Incomplete Factorization Methods for Fully Implicit Simulation of Enhanced Oil Recovery, *SIAM J Sci. Stat. Comp.* 5(3).

Bejan, A., 1979, On the Boundary Layer Regime in a Vertical Enclosure Filled with a Porous Medium, *Lett. Heat Mass Trans.*, 6, 93–102.

Bennacer, R., Tobbal, A., Beji, H. and Vasseur, P., 2001, Double Diffusive Convection in a Vertical Enclosure Filled with Anisotropic Porous Media, *Int. J. Therm. Sci.*, 40(1), 30.

Bennacer, R., Beji, H. and Mohamad, A.A., 2003, Double Diffusive Convection in a Vertical Enclosure Inserted with Two Saturated Porous Layers confining a Fluid Layer, *Int. J. Therm. Sci.*, 42(2), 141.

Berner, C., Dust, F. and McEligot, D. M., 1984, Flow around Baffles, *J. Heat Transfer*, 106, 743–749.

Bhattacharya, A., Calmidi, V.V. and Mahajam, R.L., 2002, Thermophysical Properties of High Porosity Metal Foams, *Int. J. Heat Mass Tran.*, 45(5), 1017–1031.

Boussinesq J., 1877, Essai sur La Theorie des Eaux Courants, *Mem. pres. par div. savants a 1 Acad. Sci.*, 23(1), 1–680.

Bradshaw, P., 1978 (ed.), *Turbulence, Topics in Applied Physics*, Vol. 12, 2nd Ed., Springer-Verlag, Heidelberg.

Braga, E.J. and de Lemos, M.J.S., 2002a, Free Convection in Square and Rectangular Cavities Heated from below or on the Left, *Proc. CONEM 2002*, 3rd Congresso Nacional de Engenharia Mecânica, João Pessoa, PB, Brazil, August 12–16.

Braga, E.J. and de Lemos, M.J.S., 2002b, Laminar Natural Convection in Concentric and Eccentric Annuli, *Proc. ENCIT2002*, 9th Brazilian Congress of Thermal Engineering and Sciences (on CD-ROM), Caxambu, MG, Brazil, October 13–17.

Braga, E.J. and de Lemos, M.J.S., 2002c, Natural Convection in Turbulent Regime in Concentric and Eccentric Horizontal Annular Regions, Paper AIAA-2002-3316, *Proc. of 8th AIAA/ASME*, Joint Thermophysics and Heat Transfer Conference, St Louis, Missouri, USA, June 23–27.

Braga, E.J. and de Lemos, M.J.S., 2002d, Natural Convection in Cavities Completely Filled with Porous Material, *Proc. APM2002*, 1st International Conference on Applications of Porous Media, Paper APM-164, Vol. 1, Jerba, Tunisia, June 2–8, pp. 551–560.

Braga, E.J. and de Lemos, M.J.S., 2002e, Turbulent Natural Convection in Enclosures Completely Filled with Porous Material, Paper IMECE2002-34403, 2002 ASME International Mechanical Engineering Congress, New Orleans, LA, USA, November 17–22.

Braga, E.J. and de Lemos, M.J.S., 2003a, Turbulent Heat Transfer in a Horizontal Enclosure with a Thin Porous Obstruction in the Middle, *Proc. IMECE 2003*, ASME International Mechanical Engineering Congress and Exposition, Washington, DC, USA, November 16–21.

Braga, E.J. and de Lemos, M.J.S., 2003b, Turbulent Natural Convection in Horizontal Composite Cavities, Paper IMECE2003-41456, *Proc. IMECE2003*, 2003 ASME International Mechanical Engineering Congress, Washington, DC, USA, November.

Braga, E.J. and de Lemos, M.J.S., 2004, Turbulent Natural Convection in a Porous Square Cavity Computed with a Macroscopic k-$\varepsilon$ Model, *Int. J. Heat Mass Tran.*, 47(26), 5639–5650.

Braga, E.J. and de Lemos, M.J.S., 2005, Heat Transfer in Enclosures having a Fixed Amount of Solid Material Simulated with Heterogeneous and Homogeneous Models, *Int. J. Heat Mass Tran.*, Vol. 48(23–24), 4748–4765.

Brenner, H., 1980, Dispersion Resulting from Flow through Spatially Periodic Porous Media, *Phil. Trans. Roy. Soc.*, 297, 81–133.

Breugem, W.P. and Boersma, B.J., 2005, Direct Numerical Simulations of Turbulent Flow over a Permeable Wall using a Direct and a Continuum Approach, *Phys. Fluids*, 17(2).

Breugem, W.P., Boersma, B.J. and Uittenbogaard, R.E., 2004, Direct Numerical Simulations of Plane Channel Flow Over a 3d Cartesian Grid Of Cubes, *Proc. of ICAPM2004 – International Conference on Applications of Porous Media 2004*, pp. 27–35, Évora, Portugal, May 24–27.

Brinkman, H.C., 1947, A Calculation of the Viscous Force Excerted by a Flowing Fluid on a Dense Swarm of Particles, *Appl. Sci. Research*, A1, 27–34.

Carbonell, R.G. and Whitaker, S., 1984. Heat and Mass Transfer in Porous Media, in *Fundamentals of Transport Phenomena in Porous Media*, Bear and Corapcioglu (eds), Martinus Nijhoff Publishers, Dordrecht, 121–198.

Carman, P.C., 1956, *Flow of Gases through Porous Media*, Academic Press, New York.

Catton, I., 1978, Natural Convection in Enclosures, *Proc. of 6th. Intern. Heat Transfer Conf.*, Toronto, Canada, Vol. 6, p. 13.

Chan, E.C., Lien, F.S. and Yovanovich, M.M., 2000, Macroscopic Numerical Study of Forced Convective Heat Transfer in a Back-step Channel through Porous Layer, *Proc. of NTHC2000 – ASME National Heat Transfer Conference* (on CD-ROM), ASME-HTD-I463CD, Paper NHTC2000-12118, ISBN:0-7918-1997-3, August 20–22, Pittsburgh, Pennsylvania, USA.

Chapman, D.R., 1979, Computational Aerodynamics Development and Outlook, *AIAA J*, 17, 1293–1313.

Charrier-Mojtabi, M.C., 1997, Numerical Simulation of Two – and – Three-dimensional Free Convection Flows in a Horizontal Porous Annulus Using a Pressure and Temperature Formulation, *Int. J. Heat Mass Tran.*, 40(7), 1521–1533.

Chen, C.J. and Jaw, S.-Y., 1998, *Fundamentals of Turbulence Modeling*, Taylor and Francis, New York.

Chen, B.C.-J., Sha, W.T., Doria, M.L., Schmitt, R.C. and Thompson, J.F., 1980, BODYFIT-1FE: A Computer Code for Three-Dimensional Steady-State/Transient Single-Phase Rod-Bundle Thermal-Hydraulic Analysis, ANL-80-127.

Chen, M.-H., Hsu, C.-C. and Shyy, W., 1991, Assessment of TVD Schemes for Inviscid and Turbulent Flow Computation, *Int. J. Num. Meth. Fl.*, 12, 161–177.

Chien, T.H., Chen, B.C.J, Sha, W.T. and Kim, J.H., 1983, BODYFIT-2PE and Its Application to CHF Prediction of GE $3 \times 3$ Rod Bundle, *Proc. 2nd. Int. Top. Meet. Nucl. React. Therm-Hydr.*, vol. 1, Santa Barbara, 498–507.

Chorin, A.J., 1967, A Numerical Method for Solving Incompressible Viscous Flow Problems, *J. Comput. Phys*, 2, 12–26.

Chorin, A.J., 1971, A Numerical Solution of the Navier–Stokes Equation, *Math. Comp.*, 22, 2745–2762.

Churchill, S., 1992, Free Convection in Layers and Enclosures, Chapter 2.5–8 in *Handbook of Heat Exchanger Design*, G.F. Hewitt (ed.), Hemisphere Pub. Corp., New York.

Collins, R.E., 1961, *Flow of Fluids through Porous Materials*, Reinhold, New York.

Da Silva Miranda, B.M. and Anand, N.K., 2004, Convective Heat Transfer in a Channel with Porous Baffles, *Num. Heat Transfer – Part A*, 46(5), 425–452.

Darcy, H., 1856, *Les Fontaines Publiques de la Ville de Dijon*, Victor Dalmond, Paris.

Davidson, P.A., 2004, *Turbulence – An Introduction for Scientists and Engineers*, Oxford University Press.

de Lemos, M.J.S., 1988a, An Alternating Symmetrical Coupled Gauss-Seidel procedure for Two and Three-dimensional Recirculating Flows, *ASME Winter Meeting*, Chicago, USA, November 28–December 2.

de Lemos, M.J.S., 1988b, Anisotropic Turbulent Transport Modeling for Rod-Bundle, *Int. J. Heat and Technology*, 6(1–2), 27–37.
de Lemos, M.J.S., 1990, Computation of Laminar Axi-Symmetric Recirculating Flows Using Primitive Variables and a Block-Implicit Scheme, *Proc. 3rd Braz. Therm. Sci. Meeting*, 1, 375–380.
de Lemos, M.J.S., 1992a, Computation of Buoyancy-Driven Flows Using a Block-Implicit Numerical Scheme, in *Modern Developments in Numerical Simulation of Flow and Heat Transfer*, Chen, J.L.S. and Vafai, K. (eds), ASME-HTD, vol. 194, 83–89, ASME, USA, *Proc. of the 28th National Heat Transfer Conference*, San Diego, CA, August 9–12.
de Lemos, M.J.S., 1992b, Simulation of Swirling Flow in a Model Combustor Using a Locally-Coupled Numerical Method, in *Fundamental and Applied Heat Transfer Research in Gas Turbines Engines*, Metzger, D.E. and Crawford, M.E. (eds), ASME-HTD, Vol. 226, 79–84, ASME, USA, *Proc. of Winter Annual Meeting of the ASME*, Anaheim, CA, November 8–13.
de Lemos, M.J.S., 1992c, Computation of Laminar Flows in Sub-channels of Rod-arrays using Generalized Coordinates, *Proc. of Joint ANS/ASME Conf.*, San Diego, USA, August, pp. 104–112.
de Lemos, M.J.S., 1992d, Numerical Investigation of laminar flow in ducts with Cusped Corners, *Proc. of Xth Heat Transfer Conf.*, Genova, Italy, June 25–27, pp. 223–233.
de Lemos, M.J.S., 1994a, A Locally-Coupled Numerical Method for High Performance Computational Fluid Dynamics, *Proc. of SUPERCOMP94 – 1994 Supercomputing Seminar*, CEN-SUP/UFRGS, Porto Alegre, Brazil, pp. 45–48.
de Lemos, M.J.S., 1994b, The effect of Grid Non-Orthogonality in Predicting Axial Flow and Wall Shear Stress in ducts with Cusped Corners, *Proc. of 5th Braz. Therm. Sci. Meeting*, São Paulo, December, pp. 317–320.
de Lemos, M.J.S., 1996, Computation of Heated Swirling Flows with a Fully-Coupled Numerical Scheme, *1996 Nat. Heat Transfer Conf.*, Houston, USA, August 17–20.
de Lemos, M.J.S., 1997a, Locally-Coupled Numerical Solution of Thermally-Driven Cavity Flows, *Proc. of COBEM97 – 14th Brazilian Congress of Mechanical Engineering* (on CD-ROM), Bauru, São Paulo, Brazil, December 8–12.
de Lemos, M.J.S., 1997b, Cell-Implicit Numerical Computation of Flow Field and Heat Transfer in Inclined Cavities, *Proc. of COBEM97 – 14th Brazilian Congress of Mechanical Engineering* (on CD-ROM), Bauru, São Paulo, Brazil, December 8–12.
de Lemos, M.J.S., 2000, Flow and Heat Transfer in Rectangular Enclosures using a New Block-implicit Numerical Method, *Numer. Heat Transfer, Part B – Fundamentals*, 37(4), 489–508.
de Lemos, M.J.S., 2003a, A Block-Implicit Method for Numerical Simulation of Swirling Flows in a Model Combustor, *Int. Commun. Heat Mass*, 30(3), 369–378.
de Lemos, M.J.S., 2003b, A Block-Implicit Numerical Procedure for Simulation of Buoyant Swirling Flows in a Model Furnace, *Int. J. Num. Meth Fl.*, 43(3), 281–299.
de Lemos, M.J.S., 2004, Turbulent Kinetic Energy Distribution Across the Interface Region between a Porous Medium and a Clear Fluid, *Proc. ICTEA – Int. Conf. Thermal Eng. Theory and Applications*, Paper ICTEA-TF2-01, Beirut, Lebanon, May 31–June 4.
de Lemos, M.J.S., 2005a, Turbulent Kinetic Energy Distribution across the Interface between a Porous Medium and a Clear Region, *Int. Commun. Heat Mass*, 32(1–2), 107–115.
de Lemos, M.J.S., 2005b, The Double-Decomposition Concept for Turbulent Transport in Porous Media, Chapter 1 in *Transport Phenomena in Porous Media*, Vol. 3, Ingham, D.B., Pop, I. (eds), Elsevier Science, New York, pp. 1–33.
de Lemos, M.J.S., 2005c, Mathematical Modeling and Applications of Turbulent Heat and Mass Transfer in Porous Media, Chapter 10 in *Handbook of Porous Media*, 2nd edn, Vafai, K. (ed.), Taylor and Francis Group, LLC, New York, 409–454.

de Lemos, M.J.S. and Braga, E.J., 2003, Modeling of Turbulent Natural Convection in Porous Media, *Int. Commun. Heat Mass*, 30, 615–624.

de Lemos, M.J.S. and Graminho, D.R., 2005, Turbulent Impinging Jet into a confined Porous Layer, *Proc. of IMECE2005*, Paper IMECE2005-81480, 2005 ASME International Mechanical Engineering Congress and Exposition, Orlando, Florida, USA, November 5–11.

de Lemos, M.J.S. and Magro, V.T., 2003a, Turbulent Free Convection in an Composite Enclosure, *ASME – NHTC2003 Summer Heat Transfer Conference, 2003*, Paper HT2003-47246, Las Vegas, Nevada, USA, July 20–23.

de Lemos, M.J.S. and Magro, V.T., 2003b, Nusselt Number and Temperature Distribution in an Horizontal Cavity containing a Layer of Porous Material at the Bottom, *ASME – IMECE03*, International Mechanical Engineering Congress, 2003, Paper IMECE2003-41468, Washington, DC, USA, November 15–21.

de Lemos, M.J.S. and Mesquita, M.S., 2003, Turbulent Mass Transport in Saturated Rigid Porous Media, *Int. Commun. Heat Mass*, 30, 105–113.

de Lemos, M.J.S. and Pedras, M.H.J., 2000a, Modeling Turbulence Phenomena in Incompressible Flow through Saturated Porous Media, *Proc. of 34th ASME – National Heat Transfer Conference* (on CD-ROM), ASME-HTD-I463CD, Paper NHTC2000-12120, ISBN:0-7918-1997-3, Pittsburgh, Pennsylvania, August 20–22.

de Lemos, M.J.S. and Pedras, M.H.J., 2000b, Simulation of Turbulent Flow through Hybrid Porous Medium-Clear Fluid Domains, *Proc. of IMECE2000 – ASME – Intern. Mech. Eng. Congr.*, ASME-HTD-366-5, ISBN:0-7918-1908-6, Orlando, Florida, USA, November 5–10, pp. 113–122.

de Lemos, M.J.S. and Pedras, M.H.J., 2001, Recent Mathematical Models for Turbulent Flow in Saturated Rigid Porous Media, *ASME Journal of Fluids Engineering*, 123(4), 935–940.

de Lemos, M.J.S. and Rocamora Jr., F.D., 2002, Turbulent Transport Modeling for Heated Flow in Rigid Porous Media, *Proc. 12th Intern. Heat Transfer Conf.*, 4, Kyongyu, Korea, August 23–28, pp. 791–796.

de Lemos, M.J.S. and Santos, N.B., 2004,, Laminar Heat Transfer in a Parallel Plate Channel with Solid and Porous Baffles, *Proc. of IMECE04*, 2004 ASME International Mechanical Engineering Congress, Paper IMECE2004-62362, Anaheim, CA, November 13–20.

de Lemos, M.J.S. and Sesonske, A., 1985, Turbulence Modeling in Combined Convection in Mercury Pipe Flow, *Int. J. Heat Mass Tran.*, 28(6), 1067–1088.

de Lemos, M.J.S. and Silva, R.A., 2003, Turbulent Flow around a Wavy Interface between a Porous Medium and a Clear Domain, *Proc. ASME-FEDSM2003*, Fluids Engineering Division Summer Meeting, Paper FEDSM2003-45457 (on CD-ROM), Honolulu, Hawaii, USA, July 6–11.

de Lemos, M.J.S. and Silva, R.A., 2006, Turbulent Flow over a Layer of a Highly Permeable Medium Simulated with a Diffusion-Jump Model for the Interface, *Int. J. Heat Mass Tran.*, 49(3–4), 546–556.

de Lemos, M.J.S. and Tofaneli, L.A., 2003a, Application of a Macroscopic Turbulence Model to Simulation of Flow in a Channel with equally Spaced Porous Fins, *Proc. of ASME Summer Heat Transfer Conference* (on CD-ROM), Paper HT2003-40248, Las Vegas, NV, USA, July 21–23.

de Lemos, M.J.S. and Tofaneli, L.A., 2003b, Pressure Drop Characteristics of Parallel-Plate Channel Flow with Porous Obstructions at Both Walls, *Proc. of 2003 ASME International Mechanical Engineering Congress* (on CDROM), Paper IMECE2003-41453, Washington DC, November 16–21.

de Lemos, M.J.S. and Tofaneli, L.A., 2004, Modeling of Double-Diffusive Turbulent Natural Convection in Porous Media, *Int. J. Heat Mass Tran.*, 47, 4221–4231.

de Vahl Davis, G., 1983, Natural Convection in a Square Cavity: A Benchmark Numerical Solution, *Int. J. Num. Meth. Fl.*, 3, 249–264.

de Vahl Davis, G. and Jones, I.P., 1983, Natural Convection in a Square Cavity – A Comparison Exercise, *Int. J. Num. Meth. Fl.*, 3, 227–248.

DeWiest, R.J.M., 1969, *Flow through Porous Media*, Academic Press, New York.

Dullien, F.A.L., 1979, *Porous Media Fluid Transport and Porous Structure*, Academic Press, New York.

Ergun, S., 1952, Fluid Flow through Packed Columns, *Chem. Eng. Prog.*, 48(2), 89–94.

Ferguson, W.I. and Wadsley, A.W., 1986, The Construction of Curviline Coordinate Grid Systems for Reservoir Simulation, *SPE Europian Petroleum Conference*, London, October.

Ferziger, J.H. and Periç, M., 1999, *Computational Methods for Fluid Dynamics*, 2nd edn, Springer, New York.

Forchheimer, P., 1901, Wasserbewegung durch Boden, *Z. Ver. Dtsch. Zucker-Ind*, 45, 1782–1788.

Galpin, P.F. and Raithby, G.D., 1986, Numerical Solution of Problems in Incompressible Flow: Temperature of the Temperature Velocity Coupling, *Num. Heat Transfer*, 9, 105–129.

Galpin, P.F., van Doormaal, J.P. and Raithby, G.D., 1985, Solution of Incompressible Mass and Momentum Equations by Aplication of a Coupled Equation Line Solver, *Int. J. Num. Meth. Fl.*, 56, 615–625.

Getachewa, D., Minkowycz, W.J. and Lage, J.L., 2000, A Modified Form of the k-ε Model for Turbulent Flow of an Incompressible Fluid in Porous Media, *Int. J. Heat Mass Tran.*, 43, 2909–2915.

Gibson, M.M. and Launder, B.E., 1976, On the Calculation of Horizontal, Turbulent Free Shear Flow under Gravitational Influence, *J. Heat Transfer ASME*, 98C, 81–87.

Gosman, A.D. and Ideriah, F.J.K, 1976, TEACH-T: A General Computer Program for Two-Dimensional, Turbulent, Recirculating Flows, Fluids Section, Dept. of Mech. Eng. Report (unnumbered), Imperial College, London.

Gosman, A.D, Tsui, Y.Y. and Watkins, A.P., 1984. *Calculation of Three-Dimensional Flows in Engines*, SAE paper 840229.

Goyeau, B., Songbe, J.P. and Gobin, D., 1996, Numerical Study of Double-Diffusive Natural Convection in a Porous Cavity using the Darcy-Brinkman Formulation, *Int. J. Heat Mass Trans.*, 39(7), 1363.

Gratton, L., Travkin, V.S. and Catton, I., 1994, Numerical Solution of Turbulent Heat and Mass Transfer in a Stratified Geostatistical Porous Layer for High Permeability Media, *ASME Proceedings HTD*, Vol. 41, 1–14.

Gray, W.G. and Lee, P.C.Y., 1977, On the Theorems for Local Volume Averaging of Multiphase System, *Int. J. Multiphase Flow*, 3, 333–340.

Gross, R.J., Bear, M.R. and Hickox, C.E., 1986, The Application Of Flux-Corrected Transport (FCT) to High Rayleigh Number Natural Convection in a Porous Medium, *Proc. 8th Int. Heat Transfer Conf.*, San Francisco, CA USA.

Haji-Sheikh, A., Minkowycz, W.J. and Sparrow, E.M., 2004, A Numerical Study of the Heat Transfer to Fluid Flow through Circular Porous Passages, *Num. Heat Transfer* 46(10), 929–955.

Han, N.W., Bhakta, J. and Carbonell, R.G., 1985, Longitudinal and Lateral Dispersion in Packed Beds: Effect of Column Length and Particle Size Distribution, *AICHE J.* 31, 277–288.

Harlow, F.H. and Welch, J.E., 1965, Numerical Calculation of Time-Dependent Viscous Incompressible Flow of Fluid with Free Surface, *Physics Fluids*, 8, 2182–2189.

Harten, A., 1983, High Resolution Schemes for Hyperbolic Conservation Laws, *J. Comput. Phys.*, 49, 357–393.

Harten, A., 1984, On a Class of High Resolution Total Variation Diminishing Schemes, *SIAM J. Numer. Anal.*, 21(1), 1–23.

Henkes, R.A.W.M., Van Der Vlugt, F.F. and Hoogendoorn, C.J., 1991, Natural-Convection Flow in a Square Cavity Calculated with Low-Reynolds-Number Turbulence Models, *Int. J. Heat Mass Tran.*, 34(2), 377–388.

Herring, J.R., 1977, Subgriding-Scale MODELING – An Introduction and Overview, in *Proc. Turbulent Shear Flows I*, Durst, F. *et al.* (ed.), Springer-Verlag, Heidelberg.

Hinze, J.O., 1975, *Turbulence*, 2nd edn, Mc-Graw Hill, New York.

Hirsch, C., 1990, *Numerical Computation of Internal and External Flows*, Vols 1 and 2, Wiley, New York.

Hoffmann, M.R., 2004, Application of a Simple Space-Time Averaged Porous Media Model to Flow in Densely Vegetated Channels, *J. Porous Media*, 7(3), 183–191.

Hogg, S. and Leschiziner, M.A., 1989a, *Int. J. Heat Fluid Fl.* 10(1), 16.

Hogg, S. and Leschiziner, M.A., 1989b, *AIAA J.*, 27(1), 57.

Hollands, K.G.T., Unny, T.E., Raithby, G.D. and Konicek, L., 1976, Free convection Heat Transfer Across Inclined Air Layers, *J. Heat Transfer*, Vol., 189–193.

Hortmann, M., Periç, M. and Scheider, G., 1990, Finite Volume Multigrid Predictions of Laminar Natural Convection: Benchmark solutions, *Int. J. Num. Meth. Fl.*, 11, 189–207.

Houpeurt, A., 1957, Eléments de Mécanique des Fluids dans les Milieux Poreux, *Inst. Français du Pétrole*, Paris.

House, J.M., Beckermann, C. and Smith, T.F., 1990, Effect of a Centered Conducting Body on Natural Heat Transfer in a Enclosure, *Num. Heat Transfer – Part A*, 18, 213–225.

Hsu, C.T., 1999, *J. Heat Transfer*, 121, 733.

Hsu, C.T. and Cheng, P., 1990, Thermal Dispersion in a Porous Medium, *Int. J. Heat Mass Tran.*, 33, 1587–1597.

Hsu, P.F., Howell, J.R. and Matthews, R.D., 1993, A Numerical Investigation of Premixed Combustion within Porous Inert Media, *J. Heat Transfer*, 115, 744–750.

Huang, P.C. and Vafai, K., 1994, Internal Heat Transfer Augmentation in a Channel using an Alternate Set of Porous Cavity-block Obstacles, *Num. Heat Transfer*, 25(5), 519–539.

Huget, H.G., 1985, The Evaluation and Development of Finite Volume Approximation Schemes for Fluid Flow and Heat Transfer Predictions, PhD Thesis, University of Waterloo, Canada.

Hunt, M.L. and Tien, C.L., 1988, Non-Darcyan Convection in Cylindrical Packed Beds, *J. Heat Transfer ASME*, 110, 378–384.

Hwang, J.J., 1997, Turbulent Heat Transfer and Fluid Flow in a Porous-Baffled Channel, *J. Thermophys. Heat Tr.*, 11(3), 429–436.

Incropera, F.P. and DeWitt, D.P., 1990, *Fundamentals of Heat and Mass Transfer*, 3rd edn, Wiley, Canada.

Ingham, D.B. and Pop, I. (eds), 1998, *Transport Phenomena in Porous Media*, Elsevier, Amsterdam.

Ingham, D.B. and Pop, I. (eds), 2002, *Transport Phenomena in Porous Media II*, Pergamon, Oxford.

Ingham, D.B. and Pop, I. (eds), 2005, *Transport Phenomena in Porous Media III*, Pergamon, Oxford.

Issa, R.I., 1986, Solution of the Implicitly Discretized Fluid Flow Equations by Operator Splitting, *J. Comput. Phys.*, 62, 40–65.

Jones, W.P. and Launder, B.E., 1972, The Prediction of Laminarization with Two-Equation Model of Turbulence, *Int. J. Heat Mass Tran.*, 15, 301–314.

Joshi, D.S. and Vanka, S.P., 1991, Multigrid Calculation Procedure for Internal Flows in Complex Geometries, *Num. Heat Transfer*, 20-B.

Kalita, J.C. and Dalal, D.C., 2003, Dass, Fully Compact Higher-order Computation of Steady State Natural Convection in a Square Cavity, Physical Review E 64-066703, 1–13.

Ko, Kang-Hoon and Anand, N.K., 2003, Use of Porous Baffles to Enhance Heat Transfer in a Rectangular Channel, *Int. J. Heat Mass Tran.*, 46, 4191–4199.

Kaviany, M., 1991, *Principles of Heat Transfer in Porous Media*, Springer, New York.

Kaviany, M., 1995, *Principles of Heat Transfer in Porous Media*, Springer, New York.

Kelkar, K.M. and Patankar, S.V., 1987, Numerical Prediction of Flow and Heat Transfer in a Parallel Plate Channel with Staggered Fins, *J. Heat Transfer*, 109, 25–30.

Kim, J., Kline, S.J. and Johnston, J.P., 1980, Investigation of a Reattaching Turbulent Shear Layer: Flow over a Backward-Facing Step, *ASME J. Fluids Engng*, 102, 302–308.

Knupp, P.M. and Lage, J.L., 1995, Generalization of the Forchheimer-Extended Darcy Flow Model to the Tensor Permeability case via a Variational Principle, *J. Fluid. Mech.*, 299, 97–104.

Kolmogorov, A.N., 1942, Equations of Turbulent Motion of an Incompressible Fluid, *Izv. Akad. Nauk. USSR*, Seria fizicheska, #1–2, pp. 56–58.

Kuwahara, F. and Nakayama, A., 1998, Numerical Modeling of Non-Darcy Convective Flow in a Porous Medium, *Proc. 11th Int. Heat Transf. Conf.*, 4, Kyongyu, Korea, August 23–28, pp. 411–416.

Kuwahara, F., Nakayama, A. and Koyama, H., 1996, A Numerical Study of Thermal Dispersion in Porous Media, *J. Heat Transfer*, 118, 756–761.

Kuwahara, F., Shirota, M. and Nakayama, A., 2001, *Int. J. Heat Mass Tran.* 44, 1153.

Kuznetsov, A.V., 1996, Analytical Investigation of the Fluid Flow in the Interface Region between a Porous Medium and a Clear Fluid in Channels partially filled with a Porous Medium, *Int. J. Heat and Fluid Fl.*, 12, 269–272.

Kuznetsov, A.V., 1997, Influence of the Stresses Jump Condition at the Porous-Medium/Clear-Fluid Interface on a Flow at a Porous Wall, *Int. Commun. Heat Mass*, 24, 401–410.

Kuznetsov, A.V., 1998, Thermal Nonequilibrium Forced Convection in Porous Media, in Ingham, D.B. and Pop, I. (eds), *Transport Phenomena in Porous Media*, 1st edn, Elsevier Sc., Oxford, p. 103.

Kuznetsov, A.V., 1999, Fluid Mechanics and Heat Transfer in the Interface Region between a Porous Medium and a Fluid Layer: A Boundary Layer Solution, *J. Porous Media*, 2(3), 309–321.

Kuznetsov, A.V., 2004, Numerical Modeling of Turbulent Flow in a Composite Porous/Fluid Duct utilizing a Two-layer k–ε Model to Account for Interface Roughness, *Int. J. Therm. Sci.*, 43(11), 1047–1056.

Kuznetsov, A.V. and Xiong, M., 2003, Development of an Engineering Approach to Computations of Turbulent Flows in Composite Porous/Fluid Domains, *Int. J. Therm. Sci.*, 42, 913.

Kuznetsov, A.V. and Becker, S.M., 2004, Effect of the Interface Roughness on Turbulent Convective Heat Transfer in a Composite Porous/Fluid Duct, *Int. Commun. Heat Mass*, 31(1), 11.

Kuznetsov, A.V., Cheng, L. and Xiong, M., 2002, Effects of Thermal Dispersion and Turbulence in Forced Convection in a Composite Parallel-plate Channel: Investigation of Constant Wall Heat Flux and Constant Wall Temperature Cases, *Num. Heat Transfer – Part A*, 42(4), 365–383.

Lage, J.L. and Bejan, A., 1991, The Ra-Pr Domain of Laminar Natural Convection in an Enclosure Heated from the Side, *Num. Heat Transfer – Part A*, 19, 21–41.

Lage, J.L., de Lemos, M.J.S. and Nield, D.A., 2002, Modeling Turbulence in Porous Media, in *Transport Phenomena in Porous Media*, Vol. 2, Ingham, D.B. and Pop, I. (eds), pp. 198–230.

Lane, S.N. and Hardy, R.J., 2002, Porous Rivers: A New Way of Conceptualising and Modelling River and Floodplain Flows?, Chapter 16 in *Transport Phenomena in Porous Media II*, Ingham, D. and Pop, I. (eds), 1st edn, Pergamon Press, 425–449.

Launder, B.E., 1975, Second-Moment Closure: Present ... and Future?, *Int. J. Heat Fluid Fl.*, 10(4), 282–300.

Launder, B.E. and Spalding, D.B., 1974, The numerical Computation of Turbulent Flows, *Comput. Method. Appl. M.*, 3, 269–289.

Launder, B.E., Reece, G.J. and Rodi, W., 1975, Progress in the Development of a Reynolds Stress Closure, *J. Fluid Mech.*, 68, 537–566.

Lee, K. and Howell, J.R., 1987, Forced Convective and Radiative Transfer within a highly Porous Layer Exposed to a Turbulent External Flow Field, *Proc. of the 1987 ASME-JSME Thermal Eng. Joint Conf.*, 2, 377–386.

Leonard, B.P., 1979, A stable and Accurate Convective Modeling Procedure based on Quadratic Upstream Interpolation, *Comput. Method Appl. M.*, 19, 59–98.

Leschziner, M.A., 1989, Modeling Turbulent Recirculating Flows by Finite-volume methods – Current Status and Future Trends, *Int. J. Heat Fluid Fl.*, 17(3), 186–200.

Leschziner, M.A. and Dimitriadis, K.P., 1989, Computation of Three-dimensional Turbulent Flow in Non-Orthogonal Junctions by a Branch-Coupling Method, *Comput. Fluids*, 17, 371–396.

Lesieur, M., 1997, *Turbulence in Fluids*, 3rd Rev. Ed., Kluwer Academic Publishers, Dordrecht.

Libby, P., 1996, *Introduction to Turbulence*, Taylor and Francis, New York.

Lim, I.G. and Matthews, R.D., 1993, Development of a Model for Turbulent Combustion within Porous Inert Media, *Trensp. Phenm. Therm. Eng.*, Begell House Inc. Publ., 631–636.

Lopez, J.R., Anand, N.K. and Fletcher, L.S., 1996, Heat Transfer in a Three-Dimensional Channel with Baffles, *Num. Heat Transfer – Part A*, 30, 189–205.

Lujboja, M. and Rodi, W., 1981, Prediction of Horizontal and Vertical Turbulent Buoyant Wall Jets, *J. Heat Transfer*, 103, 343–349.

Magro, V.T. and de Lemos, M.J.S., 2002a, Convecção Natural em Regime Laminar em Cavidade Contendo Material Poroso (in Portuguese), *Proc. CONEM 2002 – II Congresso Nacional de Engenharia Mecânica*, João Pessoa, PA, August. 23–28.

Magro, V.T. and de Lemos, M.J.S., 2002b, Efeito da Permeabilidade e Porosidade na Convecção Natural em Cavidade contendo Material Poroso (in Portuguese), *Proc ENCIT 2002* – Congresso Brasileiro de Engenharia e Ciências Térmicas, Caxambu, MG, Brazil, October 15–18.

Magro, V.T., Graminho, D.R. and de Lemos, M.J.S., 2003, Turbulent Heat Transport Across An Enclosure Containing a Porous Layer of Variable Thickness, *Proc. of COBEM2003, 17th International Congress of Mechanical Engineering*, São Paulo, Brazil, November 10–14.

Maliska, C.R., 1981, A Solution Method for Three-Dimensional Parabolic Flows in Non-Orthogonal Coordinates, PhD Thesis, University of Waterloo, Waterloo, Canada.

Maliska, C.R., 1995, *Computational Heat Transfer and Fluid Mechanics – Fundamentals and Generalized Coordinates* (in Portuguese), Livro Técnico, Rio de Janeiro, Brazil.

Maliska, C.R., Cunha, A.R., Livramento, M.A. and Silva, A.F.C., 1994, Tridimensional Petroleum Reservoir Simulation Using Generalized Curvilinear Grids, *Proc. of 5th Brazilian Thermal Science Meeting*, São Paulo, Brazil.

Mamou, M., Vasseur, P. and Bilgen, E., 1995, Multiple Solutions for Double-Diffusive Convection in a Vertical Porous Enclosure, *Int. J. Heat Mass Tran.*, 38(10), 1787.

Mamou, M., Hasnaoui, M., Amahmida, A. and Vasseur, P., 1998, Stability Analysis of Double Diffusive Convection in a Vertical Brinkman Porous Enclosure, *Int. Commun. Heat Mass*, 25(4), 491.

Manole, D.M. and Lage, J.L., 1992, Numerical Benchmark Results for Natural Convection in a Porous Medium Cavity, HTD-Vol 216, *Heat and Mass Transfer in Porous Media*, ASME Conference, 1992, 55–60.

Manole, D.M. and Lage, J.L., 1995, Numerical Simulation of Supercritical Hadley Circulation, within a Porous Media, Induced by Inclined Gradients, *Int. J. Heat Mass Tran.*, 38(14), 2583–2593.

Massarotti, N., Nithiarasu, P. and Carotenuto, A., 2003, Microscopic and Macroscopic Approach for Natural Convection in Enclosures Filled with Fluid Saturated Porous Medium, *Int. J. Numer. Method H.*, 13(7), 862–886.

Masuoka, T. and Takatsu, Y., 1996, Turbulence Model for Flow through Porous Media, *Int. J. Heat Mass Tran.*, 39, 2803–2809.

Merkle, C.L. and Choi, Y.H., 1987, Computation of Low Speed Flow with Heat Addition, *AIAA J.*, 25, 831–838.

Merkle, C.L., Venkateswaran, S. and Buelow, P.E.O., 1992, The Relationship between Pressure-Based and Density-Based Algorithms, AIAA Paper No. 92-0425.

Merrikh, A.A. and Lage, J.L., 2004, Effect of Distributing a Fixed Amount of Solid Constituent inside a Porous Medium Enclosure on the Heat Transfer Process, *ICAPM 2004, Proc. of Intern. Conf. Applications of Porous Media*, 51–57.

Merrikh, A.A. and Lage, J.L., 2005, Natural Convection in a Enclosure with Disconnected and Conducting Solid Blocks, *Int. J. Heat Mass Tran.*, 48 (7), 1361–1372.

Merrikh, A.A. and Mohamad, A.A., 2001, Blockage Effects in Natural Convection in Differentially Heated Enclosures, *J. Enhanced Heat Transfer*, 8(1), 55–72.

Merrikh, A.A. and Mohamad, A.A., 2002, Non-Darcy Effects in Buoyancy Driven Flows in an Enclosure Filled with Vertically Layered Porous Media, *Int. J. Heat Mass Tran.*, 45, 4305–4313.

Merrikh, A.A., Lage, J.L. and Mohamad, A.A., 2002, Comparison between Pore-Level and Porous Medium Models for Natural Convection in a Non-Homogeneous Enclosure, Fluid Flow and Transport in Porous Media: Mathematical and Numerical Treatment, Chen, Z. and Erwin, R. (eds), *Contemporary Mathematics*, 295, 387–396.

Merrikh, A.A., Lage, J.L., and Mohamad, A.A., 2005, Natural Convection in Non-Homogeneous Heat Generating Media: Comparison of Continuum and Porous-Continuum Models, *J. Porous Media*, 8(2), 149–163.

Mesquita, M.S. and de Lemos, M.J.S., 2004, Longitudinal and Transverse Mass Dispersion Coefficients for an Inifinite Porous Medium, *Proc. of ICAPM 2004 – Intern. Conf. Applications of Porous Media*, Evora, Portugal, May 24–27.

Minkowycz, W.J., Sparrow, E.M., Schneider G.E. and Pletcher, R.H., 1988, *Handbook of Numerical Heat Transfer*, Wiley, New York.

Mohamad, A.A. and Bennacer, R., 2002, Double Diffusion Natural Convection in an Enclosure Filled with Saturated Porous Medium Subjected to Cross Gradients; Stably Stratified Fluid, *Int. J. Heat Mass Tran.*, 45(18), 3725.

Monin, A.S. and Yaglom, A.M., 1975, *Statistical Fluid Mechanics*, Vol. 1 and 2, The MIT Press, Massachusetts.

Moya, S.L., Ramos, E. and Sen, M., 1987, Numerical Study of Natural Convection in a Tilted Rectangular Porous Material, *Int. J. Heat Mass Tran.*, 30, 741–756.

Muskat, M., 1946, *The Flow of Homogenous Fluid through Porous Media*, Edwards, Michigan.

Nakayama, A., 1985, A Numerical Method for Solving Momentum Equations in Generalized Coordinates (Its Application to Three-Dimensional Separated Flows), *J. Fluid Eng.*, 107, 49–54.

Nakayama, A. and Kuwahara, F., 1999, A Macroscopic Turbulence Model for Flow in a Porous Medium, *J. Fluids Eng.*, 121, 427–433.

Nakayama, A., Kuwahara, F., Sugiyama M. and Xu, G., 2001, *Int. J. Heat Mass Tran.*, 44, 4375.

Nield, D.A., 1991, The Limitations of the Brinckman–Forchheimer Equation in Modeling Flow in a Saturated Porous Medium and at the Interface, *Int. J. Heat Fluid Fl.*, 12, 269–272.

Nield, D.A. and Bejan, A., 1999, 2nd edn, *Convection in Porous Media*, Springer, New York.

Nikjook, M. and Mongia, H.C., 1991, *Int. J. Heat Fl.*, 12(1), 12.

Nisizima, S. and Yoshizawa, A., 1987, Turbulent Channel and Couette Flows Using an Anisotropic $k$-$\varepsilon$ Model, *AIAA J.*, 25(3), 414.

Nithiarasu, P., Sundararajan T. and Seetharamu, K.N., 1997, Double-Diffusive Natural Convection in a Fluid Saturated Porous Cavity with a Freely Convecting Wall, *Int. Commun. Heat Mass*, 24(8), 1121.

Ochoa-Tapia, J.A. and Whitaker, S., 1995a, Momentum Transfer at the Boundary between a Porous Medium and a Homogeneous Fluid – I: Theoretical Development, *Int. J. Heat Mass Tran.*, 38, 2635–2646.

Ochoa-Tapia, J.A. and Whitaker, S., 1995b, Momentum Transfer at the Boundary between a Porous Medium and a Homogeneous Fluid – II. Comparison with Experiment, *Int. J. Heat Mass Tran.*, 38, 2647–2655.

Ochoa-Tapia, J.A. and Whitaker, S., 1997, Heat Transfer at the Boundary between a Porous Medium and a Homogeneous Fluid, *Int. J. Heat Mass Tran.*, 40, 2691–2707.

Padila Jr., A.L.A. and de Lemos, M.J.S., 1992, Solution of Axial Flow over Bare Rod Bundles, *Proc. of Intern. Mobility Tech.*, SAE/Brazil Conf., SAE paper # 921509E, São Paulo, Brazil, October 13–14.

Park, T.S. and Sung, H.J., 1995, A Nonlinear Low-Reynolds-Number k-ε Model for Turbulent Separated and Reattaching Flows, *Int. J. Heat Mass Tran.*, 38, 2657–2666.

Patankar, S.V., 1980, *Numerical Heat Transfer and Fluid Flow*, Hemisphere, New York.

Patankar, S.V., 1981, A Calculation Procedure for Two-dimensional Elliptic Situations, *Num. Heat Transfer*, 4, 409–425.

Patankar, S.V., 1988, Recent Developments in Computational Heat Transfer, *J. Heat Trans-TASME.*, 110, 1037–1045.

Patankar, S.V. and Spalding, D.B., 1972, A Calculation Procedure for Heat, Mass and Momentum Transfer in Three Dimensional Parabolic Flows, *Int. J. Heat Mass Tran.*, 15, 1787.

Peard, T.E., Peters, J.E., Brewster, M.Q. and Buckius, R.O., 1993, Radiative Heat Transfer Augmentation in Gas-Fired Radiant Tube Burner by Porous Inserts: Effect on Insert Geometry, *Exp. Heat Transfer*, 6, 273–286.

Pedras, M.H.J. and de Lemos, M.J.S., 1999, On volume and Time averaging of Transport Equations for Turbulent Flow in Porous Media, *Proc. 3rd ASME/JSME Joint Fluids Engineering Conference* (on CDROM), ASME-FED-248, Paper FEDSM99-7273, ISBN 0-7918-1961-2, San Francisco, California, July 18–23.

Pedras, M.H.J. and de Lemos, M.J.S., 2000a, On the Definition of Turbulent Kinetic Energy for Flow in Porous Media, *Int. Commun. Heat Mass*, 27(2), 211–220.

Pedras, M.H.J. and de Lemos, M.J.S., 2000b, Numerical Solution of Turbulent Flow in Porous Media Using a Spatially Periodic Cell and the Low Reynolds k-ε Model (in Portuguese), *Proc. of CONEM2000 – National Mechanical Engineering Congress* (on CD-ROM), Natal, Rio Grande do Norte, Brazil, August 7–11.

Pedras, M.H.J. and de Lemos, M.J.S., 2001a, Macroscopic Turbulence Modeling for Incompressible Flow through Undeformable Porous Media, *Int. J. Heat Mass Tran.*, 44(6), 1081–1093.

Pedras, M.H.J. and de Lemos, M.J.S., 2001b, Simulation of Turbulent Flow in Porous Media using a Spatially Periodic Array and a Low Re Two-equation Closure, *Num. Heat Transfer – Part A Applications*, 39, 35–59.

Pedras, M.H.J. and de Lemos, M.J.S., 2001c, On Mathematical Description and Simulation of Turbulent Flow in a Porous Medium Formed by an Array of Elliptic Rods, *ASME J. Fluid Eng.*, 123(4), 941–947.

Pedras, M.H.J. and de Lemos, M.J.S., 2003, Computation of Turbulent Flow in Porous Media Using a Low Reynolds k-$\varepsilon$ Model and an Infinite Array Of Transversally-Displaced Elliptic Rods, *Num. Heat Transfer, Part A – Applications*, 43(6), 585–602.

Pedras, M.H.J. and de Lemos, M.J.S., 2004, Turbulent Dispersion in Porous Media Modeled as an Infinite Array of Transversally-Displaced Elliptic Rods, *Proc. ICAPM 2004 2nd Int. Conf. on Applications of Porous Media*, Evora, Portugal, May 24–27.

Pedras, M.H.J. and de Lemos, M.J.S., 2005, Thermal Dispersion in Porous Media as a Function of the Solid-Fluid Conductivity Ratio, *Int. J. Heat Mass Tran.* (submitted).

Pedras, M.H.J., Rocamora Jr., F.D. and de Lemos, M.J.S., 2003a, Simulation of Turbulent Thermal Dispersion in Porous Media Using a Periodic Cell with Prescribed Heat Flux at the Boundaries, *Proc. 3rd Int. Conf. Computational Heat and Mass Trans.*, Banff, Canada, May 26–30.

Pedras, M.H.J., Rocamora Jr., F.D. and de Lemos, M.J.S., 2003b, Turbulent Dispersion in a Porous Media Modeled as an Infinite Array of Longitudinally-Displaced Elliptic Rods, *Proc. of the COBEM2003 17th Int. Cong. of Mechanical Engineering*, São Paulo, Brazil, November 10–14.

Peric, M., 1990, Analysis of Pressure-Velocity Coupling on Non-Orthogonal Grids, *Num. Heat Transfer*, 17-B, 63–82.

Perry, R.H. and Chilton, C.H. (eds), 1973, *Chemical Engineers' Handbook*, 5th edn, McGraw Hill, New York.

Peyret, R. and Taylor, T.D., 1983, *Computational Methods for Fluid Flow*, Springer, New York.

Pope, S.B., 1975, A More General Effective-Viscosity Hypothesis, *J. Fluid Mech.*, 72, 331.

Pope, S.B., 1978, The Calculation of Turbulent Recirculating Flows in General Orthogonal Coordinates, *J. Comput. Phys.*, 26, 197–217.

Prakash, M., Turan, F.O., Li, Y., Manhoney, J. and Thorpe, G.R., 2001a, Impinging Round Jet Studies in a Cylindrical Enclosure with and without a Porous Layer: Part I – Flow Visualisations and Simulations, *Chem. Eng. Sci.*, 56, 3855–3878.

Prakash, M., Turan, F.O., Li, Y., Manhoney, J. and Thorpe, G.R., 2001b, Impinging Round Jet Studies in a Cylindrical Enclosure with and without a Porous Layer: Part II – DLV Measurements and Simulations, *Chem. Eng. Sci.*, 56, 3879–3892.

Prandtl, L., 1925, Bemerkungen uber die Entsthehung der Turbulenz, *ZAMM*, 1(6), 431–436.

Prandtl, L., 1945, Uber ein neus Formelsystem für die ausgebildete Turbulenz, *Nachr. Akad. Wiss.* Göttingen, Math-Phys. Klasse, p. 6.

Prasad, V. and Kulacki, F.A., 1984, Convective Heat Transfer in a Rectangular Porous Cavity-Effect of Aspect Ratio on Flow Structure and Heat Transfer, *J. Heat Transfer*, 106, 158–165.

Prinos, P., Sofialidis, D. and Keramaris, E., 2003, Turbulent Flow over and within a Porous Bed, *J. Hydraul. Eng.*, 129(9), 720–733.

Prithiviraj, M. and Andrews, M.J., 1998, Three-dimensional Numerical Simulation of Steel-and-Tube heat Exchanger: Part I – Foundation and Fluid Mechanics, *Num. Heat Transfer – Part A: Applications*, 33, 799–816.

Pullian, T.H. and Steger, J.L., 1980, Implicit Finite Difference Simulation of Three-Dimensional Compressible Flow, *AIAA J.*, 18, 159–167.

Quintard, M., 1998, Modelling Local Non-Equilibrium Heat Transfer in Porous Media, Heat Transfer 1998, *Proc. 11th Int. Heat Transfer Conf.*, Kyongyu, Korea, Vol. 1, 279–285.

Quintard, M., Kaviany, M. and Whitaker, S., 1997, Two-Medium Treatment of Heat Transfer in Porous Media: Numerical Results for Effective Properties, *Adv. Water Resour.*, 20, 77–94.

Rabi, J.A. and de Lemos, M.J.S., 2001, *Appl. Math. Comput.* 124(2), 215.

Raithby, G.D., 1976, Skew Upstream Differencing Schemes for Problems involving Fluid Flow, *Comp. Meth. Appl. Mech.*, 9, 153–164.

Raithby, G.D. and Torrance, K.E., 1974, Upstream-weighted Differencing Scheme and their Application to Elliptic Problems Involving Fluid Flow, *Comp. and Fluids*, 2, 191–206.

Rapley, C.W., 1982, The Simulation of Turbulent Flow and Heat Transfer in Bare Rod-Bundles and other Non-circular Passages, Lect. Series 1982-04, *Comp. Fluid Mech.*, von Karman Inst. for Fluid Dynamics.

Rapley, C.W., 1985, Turbulent Flow in Ducts with Cuspid Corners, *Int. J. Num. Meth. Fl.*, 5, 155–167.

Rhie, C.M., 1986, A Pressure Based Navier–Stokes solver using the Multigrid Method, AIAA Paper No. 86-0207.

Richtmyer, R.D. and Norton, K.W., 1967, *Difference Methods for Initial-Value Problems*, John Wiley & Sons, New York.

Roache, P.J., 1972, *Computational Fluid Dynamics*, Hermosa Publ., Albuquerque, New Mexico.

Rocamora Jr., F.D. and de Lemos, M.J.S., 2000a, Analysis of Convective Heat Transfer for Turbulent Flow in Saturated Porous Media, *Int. Commun. Heat Mass*, 27(6), 825–834.

Rocamora Jr., F.D. and de Lemos, M.J.S., 2000b, Laminar Recirculating Flow And Heat Transfer in Hybrid Porous Medium-Clear Fluid Computational Domains, *Proc. 34th ASME-National Heat Transfer Conference* (on CD-ROM), ASME-HTD-I463CD, Paper NHTC2000-12317, ISBN:0-7918-1997-3, Pittsburgh, Pennsylvania, August 20–22.

Rocamora Jr., F.D. and de Lemos, M.J.S., 2000c, Prediction of Velocity and Temperature Profiles for Hybrid Porous Medium-Clean Fluid Domains, *Proc. CONEM2000 – National Mechanical Engineering Congress* (on CD-ROM), Natal, Rio Grande do Norte, Brazil, August 7–11.

Rocamora Jr., F.D. and de Lemos, M.J.S., 2000d, Heat Transfer in Suddenly Expanded Flow in a Channel with Porous Inserts, *Proc IMECE2000 – ASME – Intern. Mech. Eng. Congr.*, ASME-HTD-366-5, ISBN:0-7918-1908-6, Orlando, Florida, November 5–10, pp. 191–195.

Rocamora Jr., F.D., 2001, *Heat Transport Modeling for Laminar and Turbulent Flows in Porous Media*, PhD Thesis, Instituto Tecnológico de Aeronáutica, Brazil.

Rodi, W., 1972, *The Prediction of Free Turbulent Boundary Layers by Use of a Two-Equation Model of Turbulence*, PhD Thesis, University of London.

Rubin, B. and Blunt, M.J., 1991, High-Order Implicit Flux Limiting Schemes for Black-Oil Simulating, SPE 21222, *11th SPE Symposium on Reservoir Simulation*, Anahein, February 17–20.

Rubinstein, R. and Barton, J.M., 1990, Renormalization Group Analysis of the Stress Transport Equation, *Phys. Fluids*, Part A, 2(8), 1472–1476.

Sahraoui, M. and Kaviany, 1995, Direct simulation vs Time-Averaged Treatment of Adiabatic, Premixed Flame in a Porous Medium, *Int. J. Heat Mass Tran.*, 18, 2817–2834.

Saito, M. and de Lemos, M.J.S., 2005, Interfacial Heat Transfer Coefficient for Non-Equilibrium Convective Transport in Porous Media, *Int. Commun. Heat Mass*, 32(5), 5, 667–677.

Saito, M. and de Lemos, M.J.S., 2006, A Correlation for Interfacial Heat Transfer Coefficient for Turbulent Flow over an Array of Square Rods, *J. Heat Transfer*, Vol. 128(5), 444–452.

Santos, N.B. and de Lemos, M.J.S., 2004a, Laminar Flow and Heat Transfer Analysis in a Channel with Porous Baffles, *Proc. of CONEM2004 – National Mechanical Engineering Congress* (on CD-ROM), Belém, PA, Brazil, August 10–13.

Santos, N.B. and de Lemos, M.J.S., 2004b, Effect of Porosity and Permeability on Heat Transfer Enhancement in Channels with Porous Baffles, *Proc. of ENCIT 2004 – National Thermal Sciences Meeting* (on CD-ROM), Rio de Janeiro, RJ, Brazil, November 29–December 3.

Santos, N.B. and de Lemos, M.J.S., 2006, Flow and Heat Transfer in a Parallel Plate Channel with Porous and Solid Baffles, *Num. Heat Transfer – Part A*, 49(5), 471–494.

Scheidegger, A.E., 1974, *The Physics of Flow through Porous Media*, Univ. of Toronto Press, Toronto.

Schneider, G.E. and Zedan, M., 1981, A Modified Strongly Implicit Procedure for Numerical Solution of Field Problems, *Num. Heat Transfer*, 4, 1–19.

Schumann, U., Grotzbach, G. and Kleiser, L., 1980, Direct Numerical Simulation of Turbulence, in *Prediction Methods for Turbulent Flows*, Kollman (ed.), MgGraw-Hill, New York.

Sha, W.T., 1981, A New Porous-Media Approach for Thermal-Hydraulic Analysis, *T. Am. Nucl. Soc.*, 39, 510–512.

Sha, W.T. and Thompson, J.F., 1979, Rod-Bundle Thermal Hydraulic Analysis using Boundary-Fitted Coordinate System, Rept. ANL-78-1.

Sharpe, H.N., 1993, Validation of an Adaptive, Orthogonal, Curvilinear Gridding Procedure for Reservoir Simulation, SPE 25262, *12th SPE Symposium on Reservoir Simulation*, New Orleans, USA, February 28–March 3.

Sharpe, H.N. and Anderson, D.A., 1991, Orthogonal Curvihne Grid Generation with Preset Internal Boundaries for Reservoir Simulation, *11th SPE Symposium on Reservoir Simulation*, Anahein, California, USA, February.

Shih, T.H., Zhu, J. and Lumley, J.L., 1993, A Realizable Reynolds Stress Algebraic Equation Model, NASA TM-105993.

Shyy, W., 1985, A Study of Finite Difference Approximations to Steady-State, Convention Dominated Flow Problems, *J. Comput. Phys.*, 57, 415–438.

Shyy, W., 1994, Elements of Pressure-Based Computational Algorithms for Complex Fluid Flow and Heat Transfer, *Adv. in Heat Transfer*, 25, 191–275.

Shyy, W. and Braaten, M.E., 1988, Adaptive Grid Computation for Inviscid Compressible Flows Using a Pressure Correction Method, *Proc. of the AIAA/ASME/SIAM/APS First National Fluid Dynamics Congress*, AIAA-CP888, 112–120.

Shyy, W., Chen, M.H. and Sun, C.S., 1992, Pressure-based multigrid algorithm for flow at all speeds, *AIAA 30th Aerospace Sciences Meeting*, Paper No. 92-0548, also *AIAA J.* 30, 2660–2669.

Silva, R.A. and de Lemos, M.J.S., 2002, Laminar flow in a Channel with a Layer of Porous Material considering the Non-linear Forchheimer Term (in Portuguese), *Proc. of II National Congress of Mechanical Engineering* (on CD-ROM), João Pessoa-PB, Brazil.

Silva, R.A. and de Lemos, M.J.S., 2003a, Numerical Analysis of the Stress Jump Interface Condition for Laminar Flow over a Porous Layer, *Num. Heat Transfer – Part A: Applications*, 43(6), 603–617.

Silva, R.A. and de Lemos, M.J.S., 2003b, Turbulent Flow in a Channel Occupied by a Porous Layer Considering the Stress Jump at the Interface, *Int. J. Heat Mass Tran.*, 46, 5113–5121.

Silva, R.A. and de Lemos, M.J.S., 2003c, Laminar Flow Around a Sinusoidal Interface Between a Porous Medium and a Clear Fluid, *Proc. COBEM2003 – 17th. Int. Congr. of Mech. Engineering*, Paper 1528 (on CD-ROM), São Paulo, Brazil, November 10–14.

Slattery, J.C., 1967, Flow of Viscoelastic Fluids through Porous Media, *AICHE. J*, 13, 1066–1071.

Spalding, D.B., 1972, A Novel Finite-Difference Formulation for Differential Expressions involving both First and Second Derivatives, *Int. J. Num. Meth. Eng.*, 4, 551.

Speziale, C.G., 1987, On Nonlinear $k$-l and $k$-$\varepsilon$ Models of Turbulence, *J. Fluid Mech.*, 176, 459–475.

Stone, H.L., 1968, Iterative Solution of Implicit Approximations of Multidimensional Partial Differential Equations, *SIAM J. Num. Anal.*, 5(3), 530–558.

Tang, L. and Joshi, Y.K., 1999, Application of Block Implicit Multigrid approach to Three-dimensional Heat Transfer Problems Involving Discrete Heating, *Num. Heat Transfer – Part A*, 35, 717–734.

Takatsu, Y. and Masuoka, T., 1998, Turbulent Phenomena in Flow through Porous Media, *J. Porous Media*, 1, 243–251.

Tennekes, H. and Lumley, J.L., 1972, *A First Course in Turbulence*, The MIT Press, Massachusetts.

Tofaneli, L.A. and de Lemos, M.J.S., 2002a, Escoamento Laminar Em Região Espacialmente Periódica Em Canal Contendo Obstrução Porosa (in Portuguese), *Proc. of CONEM2002 – National Mechanical Engineering Congress*, Paper CPB0275, João Pessoa, PB, Brazil, August. 12–16.

Tofaneli, L.A. and de Lemos, M.J.S., 2002b, Influência da Porosidade e da Permeabilidade de Aletas Porosas no Escoamento em Regime Laminar em Canal Entre Placas (in Portuguese), *Proc. ENCIT2002 – National Thermal Sciences Meeting*, Caxambú, MG, Brazil, October 13–17.

Travkin, V.S. and Catton, I., 1992, Models of Turbulent Thermal Diffusivity and Transfer Coefficients for a Regular Packed Bed of Spheres, *Proc. 28th National Heat Transfer Conference*, San Diego, C-4, ASME-HTD 193, pp. 15–23.

Travkin, V.S. and Catton, I., 1995, A Two Temperature Model for Turbulent Flow and Heat Transfer in a Porous Layer, *J. Fluid Eng.*, 117, 181–188.

Travkin, V.S. and Catton, I., 1998, Porous Media Transport Descriptions – Non-local, Linear and Non-linear against Effective Thermal/Fluid Properties, *Adv. Colloid Interfac.*, 76–77, 389–443.

Travkin, V.S., Catton, I. and Gratton, L., 1993, Single-Phase Turbulent Transport in prescribed Non-Isotropic and Stochastic Porous Media, *Heat Transfer in Porous Media*, ASME HTD-240, pp. 43–48.

Travkin, V.S., Hu, K. and Catton, I., 1999, Turbulent Kinetic Energy and Dissipation Rate Equation Models for Momentum Transport in Porous Media, *Proc. 3rd ASME/JSME Joint Fluids Engineering Conference* (on CD-ROM), Paper FEDSM99-7275, San Francisco, California, USA, July 18–23.

Vafai, K., 1984, Convective Flow and Heat Transfer in Variable-Porosity Media, *J. Fluid Mech.*, 147, 233–279.

Vafai, K., 1995, On the Limitations of the Brinckman-Forchheimer-Extended Darcy Equation, *Int. J. Heat Mass Tran.*, 16, 11–15.

Vafai, K. (ed.), 2005, *Handbook of Porous Media*, 2nd. Revised and Exp. Edn., Taylor and Francis, New York.

Vafai, K. and Tien, C.L., 1981, Boundary and Inertia Effects on Flow and Heat Transfer in Porous Media, *Int. J. Heat Mass Tran.*, 24, 195–203.

van Doormaal, J.P. and Raithby, G.D., 1984, Enhancements of the SIMPLE Method for Predicting Incompressible Flow, *Num. Heat Transfer.*, 7, 147–163.

Vanka, S.P., 1985, Block-Implicit Calculation of Steady Turbulent Recirculating Flows, *Int. J. Heat Mass Tran.*, 28, 2093–2103.

Vanka, S.P., 1986a, A Calculation Procedure for Three-Dimensional Steady Recirculating Flows Using Multigrid Methods, *Comput. Meth. App. Mech. Eng.*, 55, 321–338.

Vanka, S.P., 1986b, Block-Implicit Multigrid Solution of Navier-Stokes Equations in Primitive Variables, *J. Comput. Phys.*, 65, 138–158.

Vanka, S.P., 1987, Second-Order Upwind Differencing in a Recirculating Flow, *AIAA J.*, 25, 1435–1441.

Versteeg, H.K. and Malalasekera, W., 1995, *An Introduction to Computational Fluid Dynamics – The Finite Volume Method*, Longman, London.

Vinokur, M., 1974, Conservative Equations for Gas Dynamics in Curvilinear Coordinates, *J. Comput. Phys.*, 14, 105–125.

Voronoi, G., 1908, Nouvelle Applications des Parameters Continues à la Teorie des Formes Quadratiques, Deuxième Memoire, Recherches sur les Paralleloedres Primitifs, *Z. Angew. Math.*, 134, 198–287.

Wakao, N., Kaguei, S. and Funazkri, T., 1979, Effect of Fluid Dispersion Coefficients on Particle-to-Fluid Heat Transfer Coefficients in Packed Bed, *Chem. Eng. Sci.*, 34, 325–336.

Walker, K.L. and Homsy, G.M., 1978, Convection in Porous Cavity, *J. Fluid Mech.*, 87, 49–474.

Wang, H. and Takle, E.S., 1995, Boundary-Layer Flow and Turbulence near Porous Obstacles, *Bound-Lay Meteorol.*, 74, 73–88.

Warming, R.F. and Beam, R.M. 1976, Upwind Second-Order Difference Schemes and Applications in Aerodynamic Flows, *AIAA J.*, 14(9), 1241–1249.

Warsi, Z.U.A., 1993, *Fluid Dynamics: Theoretical and Computational Approaches*, CRC Press, Boca Raton.

Webb, B.W. and Ramadhyani, S., 1985, Conjugate Heat Transfer in a Channel with Staggered Ribs, *Int. J. Heat Mass Tran.*, 28(9), 1679–1687.

Whitaker, S., 1966, Equations of Motion in Porous Media, *Chem. Eng. Sci.*, 21, 291–300.

Whitaker, S., 1967, Diffusion and Dispersion in Porous Media, *AICHE. J*, 13, 420–431.

Whitaker, S., 1969, Advances in Theory of Fluid Motion in Porous Media, *Ind. Eng. Chem.*, 61, 14–28.

Whitaker, S., 1999, *The Method of Volume Averaging*, Kluwer Academic Publishers, Dordrecht.

Withington, J.P., Shuen, J.S. and Yang, V., 1991, A Time Accurate, Implicit Method for Chemically Reacting Flows at all Mach Numbers, AIAA Paper 81-0581.

Yang, K.T., 1987, Natural Convection in Enclosures, Chapter 13 in *Handbook of Single Phase Convective Heat Transfer*, Kakaç, S., Shah, R.K. and Aung, W. (eds), John Wiley & Sons, New York.

Yang, Y.-T. and Hwang, C.-Z., 2003, Calculation of Turbulent Flow and Heat Transfer in a Porous-Baffled Channel, *Int. J. Heat Transfer*, 46, 771–780.

Yanosik, J.L. and McCracken, T.A., 1979, A Nine-Point Finite Difference Reservoir Simulator for Realistic Prediction of Unfavorable Mobility Ratio Displacements, *Soc. Pet. Eng. J.*, 18, 253–262.

Zhou, X.Y. and Pereira, J.C.F., 2000, A Multidimensional Model for Simulating Vegetation Fire Spread Using a Porous media Sub-model, *Fire Mater.*, 24, 37–43.

Zhukauskas, A., 1972, Heat Transfer from Tubes in Cross Flow, *Adv. in Heat Transfer*, 8, Academic Press, New York,

# Index